LIDC Contributions on Antitrust Law, Intellectual Property and Unfair Competition

Series Editor
International League of Competition Law (LIDC), Competition Law (LIDC), Lausanne, Switzerland

In each volume of the LIDC Contributions on Antitrust Law, Intellectual Property and Unfair Competition, two topics are discussed, one focusing on competition law, the other on intellectual property and/or unfair competition.

Practitioners and academics from different industrial countries contribute national reports on each topic, followed by an analysis and a comparison of the respective national situations as well as an international report. The national reports include essential and hard-to-find case law analysis. The common trends and principles identified in the international reports serve as a basis for more general recommendations, some of which have been presented to national and international authorities.

The series is produced by the International League of Competition Law (LIDC), an association that brings together national associations for antitrust law, intellectual property and unfair competition law. Created in 1930 by a German lawyer who subsequently became Professor of Law at the University of Saint Louis in the United States, the LIDC has a long standing reputation for scientific exploration, debate and work.

Bruce Kilpatrick • Pierre Kobel •
Pranvera Këllezi
Editors

Antitrust in Data Driven Markets & Legal Framework for Influencers, Native Advertising and Control over the Use of AI in Marketing

Springer

Editors
Bruce Kilpatrick
Addleshaw Goddard LLP
London, UK

Pierre Kobel
Kobel Avocat Attorney-at-law
Geneva, Switzerland

Pranvera Këllezi
Këllezi Legal
Geneva, Switzerland

ISSN 2199-742X ISSN 2199-7438 (electronic)
LIDC Contributions on Antitrust Law, Intellectual Property and Unfair Competition
ISBN 978-3-031-07421-9 ISBN 978-3-031-07422-6 (eBook)
https://doi.org/10.1007/978-3-031-07422-6

This Springer imprint is published by the registered company Springer Nature Switzerland AG
The registered company address is: Gewerbestrasse 11, 6330 Cham, Switzerland

Preface

Like any other institution whose work is based on the open discussion of expert reports, synthesised on international reports and resolutions, the works of the LIDC (the International League of Competition Law) were affected by the pandemic. The next Congress due to be held in Brussels in 2020 was thus postponed and finally took place at the end of September 2021. Under the banner "Do disruptive innovations call for disruptive competition law enforcement?" the Congress included numerous panels and presentations related to the topics covered in this ninth edition of LIDC Contributions on Antitrust Law, Intellectual Property and Unfair Competition.

The Congress is, in the editors' opinion, a real "one of a kind" event, bringing together leading practitioners across the antitrust and intellectual property fields from a wide variety of jurisdictions across the globe, with detailed national reports on two topics chosen by the LIDC's Scientific Committee and an international comparative report on each topic.

It takes a huge amount of work and commitment from the national reporters, and particularly the international reporters, to produce their studies, and we would like to thank them for their enormous contribution, which is greatly appreciated. This year's Congress brought together delegates from across the world, with national reports from jurisdictions including Brazil, Japan and Ukraine, as well as from across the European Union, including the United Kingdom, and Switzerland. The purpose of this book is to share their learnings with a wider audience across the members of the LIDC and its constituent national organisations and with academics, practitioners and students across the world. We hope you enjoy reading it as much as we enjoyed listening to the debate and discussions that took place in Brussels.

This volume of the LIDC Contributions focuses on the data-driven society, the issue of digitalisation of trade with all related marketing methods and the recourse to artificial intelligence. Whilst these questions are wonderfully examined and presented from the antitrust viewpoint in the international report prepared by Prof. Dominik Schäfers and from the various legislation applicable to marketing methods and fair-trading laws in the international report drafted by Marta Delgado Echevarria, the novelty of the field and questions led the national reporters and the international reporters to consider the whole spectrum of applicable legislation.

The antitrust contributions of the reporters illustrate the advantages of the current broad competition legal framework when dealing with practices that are so novel, and also their limits. Market definition and market power are difficult to identify and measure when dealing with digital services and platforms. Traditional ex post enforcement mechanisms also have their limits, considering the constant need to secure non-discriminatory access to markets and practices that are in constant evolution. The reports thus describe the various attempts made by legislators and authorities, like with the Digital Market Act in the European Union, to intervene earlier in the process at compliance level instead of enforcement.

They also show how authorities in many countries are currently questioning their traditional analysis and legal framework, embarking on sector enquiries and new ways to apply or interpret the existing legal framework to tackle the issues posed by the digital economy.

As far as unfair competition is concerned, marketing practices such as native advertising, the use of influencers, comparative reviews and price discrimination are also calling for a new legal framework. The reports show the current legislative efforts made at the EU level and in various jurisdictions to tackle issues such as fake reviews. If personalised pricing can be seen as optimising, the immediate use of personal data may lead to inacceptable discrimination. This calls into question the use of personal data by trading platforms and, as a result, the need for some level of integrated approach between personal data protection and antitrust enforcement. The principle of transparency underlying personal data protection regulations is not only difficult to enforce but is also sometimes challenged by the protection of trade secrets.

This volume provides a well of information as to new trends, new legislation, regulations, decisions and enforcement policies regarding these fascinating questions.

The final reports have now been sent to national and supranational competition law enforcers, which have in the past commented very positively on LIDC reports and their value to enforcers. The works of the LIDC have been a well of practical guidance for generations of lawyers, whether they are members of the LIDC, and for regulatory authorities.

The editors would like to thank all the authors for their contributions and their patient collaboration during the editing of this book. They would like to express their sincere gratitude to the members of the Bureau, of the Council and of the Scientific Committee for their kind support and encouragement during the preparation of this book.

London, UK — Bruce Kilpatrick
Geneva, Switzerland — Pierre Kobel
Geneva, Switzerland — Pranvera Këllezi

Contents

Contributors

Lili Albert Deloitte Legal, Budapest, Hungary

Ingrid Halvorsen Barlund University of Bergen, Bergen, Norway

Sir Christopher Bellamy QC Monckton Chambers, London, UK

Friso Bostoen Consumer Competition Market, KU Leuven, Leuven, Belgium
Research Foundation – Flanders, Brussels, Belgium

Mark Burdon Faculty of Law, Queensland University of Technology, Brisbane, QLD, Australia

Johana Cau Lenz & Staehelin, Geneva, Switzerland

Laurence Charvoz Avocat, Angers, France

Benjamin Docquir Osborne Clarke, Brussels, Belgium

Marta Delgado Echevarría Jones Day, Madrid, Spain

Liliana Eskenazi Fréget Glaser & Associés AARPI, Paris, France

Judit Firniksz Firniksz Law Firm, Budapest, Hungary
Pázmány Péter Catholic University, Budapest, Hungary

Gerhard Fussenegger bpv Huegel, Vienna/Brussels, Belgium

Ronny Gjendemsjø University of Bergen, Bergen, Norway

Daniel Green Linklaters LLP, Paris, France

Thomas Hoeren Westfälische Wilhelms-Universität, Münster, Germany

Hanna Hoffmann Westfälische Wilhelms-Universität, Münster, Germany

Louis Hoffreumon Osborne Clarke, Brussels, Belgium

Barbora Jedličková TC Beirne School of Law, The University of Queensland, Brisbane, QLD, Australia

Adrian Kubat University of Vienna, Vienna, Austria

Nizar Lajnef UGGC Avocats, Paris, France

Jean-Julien Lemonnier Stephenson Harwood LLP, Paris, France

Adam Liber PROVARIS Varga & Partners, Budapest, Hungary

Mark-Oliver Mackenrodt Technical University Munich, TUM School of Management, Munich, Germany

Brenda Marshall Faculty of Law, Bond University, Robina, QLD, Australia

Lauren Mechri Fieldfisher, Paris, France

Péter Mezei Hegymegi-Barakonyi & Partner Baker & McKenzie Law Firm, Budapest, Hungary

Felipe Barros Oquendo Di Blasi, Parente & Associates, Rio de Janeiro, Brazil

Eirik Østerud University of Oslo, Oslo, Norway

Viktoria H. S. E. Robertson Vienna University of Economics and Business and University of Graz, Vienna/Graz, Austria

Dominik Schäfers Goethe Universität, Frankfurt, Germany

Arno Scharf University of Vienna, Vienna, Austria

Abbreviations

AUD	Australian dollar
AI	Artificial intelligence
Berne Convention	The Berne Convention for the Protection of Literary and Artistic Works of 9 September 1886, as amended
Brussels Convention	Brussels Convention Relating to the Distribution of Programme-Carrying Signals Transmitted by Satellite, 1974
BGBl.	Bundesgesetzblatt (Germany)
BGH	Bundesgerichtshof (Germany)
BGN	Bulgarian lev
B2B	Business to business
B2C	Business to consumers
bn	Billion
BRL	Brazilian real (reais)
c./ca.	Circa
cf.	Compare
CFI	Court of First Instance of the ECJ (before 1 December 2009)
CFREU	The Charter of Fundamental Rights of the European Union, OJ 2010 C 83, p. 389
CHF	Swiss franc
CJEU	Court of Justice of the European Union (after 1 December 2009)
CMLR	Common Market Law Review
CR_n	Concentration ratio, measuring the percentage market share held by *n* largest undertakings
DCT(s)	Digital comparison tool(s)
De minimis Notice	Commission Notice on agreements of minor importance which do not appreciably restrict competition under Article 101(1) of the Treaty on the Functioning of the European Union (de minimis), OJ 2014 C 291, p. 1

Directive 97/7	Directive 97/7 of the European Parliament and of the Council of 20 May 1997 on the protection of consumers in respect of distance contracts, OJ 1997 L 144, p. 19
Directive 96/09 or Database Directive	Directive 96/9 of the European Parliament and of the Council of 11 March 1996 on the legal protection of databases, OJ 1996 L 77, p. 20
Directive 2000/31 or Directive on electronic commerce	Directive 2000/31 of the European Parliament and of the Council of 8 June 2000 on certain legal aspects of information society services, in particular electronic commerce, in the Internal Market, OJ 2000 L 178, p. 1
Directive 2001/21 or InfoSoc Directive	Directive 2001/29 of the European Parliament and of the Council of 22 May 2001 on the harmonisation of certain aspects of copyright and related rights in the information society, OJ 2001 L 167, p. 10
Directive 2001/83	Directive 2001/83 of the European Parliament and of the Council of 6 November 2001 on the Community code relating to medicinal products for human use, OJ 2001 L 311, p. 67
Directive 2004/48	Directive 2004/48 of the European Parliament and of the Council of 29 April 2004 on the enforcement of intellectual property rights, OJ 2004 L 157, p. 45
Directive 2005/29 or Unfair Commercial Practices Directive	Directive 2005/29 of the European Parliament and of the Council of 11 May 2005 concerning unfair business-to-consumer commercial practices in the internal market and amending Council Directive 84/450, Directives 97/7, 98/27 and 2002/65 of the European Parliament and of the Council and Regulation 2006/2004 of the European Parliament and of the Council, OJ 2005 L 149, p. 22
Directive 2006/115 or Rental and Lending Rights Directive	Directive 2006/115 of the European Parliament and of the Council of 12 December 2006 on rental right and lending right and on certain rights related to copyright in the field of intellectual property, OJ 2006 L 376, p. 28
Directive 2009/24 or Software Directive	Directive 2009/24 of the European Parliament and of the Council of 23 April 2009 on the legal protection of computer programs, OJ 2009 L 111, p. 16

Directive 2012/28 or Orphan Works Directive	Directive 2012/28 of the European Parliament and of the Council of 25 October 2012 on certain permitted uses of orphan works, OJ 2012 L 299, p. 5
Directive 2014/104	Directive 2014/104 of the European Parliament and of the Council of 26 November 2014 on certain rules governing actions for damages under national law for infringements of the competition law provisions of the Member States and of the European Union, OJ 2014 L 349, p. 1
Directive 2019/1 or ECN+ Directive	Directive 2019/1 of the European Parliament and of the Council to empower the competition authorities of the Member States to be more effective enforcers and to ensure the proper functioning of the internal market (ECN+ Directive), OJ 2019 L 11, p. 3
DKK	Danish krone
DM	Deutsche mark
e.g. or eg	For example
EC	European Community
ECHR	Council of Europe, European Convention for Human Rights of 4 November 1950
ECJ	European Court of Justice (before 1 December 2009)
ECR	European Court Reports
ECtHR	European Court of Human Rights
EDPB	European Data Protection Board
EU	European Union
Ff	And following
GBP	Pound sterling (UK)
GATT 1947	General Agreement on Tariffs and Trade 1947
GATT 1994	General Agreement on Tariffs and Trade 1994
GC	General Court of the CJEU (after 1 December 2009)
GDP	Gross domestic product
GVH	Hungarian Competition Authority (Gazdasági Versenyhivatal)
Guidelines on the effect on trade concept	Commission Notice—Guidelines on the effect on trade concept contained in Articles 81 and 82 of the Treaty, OJ 2004 C 101, p. 81
Guidelines on Vertical Restraints	Guidelines on Vertical Restraints, OJ 2010 C 130, p. 1
	Commission staff working document of 25 May 2016. Guidance on the implementation/application

Guidance on the implementation of the UCP Directive	of Directive 2005/29 on unfair commercial practices accompanying the document communication from the commission to the European parliament, the Council, the European economic and social Committee and the Committee of the regions a comprehensive approach to stimulating cross-border e-commerce for Europe's citizens and businesses (swd/2016/0163 final)
ha	Hectare
Hague Agreement	Hague Agreement Concerning the International Registration of Industrial Designs, 1925
HRK	Croatian kuna (*hrvatska kuna*)
HMT	Hypothetical monopolist test
HUF	Hungarian forint (*Magyar forint*)
i.e.	Id est (that is)
Id./Idem.	The same as previously mentioned
IP	Intellectual property
Lisbon Agreement	Lisbon Agreement for the Protection of Appellations of Origin and their International Registration of October 31, 1958, as revised at Stockholm on July 14, 1967, and as amended on September 28, 1979
Locarno (Classification) Agreement	Locarno Agreement Establishing an International Classification for Industrial Designs, 1979
M	Million
m^2	Square metre
Min	Minutes
Madrid Agreement	Madrid Agreement Concerning the International Registration of Marks, 1979
MFN	Most favoured nation
NAAT-rule	The non-appreciable affectation of trade rule
Nice (Classification) Agreement	Nice Agreement Concerning the International Classification of Goods and Services for the Purposes of the Registration of Marks, 1979
Notice on the relevant market	Commission Notice on the definition of relevant market for the purposes of Community competition law, OJ 1997 C 372, p. 5
OECD	Organisation for Economic Co-operation and Development
OJ	Official Journal
p./pp.	Page(s)
para/paras	Paragraph(s)

Phonograms Convention	Convention for the Protection of Producers of Phonograms Against Unauthorized Duplication of Their Phonograms, 1971
PTL	Patent Law Treaty, 2000
Paris Convention	Paris Convention for the protection of industrial property, 1883
pt/pts	Point(s)
kg	Kilogram
R&D	Research and development
Regulation 2659/2000	Commission Regulation 2659/2000 of 29 November 2000 on the application of Article 81(3) of the Treaty to categories of research and development agreements, OJ 2000 L 304, p. 7
Regulation 1400/2002	Commission Regulation 1400/2002 of 31 July 2002 on the application of Article 81(3) of the Treaty to categories of vertical agreements and concerted practices in the motor vehicle sector, OJ 2002 L 203, p. 30
Regulation 1/2003	Council Regulation 1/2003 of 16 December 2002 on the implementation of the rules on competition laid down in Articles 81 and 82 of the Treaty, OJ 2003 L 1, p. 1
Regulation 2006/2004	Regulation 2006/2004 of the European Parliament and of the Council of 27 October 2004 on cooperation between national authorities responsible for the enforcement of consumer protection laws (the Regulation on consumer protection cooperation), OJ 2004 L 364, p. 1
Regulation 139/2004	Council Regulation 139/2004 of 20 January 2004 on the control of concentrations between undertakings, OJ 2004 L 24, p. 1
Regulation 110/2008	Regulation 110/2008 of the European Parliament and the Council of 15 January 2008 on the definition, description, presentation, labelling and protection of geographical indications of spirit drinks, OJ 2008 L 39, p. 16
Regulation 330/2010	Commission Regulation 330/2010 of 20 April 2010 on the application of Article 101(3) of the Treaty on the Functioning of the European Union to categories of vertical agreements and concerted practices, OJ 2010 L 102, p. 1
Regulation 1151/2012	Regulation 1151/2012 on quality schemes for agricultural products and foodstuffs, OJ 2012 L 343, p. 1

Regulation 608/2013	Regulation 608/2013 of the European Parliament and of the Council of 12 June 2013 concerning customs enforcement of intellectual property rights and repealing Council Regulation 1383/2003, OJ 2013 L 181, p. 15
Regulation 1308/2013	Regulation 1308/2013 of the European Parliament and of the Council of 17 December 2013 establishing a common organisation of the markets in agricultural products and repealing Council Regulations 922/72, 234/79, 1037/2001 and 1234/2007, OJ 2013 L 347, p. 671
Regulation 251/2014	Regulation 251/2014 of the European Parliament and of the Council of 26 February 2014 on the definition, description, presentation, labelling and the protection of geographical indications of aromatized wine products, OJ 2014 L 84, p. 14
Regulation 316/2014	Commission Regulation 316/2014 of 21 March 2014 on the application of Article 101(3) of the Treaty on the Functioning of the European Union to categories of technology transfer agreements, OJ 2014 L 93, p. 17
Regulation 2016/679 or GDPR or General Data Protection Regulation	Regulation 2016/679 of the European Parliament and of the Council of 27 April 2016 on the protection of natural persons with regard to the processing of personal data and on the free movement of such data, and repealing Directive 95/46/EC (General Data Protection Regulation), OJ 2016 L 119, p. 1
Regulation 2019/1150 or P2B Regulation	Regulation 2019/1150 of the European Parliament and of the Council of 20 June 2019 on promoting fairness and transparency for business users of online intermediation services, OJ 2019 L 186, p. 57
Rome Convention	International Convention for the Protection of Performers, Producers of Phonograms and Broadcasting Organizations, 1961
RON	Romanian leu
SEK	Swedish krona
SIEC	Significant Impediment of Effective Competition
Singapore Treaty	Singapore Treaty on the Law of Trademarks, 2006
SMEs	Small and medium-sized enterprises
SMP	Significant market power
SSNIP	Small but significant and non-transitory increase in price
TEC	Treaty Establishing the European Community

TFEU	Treaty on the Functioning of the European Union
TLT	Trademark Law Treaty, 1994
TRIPs	Agreement on Trade-Related Aspects of Intellectual Property Rights, Annex 1C of the Marrakesh Agreement Establishing the World Trade Organization, signed in Marrakesh, Morocco on 15 April 1994
UAH	Ukrainian hryvnia
UK	United Kingdom
US/USA	United States of America
V	Versus
Vienna (Classification) Agreement	Vienna Agreement Establishing an International Classification of the Figurative Elements of Marks, 1985
WCT/WIPO Copyright Treaty	World Intellectual Property Organization Copyright Treaty, 1996
WPPT	WIPO Performances and Phonograms Treaty, 1996

Part I

Antitrust in Data Driven Markets

International Report

1

Dominik Schäfers

1.1 Introduction

This International Report deals in a broader sense with the role of antitrust authorities regarding the digital economy. On the basis of selected issues, this Report examines in particular to what extent the applicable competition law regimes within the individual jurisdictions are suitable for responding to the challenges of digital markets.

As a first step, this article deals in general with the challenges posed by digitalisation to competition law. In particular, general developments in the legislation of the jurisdictions examined are taken into account (Sect. 1.2). Subsequently, the report deals with individual issues in detail (Sects. 1.3 to 1.11). Finally, some fundamental results are summarised (Sect. 1.12).[1]

[1] The author thanks the National Reporters for their excellent contributions; from Australia: Barbora Jedličková, Brenda Marshall and Mark Burdon; Austria: Gerhard Fussenegger and Viktoria H.S.E. Robertson; Belgium: Friso Bostoen; France: Laurence Charvoz, Liliana Eskenazi, Nizar Lajnef, Jean-Julien Lemonnier and Lauren Mechri; Germany: Mark-Oliver Mackenrodt; Hungary: Judit Firniksz and Péter Mezei; Norway: Ingrid Halvorsen Barlund, Ronny Gjendemsjø and Eirik Østerud; Switzerland: Johana Cau; and the United Kingdom: Sir Christopher Bellamy. The National Reports are based on a questionnaire that was made available to the National Reporters. The International Report has based its systematisation and summary of the National Reports on the

D. Schäfers (✉)
Attorney at law, Counsel, BLD Bach Langheid Dallmayr Rechtsanwälte Partnerschaftsgesellschaft mbB, Cologne, Germany

Entlastungsprofessur Zivilrecht, Goethe Universität Frankfurt a.M., Frankfurt am Main, Germany

Institut für Internationales Wirtschaftsrecht, Westfälische Wilhelms-Universität Münster, Münster, Germany
e-mail: dominik.schaefers@bld.de; schaefers@jur.uni-frankfurt.de; dominik.schaefers@uni-muenster.de

B. Kilpatrick et al. (eds.), *Antitrust in Data Driven Markets & Legal Framework for Influencers, Native Advertising and Control over the Use of AI in Marketing*, LIDC Contributions on Antitrust Law, Intellectual Property and Unfair Competition, https://doi.org/10.1007/978-3-031-07422-6_1

1.2 Background

Due to its disruptive potential, digitalisation poses substantial challenges to societies and legal systems worldwide. The digitalisation of markets has profoundly changed the way in which goods and services are designed, produced, marketed and sold.[2] New, often two-sided or multi-sided markets are emerging, in which the end customer's consideration is partly no longer a monetary payment but is "paid for" with personal data or attention.[3] The role of consumers is no longer limited to buying products; rather, they create new content themselves in many ways and exchange information about products online.[4] Digitalisation enables companies in an unprecedented way not only to meet the individual needs of consumers, especially through individualised advertising or comparison portals, but also to influence consumer behaviour.[5] Moreover, digitalisation has enabled the emergence of large, powerful platforms that benefit in particular from network effects, economies of scale and the scope of and access to data, and which also benefit from being integrated into their own ecosystems, and from mediating the majority of transactions between commercial and end users, acting as gatekeepers between them.[6] All of this comes with the risk that users will become dependent on powerful companies, that data protection regulations will be disregarded, that the degree of innovation in digital markets will decrease and that users will be treated unfairly.[7]

These developments naturally affect the legal framework of competition law in the individual jurisdictions. The competition law of all jurisdictions is, at least as a starting point, neutral with regard to the question of whether companies are active in "the digital or analogue world". The "traditional" legal framework can therefore also be applied to online markets. In this respect, at least from the point of view of the competition authorities or the legislator, it is advantageous that especially the prohibitions of abuse in the individual jurisdictions are often formulated in general clauses. Thus, the intervention standards are sufficiently flexible to be able to react to new technical developments and permanently accompany their rapid further development.[8]

It is, of course, obvious that the challenges posed by digitalisation for competition law cannot be overcome by resorting to traditional general clauses alone. In

---

systematics of the questionnaire. Other special features that go beyond the questionnaire and were mentioned in the National Reports were taken into account if, in the opinion of the National Reporter, they were of general significance or of particular interest. Here and there, the author has added his own critical comments.

[2] Austrian Report, Sect. 3.1; Hungarian Report, Sect. 7.1.

[3] See, on this issue, Sect. 1.4 below.

[4] Hungarian Report, Sect. 7.1.

[5] In detail *Schäfers*, Rechtsgeschäftliche Entscheidungsfreiheit im Zeitalter von Digitalisierung, Big Data und Künstlicher Intelligent, AcP 221 (2021), p. 32 et seq. with further reference.

[6] Proposal for a Regulation of the European Parliament and of the Council on contestable and fair markets in the digital sector (Digital Markets Act), COM(2020) 842 final, p. 1.

[7] UK Report, Sect. 10.1.1.

[8] See, for instance, Austrian Report, Sect. 3.2.2; Swiss Report, Sect. 9.1.

particular, the associated legal uncertainty is also unacceptable from the point of view of the rule of law. Moreover, it is to be feared that dangers to competition could not be recognised and, above all, not be combated sufficiently effectively. For example, the question arises to what extent the design of abuse proceedings should be, in particular with regard to their complexity and duration, to be a sufficiently adequate means to counter abusive behaviour in a constantly and rapidly developing high-technology sector?[9] On the other hand, there would be a danger that (exclusive) recourse to traditional legal frameworks not considering the specifics of digital markets would hinder innovation. In this respect, a quotation from the Swiss Competition Commission is convincing: "A fight with equal arms is more useful to competition than a blind application of old rules in the face of new forms of economy. It is therefore necessary to take a critical look at existing regulation."[10]

Even if not all legislators in the jurisdictions studied have yet enacted new competition law regulations explicitly tailored to digital markets, it can be said that the requirement to have "a critical look" at existing regulations has already been met in all jurisdictions. The current discourse on the application of competition law to digital markets can be characterised in general terms by a number of trends and features.[11]

Across all jurisdictions, there has been an increase in reports, sector enquiries and market studies, with which the competition authorities in particular are addressing the phenomenon of digitalisation and the associated challenges of competition law.[12] From an institutional point of view, the competition authorities are reacting to the challenges posed by digitalisation in part by setting up special departments that focus on the development of digital technologies. In the United Kingdom, for instance, there has been created a Digital Markets Taskforce in March 2020, led by the Competition and Markets Authority (CMA) with support from other authorities. Its task is to provide advice to the government's decisions on how to promote competition in digital markets to the benefit of consumers and businesses.[13] Additionally, a Digital Markets Unit (DMU) has been established within the framework of the CMA, which is to be endowed with potentially far-reaching competences in the future.[14]

In all jurisdictions, the realisation is also coming to the fore "that competition law is not an island . . . but part of a broader set of legal rules that apply to digital markets and that, ideally, should work together".[15] In this respect, data protection law, fair

[9] Cf. on this aspect for instance Austrian Report, Sect. 3.2.2; Belgian Report, Sect. 4.2; Hungarian Report, Sect. 7.5.6.

[10] Competition Commission annual report, 2016, p. 31; Swiss Report, Sect. 9.1.

[11] They do not necessarily apply to every single jurisdiction; cf. on the following systematisation in particular the Austrian Report, Sect. 3.1.

[12] See the list in the Austrian Report, Sect. 3.1; see also the respective National Reports.

[13] UK Report, Sect. 10.1.

[14] UK Report, Sect. 10.1; see for further examples French Report, Sect. 5.1; Swiss Report, Sect. 9.1.

[15] Austrian Report, Sect. 3.1.

trading law, telecommunications law, consumer protection law and also general private law, in particular contract law, are addressed. The "smooth interaction" of these different areas of law requires, among other things, that different national authorities, e.g. data protection authorities and competition authorities, cooperate more closely with each other and that there is also greater international cooperation between the authorities. Just for the sake of completeness, it should be pointed out that the interlocking of the legal areas raises not only institutional but also difficult legal-dogmatic and especially legal-theoretical problems. These aspects will be examined in more detail later on.[16]

As far as the legal framework is concerned, it is worth mentioning first of all that in some jurisdictions, in view of the challenges posed by digitalisation, legal instruments are experiencing a kind of renaissance that was not intended to regulate digital markets. In addition, some legislators have already reacted and started to enact specific legal norms tailored to digitalisation. At the EU level, the European Parliament has meanwhile approved the Commissi's proposal for the Digital Marktes Act (DMA), which primaril adresses the challenges posed by gatekeepers, and which will probably enter into force in autums 2022.[17]

At the national level, for example,[18] the German legislator has already repeatedly taken action to adapt the German competition law to some challenges of digitalisation. For instance, the legislator has clarified that the assumption of a market is not precluded by the fact that a service is provided free of charge.[19] The Austrian legislator is going to introduce a provision according to which data are incorporated as a relevant factor for the assessment of dominance in order to combat anti-competitive behaviour in digital markets.[20]

The legal framework in the United Kingdom has not yet been changed. However, based on the Furman Report, the CMA's Digital Report concluded that a specific pro-competitive regime for companies distinguished as having a strategic market status (SMS) is urgently needed.[21] The government has already responded to this proposal in a Government Response and shared key findings from the CMA Digital Report. In particular, the government agrees that the United Kingdom's existing competition tools are not sufficient to meet the challenges of dominant platforms in the digital advertising, general search and social media markets and that there has to

[16] See Sect. 1.9 below.

[17] Proposal for a Regulation of the European Parliament and of the Council on contestable and fair markets in the digital sector (Digital Markets Act), COM(2020), 842 final. Available at https://eur-lex.europa.eu/legal-content/EN/TXT/PDF/?uri=CELEX:52020PC0842&from=DE.

[18] The following examples are not exhaustive. For further details on changes to national law, see the National Reports.

[19] Section 18 (2a) Gesetz gegen Wettbewerbsbeschränkungen in der Fassung der Bekanntmachung vom 26. Juni 2013, BGBl. I p. 1750, 3245.

[20] Kartell- und Wettbewerbsrechtsänderungsgesetz 2021 (KaWeRÄG 2021), 114/ME XXVII. GP (Ministerialentwurf), introducing a new § 28a KartG.

[21] UK Report, Sect. 10.3.

be established an enforceable code of conduct to govern the behaviour of platforms funded by digital advertising as far as they have an SMS status.[22]

Finally, it should be noted that in all jurisdictions, even in those in which the legislator has not yet explicitly reacted to the competition law challenges of digitalisation, there are a large number of official decisions, especially in the areas of merger and abuse, which deal with digital markets. The most prominent decision so far is probably the Facebook decision of the German Bundeskartellamt, which is currently also being dealt with by the Court of Justice of the European Union (CJEU) within the framework of a referral procedure according to Art. 267 of the Treaty on the Functioning of the European Union (TFEU). We will return to this decision later.[23]

1.3 Defining the Term "Platform"

1.3.1 Legal Framework

It is clear from all National Reports that online platforms pose considerable challenges to the law, especially antitrust law, and that the regulation of digital platforms is a top priority from the perspective of the national competition authorities. In particular, the focus is on questions of market definition and the determination of market power as well as the possibilities of digital platforms, in particular, to abuse their (potential) market power due to their huge data stock. In this respect, the necessity arises to answer the question of what is meant by a digital platform in the first place. To that extent, it is clear that most legislation does not contain legal provisions that define or address platforms explicitly. Conclusions can insofar only be drawn from initial decisions of the authorities and courts, reports or case studies. In this context, the already mentioned advantage of the relatively vague definitions of antitrust norms within the individual jurisdictions becomes apparent once again as they allow to also get a grip on digital platforms.

Since 2017, German competition law explicitly has addressed platforms, even though the relevant section of the German Cartel Act[24] does not use the term "platform" but speaks of "multilateral markets". In spite of this first regulatory approach, the German legislator has, however, deliberately refrained from a specific definition of the term platform since this term had not yet been used uniformly in law and a uniform definition would only emerge over time.[25] Primarily, the Bundeskartellamt and the courts will thus drive the exact definition of the term platform forward in Germany. The German legislator's approach is interesting in

[22] UK Report, Sect. 10.3.1.1.; cf. as well Sect. 1.6.5 below.

[23] Section 1.9.3. below.

[24] Section 18 (3a) Gesetz gegen Wettbewerbsbeschränkungen in der Fassung der Bekanntmachung vom 26. Juni 2013, BGBl. I p. 1750, 3245.

[25] Legislative reasons, Official Gazette 18/10207, p. 47.

that it represents a kind of a compromise between the basic options of either deciding against specific digitalisation-related regulations in the interest of the flexibility of antitrust law or enacting specific norms tailored to digitalisation in the interest of legal certainty.

French law also has a legal definition for online platform providers, although this is regulated not in antitrust law but in the French Consumer Code and, in particular, provides for information obligations of platform providers towards consumers. According to the relevant section,[26] an online platform provider is *any natural or legal person offering, on a professional basis, including for free, an online communication service to the public that is based on (1) ranking or referencing contents, goods or services offered or uploaded by third parties by using computerised algorithms; (2) allowing several parties to get in contact with one another for the sale of goods, the provision of services or the exchange or sharing of content, goods or services.*

1.3.2 Characteristics

There seems to be an agreement that platforms are characterised by their multi-sided nature: the platform operator is an intermediary connecting two or more distinct but interdependent groups of users and enables direct interaction between different user groups for the provision of goods, services, digital content and information online.[27] The activity of the platform operator is limited to enabling contact between the different user groups, using the platform for different reasons.[28] Digital platforms are, e.g., e-commerce platforms, peer-to-peer platforms, social media networks and search engines.[29]

It is typical for digital platforms to collect and use a variety of data in order to provide their services.[30] The integration of different services is also a key concept in the business model of online platforms[31] because as soon as the platform operator has won customers, especially by offering them a service "for free", they demand money from the platform users on the other side of the market. For example, Amazon earns money from the fact that retailers on the Amazon Market Place have to pay a commission to Amazon, while Google provides its search service to

[26] Article L. 111-7 Code de la Consommation.

[27] Cf. the definition by the Bundeskartellamt, Decision of 8 September – B6-126/14, para 121; see also Bundeskartellamt, Working Paper, Marktmacht von Plattformen und Netzwerken, 2016, p. 14; Australian Report, Sect. 2.2; Belgian Report, Sect. 4.3.1; French Report, Sect. 5.3.2.1 (structuring platform); German Report, Sect. 6.3; Hungarian Report, Sect. 7.3.1 with reference to the Guidance on the Implementation/Application of Directive 2005/29/EC on Unfair Commercial Practices, section 5.2.1; UK Report, Sect. 10.2.1.

[28] Australian Report, Sect. 2.2.

[29] Australian Report, Sect. 2.2; UK Report, Sect. 10.2.1.

[30] UK Report, Sect. 10.2.1.

[31] UK Report, Sect. 10.2.1.

customers for free in order to generate advertising revenue on the other side of the market.[32]

The German report highlights the importance of indirect network effects to digital platforms. They occur when the benefit that the platform has for one user group is influenced by the number of users of another user group. According to the German Bundeskartellamt, indirect network effects can also be negative, for example if an additional "advertising unit" has no benefit for the user but only a disadvantage, e.g. in the form of a time loss. The indirect network effect is in any case qualified in German public authority practice and science as a sufficiently concrete criterion for distinguishing platforms from networks.[33] These platforms and networks enable users to establish exchange relationships among themselves. Such networks result in *direct* network effects since the benefit associated with the facility for the individual user also increases as the number of users on the same market side grows.[34]

1.3.3 Different Types of Digital Platforms

In some jurisdictions, several types of platforms are differentiated.

The Australian Competition and Consumer Commission (ACCC), in line with the Australian Digital Platforms Inquiry (DPI[35]), distinguishes between search engines, social media platforms and digital content aggregation platforms.[36] It is noteworthy that other digital platforms, for instance shopping platforms like Amazon, were not included in the DPI since the DPI's scope was narrow and market specific.[37] The UK report states that online marketplaces such as Amazon as well as comparison portals offer consumers the benefit of conducting a specialist search.[38] These specialised search engines are to be distinguished from general search engines from the consumer's point of view.[39] General search engines help consumers with a wider range of queries, including many who are not served by specialised providers, while specialised search provides functionality that allows consumers to compare listed

---

[32] UK Report, Sect. 10.2.1.

[33] *Fuchs*, In: Immenga/Mestmäcker (ed.), Wettbewerbsrecht, 6th Ed., Beck 2020, § 18 RWB para 71.

[34] *Fuchs*, In: Immenga/Mestmäcker (ed.), Wettbewerbsrecht, 6th Ed., Beck 2020, § 18 RWB para 71.

[35] https://www.accc.gov.au/system/files/Digital%20platforms%20inquiry%20-%20final%20report.pdf. Accessed 24 October 2021.

[36] Australian Report, Sect. 2.2.

[37] Australian Report, Sect. 2.2; it only focused on "*digital search engines, social media platforms and other digital content aggregation platforms (platform services) on the state of competition in media and advertising services markets*" (see DPI p. 537).

[38] UK Report, Sect. 10.2.1.

[39] UK Report, Sect. 10.2.1.

products across multiple dimensions.[40] However, the report emphasises that it is unlikely that there will be a formal approach distinguishing between different types of platforms or different functions within platforms. On the contrary, each situation will be judged on its own facts.[41]

The Belgian and German reports agree as a starting point that it is necessary to differentiate between two types of platforms, namely matching platforms and advertising platforms:[42]

Matching platforms, for instance Amazon Marketplace and the App Store, tend to achieve the best possible matchmaking between users from different groups. The interaction can lead to an actual transaction, especially a purchase, but this is not mandatory, as the example of dating platforms shows.[43] In addition, platforms often maintain an uneven prize structure in the sense that one side of the user does not have to pay a price for the services. Platforms that aim to bring about an actual transaction are referred to as transaction platforms.[44] These form a subgroup of matching platforms.

The German report classifies advertising platforms additionally as a subset of so-called attending platforms, which are to be differentiated from matching platforms.[45] On such platforms, offering a service to one group of users does not necessarily require the existence of another market side. Rather, it is a strategic decision to place advertisements in order to finance the other market side. One example is Spotify, a platform that offers its users the possibility to stream music titles on the basis of licences. If users opt for the free variant, the service is financed by the insertion of advertising. In this case, the advertisers are a second user group. However, customers can also choose a paid variant in which no advertisements are shown. In this case, there is no second user group, so in the view of the Bundeskartellamt, Spotify does not function as a platform.[46]

1.3.4 Market Definition

With regard to the different platform services and the different market sides, the question of market definition arises. For example, on the online sales platform Amazon, goods are sold both by Amazon itself and by third-party sellers via the Amazon Marketplace, without users being able to see a sufficient distinction between Amazon's own products and those of third-party sellers. In view of the fact that platforms always have at least two user groups, it would be conceivable that

[40] UK Report, Sect. 10.2.1.

[41] UK Report, Sect. 10.2.1.

[42] Belgian Report, Sect. 4.3.1; German Report, Sect. 6.3.

[43] German Report, with reference to BKartA, 22.10.2015, B6-57/15 – *OCPE/EliteMedianet.*

[44] German Report.

[45] German Report.

[46] Bundeskartellamt, Arbeitspapier – Marktmacht von Plattformen und Netzwerken, 2016, p. 28.

the platform services for both market sides each represent independent products and markets. In consideration of the indirect network effects, which are typical for platforms and the resulting interrelationships, it would also be conceivable to consider the platform as a whole to be a single market.

The national authorities from Germany and Belgium follow a nuanced approach in this respect: if the platform is a matching platform or a transaction platform, both sides of the market are to be classified as a single market. On the other hand, the two sides of an attention platform are to be qualified as different markets.[47] For example, the Belgian competition authority has qualified Immoweb's real estate website as a matching platform that operates on a multi-sided market with buyers on one side and sellers on the other. Accordingly, it is to be assumed that there is a single market for online real estate portals. This corresponds to approaches in the economic literature[48] and is also in line with the European Commission's approach in this regard.[49] However, the German Report also emphasises that the question must be decided on a case-by-case basis.[50]

In Austria, there is neither a legal definition of platforms nor soft law dealing with platforms. How the Australian National Capital Authority (NCA) will deal with platforms can, to some extent, be derived from a case study of its Amazon investigation and settlement. Although the authority clarifies that it would be insufficient to qualify platforms as one-sided markets in view of Amazon's trading platforms, it remains open as to whether multi-sided markets can be qualified as separate markets for the respective user types or as a single market.[51] In any case, the interdependencies between the individual market sides would have to be taken into account. In a decision against Amazon, the authority came to the conclusion that the market for online trading platforms does not include stationary shops and web shops of retailers.[52]

The Australian ACCC has already been able to take into account the two-sided nature of the market, particularly in the review of Expedia's proposed acquisition of Wotif in 2017.[53] Nevertheless, the Australian Report emphasises that in view of the current lack of cases, it is not yet possible to make statements on the definition of the relevant market. The Swiss Report underlines as well the complexity of market definition in multi-sided digital markets using the example of the Android platform,

---

[47] Belgian Report, Sect. 4.3.1; German Report, Sect. 6.3.

[48] *L. Filistrucchi, D. Geradin, E. van Dammer, P. Affeldt*, Market definition in two-sided markets: theory and practice. Journal of Competition Law and Economics 2014 (2), pp. 293–339.

[49] The approach is probably also being followed in Switzerland, as the Report of the Federal Council on the main framework conditions for the digital economy, 11 January 2017, p. 160f. also refers to *L. Fillistrucchii* et al.

[50] German Report, Sect. 6.3.

[51] Austrian Report, Sect. 3.2.5.

[52] Austrian Report, Sect. 3.2.5.

[53] ACCC Public Register (n.d.) Expedia Inc – proposed acquisition of Wotif.com Holdings Limited, available at http://registers.accc.gov.au/content/index.phtml/itemId/1182044/fromItemId/751046. Accessed 25 October 2021.

on which at least five parties interacted in addition to Google. However, the Swiss competition authority does not seem to have developed any further criteria as to how the market definition is to be carried out in individual cases.[54]

In summary, it is surprising that, given the importance of market definition in competition law and the role of digital platforms, there is so little agreement on how a market definition should be carried out in individual cases.

1.4 Market for Free Services

1.4.1 General

As far as can be seen, in all jurisdictions, supply markets are based on the substitutability of the product from a demand-side point of view (demand market concept). According to the small but significant non-transitory increase in prices (SSNIP) test practised in this respect, it depends on whether the demanders would turn to an alternative product if the supplier were to increase the price of the offered product by 5–10 per cent. If this question is answered in the affirmative, the alternative product belongs to the supplier's market. Otherwise, the alternative product belongs to a different market.[55]

Especially in digital markets, however, suppliers often do not demand monetary consideration for their own performance. This is especially true in the two-sided markets already mentioned. For example, the use of the Internet search engine Google or the use of Instagram or Facebook is free of charge for one user group of the platforms, while another user group, especially advertisers, provides a charge for distributing their product, for instance advertisement. In this respect, the question arises whether the market quality is to be affirmed only for the paid side or also for the free side. Even if this question is closely connected to the question of market definition in two-sided markets,[56] it can in principle also arise in one-sided markets.[57]

[54] Swiss Report, 9.2.2; cf. also Report of the Federal Council on the main framework conditions for the digital economy, 11 January 2017, p. 162f., retrievable (in French) under: https://www.newsd.admin.ch/newsd/message/attachments/46894.pdf.

[55] Cf. in general *Lindsay*, EC Merger Regulation: Substantive Issues, Sweet Maxwell 2003, p. 75 et seq.

[56] Section 1.4.3 above; Norwegian Report, Sect. 8.3.2.

[57] German Report, Sect. 6.4.

1.4.2 Fundamental Acceptance of the Concept of Markets for Free Services

It seems generally accepted in all jurisdictions examined that the fact that a service is offered free of charge does not prevent the assumption of a market.[58] As examples, markets for supplying social media services and supplying general search services are mentioned. This view also corresponds to the view of the European Commission, e.g. in *Facebook/WhatsApp*.[59] It can be decisive for the success of a platform that the operator gains as many customers as possible on one side of the market in order to be able to achieve high network effects on the other side of the market. Users whose benefit from the platform does not depend or depends only to a limited extent on the number of users on the other side of the market will be won over in particular by low prices. In this respect, it may make economic sense for the platform operator not to demand any monetary consideration at all.[60]

A corresponding legal regulation has so far only been found in Germany. There, the legislator expressly clarified in 2017 that it does not prevent the assumption of a market that a service is provided free of charge.[61] This was also in line with the Bundeskartellamt's previous practice.[62] In contrast, the Oberlandesgericht (OLG) Düsseldorf still assumed in 2015 that in the case of hotel booking portals, only the paid side constitutes a market.[63] The Bundeskartellamt has already applied the new provision, for example, to a platform that provides services with regard to event tickets. The relationship between the platform and advance booking agencies was qualified as a market, although the service was free of charge.[64] In the Facebook case, the authority also assumed that the social network services qualified as a market despite being free of charge.[65]

In other jurisdictions, national law does not provide for a legal provision comparable to the German regulation. However, as already mentioned, the possibility of a market for free services is also recognised here in principle. In particular, corresponding statements can sometimes be found in publications of the NCA,

---

[58] Australian Report, Sect. 2.3; Austrian Report, Sect. 3.2.6 ("undisputed"); Belgian Report, Sect. 4.3.1; French Report, Sect. 5.4.; German Report, Sect. 6.4; Hungarian Report, Sect. 7.4; Norwegian Report, Sect. 8.3.2; Swiss Report, Sect. 9.2.2; UK Report, Sect. 10.2.2.

[59] COMP/M.7217.

[60] German Report, Sect. 6.4; cf. also Norwegian Report, Sect. 8.3.2.

[61] Section 18 (2a) Gesetz gegen Wettbewerbsbeschränkungen in der Fassung der Bekanntmachung vom 26. Juni 2013, BGBl. I p. 1750, 3245.

[62] Bundeskartellamt, Fallbericht vom 25. Juni 2015, Az. B6-39/15 – Online-Immobilienplattformen; Beschluss vom 22. Oktober 2015, Az. B6-57/15 – Online-Datingplattformen.

[63] OLG Düsseldorf, 9 January 2015, VI Kart 1/14 (V), para. 43.

[64] Bundeskartellamt, Decision of 4 December 2017 – B6-132/14-2, para 144; Bundeskartellamt, Decision of 23 November 2017 – B6-35/17, para. 123.

[65] Bundeskartellamt, Decision of 6 February 2019 – B6-22/16, paras. 239 ff; BGH, Judgment of 23 June 2020 – KVR 69/19, paras. 27 ff. – *Facebook* (interim decision).

and/or the authorities have assumed that a market exists in their case despite the fact that a service is free of charge. It is even assumed that free services of digital platforms do not pose any specific challenges to competition law since corresponding services were already known "in the analogue age".[66] The possibility of a market for free services is justified in particular by the fact that the service is—actually—not for free at all since the user provides another form of non-monetised consideration, in particular through the provision of attention,[67] personal data, preferences and user-generated content.[68] According to the Hungarian Competition Authority (Gazdasági Versenyhivatal (GVH)), even circumstances such as the limiting of the scope of future decisions can be considered a value that is provided by the user.[69] Given that there is a widespread agreement that users provide significant economic value to platforms, it is even suggested that *the zero-price paid is in fact too high and consumers could be extracting a greater value in return for their data.*[70]

Considering that the concept of markets for free services does not seem to pose any major difficulties for the jurisdictions examined, the question may arise whether it is at all advantageous for the legislator to enact a corresponding legal regulation, as in Germany. As the German Report emphasises, Section 18 (2a) German Cartel Act[71] simplifies the application of the law since the assumption of a market, for instance, does not depend on the discussion of whether the user, who does not pay a monetary consideration, might provide another service. In other words, the assumption of a market for free services does not require any special justification.[72] However, this does not mean that under German law, every free service also justifies the assumption of a relevant market. Rather, it is necessary that the free service is offered for economic reasons or as part of a strategy that is at least indirectly profit oriented. In this regard, a private scholarship, for instance, may not be qualified as a market.[73] The German legislator emphasises in particular that the assumption of a market necessarily presupposes an exchange relationship. This makes it clear that the term "free" is only to be understood in the sense that the user does not necessarily owe a monetary consideration. However, some kind of exchange must take place. Thus, the criteria outlined above for determining the existence of a market for free services are therefore likely to remain relevant for German law as well.

66 French Report, Sect. 5.4; UK Report, Sect. 10.2.2.

67 Cf. for Instance UK Report, Sect. 10.2.2.

68 Swiss Report, Sect. 9.2.2; Hungarian Report, Sect. 7.4.1; Norwegian Report, Sect. 8.3.2; UK Report, Sect. 10.2.2.

69 Hungarian Report, Sect. 7.4.

70 UK Report, Sect. 10.2.2 with reference to *Harford*, 'Treat social media like email and search engines', Financial Times, 27 April 2018.

71 Gesetz gegen Wettbewerbsbeschränkungen in der Fassung der Bekanntmachung vom 26. Juni 2013, BGBl. I p. 1750, 3245.

72 German Report, Sect. 6.4.

73 German Report; Sect. 6.4; Legislative reasons, Official Gazette 18/10207, p. 48.

1.4.3 Methodological Issues

Even if the possibility of a market for free services is recognised, there are at least considerable problems with regard to the method of market definition. As already seen above, the SSNIP test usually used in this respect cannot be applied (without modifications) because a hypothetical 5-per-cent price increase of a non-monetised service is logically excluded. The application of the SSNIP test to multi-sided markets is generally made even more complex by the fact that the determination of prices for each side depends on the determination of prices on the other side and therefore cannot be considered separately for both parties.[74]

The Belgian and the Norwegian reports suggest adapting the SSNIP test and—as already practised by the European Commission in *Google Android*[75]—to focus on shifts in consumer demand in response to a quality degradation rather than a price increase (significant non-transitory decrease in quality (SSNDQ) test).[76]

In practice, however, such a test is associated with considerable legal uncertainties because it is unclear on the basis of which criteria "quality" can be determined with legal certainty.[77] In view of these uncertainties, the Belgian NCA has already called for updated guidelines.[78] In Australia, there is the peculiarity that although the assumption of a market for free services would be compatible with the wording of the law, the then Trade Practices Tribunal in the QMA case in 1976 expressly relied on hypothetical price changes for the determination of the market.[79] However, one hopes for clarifying words from the Federal Court in the case *Dialogue Consulting Pty Ltd. v. Instagram Inc & Ors*.

Since the question of a market definition depends on how consumer demand changes in relation to any change in "free" goods, the call of the president of the Belgian NCA for a "more behavioural economic approach" (notion by author) can be endorsed, who, in view of non-price markets, calls for greater consideration of behavioural economics in competition law.[80]

---

[74] Swiss Report, Sect. 9.2.2, with further reference.

[75] CASE AT.40099, paras 284 – 322 and 483 – 566.

[76] Belgian Report, Sect. 4.3.1; Norwegian Report, 8.3.2.

[77] Belgian Report, Sect. 4.3.1 with further reference.

[78] Belgian Report, Sect. 4.3.1; BCA, Decision of 7 November 2016, ABC-2016-I/O-31-AUD, *Immoweb*, paras 22–27.

[79] Australian Report, Sect. 2.3.

[80] Belgian Report, Sect. 4.3.1; with reference to *Steenbergen*, On competition policy and online markets, European Competition and Regulatory Law Review 2019(1), VII.

1.5 Market Power and Abuse of Dominant Position

1.5.1 General

Usually, the criteria for determining market power are market shares and barriers to entry for competitors. These criteria can also be taken into consideration in digital markets. However, they are not as meaningful there as in traditional "analogue" markets. For example, the criterion of turnover fails if a company offers free services. It needs to be kept in mind as well that digital markets are often very dynamic. Even if a company currently has a high market share, this can change within a very short time.

Nevertheless, with the exception of Germany, the legislators within the legal systems examined have not yet taken action and developed criteria that would have to be taken into account when determining market power on digital markets. However, changes to the legal framework are being considered at the suggestion of the competition authorities, for example in Switzerland. In view of a statement by the Competition Commission, the Swiss legislator declared *that it may be necessary and useful to adapt the merger notification criteria so that the authorities can examine mergers or acquisitions of young internet platforms that could possibly impact competition.*[81] In addition, the Swiss Competition Commission pointed out in a 2016 report *that the turnover-based thresholds in merger control could lead to a situation where mergers are not controlled, even though, in relation to customer data, a dominant position exists.*[82] Subsequently, the Swiss government has stated that it could be useful to adapt the merger notification criteria to ensure that competition authorities can also intercept mergers or acquisitions of young Internet platforms that potentially affect competition.[83]

In Hungary, a new merger notification regime had been introduced in 2017. It allows the GVH to exercise control over some lower-value transactions, like start-ups or newly established players on the digital markets with a high development and innovation potential.[84] The Norwegian NCA is well-equipped in the digital sector with *ex ante* enforcement tools against concentrations below the general turnover thresholds for mandatory notification and against acquisitions of non-controlling shareholdings.[85]

The Australian ACCC sees the need for far-reaching reforms to merger control law. There should be a new test, predicating on a lower probability of competitive harm than that which applies to acquisitions in the broader economy. Furthermore, the application of this test to any given digital platform should depend on the size

[81] Swiss Report, Sect. 9.2.2.

[82] Swiss Report, Sect. 9.2.2. with reference to Competition Commission annual report, 2016, p. 28.

[83] Swiss Report, Sect. 9.2.2.

[84] Hungarian Report, Sect. 7.3.2.3.

[85] Norwegian Report, Sect. 8.2.

and scope of the platform's services in Australia and whether it acts as a "gateway".[86]

The current Austrian law, especially in connection with assessing market power, leaves sufficient room for manoeuvre and does not need to be adapted. The problem of applying the current antitrust law to digital cases is not that the existing legal framework is insufficient but that digital cases require more effort and that the authorities are not sufficiently equipped in this respect.[87]

Apparently, the national competition authorities have not yet issued guidelines or other soft law instruments dealing specifically with market power in digital markets. However, with the exception of the Norwegian and Hungarian competition authorities, the national authorities have at least dealt with the phenomenon of market power in digital markets in reports and opinions. Some competition authorities have also been confronted with the issue in case practice in the context of both merger control and abuse proceedings. In this respect, it is noteworthy that the Australian ACCC recently referred to a case from 1976, which cited rather static criteria for the assessment of market power.[88] The Australian Report emphasises that the authority failed to adapt its guidelines to digital markets on this case. Thus, there are concerns as to whether this approach is appropriate for digital markets.[89] The Hungarian Report outlines the fact that despite numerous guidelines and other soft law instruments dealing with digital markets, none of them specially addressed the abuse of market power so that there is some kind of a *vacuum regarding judicial interpretations of market power definitions concerning digital markets*, especially since no cases have yet been decided by the authorities.[90]

1.5.2 Relevant Criteria

As already mentioned, German law is the only legal system to date that provides explicit regulations dealing with the question of which criteria can be relevant for the determination of market power in digital markets. According to German law, market shares remain the starting point for determining the market position of a company, although this criterion is of comparatively minor importance. In order to determine market shares on digital markets, the Bundeskartellamt uses alternative criteria in its case practice, e.g. the used number of search queries with regard to search engines or the number of registered members and individuals with regard to dating platforms.[91] In addition, the provision of Section 18 (3a) German Cartel Act now expressly contains five further criteria that are to be taken into account, in particular, in the case

[86] Australian Report, Sect. 2.6.

[87] Austrian Report, Sect. 3.2.4.

[88] Australian Report, Sect. 2.4.

[89] Australian Report, Sect. 2.4.

[90] Hungarian Report, Sect. 7.3.2.

[91] German Report, Sect. 6.2.1.

of multi-sided markets (meaning platforms) and networks. The provision, which is not exhaustive, mentions (*i*) direct and indirect network effects, (*ii*) the parallel use of networks (so-called multi-homing) and the cost of switching between platforms, (*iii*) economies of scale, (*iv*) access to data and (*v*) competitive pressure through innovation. Finally, Section 18b (3b) German Cartel Act, which addresses companies that are active as intermediaries in multi-homed markets, did come into force in 2021. According to this legal provision, when assessing the market position of such companies, the importance of the intermediary services it provides for access to procurement and sales markets must also be taken into account in particular.

Even if other jurisdictions do not contain any legal provisions as in Germany, similar criteria can be found, in particular, in the reports of the competition authorities and their case practice. When it comes to barriers to entry, in other jurisdictions, criteria such as strong network effects,[92] barriers to switching,[93] the existence of economies of scale,[94] access to user data[95] or the vulnerability of the undertaking's position in the market in relation to innovation cycles[96] are also considered to be essential. Importance is attached not only to the number of data an undertaking possesses but in particular also to the quality of the data.[97] For example, the Hungarian GVH, referring to the European Commission's *Apple/ Shazam*[98] EU case, emphasised that the most valuable data are those that can be used to predict consumer behaviour.[99] In this respect, it becomes clear that some kind of a "more behavioural economic approach" already mentioned above[100] may be important not only in the context of determining the relevant market but also in connection with the determination of market power. This is also made clear in the UK Report, where, inter alia, consumer behaviour and the power of defaults, i.e. default behaviour by consumers and the way in which choices are presented to consumers, are qualified as important criteria.[101]

In the United Kingdom, the CMA noted that metrics such as level of concentrations, user numbers, traffic generated or time spent on each service could be considered by assessing market power. This is consistent with the Belgian Report.[102] Also important are barriers to entry and the degree of control that some platforms have over (i) the relationship between buyers and sellers and (ii) the access

---

92 Austrian Report, Sect. 3.2.4; Belgian Report, Sect. 4.3.1; French Report, Sect. 5.3.1.

93 Austrian Report, Sect. 3.2.4.

94 Belgian Report, Sect. 4.3.1; French Report, Sect. 5.3.1.

95 Austrian Report, Sect. 3.2.4.

96 Swiss Report, Sect. 9.2.2.

97 Hungarian Report, Sect. 7.3.2.2; s. also German Report, Sect. 6.2.1 with reference to the German Legislative reasons, Official Gazette 18/10207, p. 51.

98 Case M.8788 - *Apple/Shazam*.

99 Hungarian Report, Sect. 7.3.2.2.

100 Section 1.4.3 above.

101 UK Report, Sect. 10.2.

102 Belgian Report, Sect. 4.3.1.

that advertisers have to potential buyers.[103] The CMA has also used profitability analysis and has identified the following conditions as creating barriers to entry and undermining competition: (i) network effects and economies of scale, (ii) consumer behaviour and the power of defaults, (iii) unequal access to user data, (iv) problems relating to a lack of transparency and (v) vertical integration and conflicts of interest.[104]

The Norwegian NCA is currently, especially in merger cases, increasing reliance on other analytical tools than structural market factors to measure market power, whereby the authority is inspired in particular by US economic theory and UK merger practice. The NCA bases its decisions especially on customer surveys, analyses of closeness of competition, diversion ratios and price pressure indexes.[105]

1.5.3 Relative Market Power or Economic Dependency

In several of the legal systems examined, the national cartel authorities have powers to take action on the basis of specific competition law standards against companies that do not hold a dominant position. Norway is a notable exception in this respect. The Norwegian Competition Act does not provide for any general possibilities to initiate abuse proceedings against companies that do not have a dominant market position. Although the government would have the legal possibilities to implement corresponding, market-specific regulations, such legal changes are so far only on the agenda for the grocery market but not for digital markets.[106]

What most of the regulations have in common is that the company that is not necessarily in a dominant position holds a position by virtue of which another company is "dependent" on it in the broadest sense.

Thus, under Austrian law, a company is also considered dominant when it has a predominant position in relation to its customers or suppliers. This is the case when customers or suppliers are obliged to maintain business relations with the dominant company in order to avoid suffering significant economic losses.[107] The French NCA can sanction the abusive exploitation of the existence of a situation of economic dependency if the abuse has an actual or potential effect on the functioning of the structure of the competition of the market. In this context, the situation of economic dependency requires that it is impossible for the dependent undertaking to find a technically and commercially equivalent solution.[108] Although the relevant provision of the Code de Commerce itself was initially introduced to protect suppliers against large retailers, this hitherto rather neglected instrument is

[103] UK Report, Sect. 10.2.

[104] UK Report, Sect. 10.2.

[105] Norwegian report, Sect. 8.3.2.

[106] Norwegian Report, Sect. 8.3.7.

[107] Austrian Report, Sect. 3.2.10.

[108] Article L.420-2§2 of the French Commercial Code.

experiencing a renaissance, so to speak, as it is considered suitable to provide good services to assess unfair trading conditions in the field of online platforms and marketplaces.[109] For example, the French NCA has already issued two sanction decisions against Apple and Google based on the relevant regulation in the past year.[110]

The Belgian legislator has taken the outlined French regulation as a model and enacted a law on the abuse of economic dependency in 2020. According to this law, the competition authority can also take action against a company for abusive behaviour if it does not have absolute market power but rather a case of economic dependency exists. This is qualified as a sub-case of relative market power. This regulation was not intended for the digital economy but could play a central role here.[111]

German law also provides for comparable regulations in Section 20 German Cartel Act.[112] According to Section 20 (1) sentence 2, certain prohibitions intended for dominant companies also apply to companies with relative market power. Relative market power of the company requires that other companies are dependent on it as suppliers or customers in such a way that there are no sufficient and reasonable possibilities to switch to third companies. Section 20 (1) sentence 2 German Cartel Act, which was not introduced until 2021, is specifically tailored to digital markets and also provides for prohibitions for undertakings that are active as intermediaries in multi-sided markets, insofar as other undertakings are dependent on their intermediary services in such a way that sufficient and reasonable sales opportunities do not exist. It is worth mentioning in this respect that, according to Section 20 (1a) German Cartel Act, which was also introduced in 2021, dependency within the meaning of Section 20 (1) German Cartel Act can also result from the fact that a company is dependent on access to data controlled by another company for its own activities. The denial of access to such data for a reasonable fee may be abusive. Furthermore, German law allows abuse proceedings for unfair obstruction against companies that have superior market power over small and medium-sized competitors (Section 20 (3) German Cartel Act). According to Section 20 (3a) German Cartel Act, which is also tailored to digital markets and came into force in 2021, an unfair obstruction in this sense can also be that the company with superior market power obstructs the independent achievement of network effects by competitors. As the German Report explicitly emphasises, it can be assumed that the provision of Section 20 German Cartel Act will have considerable practical significance for the protection of competition in digital markets.

The concept of relative market power will also come into force in the Swiss legal system at the end of 2021, whereby the Swiss legislator has been inspired by the

[109] French Report, Sect. 5.9.1.

[110] French Report, Sect. 5.9.1. with further reference.

[111] Belgian Report, Sect. 4.3.2.4.

[112] Gesetz gegen Wettbewerbsbeschränkungen in der Fassung der Bekanntmachung vom 26. Juni 2013, BGBl. I p. 1750, 3245.

German regulations in this respect. In contrast to German law, "the amendment does not explicitly include relative dominance for intermediation on platforms [. . .] and data".[113] The Swiss Report also points out that the Competition Commission and the Swiss courts will develop their own interpretation practice, which currently leads to a considerable degree of legal uncertainty.[114]

According to Section 46 (4)(a)(ii) of the Australian CCA, buying and selling powers are also to be taken into account when assessing the existence of significant market power. This makes it possible, under certain conditions, to assume a misuse of market power under section 46 even in the case of bargaining power. The possibility to take into account vertical and other non-horizontal relations could thus become significant for the determination of market power, especially in the digital world.[115]

1.5.4 Other Legal Instruments

Given the fact that market definition and the proof of a dominant position are associated with considerable difficulties, especially in digital markets; that abuse proceedings are too costly and lengthy; and that the associated ex post assessment of conduct may not be suitable for averting damage from the fast-moving digital markets,[116] there seems to be an increasing focus on legal instruments that allow intervention even if a company does not have a dominant market position. In this respect, more general institutes and instruments beyond the core antitrust law, such as provisions of fair trading law or duties of good conduct under commercial law, are coming into focus as well.[117]

For instance, French competition law allows certain abusive practices (e.g. abruptly terminating established commercial relations) committed by undertakings towards their partners to be sanctioned regardless of whether the company has a dominant position and regardless of the effects of the abusive practice on the market. The potential importance of these instruments for digital markets becomes clear when one realises that the Paris Commercial Court has recently imposed sanctions on Booking, Expedia and Amazon on the basis of such.[118] However, it is not the French NCA that is responsible in this respect but the Commercial Court. In Austria, to give another example, sanctions by the Cartel Court can be considered if a company violates provisions on good business

[113] Swiss Report, Sect. 9.2.5.

[114] Swiss Report, Sect. 9.2.5.

[115] Australian Report, Sect. 2.4.

[116] Belgian Report, Sect. 4.3.2.4; French Report, Sect. 5.9.2.

[117] Cf. for instance Austrian Report, Sect. 3.2.10; French Report, Sect. 5.9.2.

[118] French Report, Sect. 5.9.2.

behaviour, provided the violation is likely to threaten performance-related competition.[119]

The Hungarian Competition Authority GVH has competences in both antitrust and consumer protection. Especially in digital markets, the advantage is that the authority can act independently of proof of a dominant position or abuse. Due to its competences in consumer protection law, the GVH was able to investigate comparable questions as other majorcompetition law enforcing jurisdictions like the European Union and Germany, whereby the advantage in this respect is that consumer protection procedures can be carried out significantly faster.[120]

1.6 Undertakings with Excessive Market Power

1.6.1 General

The activities of companies with excessive market power are followed with great concern. The focus here is primarily on Facebook, Google and their activities in the online advertising market.

According to the investigations of the competition authorities in Australia, France and the United Kingdom, Facebook and Google play a central role in the market for online advertising.[121] Various reasons are given for the strong position of both companies. In general, they offer specific benefits, such as targeted advertising, under more favourable conditions than those offered by other providers, especially of traditional advertising.[122] Google can permanently improve its algorithms based on "click-and-query-data" from the broad user base and now even trace users offline.[123] In addition, compared to their competitors, both companies have significantly larger data volumes, which are also constantly updated, and are active at both publishing and technical intermediation levels. Moreover, almost all social platforms, including Facebook in particular, make the use of their platforms conditional on users agreeing to personalised advertising (take it or leave it). In particular, the lack of alternatives forces users to disclose more data about themselves than they would usually be willing to disclose.[124] Overall, the large number of users of Facebook and Google are a source of market power due to network effects and economies of scale. Additionally, the data possessed by these companies can be a

[119] Austrian Report, Sect. 3.2.10.

[120] Hungarian Report, Sects. 7.1 and 7.3.2.

[121] Australian Report, Sect. 2.9; French Report, Sect. 5.5; UK Report, Sect. 10.1.2; The Bundeskartellamt's sector enquiry has not yet been completed. It aims in particular to better understand the significance of various technical services such as data collection and processing, targeting and benchmarking and auctioning of advertising spots, cf. German Report, Sect. 6.5.

[122] Australian Report, Sect. 2.9.

[123] UK Report, Sect. 10.1.2.

[124] CMA Digital Report, paragraphs 31 to 36.

source of exclusionary und exploitative behaviour.[125] There are inter alia concerns about the possibility for leverage into other areas enabled by the Google and Facebook ecosystems and, particularly for Google, the possibility to self-preference its own activities and to reinforce its market power.[126] There are also complaints about a lack of transparency, which affects not only the consumers but also the advertising companies on the other side of the market. For instance, due to a lack of transparency in the operation of online-automated bidding for digital advertising inventory, advertisers often do not know whether they are receiving value for money.[127]

Against the background of the excessive market power especially held by Google and Facebook, and also by other companies such as Apple, Microsoft and Amazon, concerns are expressed that the current law is still up to the challenges posed by digital platforms.[128] In addition, there are institutional concerns, which have already been mentioned above.[129] In particular, it is argued that *ex post* control by way of abuse proceedings is too cumbersome and lengthy to be able to avert damage from digital markets. In this respect, there is, for example, a plea for greater recourse to interim measures and a lowering of the burden of proof, although concerns about the rule of law are also raised in this respect.

1.6.2 European Level: Digital Markets Act

At the EU level, the DMA aims to address the challenges posed by powerful digital platforms, particularly to ensure contestable and fair markets in the digital sector across the European Union, where so-called gatekeepers are present. According to Article 3 DMA, a provider of core platform services shall be designated as gatekeeper if (i) it has a significant impact on the internal market; (ii) it operates a core platform service which serves as an important gateway for business users to reach and users; and (iii) it enjoys an entrenched and durable position in its operations or it is foreseeable that it will enjoy such a position in the near future.

The DMA provides for a total of 18 requirements and prohibitions that gatekeepers must comply with and introduces a new ex ante regulatory instrument. In this respect, the DMA aims to prevent, among other things, the behaviour of powerful companies described above, which is considered particularly problematic. For instance, a gatekeeper shall refrain from combining personal data sourced from its core platform services with personal data from any other services offered by the gatekeeper or with personal data from third-party services.

---

[125] Belgian Report, 4.3.2.2.

[126] Australian Report, Sect. 2.9; Belgian Report, Sect. 4.3.2.1; Norwegian Report, Sect. 8.3.5; UK Report, Sect. 10.1.2.

[127] UK Report, 10.1.2.

[128] Cf. for instance Belgian Report, Sect. 4.3.2.4; UK Report, Sects. 10.1 and 10.3.

[129] Section 1.2 above.

1.6.3 German Law

The German legislator has already reacted to some of the challenges posed by undertakings with excessive market power. According to Section 19a of the German Cartel Act,[130] the Bundeskartellamt can determine that an undertaking that is active to a considerable extent on platform markets has an "overriding cross-market importance" for competition. The provision intends to enable effective control over digital corporations. It aims at companies that not only have a dominant position in individual platform or network markets but also have the resources and opportunities to continuously expand their own business activities.[131] If these conditions are met, the Bundeskartellamt can, in a second step, prevent certain strategies of such companies, including, inter alia, self-preferencing, impeding competitors, committing abuse by using data, enabling certain deficiencies in the information policy to prevent other companies and so on.[132] Furthermore, the addressees of the standard bear the burden of proof that their conduct is objectively justified.

It would go beyond the scope of this article to compare the current German regulatory concept with the concept of the DMA. However, two major differences should be pointed out: whereas under German law a gatekeeper status can only be established in the context of a formal administrative procedure, under Art. 3(3) DMA, companies are obliged to notify the Commission of their gatekeeper status on their own initiative. In particular, however, under the current German legal situation, there are no ex ante obligations on the companies concerned even after they have qualified as gatekeepers. Rather, the Bundeskartellamt can only prohibit the companies from certain conduct in specific individual cases.

1.6.4 French Law

Various French regulators have also recently endeavoured to find new expressions of the market power of digital platforms. Thus, the concept of a structuring platform is proposed.[133]

According to this, market power only becomes a trigger for intervention by public authorities instead of a legal standard to demonstrate an abuse of dominant position.[134] In this respect, of course, criteria are also required by which a structuring platform can be identified. So far, the Financial Conduct Authority (FCA) and the

[130] Gesetz gegen Wettbewerbsbeschränkungen in der Fassung der Bekanntmachung vom 26. Juni 2013, BGBl. I p. 1750, 3245.

[131] *Paal*, IN: Gersdorf/Paal (ed.), BeckOK Informations- und Medienrecht, 36th Edition, Beck 2021, § 19a GWB para 1.

[132] Cf. as well German Report, Sect. 6.2.9.

[133] French Report, Sect. 5.3.

[134] French Report, Sect. 5.3.1.

French electronic communications regulator ARCEP have made proposals that are comparable to the criteria of a gatekeeper within the meaning of Art. 3 DMA.[135] The French ARCEP suggests a definition for structuring digital platforms that relies on a body of primary evidence (e.g. bottleneck power of the platform, size of the platform defined in terms of the number of unique users of the platform) and secondary evidence, e.g. access to significant volumes of qualitative data. According to the French Report, the proposed approach differs from a "classic" dominance assessment insofar as a *definition essay with an operational scope* is proposed.[136]

The extent to which the status of a company as a structuring platform under French competition law entails further obligations cannot be assessed here. In this context, the French Report outlines the fact that the FCA has already applied its traditional decisional practice to "structuring" platforms". The authority has stated that "[t]*he principle of commercial freedom also applies to so-called 'structuring' platforms, [. . .] provided that the rules governing access or maintenance of economic operators to or on these platforms do not have the purpose or effect of restricting competition*".[137] The FCA went on to say that the structuring platform is *free to define rules providing additional protection for Internet users compared with what is otherwise required by the regulations* but must not *direct the economic model of the economic operators listed on its platform, limit their freedom of enterprise, and influence the quality and diversity of the offer open to Internet users* by means of anti-competitive practices.[138]

1.6.5 UK Law

As already seen,[139] in the United Kingdom, there is widespread agreement that a special pro-competitive regime is needed for companies holding a strategic market status. In particular, current tools such as the special responsibility of dominant companies were insufficient to meet the challenges of these companies.

In its Taskforce Report, the Digital Markets Taskforce made some suggestions on key issues that the new law should address. A central concern of the proposed regime is the concept of strategic market status. The Task Force recommends that this status

---

[135] Proposal for a Regulation of the European Parliament and of the Council on contestable and fair markets in the digital sector (Digital Markets Act), COM(2020) 842 final.

[136] French Report, Sect. 5.3.2.2.

[137] Decision No. 21-D-07 of 17 March 2021 on a request for interim measures filed by Interactive Advertising Bureau France, Mobile Marketing Association France, Union Des Entreprises de Conseil and Achat Media, and Syndicat des Régies Internet associations in the sector of mobile applications advertising on iOS, para. 136; French Report, Sect. 5.6.3.

[138] Decision No. 21-D-07 of 17 March 2021 on a request for interim measures filed by Interactive Advertising Bureau France, Mobile Marketing Association France, Union Des Entreprises de Conseil and Achat Media, and Syndicat des Régies Internet associations in the sector of mobile applications advertising on iOS, paras. 137 and 138; French Report, Sect. 5.6.3.

[139] Section 1.2 above.

requires that a firm has *substantial, entrenched market power in at least one digital activity, providing the firm with a strategic position in respect of that activity.*[140] The temporary or transitory possession of market power is not so important; it is rather crucial that the position becomes – as provided under the DMA – *entrenched*, which means that it is expected to *persist over time and is unlikely to be competed away in the short term.*[141] If an undertaking holds entrenched market power, it can be assumed that prices on the market concerned will be permanently higher, while quality, investment and innovation will be permanently lower, to the long-term detriment of consumers.[142]

As far as a company has a strategic market status, it will in future have to comply with a legally enforceable code of conduct. This code of conduct will provide a set of *ex ante* principles. It should aim to prevent consumers from being exploited and the company from engaging in a behaviour that undermines competition.[143] Interestingly, the code will be tailor-made for each individual company with structured market power.

1.6.6 Other Jurisdictions

In Hungary, the sector-specific National Regulatory Authority (NRA) responsible in this regard has endeavoured to define a list of companies with significant market power in two Broadband Market Resolutions and has established rules according to which the companies concerned must guarantee access to their infrastructure so that competitors without the corresponding infrastructure can also enter the market.[144] The Benelux NCAs also advocate the introduction of ex ante instruments and even suggest that the criterion of abuse be dispended.[145] In Australia, the News Media and Digital Platforms Mandatory Bargaining Code has now come into force. The Code aims to redress imbalances between digital platforms and news companies.[146]

---

[140] Appendix B § 56, Taskforce Report; UK Report, Sect. 10.3.1.2.

[141] Taskforce Report, paras 28-30; UK Report, Sect. 10.3.1.2.

[142] UK Report, Sect. 10.3.1.2.

[143] UK Report, Sect. 10.3.1.2.

[144] The NRA's Broadband Decisions are available at http://nmhh.hu/dokumentum/191571/3a_piac_hatrozat_17915_64.pdf and http://nmhh.hu/dokumentum/191577/3b_piac_hatrozat_17920_64.pdf (accessed 25 October 2021); cf. also Sect. 1.11 below.

[145] Belgian Report, Sect. 4.3.2.4.

[146] Australian Report, Sect. 2.1.; see as well UK Report, Sects. 10.3 and 10.3.1.

1.7 Comparison Portals

Comparison portals are generally viewed positively in various jurisdictions, from the perspective of both competition law and consumer protection law.[147] In particular, comparison portals can contribute to consumers making more informed decisions in line with their needs due to increased transparency. At the same time, the competitive pressure is increased and competition is strengthened. Of course, this is only true if the comparison portals follow certain codes of conduct and meet certain quality criteria, especially if they are neutral. Otherwise, the effects of comparison portals on competition can be negative and harm consumers.

Sector enquiries, market studies and proceedings of the authorities have shown that comparison portals often do not meet these requirements. For example, in a case against Trivago, the Australian ACCC revealed that it had an influence on the rating of a hotel whether it had paid a cost-per-click fee. Trivago had also not compared the same room categories, for example, without making this sufficiently clear to consumers.[148] Comparable practices were also uncovered by the market enquiry of the German Bundeskartellamt and the Hungarian GVH.[149] Other shortcomings include the fact that platforms often cooperate with each other, e.g. by using the same databases, which gives consumers the erroneous impression that the result of one comparison portal has been confirmed by the (independently obtained) result of another comparison portal. The market coverage of the comparison portals is also low in many cases, with precisely those companies that were not willing to pay a fee to the comparison portal operator being excluded from the comparison. In addition, it is sometimes unclear on the basis of which criteria services are compared with each other at all, or complex services are compared in an under-complex manner on the basis of only one criterion, which does not allow for a meaningful result.[150] All these shortcomings ultimately result in consumers not being properly informed and not being able to make a sufficiently informed decision. This is aggravated by the fact that consumers who have consulted a comparison portal will often be convinced that they are making a particularly informed decision that is in their best interests.

Differences exist in the various jurisdictions with regard to the question of the extent to which the practices outlined constitute violations of the law and also, in particular, can be enforced by the competition authorities or at least other authorities. According to the Australian Report, the current Australian legal framework is sufficient and has, in practice, already proven itself to get the outlined practices under control.[151] The French Consumer Code provides specific information obligations for operators of online platforms, the breach of which can result in an

[147] The Norwegian Report, Sect. 8.3.6 offers some serious criticism. However, this criticism seems to refer only to comparison portals in markets that are oligopolistic.

[148] Australian Report, Sect. 2.7.3.

[149] German Report, Sect. 6.9; Hungarian Report, Sect. 7.5.3.1.

[150] Hungarian Report, 7.5.3.1.

[151] Australian Report, Sect. 2.7.3.

administrative fine; in this respect, the Direction générale de la concurrence, de la consommation et de la répression des fraudes (DGCCRF), not the FCA, is responsible.[152] Under Hungarian law, the GVH is also responsible for enforcing fair trading law and consumer protection law; in addition, criminal and civil sanctions are possible in the case of misleading commercial practices, whereby consumer class actions are also conceivable.

The CMA also has comprehensive competences, especially in consumer protection law. In CMA's Digital Comparison Tools Market Study, the authority made a series of recommendations to ensure that consumers are treated fairly and that comparison portals, in particular, are clear, accurate, responsible and easy to use (CLEAR).[153] The CMA has also used its powers to enforce consumer protection rules on many occasions, for example taking action against misleading sales practices in the hotel booking sector.

In contrast, there is probably a deficit in legal enforcement in Germany in particular.[154] Although the Bundeskartellamt may conduct sector enquiries that concern aspects of consumer protection law, the authority has no competence to sanction such conduct in any form. In principle, some of the uncovered practices may violate the Unfair Competition Act.[155] However, as the German Report also points out, such violations can be asserted not by a public authority but only by way of private enforcement. In this respect, of course, it must be taken into account that individual consumers cannot derive any claims from the Unfair Competition Act. One could still think of a claim for damages under private law by affected consumers who have made an uninformed decision. However, it is questionable to what extent consumers should have any knowledge at all that they have become "victims" of a corresponding practice. Even if they have such knowledge, the damages they suffer are likely to be small in many cases, so a lawsuit will often fail due to the consumers' rational lack of interest. To a certain extent, only the "Musterfeststellungsklage" to §§ 611 ff of the German Code of Civil Procedure (ZPO),[156] a certain form of class action, could provide a remedy.

[152] French Report, Sect. 5.8.2.

[153] UK Report, Sect. 10.5.1.

[154] Cf. in this respect *Podszun/Busch/Henning-Bodewig*, Consumer Law in Germany: A Shift to Public Enforcement?, EuCML 2019, 75.

[155] Gesetz gegen den unlauteren Wettbewerb in der Fassung der Bekanntmachung vom 3. März 2010, BGBl. I p. 254.

[156] Zivilprozessordnung in der Fassung der Bekanntmachung vom 5. Dezember 2005, BGBl. I p. 3202; 2006 I, p. 431; 2007 I, p. 1781.

1.8 Most Favoured Nation Clauses

1.8.1 General

Most favoured nation (MFN) clauses, also called best price clauses or parity clauses, have already been the subject of official investigations and decisions in most jurisdictions and have in some cases already occupied the courts and legislators. They play an important role in digital platforms, especially in comparison portals where customers can also conclude a contract with the providers, such as Expedia, Booking and HRS, or on Amazon. In digital markets, MFN clauses are typically designed in such a way that the provider offering a service or good on one platform is not authorised to offer the same service on another distribution channel at a lower price or on better terms. The other distribution channels can be competing platforms, the supplier's own homepage or even offline distribution, depending on the design of the clause.

MFN clauses can take different forms depending on the context. In many jurisdictions, a fundamental differentiation between wide and narrow MFN clauses is recognised. This differentiation is linked to the extent to which the contracting parties are prohibited from offering their products through another distribution channel on better terms. A narrow MFN clause is spoken of if the prohibition only refers to the supplier's own distribution channels; e.g., a hotel is not allowed to offer rooms on its own home page at more favourable conditions. A wide clause is used if the prohibition also extends to distribution via competing platforms.[157]

In this respect, too, the terminology is not uniform and a clear-cut classification of an MFN clause as wide or narrow is not possible. As the German Report mentions by way of example, German authorities and courts have described MFN clauses as "wide" and subsequently invalid (see below), which were significantly narrower than an MFN clause in a Czech case against Booking.[158] MFN clauses sometimes go beyond a mere prohibition to offer the product on other distribution channels at more favourable conditions. For example, Amazon has given a commitment to the European Commission that it will not use MFN clauses that oblige competitors to inform Amazon about competitors' business models, promotions, prices, etc.

As far as can be seen, there is consensus within the various jurisdictions that MFN clauses can both violate the prohibition of cartels and constitute an abuse of market power. On the one hand, MFN clauses can increase competition in vertical relationships. In addition, comparison portals whose operators often use MFN clauses also have positive effects on competition. Finally, platform operators may have a legitimate interest in preventing providers from using their platform only to attract the attention of customers in order to offer them the product via another distribution channel under more favourable conditions (so-called free riding). With

[157] Cf. for instance Swiss Report, Sect. 9.2.2; Belgian Report, Sect. 4.3.2.3; German Report, Sect. 6.11; Hungarian Report, Sect. 7.5.2.2; Swiss Report, Sect. 9.2.2; UK Report, Sect. 10.6.

[158] German Report, Sect. 6.11.

regard to the advantages that MFN clauses can have, they tend to be assessed positively in Australia, for example.[159] On the other hand, it should be noted that MFN clauses can restrict competition between different platforms and also represent a considerable interference in the pricing freedom of the obligated suppliers.

The way in which competition authorities and the courts deal with MFN clauses within individual jurisdictions can only be outlined with broad brushstrokes, as the practice even within the jurisdictions still seems to be in flux.

However, the rule of thumb that broad MFN clauses are generally inadmissible while narrow MFNs may be admissible seems to be recognised across jurisdictions. It is true that narrow MFN clauses can be subsumed under the prohibition of cartels (for instance Art. 101 (1) TFEU). In view of the already mentioned interest of platform operators to prevent free riding, the use of an MFN clause can, however, be objectively justified. On the level of EU law, an MFN clause can also fall under Block Exemption Regulation 330/2010.[160] However, this justification does not apply if broad clauses are used. The rule of thumb corresponds in principle to the results of a monitoring exercise by the EU Commission and nine national competition authorities concerning the hotel sector, which was published in 2017.[161] The practice of the authorities and jurisdictions outlined in the individual national reports also seems to be oriented towards it—even beyond the hotel sector.[162]

Of course, an assessment of the respective clause in concrete individual cases is necessary.[163] That even narrow MFN clauses can also be potentially inadmissible is proven by the example of the use of narrow MFNs by Booking, which is also mentioned in the German Report. Contrary to the outlined rule of thumb, the Bundeskartellamt had assumed that narrow MFN clauses also violated Article 101 (1) TFEU without being objectively justified by efficiencies under Article 101 (3) TFEU. This decision was initially confirmed by the Düsseldorf Higher Regional Court (OLG Düsseldorf) in an interim decision but was later reserved. At the moment, the Bundesgerichtshof BGH (Supreme Court) is dealing with the question.

The Hungarian NCA also had concerns about a narrow MFN clause in the online food ordering market. The Hungarian Report emphasises that, in the opinion of the NCA, only such narrow clauses that extend to the supplier's own online sales but not

[159] Australian Report, Sect. 2.9.

[160] Commission Regulation (EU) No 330/2010 of 20 April 2010 on the application of Article 101(3) Treaty on the Functioning of the European Union to categories of vertical agreements and concerted practices, OJ from 24 April 2016 L102/1.

[161] EC Monitoring Exercise carried out in the online hotel booking sector by EU competition authorities in 2016, available at https://ec.europa.eu/competition/ecn/hotel_monitoring_report_en.pdf (accessed 25 October 2021).

[162] Austrian Report Sect. 3.2.8; Belgian Report, Sect. 4.3.2.3; French Report, Sect. 5.7.1; German Report, Sect. 6.11; Hungarian Report, Sect. 7.5.2.2; Swiss Report, Sect. 9.2.2; UK Report, Sect. 10.6.

[163] Swiss Report, Sect. 9.2.2.

to offline sales are permissible.[164] The Australian competition authority, which does not seem to adhere strictly to the dogmatic distinction between narrow and broad MFN clauses, does not seem to have found a uniform approach to MFNs across different sectors.[165]

1.8.2 MFN-Related Specific Regulation

It is worth mentioning that some jurisdictions provide special regulations for MFN clauses, which are sometimes stricter than the principles outlined above.

In 2018, the Belgian Parliament enacted a law regarding the freedom of tariffs by hoteliers in contracts with platform operators for online reservation.[166] According to this law, operators of online platforms are *per se* prohibited from restricting the freedom of hotel owners to set prices in any form—and thus also through MFNs—provided that the hotel property is located in Belgium.[167]

The use of MFN clauses in the hotel industry is also the focus of the legislator in Switzerland. Thus, the Swiss Parliament has recently approved a motion that would introduce a general ban on all MFN clauses in the hotel industry. According to the Swiss Report, this should lead to a legislative prohibition of all price parity clauses in the future, regardless of their nature.[168]

The French Commercial Code prohibits an economic operator from requiring its partners to grant it the same advantages as its competitors.[169] Based on this, the use of MFN clauses by travel and hotel booking platforms in particular has been sanctioned. In addition, any producer, trader or service provider shall be held liable and obliged to remedy the damage caused in the course of business negotiation, the conclusion or the execution of a contract for submitting or attempting to submit a trading partner to obligations creating a significant imbalance in the rights and obligations of the parties.[170] Based on this standard, for example, a broad MFN clause used by Booking was declared null and void by the Paris Commercial Court.[171] Finally, with Art. L-311-5 of the French Tourism Code, a sectoral per se ban on MFN clauses in the hotel sector was enacted. According to this legal provision, the hotelier retains the freedom to grant the client any discount or pricing

[164] Hungarian Report, Sect. 7.5.2.2.

[165] Australian Report, Sect. 2.9.

[166] Wet betreffende de tariefvrijheid van exploitanten van toeristische logies in de contracten afgesloten met platform-operators voor online reservatie, Official Gazette 2018, N. 188, p. 62710.

[167] Belgian Report, Sect. 4.3.2.3.

[168] Swiss Report, Sect. 9.2.2.

[169] Art. L. 442-3 Code de Commerce.

[170] Art. L.442-1 Code de Commerce.

[171] French Report, Sect. 5.7.2.2.

advantage of any kind, any clause to the contrary being deemed unwritten. The impact of this law is considered positive by the Comité des affaires économiques.[172]

The fact that individual legislators prohibit the use of MFN clauses in whole or in part is in line with potential legal developments at the EU level. Thus, according to Art. 5 lit. (b) DMA, a gatekeeper shall allow business users to offer the same products or services to end users through third-party online intermediation services at prices or conditions that are different from those offered through the online intermediation services of the gatekeeper. These developments at the national level are thus already at least partly in line with the above-mentioned[173] call for increased *ex ante* regulation.

1.9 Interaction Between Competition Law and Data Protection Regulation

1.9.1 General

Competition law and data protection law are materially separate legal matters, with each pursuing independent objectives, at least as their starting point.[174] While competition law protects competition and is intended to prevent companies from engaging in anti-competitive behaviour, data protection law generally aims to protect the rights and interests of individuals who are affected by the collection, evaluation and use of personal data (right to privacy, right to informational self-determination).[175] This substantive distinction between competition law and data protection law is reflected in a formal-institutional sense, in that the enforcement of data protection law in all jurisdictions is not one of the (core) competences of the competition authorities but that special data protection authorities are responsible in this respect.[176] According to Art. 51 of the General Data Protection Regulation (GDPR), the establishment of independent data protection authorities in the Member States is even mandatory.

However, both areas of law can overlap. The reason for this, to put it simply, is that they are partly linked to the same factual aspects. As data protection law is generally linked to the collection, evaluation, use and disclosure of personal data, these factual aspects raise, as already seen, specific competition law questions and concerns. For instance, the special standing of digital platforms is based precisely on

[172] French Report, Sect. 5.7.2.3.

[173] Section 1.6 above.

[174] Cf. for instance Swiss Report, Sect. 9.2.1.

[175] It is worth noting that the right to privacy is not constitutionally guaranteed in Australia, cf., Australian Report, Sect. 2.8.1.

[176] Australian Report, Sect. 2.8; Austrian Report, Sect. 3.2.3; Belgian Report, Sect. 4.3.2.2; French Report, Sect. 5.2.; Hungarian Report, Sect. 7.2; Norwegian Report, Sect. 8.3.4; Swiss Report, Sect. 9.2.1; UK Report, Sect. 10.4.3.

the fact that they and that can collect, evaluate, combine and use data otherwise, that data has monetary value, and that ownership and use of data can create or strengthen market power.[177] In addition, data protection aspects have a link to consumer welfare because consumers value data protection aspects as an important product characteristic.[178]

1.9.2 Starting Point: Complementary Relationship

The relationship between data protection law and competition law is increasingly coming into focus in most jurisdictions. One central question is whether a violation of data protection law can constitute an abuse of market power. If this question is answered in the affirmative, the result is that—contrary to the outlined material and institutional distinction between data protection law and competition law—to a certain extent the NCAs would also be responsible for sanctioning data protection law infringements.

A clear picture of opinion on this question within the various jurisdictions is not yet apparent from the National Reports. In principle, there seems to be agreement that data protection authorities and competition authorities—as the Norwegian report aptly points out—have complementary competences.[179] The enforcement of data protection law is the very own and, in principle, exclusive task of data protection authorities. However, competition authorities can intervene if a matter not only essentially concerns data protection issues but also raises specific competition law concerns, i.e. if the collection, use and analysis of data lead to anti-competitive conducts or agreements or constitute an abuse of market power. If mergers are at issue, the accumulation of data would have to give rise to competitive concerns.[180]

1.9.3 Current Approaches

According to what has been said so far, it is decisive when data protection law issues raise competition law concerns beyond data protection law.

As regards the further specification of these general requirements, the French Report refers to a joint study conducted by the FCA with the Bundeskartellamt in 2016. According to this study, data processing can always trigger competition concerns if *privacy policies are liable to affect competition, notably when they are implemented by a dominant undertaking for which data serves as a main input of its products or services and A close link between the dominance of the company, its*

[177] Austrian Report, Sect. 3.2.2.

[178] Hungarian Report, Sect. 7.2; Australian Report, Sect. 2.8.2.

[179] Norwegian Report, Sect. 8.3.4.

[180] In this sense, cf. Austrian Report, Sect. 3.2.3; Belgian Report, Sect. 4.3.2.2; French Report, Sect. 5.6.1; Norwegian Report, Sect. 8.3.4; Swiss Report, Sect. 9.2.1.

data collection processes and competition on the relevant markets could justify the consideration of privacy policies and regulations in competition proceedings.[181]

In its Facebook case, the Bundeskartellamt took a primarily data-protection-oriented approach and took the view that the abuse of a dominant position could essentially be argued on the ground that Facebook had disregarded the requirements of the GDPR when combining user data and device-related data collected on the Facebook website with "off Facebook data".[182] In contrast, the Bundesgerichsthof (BGH, Federal Supreme Court) preferred a more competition-oriented approach in order to keep the overlaps with data protection law low.[183] Thus, the court primarily focused on the fact that the lack of an option for users to choose to have only the data generated when using Facebook itself used for personalised advertising restricts the users' freedom of choice. Since it is generally part of the business model of social networks to collect data in order to place advertisements, a company that collects data in violation of the GDPR gains an advantage over its competitors that behave in a legally compliant manner. In particular, barriers to entry would be erected because competitors acting in a data-protection-compliant manner would not be able to offer personalised advertising in the same way as Facebook does and would consequently also find it difficult to finance their network. Moreover, in the context of the balancing of interests required under the German Cartel Act,[184] conduct in breach of data protection was only of secondary importance. As regards the relationship between competition law and data protection law, the German Report outlines that the BGH refers to the infringement of the GDPR only as one factor in the balancing of interests. A violation of the GDPR is not a necessary requirement for a competition law infringement, while an infringement of competition law is possible even when the requirements of the GDPR are met. The OLG Düsseldorf must now decide the case in the main proceedings. On 24 March 2021, the OLG Düsseldorf made a referral to the CJEU according to Art. 267 TFEU. With regard to the relationship between data protection law and competition law, the OLG submitted the following questions, among others, to the CJEU:

(a) Can a national cartel authority of a Member State, such as the Bundeskartellamt, which is not a supervisory authority within the meaning of Art. 51 et seq. GDRP and which examines an infringement of the prohibition of abuse under antitrust law by a dominant undertaking which does not consist in an infringement of its data processing conditions and their implementation against the GDPR, make findings, for example in the context of the balancing of interests, as to whether

[181] Bundeskartellamt and French Competition Authority, Joint Study on Competition Law and Data, 10 May 2016, pp. 23–24.

[182] See in detail German Report, Sect. 6.1; Bundeskartellamt, Decision of 6 February 2019 – B6-22/16.

[183] See in detail German Report, Sect. 6.1; BGH, Judgment of 23 June 2020 – KVR 69/19.

[184] Section 19 (1) Gesetz gegen Wettbewerbsbeschränkungen in der Fassung der Bekanntmachung vom 26. Juni 2013, BGBl. I p. 1750, 3245.

the data processing conditions of that undertaking and their implementation comply with the GDPR?

(b) If so, does this also apply with regard to Article 4(3) TEU if, at the same time, the competent lead supervisory authority pursuant to Article 56(1) GDPR subjects the data processing conditions of that undertaking to an investigation procedure?

In an opinion, the French NCA clarified its view and stated that data protection has an impact on competition and is therefore likely to be taken into account by the French NCA in its analyses.[185] In Austria, according to the NCA, violations of other fields of law must be taken into account as well when assessing abuse of a dominant position.[186] In contrast, the Belgian Report explicitly raises concerns about the practice of the Bundeskartellamt and assumes that the Belgian NCA would not intervene in a comparable case: exploitative cases were significantly more difficult to bring than exclusionary cases. In addition, data protection aspects were more likely to be affected, and the Belgian NCA preferred the approach of the European Commission of putting the data protection authorities responsible in such a case.[187] The Norwegian NCA has also expressed scepticism as to whether competition law is an adequate instrument to enforce data protection law.[188]

In Hungary, the enforcement of data protection law is also the exclusive responsibility of the data protection authority (Hungarian Data Protection Authority (HDPA)). However, since data protection could be considered part of consumer welfare and data-related practices could also constitute unfair commercial practice, the authority wants to resort to fair trading law instruments.[189] The Hungarian NCA has not yet applied antitrust law to the data-processing activities of social networks.[190]

The Australian Report emphasises that the Australian federal data protection authority (Office of the Australian Information Commissioner (OAIC)) has so far hardly taken any action against platforms and in particular has not taken into account the systemic concerns associated with the collection and use of device-related data across different platforms. The Australian Privacy Act is also "woefully under-litigated", and the central concept of "personal information" is unclear. However, the ACCC is increasingly focusing on the protection of consumer privacy. In its final report, the ACCC pleads for far-reaching changes to the Australian information privacy law framework in order to be able to take appropriate action against the market power of digital platforms.[191]

[185] French Report, Sect. 5.2.

[186] Austrian Report, Sects. 3.2.2, 3.2.3.

[187] Belgian Report, Sect. 4.3.2.2.

[188] Norwegian Report, Sect. 8.2.

[189] Hungarian Report, Sect. 7.2.

[190] Hungarian Report, Sect. 7.2.

[191] Australian Report, Sect. 2.8.1.

1.9.4 Need for Further Coordination

It should be borne in mind that the tensions between the two areas of law go beyond the just outlined question of whether NCAs can also take action based on a breach of data protection law. The question of appropriate coordination of competition law and data protection law has other dimensions as well.[192]

In particular, an uncoordinated coexistence of data protection law and competition law can lead to the two legal areas obstructing each other; e.g., the enforcement of data protection law standards has adverse effects from a competition law perspective and is therefore undesirable or vice versa. Such obstructions between antitrust law and data protection law are generally due to the fact that there can also be conflicts of objectives between the two areas of law despite partly identical objectives.[193] The UK Report impressively demonstrates the fact that such conflicts of objectives are already relevant in practice. For instance, the CMA is concerned that Google and Facebook are abusing the GDPR to deny third parties access to relevant data but are using these data themselves within their own ecosystem.[194] Google has recently announced that it will discontinue its support for third-party cookies on the Chrome browser for data protection reasons, with the support of the Information Commissioner's Office (ICO). This may be desirable from the singular perspective of data protection law, but—at least in the view of the CMA—it leads to the restriction of publishers' ability to offer personalised advertising while Google continues to have access to the data of Google Search and YouTube users.[195] Examples show that the question of the relationship between competition law and data protection law goes far beyond a clear demarcation of the two areas of law. Even if a material demarcation should succeed, the difficult question of the relationship between the two areas of law is not answered. In particular, the effects of one on the other area of law must not be denied and neglected in practice.

Given the importance of data in the digital economy and the power of digital platforms, the discussion should rather focus on exploring how data protection law and competition law can be meaningfully intertwined. This requires a more detailed analysis, especially of the extent to which both areas of law, despite their different meta-goals, also pursue common (intermediate) goals and the extent to which the respective instruments of one area of law can, if necessary, also be used in a supportive manner in order to be able to achieve the (intermediate) goals of the other area of law. In addition, special emphasis must be placed on uncovering the conflicts of objectives between the two areas of law and analysing in which context

[192] On the coordination of different areas of law and the associated methodological and doctrinal issues, cf. *Schäfers*, Korrelative Systeminterferenzen, Mohr Siebeck 2021; cf. as well *Hacker*, Datenprivatrecht: Neue Technologien im Spannungsfeld von Datenschutzrecht und BGB, Mohr Siebeck, 2020.

[193] Cf. on this concept in detail Schäfers, Korrelative Systeminterferenzen, Mohr Siebeck 2021, p. 250 ff.

[194] UK Report, Sect. 10.1.2.

[195] UK Report, Sect. 10.4.1.

data protection law and competition law obstruct each other and for what reasons. Finally, the coordination of different areas of law must also be secured institutionally by means of substantial cooperation between legislators and authorities. We should think of informal cooperation not only between data protection and competition authorities but also of legally binding procedural regulations that guarantee such cooperation. For example, a mandatory consultation of the respective other authority in administrative and judicial proceedings relating to respective other regulatory matters; the corresponding information rights of the authorities; the establishment of special, competent panels of judges at courts; etc. should be considered.

Most of the jurisdictions already seem to address such cooperative considerations, albeit only through non-binding agreements. The French DGCCRF and the Commission Nationale de l'Informatique et des Libertés (CNIL) signed, for instance, a cooperation protocol in order to strengthen the entities' collaboration. The Austrian NCA announced to join forces with the national telecommunications regulator for the purpose of monitoring digital platforms.[196] In the United Kingdom, the CMA, the ICO and Ofcom have jointly founded the Digital Regulation Cooperation Forum (DCRF), whose role consists primarily of supporting cooperation and coordination between the mentioned authorities to advance a coherent regulatory approach.[197]

Overall, it can be said that the analysis of the relationship between data protection law and competition law is only just beginning. In my view, the consistent exploration of this relationship seems elementary in order to be able to adequately meet the challenges of digitalisation. Just for the sake of completeness, it should be pointed out that with regard to the collection, evaluation and use of data, other areas of law also collide with both data protection law and competition law. Above all, one should think of fair trading law, consumer protection law and general private law, especially general contract law. Legal science and practice are thus faced with an extremely difficult coordination task that will probably occupy researchers and practitioners for many years to come.

1.10 General Terms and Conditions, Consumer Protection

Related to the questions dealt with under Sects. 1.6 and 1.8 above and not easy to differentiate from them is the issue of the extent to which competition authorities also take action against general terms and conditions, especially with regard to data collection and consumer protection law.

In the Swiss legal system, there is so far no case dealing with the use of inadmissible general terms and conditions regarding data collection.[198] The same

[196] Austrian Report, Sect. 3.1.

[197] UK Report, Sect. 10.4.1.

[198] Swiss report, Sect. 9.2.1.

applies to Hungary.[199] However, the Hungarian GVH is also very active in the enforcement of consumer protection law as far as mobile service providers are concerned.[200] In Norway, the use of general terms and conditions for data collection is primarily a task for data protection and consumer protection authorities.[201] The Austrian NCA was initially authorised as the Austrian authority for the cooperation of EU authorities in consumer protection to remedy intra-EU infringements of certain consumer protection regulations insofar as collective interests were affected. However, the authority lost its competence at the beginning of 2020.[202] The French NCA has specified the legality requirements for the terms of use of Google's services in the advertising market on several occasions since 2010. The authority placed particular emphasis on the predictability and non-discriminatory application of the terms of use.[203]

The CMA is authorised to conduct market enquiries under the specific aspect of consumer protection law. It has also taken a closer look at the default settings and the design of the privacy terms and conditions of digital platforms.[204] According to the CMA, default settings have a far-reaching influence on the shape of competition in both search and social media, because they affect the way in they which users use the search engine and, especially in the case of social networks, they have a considerable influence on the possibilities of collecting data about their customers. The CMA also criticised the design of the privacy terms and conditions, which are too long and complicated, with the result that users simply give their consent.[205] Another criticism is that often there is no possibility for consumers to use a service without being exposed to personalised advertising.[206] Even if they have the option, they need clear and understandable information about what exactly is happening with their data. In CMA's view, these conditions are often not met. As a result, many consumers use services that they would not have otherwise used if they were only properly informed.[207] Against this backdrop, the CMA Digital Report recommends a "Fairness by Design obligation" to maximise users' ability to make informed choices.[208]

In its DPI Final Report, the Australian ACCC also criticised a discrepancy between the expectations of consumers and the actual terms and conditions

[199] Hungarian Report, Sect. 7.6.2.

[200] Hungarian Report, Sect. 7.6.2.

[201] Norwegian Report, Sect. 8.3.4.

[202] Austrian Report, Sect. 3.2.11.

[203] Decision No. 21-D-07 of 17 March 2021 on a request for interim measures filed by Interactive Advertising Bureau France, Mobile Marketing Association France, Union Des Entreprises de Conseil and Achat Media, and Syndicat des Régies Internet associations in the sector of mobile applications advertising on iOS, paras. 137 and 138; French Report, Sect. 5.6.3.

[204] UK Report, Sect. 10.4.4.

[205] UK Report, Sect. 10.4.4.

[206] UK Report, Sect. 10.4.2.

[207] UK Report, Sects. 10.4.2., 10.4.3.

[208] UK Report, Sect. 10.4.3.

underlying their agreements with digital platforms, particularly with regard to what and how much data are collected by the platforms and how and to what extent they are used and passed on to third parties.[209] The fact that consumers nevertheless accepted the terms and conditions despite fundamental reservations about data protection law vis-à-vis platforms (privacy paradox) was due in particular to the lack of bargaining power of consumers, who had no choice but to join the platform, as well as information asymmetries between the platforms and the consumers.[210] In reviewing the terms and conditions of digital platforms, the ACCC specifically identified the following practices of digital platforms that are of concern: terms and conditions are presented on a take-it-or-leave-it basis; clickwrap agreements contain unobtrusive links to online documents that permit the collection, use and disclosure of vast amounts of user data; data use consents are bundled for different uses and different services; and privacy policies are long, complex, vague and difficult to navigate.[211] Since the practices outlined will further increase the imbalance between the parties, it was deemed necessary to re-evaluate the consumer protection legal framework.[212] In this respect, the peculiarity of Australian law has to be taken into account as well as the fact that the ACCC, together with other authorities, has the competence to administer also consumer protection law. Although the DPI Final Report came to the conclusion that consumer protection law provides several effective instruments against questionable practices by digital platforms, the DPI Final Report and DPSI2020 nevertheless provide for certain tightening up or innovations in consumer protection law that deal specifically with unfair contract terms and commercial practices. Changes to the Australian information privacy law framework are also needed to counter the market power of digital platforms. These changes should aim, among other things, to provide consumers with information to enable them to decide whether services meet their privacy preferences and to address bargaining imbalances in relation to the collection, use and disclosure of consumers' personal information.[213]

Since 2017, the German Bundeskartellamt has had for the first time the power to conduct sector enquiries also in relation to economic consumer protection. The competence also extends in particular to the law of unfair competition and the law on general terms and conditions. In this respect, however, it must be considered that the Bundeskartellamt has no power to sanction violations of consumer protection regulations or to enforce consumer protection regulations. Rather, consumer protection standards in Germany are largely carried out by way of private enforcement. As already mentioned above, the Bundeskartellamt had, though, just accused Facebook of using invalid general terms and conditions and thus violating the GDPR.[214]

209 Australian Report, Sect. 2.7.1.

210 Australian Report, Sect. 2.7.1.

211 Australian Report, Sect. 2.7.1.

212 Australian Report, Sect. 2.7.1.

213 Australian Report, Sect. 2.8.2.

214 Section 1.9.3. above.

1.11 Telecommunication Sector

1.11.1 General

For large digital platforms in particular, there is currently a considerable incentive to integrate vertically into the telecommunications supply chain as they are dependent on telecommunications networks and therefore have an interest in reliable and far-reaching telecommunication services.[215] The NCAs, which in principle supervise every activity of companies regardless of the economic sector, also have competences in the telecommunications sector. In addition, sector-specific authorities (sometimes several) are responsible in the individual jurisdictions as well. The responsibilities of sector-specific authorities and competition authorities may overlap, but the activities of the different authorities are primarily complementary.[216]

1.11.2 Consumer Protection

Regarding the telecommunication sector, consumer protection issues seem to play an important role, especially in Australia and Hungary.

In Australia, the telecommunications industry is monitored by three different authorities. In addition, the Telecommunications Industry Ombudsman (TIO), a free service to consumers and small businesses, has been established with the particular task of developing solutions to minor disputes.[217] Despite the establishment of the TIO, the ACCC had declared in its DPI Final Report that there were no effective dispute resolution processes for consumers and businesses in their dealings with digital platforms. The ACCC suggested that the ACMA should develop specific standards with best practice principles, e.g. concerning internal dispute resolution and accessibility or collection of information.[218] The Australian government has shown itself to be open to the proposals in principle but has also committed itself to developing a pilot external dispute resolution scheme in consultation with major digital platforms, consumer groups and relevant government agencies.[219]

In Hungary, the GVH attaches importance to effectively enforcing consumer protection standards against mobile phone providers, e.g. with regard to unlawful pricing and discounting practices or unlawful statements on market leadership and other misleading advertising.

---

[215] Australian Report, Sect. 2.5.1 with reference to DPSI2020, p. 79 et seq.

[216] Cf. for instance Norwegian Report, Sect. 8.3.3.

[217] Austrian Report, Sect. 3.2.9.

[218] Australian Report with reference to DPSI2020, p. 508 et seq.

[219] Austrian Report, Sect. 3.5.1 with further reference.

1.11.3 Mergers and Abuse of Dominant Position

In Germany, Hungary, Norway and Switzerland, the telecommunication sector was the focus of the competition authorities, especially with regard to mergers and abusive behaviour by dominant undertakings.[220] The case practice within the various jurisdictions proves in this respect that in particular commitment decisions can be an adequate means to ensure the rapid spread of high technology through the possibility of mergers on the one hand, but on the other hand to eliminate associated competition law concerns.[221]

The German Bundeskartellamt had investigated, for instance, a cooperation agreement and the creation of a joint venture between leading Internet network operators *EWE* and *Telekom Deutschland*, which wanted to cooperate on expanding the fast fibre optic network.[222] The Bundeskartellamt approved the merger, requiring the two companies to submit commitments. Among other things, it had to be guaranteed that a minimum number of households would be connected to the fast Internet by 2023 and that the network would be expanded to a certain extent in rural areas. In addition, the companies had to undertake to participate independently in future public tenders and not to involve the joint venture in the tenders.

In Hungary, the sector-specific NRA responsible in this regard has endeavoured to define a list of companies with significant market power in two Broadband Market Resolutions and has established rules according to which the companies concerned must guarantee access to their infrastructure so that competitors without the corresponding infrastructure can also enter the market.[223] With regard to the prices at which undertakings with significant market power are obliged to grant access to their networks, the NRA applied for the first time the bottom-up pure long-run incremental cost (LRIC) calculation methodology recommended by the EU Commission.[224] The case *DIGI/Invitel* concerned the merger of two major Hungarian companies in the market for wired telecommunications. In view of the extensive networks of both companies, it was to be feared that the merger would lead to a significant impairment of competition. Interestingly, the GVH did not oblige DIGI to grant other companies access to DIGI's infrastructure. Rather, DIGI had to sell the relevant networks. According to the Hungarian Report, this is a deviation from the usual practice within the European Union, where GVH is probably trying to

[220] Cf. for instance German Report, Sect. 6.8; Hungarian Report, Sect. 7.6.1, Norwegian Report, Sect. 8.3.3, Swiss Report, Sect. 9.2.4.

[221] German Report, Sect. 6.8.

[222] Bundeskartellamt, Commitment of 4 December 2019 – B7-21/18.

[223] The NRA's Broadband Decisions are available at http://nmhh.hu/dokumentum/191571/3a_piac_hatrozat_17915_64.pdf and http://nmhh.hu/dokumentum/191577/3b_piac_hatrozat_17920_64.pdf (accessed 14 September 2021); cf. already Sect. 1.5.2 above.

[224] Hungarian Report, Sect. 7.6.1.

encourage competitors to build their own parallel networks rather than to cooperate with each other.[225]

The Norwegian NCA also sees it as one of its priority tasks to strengthen competition in the telecommunication sector.[226] In particular, the authority has intervened several times against mergers in the telecommunication sector as the mergers would have had anti-competitive effects. In each case, however, the competition authority made use of commitment decisions and was thus able to accept the mergers. In addition, the authority has concentrated on proceedings against *Telenor*, which was alleged to have abused its dominant position by using pricing structures that distort competition.[227]

1.12 Summary

The National Reports were dedicated to the overarching question of whether disruptive technologies call for disruptive competition law enforcement. Based on the analysis of the reports from the various jurisdictions, this question can be generally answered neither in the affirmative nor in the negative. In particular, a blanket answer to this question that claims to apply to all jurisdictions is out of the question, if only in view of the different structures of the individual legal systems.

At any rate, competition law seems to be well positioned to meet the challenges of digitalisation, at least as a starting point. It has been shown in various contexts that the fact that the central competition law norms, in particular the prohibition of the abuse of market power, are structured as general clauses is an advantage in this respect. Competition law proves to be sufficiently flexible as it is able to react appropriately to the many challenges of disruptive technologies. In particular, a certain restraint on the part of the legislator also has the advantage of allowing authorities to gradually gain experience in dealing with new technologies under competition law, e.g. in the context of initial proceedings against digital platforms, the conduct of market enquiries, etc. This experience can possibly be better incorporated into regulation in a decentralised and flexible manner, e.g. through soft law instruments than through lengthy changes to the legal framework that may soon be outdated by case practice. For example, the initially problematic question of whether markets for free services should be recognised could be dealt with relatively quickly and uniformly within all jurisdictions without the legislators having to take action. In view of this unanimous view within different jurisdictions, it seems doubtful, for example, to what extent the introduction of a corresponding legal regulation in Germany was necessary and beneficial at all. Also, with regard to the basic criteria that can be used to determine market power in digital markets,

[225] Hungarian Report, Sect. 7.6.1.

[226] Norwegian Report, Sect. 8.3.3.

[227] Norwegian Report, Sect. 8.3.3.

especially the competition authorities in most jurisdictions have developed criteria that are largely in line with the explicit German regulation.

With regard to the principle of legal certainty and democratic legitimacy, it is certainly recommendable that the legislators take action and successively adapt the legal framework. Legal certainty is especially important for (young) companies in digital markets and generally promotes innovation. Sometimes, however, it is doubtful whether new regulations contribute to legal certainty. For example, although the German legislator has explicitly addressed platforms, the central question of the definition of platforms has been left open. The section of the German Cartel Act addressing markets for free services also does not answer the question of how to define such markets, which is considered extremely difficult in many jurisdictions. Nevertheless, such legislative initiatives can be understood as a signal to (powerful) companies on the digital markets that the legislator has their activities on its radar.

Despite this fundamentally optimistic assessment of the existing legal framework, the National Reports have also shown that new conceptual and methodological approaches, as well as completely new regulatory approaches, are required in a wide variety of contexts in order to be able to deal with the new technologies. This relates, for example, to the question of market definition on multilateral markets as well as on markets for free services. In this respect, the demand for greater consideration of the findings of behavioural economics seems to me to be of particular relevance. If one takes into account that the advantage of collecting, combining and evaluating data consists, among other things, precisely in being able to foresee the idiosyncratic needs and weaknesses of consumers and to influence their behaviour, it is unavoidable that the relevant findings of behavioural economics should also be taken into account on the part of the authorities and legislators.

In addition, there are widespread concerns about the extent to which abuse proceedings in particular are suitable for overcoming challenges in digital markets. From an institutional point of view, abuse proceedings appear to be too lengthy and cumbersome in view of the rapid developments in digital markets. The associated ex post regulation is probably not suitable for providing sufficient protection in digital markets. Whether it is also appropriate against the background of the rule of law to make greater use of interim measures or to change the burden of proof is discussed controversially. It is also evident, especially in digital markets, that proving a dominant position can be associated with considerable difficulties or that intervention by the competition authorities may be necessary even if a company does not (yet) hold a dominant position. In this respect, however, it has been shown that in many jurisdictions, legal instruments already exist to regulate companies without a dominant market position.

However, companies with excessive market power pose particular challenges to competition law. In this respect, considerable concerns are being expressed in several jurisdictions that the current legal framework is still up to these challenges. At the EU level, the DMA should be taken into account, which intends to introduce specific obligations for so-called gatekeepers and would result in *ex ante* regulation of these companies. In the United Kingdom, too, there is a plea for ex ante regulation for companies with a so-called structural market status. The German legislator has already introduced an independent standard for companies with a so-called overriding cross-market importance. Unlike the drafts at the EU level and in the United Kingdom, however, the German regulation does not lead to ex ante regulation of these companies. In France, regulators have recently developed the concept of structuring platforms to find new expressions of market power of digital platforms. This concept is based on the criteria of the DMA for gatekeepers.

Finally, one of the possibly most far-reaching challenges to competition law through digitalisation is coordinating competition law with other areas of law, especially data protection law. The process of addressing this issue is only just beginning.

2 Australia

Barbora Jedličková, Brenda Marshall, and Mark Burdon

2.1 Introduction

The Australian competition law regime is primarily governed by the Competition and Consumer Act 2010 (Cth) (CCA).[1] Australian competition law is enforced via public and private enforcement, with the Australian Competition and Consumer Commission (ACCC) as the public enforcer of federal competition and consumer laws.[2] The ACCC has investigative and information-gathering powers under Part XID of the CCA; it resolves matters either by administratively encouraging consultation or negotiation to settle disputes[3] or via litigation. Nevertheless, with Australia being a common law jurisdiction, only the court has the power to declare whether particular conduct contravenes the CCA and to make findings of liability. The

[1] The Act was previously named the Trade Practices Act 1974 (Cth).

[2] The ACCC was established by the Competition Policy Reform Act 1995 (Cth). It replaced the existing Trade Practices Commission.

[3] See ACCC (2014) Compliance & enforcement policy, available at http://www.accc.gov.au/about-us/australian-competition-consumer-commission/compliance-enforcement-policy. Accessed 10 September 2021.

B. Jedličková (✉)
TC Beirne School of Law, The University of Queensland, Brisbane, Australia
e-mail: b.jedlickova@law.uq.edu.au

B. Marshall
Faculty of Law, Bond University, Robina, Australia
e-mail: bmarshal@bond.edu.au

M. Burdon
School of Law/Digital Media Research Centre, Queensland University of Technology, Brisbane, Australia
e-mail: m.burdon@qut.edu.au

B. Kilpatrick et al. (eds.), *Antitrust in Data Driven Markets & Legal Framework for Influencers, Native Advertising and Control over the Use of AI in Marketing*, LIDC Contributions on Antitrust Law, Intellectual Property and Unfair Competition,
https://doi.org/10.1007/978-3-031-07422-6_2

ACCC is empowered to institute proceedings in court for the declaration of an infringement of the CCA and for the recovery of a pecuniary penalty on behalf of the Commonwealth and may apply for injunctions, damages and a range of orders.[4] The ACCC can also be instructed to conduct inquiries and can initiate studies of markets of its own accord.

Australia has been active with regard to competition law in the digital economy, with the ACCC entrusted to lead and overview several inquiries. This inquiry boom commenced with the ACCC being directed by the Treasurer to conduct a Digital Platforms Inquiry (DPI) on the 4th of December 2017.[5] The ACCC issued its final report on this matter on 26 July 2019 (DPI Final Report). The DPI has been followed by the Digital Advertising Services Inquiry (DASI) [6] and Digital Platform Services Inquiry (DPSI),[7] both announced in February 2020. While DASI's focus is specifically limited to the 'markets for the supply of digital advertising technology services and digital advertising agency services',[8] the DPSI's scope is broad, covering digital platform services.[9] The DPSI consists and will consist of a number of interim reports, with the scope of the new interim report being announced every 6 months in the period between 2020 and 2025. The ACCC's first DPSI interim report was published in September 2020 ('DPSI2020') and the second in March 2021 ('DPSI2021'). The DPSI2020 investigates 'competition and consumer issues associated with online private messaging services, updates previous findings reached by the ACCC as regards social media and online search services and also identifies some common concerns across different types of platforms',[10] while the DPSI2021

---

[4] See ss 76-77 of the CCA.

[5] Some less significant studies were conducted prior to the DPI. In 2015, the ACCC commissioned a study on 'The Sharing Economy and the Competition and Consumer Act' (the 'Deloitte Study'): Deloitte Access Economics (2015) The sharing economy and the Competition and Consumer Act 2015, ACCC, available at https://www.accc.gov.au/system/files/Sharing%20Economy%20-%20Deloitte%20Report%20-%202015.pdf. Accessed 10 September 2021. Also, see discussions in Australian Government Productivity Commission, Digital Disruption: What do government need to do? 2016, p. 61, available at http://www.pc.gov.au/research/completed/digital-disruption. Accessed 10 September 2021; and also I. Harper, P. Anderson, S. McCluskey and M. O'Bryan QC (2015) Competition Policy Review: Final Report, 2015, p. 26, available at https://treasury.gov.au/publication/p2015-cpr-final-report Accessed 10 September 2021.

[6] The final report due 31 August 2021.

[7] The final report due 31 March 2025.

[8] ACCC (2020) Issues Paper Ad Tech Inquiry, available at https://www.accc.gov.au/system/files/Ad%20tech%20inquiry%20-%20issues%20paper.pdf. Accessed 10 September 2021.

[9] See, ACCC (2020) Ministerial Direction 'Digital Platform services inquiry', available at https://www.accc.gov.au/focus-areas/inquiries-ongoing/digital-platform-services-inquiry-2020-2025/ministerial-direction. Accessed 10 September 2021.

[10] ACCC (2020) Digital Platform Services Inquiry: Interim Report, available at <https://www.accc.gov.au/publications/serial-publications/digital-platform-services-inquiry-2020-2025/digital-platform-services-inquiry-september-2020-interim-report. Accessed 10 September 2021.

considers ‘competition and consumer issues associated with the distribution of mobile apps to users of smartphones and other mobile devices’.[11]

The third interim report, which examines ‘the provision of web browsers and general search services to Australian consumers and the effectiveness of choice screens in facilitating competition and improving consumer choice’, was published at the end of October 2021.[12] The fourth interim review analyses ‘potential competition and consumer issues in the provision of general online retail marketplaces’ and was published in April 2022.[13]

Despite the broad title of the inquiry, ‘Digital Platforms Inquiry’, the DPI’s scope was narrow and market-specific; it focussed only on ‘digital search engines, social media platforms and other digital content aggregation platforms . . . on the state of competition in media and advertising services markets’.[14] However, this narrow scope did not stop the ACCC from recommending an economy-wide application of some of the findings around privacy protection.[15] Indeed, a significant feature of the DPI is that it involved issues related not only to Australian competition and consumer laws but also to data and privacy protection, which are areas otherwise outside of the ACCC’s authority, and their intersections. The DPI’s scope reflects the ACCC’s broad portfolio as well as its ability to incorporate public interest issues beyond its usual scope of competition and consumer laws.

Since the DPI Final Report, the ACCC has enacted a new Digital Platforms Branch (DPB) that overviews competition and consumer law issues arising from the digital economy, in particular digital platforms, thus proving that matters of the digital economy are one of ACCC’s current enforcement priorities.[16] The DPB monitors the activities of digital platforms, takes enforcement actions in that space and conducts inquiries, including the DASI and DPSI.

---

[11] ACCC (2021) Digital Platform Services Inquiry: Interim Report No. 2, available at https://www.accc.gov.au/publications/serial-publications/digital-platform-services-inquiry-2020-2025/digital-platform-services-inquiry-march-2021-interim-report. Accessed 10 September 2021.

[12] ACCC (2021) September 2021 Interim Report, available at https://www.accc.gov.au/focus-areas/inquiries-ongoing/digital-platform-services-inquiry-2020-2025/september-2021-interim-report. Accessed 10 September 2021. Also highlights that the ACCC will provide ‘its advice to the Australian Government on Google’s rollout of search engine choice options on new Android devices in Europe.’

[13] ACCC (2021) March 2022 Interim Report https://www.accc.gov.au/focus-areas/inquiries-ongoing/digital-platform-services-inquiry-2020-2025/march-2022-interim-report. Accessed 10 September 2021.

[14] Ministerial Direction (2017) Inquiry into digital platforms, available at https://www.accc.gov.au/focus-areas/inquiries-ongoing/digital-platforms-inquiry/terms-of-reference. Accessed 10 September 2021

[15] See, DPI, pp. 452-453.

[16] ACCC (2020) ACCC 2020 Compliance and Enforcement Priorities, available at https://www.accc.gov.au/media-release/accc-2020-compliance-and-enforcement-priorities. Accessed 10 September 2021.

Another successful outcome of the DPI is the introduction of the News Media and Digital Platforms Mandatory Bargaining Code in February 2021.[17] The aim of the Code is to 'address the imbalance in bargaining power between leading digital platforms and Australian news businesses. It allows news media businesses to bargain collectively or individually with designated digital platforms about payment for the inclusion of news on their services.'[18]

2.2 Platform

The term 'platform' is central when dealing with the digital economy. While Australia has yet to define the term in its legislation or cases,[19] the ACCC's view on this term is clear from the DPI Final Report, which contains a definition of the term for the purposes of inquiry. The DPI Final Report refers to 'multi-sided platforms', which 'are applications that serve multiple groups of users at once, providing value to each group based on the presence of other users'.[20] It further explains that:

> The multiple sides of these platforms consist of groups of individuals who use the platforms for different reasons. For example, one side of a platform may consist of individuals who use its search services to find content or products, while another side may consist of businesses wanting to advertise to targeted groups of those individuals.[21]

For the purposes of the DPI, the ACCC considered and evaluated the meanings of three kinds of digital platforms: search engines, social media platforms and digital content aggregation platforms:

> (i) *Search engines* are software systems designed to search for information on the World Wide Web, generally returning a curated, ranked set of links to content websites. Search engines operate in an automated fashion using sophisticated algorithms to collect information (commonly known as 'crawling') and to provide search results...
>
> (ii) *Social media platforms* are online services that allow users to participate in social networking, communicate with other users and share and consume content generated by other users (including professional publishers). Social media platforms generally display content for consumption as linear 'feeds', curated by algorithms or displayed chronologically . . . Platforms may also offer additional functions, including instant messaging services.
>
> (iii) *Digital content aggregation platforms* are online intermediaries that collect information from disparate sources and present them to consumers as a collated, curated product

[17] Legislation enacting the Code, Treasury Laws Amendment (News Media and Digital Platforms Mandatory Bargaining Code) Act 2021, was passed on 25 February 2021.

[18] News media bargaining code I ACMA, htpps://www.acma.gov.au/news-media-bargaining-code. Accessed 30 October 2021.

[19] A brief explanation of the term platform is contained in The *Corporations Act 2001* (Cth) s 9, which defines platforms to include 'a website or other electronic facility'.

[20] DPI Final Report, p.41.

[21] DPI Final Report, p.41.

> . . . Users may be able to customise, filter or search their aggregation results. Examples include Google News, Apple News and Flipboard. [22]

The DPI did not include other digital platforms such as online shopping platforms, for instance Amazon.

The DPI Final Report explained that some of the digital platforms it examined also involved other services. In particular, the ACCC noted that Facebook offered online sales of goods and services while Google included services such as mapping, email and cloud storage.[23]

Despite specific challenges that 'two-sided platforms' can comprise for the purposes of the application of competition law, the ACCC has suggested that the existing Australian 'legislative framework has been sufficiently flexible to date in accommodating the consideration of two-sided platforms and the issues they create for market analyses'.[24] Indeed, the ACCC has already been considering the two-sided nature of online platforms and their implications when applying competition law, as was the case in the review of Expedia's proposed acquisition of Wotif in 2014.[25] Nevertheless and as partially discussed below, there is a current lack of cases to make any conclusion on potential issues that the analysis of platforms can contain with regard to Australian competition law, especially the definition of the relevant market.

2.3 Relevant Market and Free Services

Market definition is essential to an assessment of whether or not competition provisions have been contravened under the CCA. For the CCA to effectively capture potential anti-competitive conduct occurring in the digital world it is essential to, amongst other things, acknowledge the existence of the relevant markets for free services. In the DPI, the ACCC expressed that free services can form relevant markets by referring to markets for supplying social media services and supplying general search services, arguing that Facebook and Google had significant market power in their respective markets.[26]

[22] DPI Final Report, p.41.

[23] DPI Final Report, p.41.

[24] OECD, 'Two-Sided Markets', Policy Roundtables, DAF/COMP(2009)20, 17 December 2009, p. 80.

[25] ACCC Public Register (n.d.) Expedia Inc – proposed acquisition of Wotif.com Holdings Limited, available at http://registers.accc.gov.au/content/index.phtml/itemId/1182044/fromItemId/751046. Accessed 10 September 2021.

[26] See, e.g. DPI, pp. 64, 77. The DPI did not provide a detailed analysis of setting these 'relevant markets'.

Whether free services can form relevant markets has not been tested by the courts as yet. In the case of *ASX v Pont Data*,[27] the court confirmed that the digital flow of information could constitute a relevant market. However, this market did not involve free services.

The CCA, especially in its definition of the market, does not preclude the courts from agreeing with the ACCC in this matter. A market is defined in Section 4E CCA as:

> a market in Australia and, when used in relation to any goods or services, includes a market for those goods or services and other goods or services that are substitutable for, other otherwise competitive with, the first-mentioned goods or services.[28]

While the wording of the CCA definition of a market does not preclude the court from finding the existence of relevant markets for free services, its interpretation could impose a hurdle in that matter. In particular, the elaborated interpretation of Section 4E by the (then) Trade Practices Tribunal in the *QCMA* case in 1976,[29] which has been used extensively by the courts, involves references to price and/or profit in ways in which the market for free services does not fulfil.

This passage from *QCMA* states that the market is:

> the area of close competition between firms or ... the field of rivalry between them. ... Within the bounds of a market there is substitution – substitution between one product and another and between one source of supply and another, in response to *changing prices*. So a market is the field of actual and potential *transactions* between buyers and sellers amongst whom there can be strong substitution, at least in the long run, if given a sufficient *price incentive* ... in determining the outer boundaries of the market we ask a quite simple but fundamental question: if the firm were to 'give less and *charge more*' would there be, to put the matter colloquially, much of a reaction.[30]

A case that could potentially introduce some clarification in this matter is *Dialogue Consulting Pty Ltd v Instagram Inc & Ors*, filed at the Federal Court in April 2019, which revolves around the misuse of market power and unconscionable conduct.

---

[27] *ASX Operations Pty Ltd v Pont Data Australia Pty Ltd (No 2)* (1990) 21 FCR 385, appealed (1991) 27 FCR 492; *ASX Operations Pty Ltd v Pont Data Australia Pty Ltd (No 1)* (1990) 27 FCR 26, (1990) 27 FCR 460 at 486.

[28] Market is defined separately in relation to mergers as a market for goods or services in Australia or a state or a territory or a region of Australia. Section 50(6) CCA.

[29] *Re Queensland Co-Op Milling Association Limited and Defiance Holdings Limited* (QCMA) (1976) 8 ALR 481.

[30] *Re Queensland Co-Op Milling Association Limited and Defiance Holdings Limited* (QCMA) (1976) 8 ALR 481, 518. (Emphasis added.)

2.4 Market Power and Data

One of the questions that arise regarding the digital economy is whether the collection and usage of data can be a relevant factor for evaluating market power. Most importantly, the assessment of market power is essential for reviews of mergers and acquisitions and as part of the consideration as to whether the misuse of market power provision, Section 46 of the CCA, was contravened. The CCA provides some guidance on the assessment of market power with regard to the misuse of market power in Section 46(4). It states that:

(a) regard must be had to the extent to which the conduct of the body corporate or of any of those bodies corporate in that market is constrained by the conduct of:
  (i) competitors, or potential competitors, of the body corporate or of any of those bodies corporate in that market; or
  (ii) persons to whom or from whom the body corporate or any of those bodies corporate supplies or acquires goods or services in that market; and
(b) regard may be had to the power the body corporate or bodies corporate have in that market that results from:
  (i) any contracts, arrangements or understandings that the body corporate or bodies corporate have with another party or other parties; or
  (ii) any proposed contracts, arrangements or understandings that the body corporate or bodies corporate may have with another party or other parties.

Subsection 46(4)(a)(ii) expressly refers to buying and selling powers as aspects for determining the existence of significant market power. In general, for an entity to have significant market power, it must be able to substantially control the market. This is achieved not only with significant horizontal market power but also by the possession of significant bargaining power (buying and selling power). The fact that buying and selling powers are considered for determining significant market power means that even bargaining power could be misused under Section 46 as long as significant market power and other Section 46 requirements are shown to exist.[31] This kind of consideration of vertical and other non-horizontal relations could be relevant for the determination of market power in the digital world.

Although Section 46(4) does not directly refer to the collection of data as a potential factor for assessing market power, which is relevant to the digital markets, its wording does not preclude it either. *ASX v Pont Data* clarified that trading with and profiteering from digitalised information and its ownership could be important

[31] As was the case in *ACCC v Australian Safeway Stores Pty Ltd* (2003) 129 FCR 339; and *TPC v Carlton United Breweries Ltd* (1990) 24 FCR 532. These cases, decided under the old provision of s 46, remain relevant for the current provision because the new provision 46 contains the same aspects for determining the degree of market power.

for determining the relevant market and market power. However, this case did not include the collection of 'free' data.

The ACCC considers dynamic market characteristics when making competition and market power assessments.[32] Nevertheless, it missed an opportunity to expand on these factors and update them with regard to the features of the digital economy in its non-binding Guidelines on Misuse of Market Power from 2018.[33] Instead, the ACCC referred to a regularly quoted case with regard to factors that can influence the degree of competitive constraint faced by a firm that are likely to be relevant to the ACCC's assessment of the existence of substantial market power. Those are, in particular, the structural factors as set out by the Tribunal in the 1976 *QCMA* decision:

(i) the number and size distribution of independent sellers, especially the degree of market concentration;
(ii) the height of barriers to entry, that is the ease with which new firms may enter and secure a viable market;
(iii) the extent to which the products of the industry are characterised by extreme product differentiation and sales promotion;
(iv) the character of 'vertical relationships' with customers and with suppliers and the extent of vertical integration; and
(v) the nature of any formal, stable and fundamental arrangements between firms which restrict their ability to function as independent entities.[34]

These factors focus on static market structures, which may not be as relevant for digital markets. Therefore, previous concerns that the current approach by the ACCC and courts to the assessment of market power continues to place too much reliance on static market definitions and existing levels of market concentration remain.[35] The ACCC's approach to the determination of market power, which focuses on static market structures and utilises the competitive restraints test as applied in *QCMA*, is also used in digital inquiries, including the recent DPSI2020.[36]

[32] For example, in relation to the review of Expedia's acquisition of Wotif [ACCC Public Register (n.d.) Expedia Inc - proposed acquisition of Wotif.com Holdings Limited, available at http://registers.accc.gov.au/content/index.phtml/itemId/1182044/fromItemId/751046. Accessed 10 September 2021.] the level and pace of dynamic change in the market was a key factor in the ACCC's decision not to challenge the acquisition.

[33] ACCC (2018) Guidelines on misuse of market power, available at https://www.accc.gov.au/publications/guidelines-on-misuse-of-market-power. Accessed 10 September 2021.

[34] *Re Queensland Co-Op Milling Association Limited and Defiance Holdings Limited (QCMA)* (1976) 8 ALR 481, 515[40]; (1976) ATPR 40–012, 17,24.

[35] I. Harper, P. Anderson, S. McCluskey and M. O'Brya QC (2015) Competition Policy Review: Final Report, available at https://treasury.gov.au/publication/p2015-cpr-final-report. Accessed 10 September 2021.

[36] See, e.g., pp 20-34, where the ACCC assesses key online private messaging services provided by Facebook and Apple using the competitive restraint test discussed in *QCMA*. Although the ACCC applies this 'static' approach, it acknowledges the dynamic nature of platform markets (see, p. 28).

While the 2018 Guidelines on Misuse of Market Power do not refer to data collection, the ACCC's view on data collection as being a relevant factor for assessing market power is obvious in the DPI Final Report, where the ACCC's view is in line with findings from other competition authorities, such as the Bundeskartellamt. The ACCC argued in the DPI Final Report that large digital platforms, in particular Google and Facebook, possess significant market power due to a number of factors, one being their accumulation of users' data.[37] The ACCC found that they obtained the data not only directly and from other sites they operated (such as YouTube, Google Maps, Instagram, Messenger) but also from third-party sites when users used Facebook's or Google's devices and advertising systems.[38]

2.5 Misuse of Market Power

When considering the various issues that appeared in the DPI and in discussions held across the globe, including in the European Union, with regard to competition law and the digital economy, misuse of market power (or, as a similar area of EU law calls it, 'abuse of dominant position') is an area of competition law that possesses various issues. These include the understandings and interpretations of not only the terms 'platform', 'market' and 'market power' but also the core element of what constitutes a misuse, in other words abuse, of market power.

The provision on the misuse of market power, Section 46 of the CCA, was changed significantly in November 2017 to introduce an effect-based test. The provision now prohibits a corporation with substantial market power from engaging in conduct that has the purpose or effect of substantially lessening competition. Previously, the provision prohibited firms with a substantial degree of market power from *taking advantage* of their market power for an anti-competitive purpose, where these purposes included:

(i) eliminating or substantially damaging a competitor ... in that or any other market;
(ii) preventing the entry of a person into that *or* any other market; or
(iii) deterring or preventing a person from engaging in competitive conduct *in that or any other market.*

The new provision, Section 46, has not yet been tested thoroughly in court.[39] Both the new and old provisions on the misuse of market power prohibit only

[37] DPI Final Report, pp. 73, 84-85.

[38] DPI Final Report, pp. 73, 84-88. In terms of the extensiveness of the data, Facebook and Google are leading platforms in that sense in Australia, whereas other websites or publishers have not been able to collect such an extensive amount of data (see, p. 86).

[39] However, in *Unlockd Limited v Google Asia Pacific Pte Limited* [2018] FCA 826, Unlockd, an advertising technology start-up, succeeded in obtaining interim injunctive relief against Google in circumstances where Unlockd claimed that Google had threatened to ban Unlockd's app from its

exclusionary conduct. Other forms of abuse, which are covered in the EU law on abuse of dominant position, in particular exploitative practices, are not dealt with under Section 46 in Australia. However, some areas of Australian consumer law, such as unconscionable conduct, can be used to tackle other issues arising from a dominant position.

Interestingly, with regard to the DPI, the ACCC referred to 'special responsibility' in relation to the conduct of dominant digital platforms, in particular Google and Facebook, in the 'Introductory Overview' of the DPI Final Report.[40] This appears to be confusing, considering that, unlike in EU competition law, neither the new nor the old provisions or the interpretation of Section 46 by the courts refer to 'special responsibility', and thus special responsibility is not analysed when determining the existence of misuse of market power.

However, when looking at the potential issues that the DPI Final Report identified with regard to the practices conducted by Google and Facebook, it is obvious that the connotation of 'special responsibilities' was used analogously in order to reflect the ACCC's 'findings' that market power or the dominant position of a 'giant' digital platform can be 'misused' or 'abused' in a way that can contravene not only Australian competition law but also the Australian Consumer Law (ACL), which is Schedule 2 to the CCA. These issues are:

(i) whether access restrictions imposed by a digital platform on a third-party app developer raise issues under section 46 of the CCA
(ii) whether representations made by Google to some users about the control users have over Google's collection of location data, raise issues under the ACL
(iii) whether representations by Google about its privacy policy, and the level of disclosure about subsequent privacy policy changes that enabled Google to combine or match different sets of user data, raise issues under the ACL
(iv) whether representations made by Facebook (and/or its related entities) in relation to the nature of its services and the scope of its terms and conditions, including terms and conditions that allowed user data to be shared with third parties, raise issues under the ACL

Google Play Store in contravention of s 46. The Federal Court made interim orders preventing Google from proceeding with the ban, noting that the likely effect of excluding Unlockd from distributing its app through the Google Play Store would be that Unlockd would cease to exist as a business. Subsequently, Unlockd entered voluntary administration, blaming the dispute with Google, whereupon the action against Google was discontinued by the liquidators. In November 2019, media reports suggested that the ACCC was preparing to launch legal action against Google in relation to the demise of Unlockd, alleging a breach of s 46: see, P. Smith (2019) ACCC to sue Google over Unlockd, available at https://www.afr.com/technology/accc-to-sue-google-over-unlockd-20191030-p535u9. Accessed 10 September 2021. To date, the case has not been commenced.

[40] DPI Final Report, p.1.

(v) whether terms of use and privacy policies used by Facebook (and/or its related entities) contain unfair contract terms.[41]

Of these potential issues, only one is relevant to the misuse of market power, while four are covered by the ACL. It is interesting to note that some of these ACL issues would be covered by competition law in (some) EU countries, for instance Germany, and this, indeed, raises the question of the objects of competition law and consumer law when dealing with business behaviour in the digital world. This, in turn, also raises the question as to the role of a competition authority (and not the consumer one) in the digital economy.

Unlike many other competition regulators across the globe, the ACCC's remit encompasses competition law and consumer law. The ACCC considers the goals and functions of these two areas of law to be 'mutually-reinforcing'.[42] For its part, competition law facilitates competitive markets so as to increase consumer choice and enhance consumer welfare, while consumer law protects consumers' ability to make free and informed choices that maximise their own utility, which, in turn, promotes competitive markets.[43]

The DPI Final Report noted that consumer transactions with digital platforms are impacted by significant information asymmetries and imbalances in bargaining power.[44] The former compromises the ability of consumers to make informed choices, the latter leaves consumers open to exploitation and both can lead to market failure. It follows, therefore, that consumer law is as critical as competition law in addressing issues of market inefficiencies and market power arising in digital markets.[45]

2.5.1 Sector-Specific Regulation: Telecommunications Industry

The DPSI2020 identified a current trend on the part of large digital platforms to integrate vertically into the telecommunications supply chain.[46] This is an efficient

[41] DPI Final Report, p. 38. Since the publication of the DPI Final Report, the ACCC initiated only one court proceeding, against Google, for misleading consumers on the collection and use of location data. ACCC (2019) Google allegedly misled consumers on collection and use of location data, available at https://www.accc.gov.au/media-release/google-allegedly-misled-consumers-on-collection-and-use-of-location-data. Accessed 10 September 2021.

[42] DPI Final Report, pp. 139-140.

[43] DPI Final Report, pp. 139-140.

[44] DPI Final Report, p. 384.

[45] DPI Final Report, pp. 139-140. In Australia, the ACL regulates the unilateral behaviour of all corporations in relation to how they deal with consumers and other businesses. The ACL prohibits misleading or deceptive conduct (Part 2-1), false or misleading representations (Part 3-1) and unconscionable conduct (Part 2-2), and provides for unfair contract terms in standard form consumer and small business contracts to be declared void and unenforceable (Part 2-3).

[46] DPSI2020, p. 79.

strategy for these platforms, given their heavy reliance on telecommunications networks and corresponding incentives to ensure that telecommunications services are reliable and far-reaching.[47] It allows the platforms to achieve lower costs and greater control over the quality and capacity of telecommunications services.[48]

In Australia, the expansion of digital platforms into telecommunications infrastructure and companies raises the spectre of the sector-specific competition regulation that applies to the telecommunications industry. This operates alongside the general misuse of market power prohibition in Section 46 CCA.

The sector-specific misuse of market power provision in Part XIB CCA, Section 151AJ, applies to telecommunications carriers and carriage service providers only.[49] Section 151AJ—a variant of both the old and new Section 46—provides that a telecommunications carrier or carriage service provider that has a substantial degree of power in a telecommunications market is prohibited from taking advantage of that power with the effect or likely effect of substantially lessening competition in that or any other telecommunications market. Additionally, Part XIC CCA allows the ACCC to 'declare' a carriage service or service that facilitates the supply of a carriage service.[50] Once a service is declared, the carrier must allow carriage service providers to obtain access to the service.[51]

In terms of consumer protection, the telecommunications industry is well served by the following three 'watchdogs'. First is the ACCC, which not only regulates telecommunications infrastructure and competition but also enforces the ACL to protect consumers and businesses. Second is the Australian Communications and Media Authority (ACMA), which works closely with industry players to ensure that appropriate and active self-regulation is developed and compliance with relevant codes and standards is met.[52] Third is the Telecommunications Industry Ombudsman (TIO), which provides free service to consumers and small businesses to facilitate the resolution of minor disputes.[53]

---

[47] DPSI2020, p. 80.

[48] DPSI2020, p. 80.

[49] In the Australian telecommunications industry, carriers operate the telecommunications facilities and infrastructure, and carriage service providers use carriers' facilities to provide telecommunication services to businesses and consumers. See Department of Infrastructure, Transport, Regional Development and Communications (n.d.) A competitive telecommunications regime for Australia, available at https://www.communications.gov.au/policy/policy-listing/competitive-telecommunications-regime-australia. Accessed 10 September 2021.

[50] See CCA Part XIC, Division 2.

[51] See CCA Part XIC, Division 4.

[52] An example of one such code is the Telecommunications Consumer Protections Code, which includes standards for advertisements, contracts and billing: see, Australian Communications and Media Authority (2021) Telecommunications Consumer Protections Code, available at https://www.acma.gov.au/telecommunications-consumer-protection-code. Accessed 10 September 2021.

[53] See, Telecommunications Industry Ombudsman (n.d.), available at https://www.tio.com.au/. Accessed 10 September 2021.

No doubt mindful of the role of the TIO, the ACCC's primary concern in the DPI was the absence of effective dispute resolution processes for consumers and businesses in their dealings with digital platforms.[54] The DPI Final Report recommended that the ACMA develop minimum internal dispute resolution standards to apply to those digital platforms that supply services in Australia and have over one million active users per month nationally.[55] At a minimum, these standards should require large digital platforms to comply with best-practice principles of internal dispute resolution design relating to visibility, accessibility, responsiveness, objectivity, confidentiality, accountability, collection of information, resourcing, no charging and continual improvement.[56]

In the event that complaints or disputes are not resolved internally, the DPI Final Report also recommended the establishment of an independent ombudsman scheme to resolve complaints and disputes between businesses and digital platforms and also consumers and digital platforms.[57] While the scope of the ombudsman scheme would be determined by the ACMA following broad consultation with relevant stakeholders, it was anticipated that the scheme would cover, at the least, the following matters: complaints or disputes from businesses relating to the purchase or performance of advertising services from digital platforms and complaints or disputes from consumers, including in relation to scams and the removal of such content.[58] The ombudsman would have the power to compel information, make decisions that are binding on digital platforms, require digital platforms to take down scam content and order compensation in appropriate cases.[59] The ACCC considers that the TIO may be an appropriate body to implement the scheme and has asked the ACMA and the TIO to investigate the feasibility of the TIO taking on the role, before considering whether a stand-alone ombudsman should be created to resolve complaints about digital platforms.[60]

The Australian government has expressed in-principle support for the above recommendations but has also committed to developing a pilot external dispute resolution scheme in consultation with major digital platforms, consumer groups and relevant governmental agencies.[61] The outcomes of the pilot scheme, along with any parallel improvements in internal dispute resolution processes generally, will inform consideration of whether to establish a Digital Platforms Ombudsman to resolve

[54] DPI Final Report, p. 507.

[55] DPI Final Report, p. 509 (Recommendation 22). This recommendation was also endorsed in the DPSI2021, p. 12.

[56] DPI Final Report, pp. 508-509.

[57] DPI Final Report, p. 510 (Recommendation 23). This recommendation was also endorsed in the DPSI2021, p. 12.

[58] DPI Final Report, pp. 509-510.

[59] DPI Final Report, p 510.

[60] DPI Final Report, p. 510.

[61] Australian Treasury (2019) Regulating in the Digital Age: Government Response and Implementation Roadmap for the Digital Platforms Inquiry, available at https://treasury.gov.au/sites/default/files/2019-12/Government-Response-p2019-41708.pdf. Accessed 10 September 2021.

complaints and disputes between digital platforms and the individual consumers and businesses using their services.

2.6 Acquisition of Market Power

The processes whereby digital platforms may gain market power through mergers and acquisitions were the focus of specific scrutiny in the DPI.[62] DPSI2020 pursued the theme, tying the expansion of digital platforms to concerns about the propensity of larger platforms to build 'ecosystems' of products and services that interoperate with each other.[63] While this practice may benefit consumers by allowing them to access multiple services on the same platform and enjoy improvements in product quality stemming from the bundling of hardware and software, it also has potential anti-competitive effects as rival platforms will need to incur significant costs and offer a greater range of services to attract consumers away from entrenched ecosystems.[64]

The current provision of the CCA dealing with mergers and acquisitions, Section 50, sets out the legal test for considering a proposed acquisition of shares or assets. In general terms, Section 50 CCA prohibits acquisitions that would have the effect, or be likely to have the effect, of substantially lessening competition in any market. Concern was expressed in the DPI Final Report that acquisitions in digital markets, or acquisitions in markets involving emerging technologies, may have the effect or likely effect of substantially lessening competition if the transaction involves the acquisition of nascent competitors (i.e. start-ups) by established platforms or the acquisition of any business with access to a large volume and scope of data or that otherwise collects and monetises data.[65]

The merger law case *Vodafone Hutchison*[66] no doubt reinforced the ACCC's view that current merger laws are not effectively preventing the acquisition of new entrants by established firms.[67] In 2019, the ACCC opposed a proposed merger between TPG Telecom Ltd and Vodafone Hutchison Australia Pty Ltd largely on the basis that it would reduce competition and contestability in the market for mobile

[62] See DPI Final Report, p. 30 (Recommendations 1 and 2).

[63] DPSI2020, p. 82.

[64] DPSI2020, p. 85.

[65] DPI Final Report, p. 105. This concern has been echoed by the UK, Australian and German competition authorities in a joint statement on the need for rigorous merger enforcement: see, Competition and Markets Authority, ACCC and Bundeskartellamt (2021) Joint statement on merger control enforcement, available at https://www.accc.gov.au/publications/joint-statement-on-merger-control-enforcement. Accessed 10 September 2021.

[66] *Vodafone Hutchison Australia Pty Ltd v ACCC* [2020] FCA 117.

[67] Indeed, in the DPI Final Report, the ACCC indicated that it may lobby for the introduction of a rebuttable presumption when it challenges an acquisition, such that the starting point for court proceedings is that the acquisition is presumed likely to substantially lessen competition, in the absence of clear and convincing evidence to the contrary: DPI Final Report, p. 105.

phone services.[68] However, Vodafone successfully applied for a declaration from the Federal Court that the merger would not be in breach of the CCA.[69]

Given the prevailing concerns, the DPI Final Report recommended amending Section 50(3) CCA to include two additional factors that must be taken into account when assessing whether a proposed acquisition has the effect or likely effect of substantially lessening competition. These factors are (i) the likelihood that the acquisition would result in the removal of a potential competitor and (ii) the nature and significance of the assets, including data and technology, being acquired directly or through the body corporate.[70] There is no bar at present to considering these matters under Section 50(3) as the list of factors currently set out therein is expressly non-exhaustive. However, the DPI Final Report cited the need to 'highlight the significance' of these two factors by articulating them in the CCA.[71]

While the two additional factors are intended to apply to all proposed acquisitions, not simply those involving digital platforms, they clearly reflect the ACCC's view that 'strategic' and 'killer' acquisitions by Google and Facebook have contributed to the market power that these platforms currently hold.[72] They also speak to the ACCC's concern that, where a proposed acquisition will combine sets of data between the target and the acquirer, competitors may not be able to match the scale or scope of those combined data sets, which may then have the effect of raising barriers to entry or foreclosing access to markets.[73] The tension here is that it is often the precise goal of digital start-ups to be acquired by a large digital platform. If the founders/investors of the target company anticipate that it will be harder to sell their

---

[68] ACCC (2019) ACCC opposes TPG-Vodafone merger, available at https://www.accc.gov.au/media-release/accc-opposes-tpg-vodafone-merger. Accessed 10 September 2021.

[69] *Vodafone Hutchison Australia Pty Ltd v ACCC* [2020] FCA 117.

[70] DPI Final Report, p. 105 (Recommendation 1).

[71] DPI Final Report, p. 106.

[72] DPI Final Report, pp. 74-76 (Google) and 80-81 (Facebook). For clarity, a 'strategic' acquisition is an acquisition that benefits the acquiring digital platform through resulting advantages of scope and a reduction in potential competition. A commonly cited example is Facebook's acquisition of Instagram in 2012. A 'killer' acquisition is a specific type of strategic acquisition where the target is a smaller innovative company and the acquiring digital platform discontinues the target's innovative activities to eliminate potential rivalry: DPI Final Report, pp. 75, 105.

[73] DPI Final Report, p. 108. By way of example, on 18 June 2020, the ACCC released its Statement of Issues detailing its preliminary concerns about the proposed acquisition of Fitbit Inc, which is active in Australia via its subsidiary, Fitbit (Australia) Pty Ltd, by Google LLC (Google). In the ACCC's view, the proposed acquisition may provide Google with access to large volumes of consumer health data (in addition to the substantial amounts of individual user data already held by Google), further entrench Google's dominant position and adversely affect competition in several digital advertising and health markets. See, ACCC (2020) Statement of Issues: Google LLC – Proposed Acquisition of Fitbit Inc, available at https://www.accc.gov.au/system/files/public-registers/documents/Google%20Fitbit%20-%20Statement%20of%20Issues%20-%2018%20June%202020.pdf. Accessed 10 September 2021.

interest to a major platform, the recommendation could inadvertently dampen competition and innovation.[74]

The DPI Final Report also recommended that 'large' digital platforms agree to a notification protocol under which they would provide advance notice of any proposed acquisitions that could potentially impact competition in Australia. The protocol would specify the types of acquisitions requiring notification (including any threshold transaction value) and the minimum advance notification period to enable the ACCC to assess the proposed acquisition.[75] In effect, this recommendation would introduce a mandatory notification regime for 'large' digital platforms, principally Google and Facebook (as the ACCC has not clarified which other platforms are considered 'large'). As such, it represents a partial departure from Australia's current merger review process, which does not include mandatory notification. However, the ACCC already expects notification of all proposed acquisitions that it considers sensitive.[76] To this extent, the recommendation would simply codify and extend the notification commitment.

Overall, the DPI Final Report recommendations relating to mergers and acquisitions represent a shift towards increased scrutiny and intervention in relation to acquisitions in the digital/technology space. In December 2019, the Australian government responded to these recommendations by committing to the introduction of a notification protocol and agreeing to undertake further public consultation on changes to Section 50 CCA.[77] After 20 months of reflection, the ACCC has flagged the need for more sweeping merger reforms, including a proposal for 'a tailored test for acquisitions by certain digital platforms'.[78] According to the ACCC, the new test should be predicated on a lower probability of competitive harm than that which applies to acquisitions in the broader economy, and the application of this test to any given digital platform should depend on the size and scope of the platform's services in Australia and whether it acts as a 'gateway' in terms of controlling how other businesses interact with consumers.[79]

---

[74] Gilbert & Tobin (2020) Competition & Consumer Regulation: Review of 2019 & Insights for 2020, available at https://www.gtlaw.com.au/insights/australian-competition-consumer-law-insights-2020. Accessed 10 September 2021.

[75] DPI Final Report, p. 109 (Recommendation 2).

[76] ACCC (2017) Merger Guidelines, available at https://www.accc.gov.au/system/files/Merger%20guidelines%20-%20Final.PDF. Accessed 10 September 2021.

[77] Australian Treasury (2019) Regulating in the Digital Age: Government Response and Implementation Roadmap for the Digital Platforms Inquiry, available at https://treasury.gov.au/sites/default/files/2019-12/Government-Response-p2019-41708.pdf. Accessed 10 September 2021.

[78] The ACCC's views have been expressed by its Chair: see, R. Sims (2021) Protecting and promoting competition in Australia, available at https://www.accc.gov.au/speech/protecting-and-promoting-competition-in-australia. Accessed 10 September 2021.

[79] Ibid.

2.7 Consumer Protection

2.7.1 Data Collection

The collection and use of consumer data is central to the business model of most digital platforms and generally results in significant benefits to consumers, including improved services, new product recommendations and the provision of free content.[80] However, the DPI Final Report noted a general mismatch between the expectations of consumers and the terms of their agreements with digital platforms regarding such matters as how much consumer data are collected by platforms,[81] how those data are used and the extent to which they are disclosed to third parties.[82]

The apparent disconnect between the privacy concerns expressed by consumers and their actual behaviour in continuing to deal with digital platforms on terms that do not prioritise privacy has been labelled the 'privacy paradox'.[83] However, this so-called 'paradox' ignores the fact that, in dealing with digital platforms, the typical consumer is disadvantaged by both a lack of bargaining power and acute information asymmetries.[84] The DPI Final Report emphasised that the ubiquity of digital platforms in the daily lives of consumers means that many have no real choice but to join or use these platforms and accept the terms of use offered.[85] Moreover, the terms often involve 'concealed data practices'—i.e., the extent of the terms and related data practices and the consequences of those data practices are hidden from consumers.[86]

The ACCC's review of digital platforms' terms and policies found the following practices to be commonplace:[87] terms are put forward on a take-it-or-leave-it basis; clickwrap agreements contain unobtrusive links to online documents that permit the collection, use and disclosure of extensive amounts of user data;[88] consents to the

---

[80] See, R. Walters, B. Zeller and L. Trakman, Personal Data Law and Competition Law – Where is it Heading?, European Competition Law Review 2018(39), pp. 505-514.

[81] Relevant here are developments in voice-activated services and augmented and virtual reality technology, which allow for more extensive data collection by digital platforms: DPSI2020, pp. 88, 99.

[82] DPI Final Report, p. 448.

[83] DPI Final Report, p. 384.

[84] DPI Final Report, p. 448.

[85] DPI Final Report, p. 400.

[86] See, K. Kemp, Concealed Data Practices and Competition Law: Why Privacy Matters, August 7, 2019, available at https://www.competitionpolicyinternational.com/concealed-data-practices-and-competition-law-why-privacy-matters/. Accessed 10 September 2021.

[87] See, DPI Final Report, pp. 400, 402. The DPSI2020 found these practices to be prevalent in the context of online messaging services as well: pp. 35-43.

[88] This includes data practices of particular concern to consumers, such as location tracking, online tracking for targeted advertising purposes and disclosure of user data to third parties: DPI Final Report, p. 421. Similar concerns were noted in the DPSI2020: p. 50. The DPSI2020 also identified further potential for consumer harm arising from scams, prominence of non-organic search results, and price discrimination by online retailers: pp. 56, 62, 100.

use of data are bundled across different uses and different services; and privacy policies are long, complex, vague and difficult to navigate.

The DPI Final Report and DPSI2020 both affirmed that these practices leverage bargaining power imbalances between consumers and digital platforms and deepen the information asymmetries between them.[89] Given the pervasiveness of data collection and use by both digital platforms and other businesses, and the consequential potential for significant detriment to consumers, it was deemed necessary to re-evaluate the current protections provided to consumers by the ACL in relation to these practices.[90]

2.7.2 ACL Provisions

In addition to its role as Australia's competition regulator, the ACCC has joint responsibility with state and territory consumer protection agencies for administering the ACL.

The DPI Final Report recognised that the ACL contains several effective tools for addressing digital platforms' practices (including 'concealed data practices') that give rise to consumer protection issues.[91] As is the case for all businesses subject to the ACL, digital platforms must not engage in conduct that is misleading or deceptive or is likely to mislead or deceive[92] or make false or misleading representations about their goods or services.[93] This includes, for example, incorrect or misleading statements about how user data are collected, used or shared.[94] Digital platforms are also prohibited from engaging in unconscionable conduct in

[89] DPI Final Report, p. 400; DPSI2020, p. 44.

[90] DPI Final Report, p. 498. In addition, the DPSI2020 recommended an enforceable code which would require greater transparency on the part of digital platforms about their consent requirements, data sharing practices, and willingness to provide consumers with opt-out controls: p. 44.

[91] DPI Final Report, pp. 436-437.

[92] ACL s 18. An intention to mislead is irrelevant to a contravention of s 18: *Parkdale Custom Built Furniture Pty Ltd v Puxu Pty Ltd* (1982) 149 CLR 191.

[93] ACL ss 29, 33 and 34.

[94] The ACCC recently commenced Federal Court proceedings against Google for allegedly misleading consumers about the way it collected, kept and used location data when certain Google Account settings were enabled or disabled. The ACCC alleges that Google did not properly disclose that both the 'Location History' setting and the 'Web & App Activity' setting had to be switched off if consumers did not want Google to collect, keep and use their location data. As a result, the ACCC alleges that Google dealt with highly sensitive and valuable personal information about consumers' locations without the consumers making an informed choice as to whether to share this data with Google. See, ACCC (2019), Google allegedly misled consumers on collection and use of location data, available at https://www.accc.gov.au/media-release/google-allegedly-misled-consumers-on-collection-and-use-of-location-data. Accessed 10 September 2021.

connection with the supply or acquisition of goods or services.[95] Additionally, as the ACCC considers digital platforms' consumer-facing terms of use and privacy policies to be standard form contracts,[96] digital platforms must comply with the unfair contract term provisions in the ACL.[97]

Nevertheless, in order to keep pace with the future, particularly as technology creates new opportunities for unfair conduct, the DPI Final Report and DPSI2020 focussed on the strengthening or introduction of ACL provisions dealing with unfair contract terms and unfair trading practices.

2.7.2.1 Unfair Contract Terms

Due to the bargaining power imbalance in the relationship between digital platforms on one side and consumers or small businesses on the other, the latter parties are unable to negotiate terms of service, including terms relating to the collection, use and disclosure of personal data. This imbalance in bargaining power allows digital platforms to include potentially unfair contract terms in their terms of use or privacy policies.[98]

A review by the ACCC of the standard terms governing the supply of advertising services by large digital platforms identified the widespread use of several potentially unfair clauses. In the main, these involve broad discretions for the platforms to remove content for any reason, suspend or terminate a user's account for any reason, vary terms without notice, impose a prohibitive dispute resolution process and specify confidentiality or publicity limitations.[99]

At present, it is not a contravention of the ACL to include unfair terms in contracts, meaning that no penalties can be sought for this conduct per se; rather, if a particular term is declared unfair, it is simply void.[100] However, there is some potential for ACL penalties for unfair contract terms to be accessed indirectly. For example, by including unfair terms in a contract (e.g. relating to 'concealed data practices'), the digital platform may be engaging in misleading or deceptive conduct in contravention of Section 18 ACL. Also, relying on unfair contract terms may amount to unconscionable conduct in contravention of Section 20 or 21 ACL. Breaches of these provisions give rise to a wide range of remedies, including the imposition of pecuniary penalties.[101] In addition, any loss a consumer or small business suffers as a result of an unfair contract term may be subject to a

---

[95] ACL Part 2-2. Although the concept has not been defined precisely in Australia, conduct is generally 'unconscionable' when it goes against good conscience by reference to the norms of society: *ACCC v Lux Distributors Pty Ltd* [2013] FCAFC 90.

[96] DPI Final Report, p. 437.

[97] ACL Part 2-3.

[98] DPI Final Report, p. 497.

[99] DPSI2020, pp. 70-71.

[100] ACL s 23.

[101] ACL s 224. Currently, maximum penalties stand at the greater of AUD 10 million, three times the benefit obtained from the breach, or 10% of annual turnover connected with Australia.

compensation order, which can be claimed by the aggrieved party or the ACCC on their behalf.[102]

Not surprisingly, the ACCC's view is that the current unfair contract terms regime does not itself provide sufficient deterrence.[103] Accordingly, the DPI Final Report and DPSI2020 have both recommended that the ACL be amended so that the inclusion of unfair terms in standard form consumer or small business contracts is prohibited and attracts the imposition of civil pecuniary penalties.[104] In response, the Australian government committed to consultations on strengthening the unfair contract term provisions of the ACL[105] and in August 2021 released exposure draft legislation giving effect to the recommendation in full, together with other enhancements.[106] If the legislation is passed, the changes to the unfair contract terms regime will have implications on an economy-wide scale for all firms that deal with consumers and small businesses, not only those that operate on digital platforms.

2.7.2.2 Unfair Trading Practices

The DPI Final Report identified various business practices that cause detriment to consumers but are not expressly prohibited by the ACL. These practices include[107] collecting and/or disclosing consumer data without express informed consent; failing to comply with reasonable data security standards, including failing to put in place appropriate security measures to protect consumer data; inducing consumer consent to data collection and use by relying on long and complex contracts or all-or-nothing clickwrap consents and providing insufficient time or information to enable consumers to properly consider the contract terms; unilaterally changing the terms on which goods or services are provided to consumers without reasonable notice and without the ability for consumers to consider the new terms, including in relation to subscription products and contracts that automatically renew; and seeking to dissuade consumers from exercising their contractual or other legal rights, including requiring the provision of unnecessary information in order to access benefits.

[102] ACL s 237.

[103] DPI Final Report, p. 441.

[104] DPI Final Report, p. 497 (Recommendation 20); DPSI2020, p. 74.

[105] Australian Treasury (2019) Regulating in the Digital Age: Government Response and Implementation Roadmap for the Digital Platforms Inquiry, available at https://treasury.gov.au/sites/default/files/2019-12/Government-Response-p2019-41708.pdf. Accessed 10 September 2021. On 13 December 2019, the Australian Government announced a consultation on 'Enhancements to Unfair Contract Term Protections': see Australian Treasury (2020), available at https://treasury.gov.au/consultation/enhancements-unfair-contract-term-protections. Accessed 10 September 2021.

[106] See, Australian Treasury (2021) Exposure Draft, Treasury Laws Amendment (Measures for a later sitting) Bill 2021: Unfair contract terms reforms, available at https://treasury.gov.au/sites/default/files/2021-08/c2021-201582_edl.pdf. Accessed 10 September. Other reforms proposed in the Bill include a rebuttable presumption that a contract term is unfair if the same or a substantially similar term (in a party's own contracts or other contracts in the industry) has previously been declared unfair.

[107] DPI Final Report, p. 498. Similar practices were also outlined in the DPSI2020, pp. 35-43.

To address such instances of problematic conduct, the DPI Final Report recommended that the ACL be amended to include an economy-wide prohibition against unfair trading practices.[108] While similar laws exist in the European Union and the United States,[109] this would be a new provision in Australia and should be 'carefully developed such that it is sufficiently defined and targeted, with appropriate legal safeguards and guidance'.[110] It is anticipated, for example, that the new prohibition would focus on practices that cause significant detriment to consumers or small businesses, thereby protecting both groups; that operate in addition to existing ACL protections, including prohibitions against misleading or deceptive conduct and unconscionable conduct; and that attract the same maximum civil pecuniary penalties that currently apply under the ACL.[111]

The Australian government has noted the recommendation in favour of a prohibition against unfair trading practices but has not specifically responded to it, preferring instead to reference the separate work on this issue being undertaken by Consumer Affairs Australia and New Zealand (CAANZ).[112]

2.7.3 Comparator Websites

The ACCC recognises that comparator websites have the potential to be of benefit to Australian consumers and to promote competition in the markets for the goods or services to which they relate. They achieve this by providing consumers with a convenient means of comparing deals offered by competing businesses and by assisting new market entrants to compete more effectively.[113]

[108] DPI Final Report, p. 498 (Recommendation 21). In lending support to this recommendation, the DASI interim report identified consumers' relationships with 'ad tech' providers as deserving of protection by a prohibition against unfair trading practices. Even though consumers are not customers of ad tech services, the data produced by their online activities are an essential input into many ad tech services. The potential for consumer harm arises when consumers are not sufficiently informed about, or do not have sufficient control over, how their data are collected and used for ad targeting purposes. See ACCC (2020) Digital Advertising Services Inquiry: Interim Report, available at https://www.accc.gov.au/focus-areas/inquiries-ongoing/digital-advertising-services-inquiry/interim-report. Accessed 10 September 2021.

[109] DPI Final Report, p. 498.

[110] DPI Final Report, p. 498.

[111] See, Gilbert & Tobin (2020) Competition & Consumer Regulation: Review of 2019 & Insights for 2020, available at https://www.gtlaw.com.au/insights/australian-competition-consumer-law-insights-2020. Accessed 10 September 2021.

[112] Australian Treasury (2019) Regulating in the Digital Age: Government Response and Implementation Roadmap for the Digital Platforms Inquiry, available at https://treasury.gov.au/sites/default/files/2019-12/Government-Response-p2019-41708.pdf. Accessed 10 September 2021.

[113] ACCC (2014) The Comparator Website Industry in Australia, available at https://www.accc.gov.au/system/files/926_Comparator%20website%20industry%20in%20Australia%20report_FA.pdf. Accessed 10 September 2021.

However, the ACCC has also identified two primary areas of concern with respect to comparator websites.[114] The first relates to the material on these websites, broadly encompassing representations about the extent of the comparison service, representations about the savings to be achieved from using the comparison service and representations that the comparison service is independent, impartial or unbiased.[115] The second area of concern involves the 'behind-the-scenes' conduct of comparator websites. In this regard, the ACCC has expressed misgivings as to whether undisclosed commercial relationships between comparator websites and their suppliers are impacting the recommendations presented to consumers and whether the information displayed on comparator websites is being updated in a timely fashion.[116]

As it stands, Australia's regulatory framework is well positioned to deal with the concerns associated with comparator portals. The ACL prohibits businesses from engaging in misleading or deceptive conduct[117] and from making a range of false or misleading representations.[118] All of the aforementioned concerns would likely constitute breaches of one or more of the existing provisions.

Consider, for example, the Full Federal Court's decision in *Trivago NV v ACCC*,[119] which upheld the first instance judgment[120] that Trivago, a hotel comparator website, had misled consumers in its television advertising and on its website about hotel room rates in breach of Sections 18, 29(1)(i) and 34 ACL. Those provisions prohibit misleading or deceptive conduct, false or misleading representations with respect to the price of goods or services and conduct that is liable to mislead the public as to the nature or characteristics of any services, respectively.

The Full Federal Court unanimously agreed that:

---

[114] ACCC (2014) The Comparator Website Industry in Australia, available at https://www.accc.gov.au/system/files/926_Comparator%20website%20industry%20in%20Australia%20report_FA.pdf. Accessed 10 September 2021.

[115] ACCC (2014) The Comparator Website Industry in Australia, available at https://www.accc.gov.au/system/files/926_Comparator%20website%20industry%20in%20Australia%20report_FA.pdf. Accessed 10 September 2021.

[116] ACCC (2014) The Comparator Website Industry in Australia. Accessed 10 September 2021, pp. 18-19. Interestingly, it appears that Australian consumers are already attuned to these issues, with market research indicating that while most consumers make use of comparator websites, the majority do so for research purposes only and not as a purchasing channel: ACCC, *The Comparator Website Industry in Australia* (Commonwealth of Australia, 2014) p. 11. The ACCC has also published a webpage to assist consumers in using comparator websites effectively: see, ACCC (n.d.) Comparator Website, available at https://www.accc.gov.au/consumers/online-shopping/comparator-websites. Accessed 10 September 2021.

[117] See ACL s 18.

[118] See ACL Part 3-1, Division 1.

[119] [2020] FCAFC 185.

[120] *ACCC v Trivago NV* [2020] FCA 16.

(i) Trivago had misled consumers by representing that its website would identify the cheapest rates available for a hotel room in response to a user's search. In fact, Trivago typically did not rank hotel rooms based on the cheapest rates, but rather according to which hotels paid it the highest 'cost-per-click' fee.[121]

(ii) Trivago's hotel room rate comparisons, through the use of strike-through prices or text in different colours, misrepresented that the comparison was between prices offered for the same room category in the same hotel. In fact, the comparison was often between prices offered for a standard room and a luxury room at the same hotel, which created a false impression of savings.[122]

(iii) Trivago had misled the public to believe that its website provided a price comparison service that was impartial, objective and transparent, particularly since it was not made clear to consumers that Trivago was receiving a 'cost-per-click' fee.[123]

In commenting on the decision, the ACCC said that the case serves as 'an important warning to comparison sites that they must not mislead consumers about the results they recommend'.[124] The announcement of the DASI,[125] and DPSI,[126] sends the same message.

2.7.4 Price Parity Clauses

Price parity clauses (also known as best price clauses) provide assurance to downstream platforms that the terms on which a supplier offers goods or services to them are at least as favourable as the terms the supplier offers to other purchasers. Such clauses encourage a downstream platform to invest in promoting the supplier's goods or services by eliminating the risk of the supplier free riding on that investment and undercutting the platform's price. A classic example here involves a consumer who searches for accommodation through an online travel site and then

---

121 [2020] FCAFC 185, [220]-[221]. Trivago operated its business such that it was paid by the featured hotels for each click it generated for them. This incentivised Trivago to promote the hotels that paid the highest amount per click, as opposed to those that provided the best value for customers.

122 [2020] FCAFC 185, [257].

123 [2020] FCAFC 185, [268].

124 ACCC (2020), Trivago loses appeal after misleading consumers over hotel ads, available at https://www.accc.gov.au/media-release/trivago-loses-appeal-after-misleading-consumers-over-hotel-ads. Accessed 10 September 2021.

125 See, ACCC (n.d.) Digital advertising services inquiry, available at https://www.accc.gov.au/focus-areas/inquiries-ongoing/digital-advertising-services-inquiry. Accessed 10 September 2021.

126 See, ACCC (n.d.) Digital advertising services inquiry, available at https://www.accc.gov.au/focus-areas/inquiries-ongoing/digital-advertising-services-inquiry. Accessed 10 September 2021.

makes the booking directly with the hotel in the expectation of obtaining a cheaper price.[127]

In Australia, price parity clauses are widely considered to increase competition in vertical relationships and feature commonly in contracts between online price comparator websites and their suppliers. However, concerns have been raised internationally, particularly in Europe, that price parity clauses in fact reduce competition by ensuring that the price available from one downstream platform will not be higher than the price available from the supplier directly or from any other platform.[128]

Awareness of these concerns prompted the ACCC, in September 2015, to request submissions from Australian accommodation providers about the use and extent of price parity clauses in their contracts with online travel sites such as Booking.com and Expedia.[129] The consultation process was intended to allow the ACCC to assess any competition issues associated with those clauses in the Australian context.[130] Twelve months later, the ACCC issued a media release commending Booking.com and Expedia for agreeing to changes that would 'increase the incentive for them to compete with each other and allow consumers to shop around to get the best deal'.[131] Specifically, Booking.com and Expedia agreed to the removal of contractual clauses requiring Australian accommodation providers to offer room rates equal to or lower than those offered on any other online travel site or on an accommodation provider's offline channels, to make all remaining room inventory available and to offer the same number and same type of rooms offered to any other online travel site.[132] Interestingly, however, the changes did not prevent Booking.com or Expedia from continuing to require that Australian accommodation providers not offer rates below those offered to online travel sites.

[127] M. Corrigan and J. McMahon (2016) Expedia and Booking.com's narrow price parity clauses OK, say ACCC – for now, available at https://www.claytonutz.com/knowledge/2016/september/expedia-and-booking-coms-narrow-price-parity-clauses-ok-say-accc-for-now. Accessed 10 September 2021.

[128] See, M. Corrigan and J. McMahon (2016) Expedia and Booking.com's narrow price parity clauses OK, say ACCC – for now, available at https://www.claytonutz.com/knowledge/2016/september/expedia-and-booking-coms-narrow-price-parity-clauses-ok-say-accc-for-now. Accessed 10 September 2021.

[129] See, ACCC (2015) ACCC consults on online travel agents, available at https://www.accc.gov.au/update/accc-consults-on-online-travel-agents. Accessed 10 September 2021.

[130] See, ACCC (2015) ACCC consults on online travel agents, available at https://www.accc.gov.au/update/accc-consults-on-online-travel-agents. Accessed 10 September 2021.

[131] ACCC (2016) Expedia and Booking.com agree to reinvigorate price competition by amending contracts with Australian hotels, available at https://www.accc.gov.au/media-release/expedia-and-bookingcom-agree-to-reinvigorate-price-competition-by-amending-contracts-with-australian-hotels. Accessed 10 September 2021.

[132] See, ACCC (2016) Expedia and Booking.com agree to reinvigorate price competition by amending contracts with Australian hotels, available at https://www.accc.gov.au/media-release/expedia-and-bookingcom-agree-to-reinvigorate-price-competition-by-amending-contracts-with-australian-hotels. Accessed 10 September 2021.

The ACCC's equanimity over the latter outcome stands in contrast to its protracted legal dispute with Australia's largest travel agency company, Flight Centre, which was adjudicated by the High Court in 2016.[133] In brief, the ACCC alleged that Flight Centre had engaged in anti-competitive, price-fixing conduct in breach of Section 45 CCA by attempting to reach a price parity agreement with Singapore Airlines, Malaysia Airlines and Emirates that would prevent those airlines from selling tickets directly to consumers at a lower price than those available through Flight Centre.[134] The High Court found for the ACCC on appeal,[135] confirming zero tolerance for anti-competitive conduct that arises under the guise of price parity clauses.

2.8 Data Protection

Australia has a comprehensive system of information privacy law,[136] historically based upon the 1980 Organisation for Economic Co-operation and Development (OECD) guidelines on the transborder exchanges of personal data.[137] The key Australian law, Privacy Act 1988 (Cth), is similar in overarching structure to the General Data Protection Regulation (GDPR) as it is based on the same types of principled protections.[138] However, there are some important differences in application, particularly regarding the involvement of regulatory activities that cross over information privacy law and competition law considerations. Actions taken against major platform companies have thus far been limited, especially by the federal Australian information privacy law regulator, the Office of the Australian Information Commissioner (OAIC). The limited actions against major platforms have not regarded the type of systemic concerns that arise from large-scale collection and use of device-related data across different platforms or services, as highlighted by the Bundeskartellamt decision. However, there are recent signals that this limited regulatory approach is changing, both by the OAIC and by the fact that the ACCC

[133] *ACCC v Flight Centre Ltd (No 2)* [2013] FCA 1313; *Flight Centre Ltd v ACCC* [2015] FCAFC 104; *ACCC v Flight Centre Travel Group Ltd* [2016] HCA 49.

[134] *ACCC v Flight Centre Travel Group Ltd* [2016] HCA 49, [25].

[135] *ACCC v Flight Centre Travel Group Ltd* [2016] HCA 49, [93] (Kiefel and Gageler JJ), [148] (Nettle J), [186] (Gordon J), French CJ dissenting at [24].

[136] The term information privacy has largely been used in Australia instead of data protection. However, as noted below, the DPI uses 'data protection' and 'privacy' separately, thus mirroring the general EU position. The term 'comprehensive' in the chapter refers to a regulatory structure predicated on one overarching information privacy law that is regulated by a specific information privacy regulator. See M. Burdon, Digital Data Collection and Information Privacy Law, Cambridge University Press, 2020.

[137] *Guidelines on the Protection of Privacy and Transborder Flows of Personal Data* (OECD, 1980).

[138] See L. A. Bygrave, Data Privacy Law: An International Perspective, Oxford University Press 2014.

is increasingly active in the promotion of consumer privacy protection, as exemplified by the DPI and the advent of the newly implemented Consumer Data Right (CDR). These changes potentially signify a movement away from the purely comprehensive system of information privacy law to a more hybrid structure that is supplemented by the application of information privacy law in different jurisdictions, particularly the GDPR. The supplementation process is likely accelerated by the DPI.

2.8.1 Background to the Australian Information Privacy Law Framework

The Privacy Act implements into Australian law two supranational instruments: the 1980 OECD Guidelines and Article 17 of the *International Covenant on Civil and Political Rights*. However, the Privacy Act's ambit is very much on the former rather than the latter, as a general right to privacy is not legally or constitutionally recognised in Australia. The Australian approach to privacy law protection, including information privacy, is thus fundamentally different from that of the European Union, wherein data protection is viewed as a foundational right of EU citizenship.[139] Instead, the Privacy Act provides a range of largely procedural, principled protections that are not intended to have the same substantive, rights-based application of the GDPR. The limited focus of the Privacy Act thus emanates in several important ways, including the historically self-constrained role that the OAIC has largely imposed on itself. This latter point is perhaps the key reason for the absence of systemic actions against the potentially infringing data-collecting activities of platforms.

The basic purpose of the Act remains the same for both government agencies and private sector organisations – the Act regulates how entities are permitted to handle and manage personal information. However, the Act's coverage of the private sector remains a significant problem as it excludes coverage of different types of entities, most notably small and medium-sized enterprises with an annual turnover of less than three million dollars.[140] Approximately 94% of Australian businesses are excluded from the Act.[141] The Privacy Act also forms the basis of separate state-based legislation that regulates certain state government agencies. As such, federal and state information privacy laws share the same foundations both in terms of content and largely in application.

The Privacy Act is woefully under-litigated, and many of the Act's foundational terms are still to be judicially considered, unlike state-based laws. Where the Act has

[139] See e.g. Article 8 of the *Charter of Fundamental Rights of the European Union OJ [2010] C 83/02 2010.*

[140] S6D The Privacy Act 1988 (Cth). Certain exceptions exist to the exemption, such as SMEs that trade in personal information or an SME that would be classed as a health service provider.

[141] See DPI 454. The ACCC estimates that the Act only covers approximately 100,000 entities.

been judicially considered, it has given rise to confusion, most notably involving the Act's definition of personal information in the Federal Court decision of *Privacy Commissioner v Telstra*.[142] Like the GDPR, this is a crucial question to resolve because the Act only covers acts and practices in relation to the handling of personal information.[143] Consequently, any information that is not personal information will not be covered by the Act. It is thus a threshold issue. Under Section 6, personal information is 'information or an opinion about an identifiable or reasonably identifiable individual'.

The Privacy Act's definition of personal information is substantively different from the GDPR's definition of personal data. Under Article 4(1) of the GDPR, personal data are 'any information relating to an identified or identifiable natural person'. The semantic change of 'about' and 'relating to' signifies different policy emphases about the breadth accorded to information privacy protection in each jurisdiction. In Australia, the breadth is constrained, and in the European Union, it is broadened, again in line with rights-based perspectives about the regulatory purpose of data protection. Recently implemented legislation involving attempts to broaden the scope of personal information (so that it captures device or system metadata for different statutory purposes to the Privacy Act) have attempted to obviate the confusion caused by the *Telstra* case by introducing a 'relates to' definition similar to the GDPR into Australian information privacy law, including the CDR and Australia's COVID contact tracing app, COVIDSafe.[144]

Other differences between the Privacy Act and the GDPR also exist. The Privacy Act implicitly regulates personal information processing and does not have a model of regulated process parties like the GDPR. Instead, information privacy obligations emanate from collecting, holding or using/disclosing personal information rather than from the processing of personal data, even though collection effectively begins a processing cycle of principled protection for most entities. These privacy principles are typical to most information privacy or data protection laws and are called the Australian Privacy Principles (APPs). The principles are inherently linked. The principled protections begin at the point of data collection (APPs 1–3, 5) and end with the destruction or de-identification of no longer required personal information (APP 11). In the interim, data collection organisations have a range of obligations to fulfil:

(i) Personal information can only be used or disclosed for a defined purpose (APP 6 and APP 7 regarding direct marketing).
(ii) The individual must be notified about the purposes of collection and subsequent use (APP 5).
(iii) There must be a defined need for the collection (APP 3).

142 *The Privacy Commissioner v Telstra Corporation Limited [2017] FCAFC 4 (19 January 2017* (2017)

143 S13(1) The Privacy Act 1988 (Cth).

144 S94D(5) Privacy Amendment (Public Health Contact Information) Act 2020.

(iv) Individuals also have a range of interaction mechanisms that seek to ensure the maintenance of individual control by being able to affirm the accuracy and currency of collected personal information (APP 10, 12-13)
(v) Personal information must be held securely (APP 11).

2.8.2 The Platform Context

A continual criticism about the OAIC has been its perceived lack of firm regulatory activity, including in the platform context. The Privacy Act was significantly amended in 2014 to give the OAIC greater investigatory powers, including the option of imposing higher civil penalties for serious privacy infringements of up to AUD 1.7 million dollars. However, this sanction has been under-utilised, and most of the OAIC's regulatory activities, in both platform and offline contexts, have focussed on serious data breaches, which have culminated in the implementation of a mandatory data breach reporting regime in 2017. The OAIC's investigatory focus in relation to information privacy problems regarding security is based on an ostensible data breach context.[145] Accordingly, the small number of actions undertaken by the OAIC against online data collectors or platforms has included discrete, highly publicised concerns, often involving other jurisdictional data protection investigations, rather than a serious consideration of systemic issues involving large-scale personal data collections across platforms. Even when such investigatory actions have been undertaken, such as the Google Streetview WiFi collection[146] or the Ashley Madison data breach,[147] the results have either been in contradiction to findings in other jurisdictions or resulted in conciliatory outcomes.

There are recent signals that the OAIC is taking a different approach, as exemplified by its belated legal action against Facebook regarding the impacts of the Cambridge Analytica scandal on Australia.[148] At face value, the OAIC action bears the same limited hallmarks of its previous investigations as it focuses on privacy infringements arising from unauthorised disclosures under APP 6 and breaches of security under APP 11. However, unlike previous investigations, in

---

[145] J. Signato and M. Burdon, The Privacy Commissioner and Own-Motion Investigations into Serious Data Breaches: A Case of Going through the Motions?, University of New South Wales Law Journal (2015) 38(3), pp. 1145-1185.

[146] M. Burdon and A. McKillip, The Google Street View Wi-Fi Scandal and Its Repercussions for Privacy Regulation, Monash University Law Review (2013) 39, p 702.

[147] Joint investigation of Ashley Madison by the Privacy Commissioner of Canada and the Australian Privacy Commissioner and Acting Australian Information Commissioner [2016] (commissioner initiated investigation).

[148] OAIC (2020) Commissioner launches Federal Court action against Facebook, available at https://www.oaic.gov.au/updates/news-and-media/commissioner-launches-federal-court-action-against-facebook/. Accessed 10 September 2021.

documents submitted to the Federal Court, the OAIC is construing harm as both an individually focussed element and a broader systemic consideration.[149]

The OAIC's action against Facebook consequently has a much broader systemic reach than previous regulatory investigations as it is specifically relevant to Facebook's status as a massive platform data collector that can shape perceptions about the applicable utility of Australian privacy law. The basis of the action is about individually focussed protections through the application of pertinent APPs but it is done so through the lens of the digital platform context and the market powers that flow. Why was then a change in regulatory direction in 2020, considering that the OAIC was provided with stronger regulatory powers in 2014 that are specifically relevant to systemic considerations? The answer to that question relates to the information-privacy-related developments that flow from the DPI report and the greater impetus that the ACCC is attaching to consumer privacy protection as a demonstrable social good in its own right and as part of a broader digital economy imperative.

Further cementing the OAIC's action against Facebook, the DPI Final Report broadly considers the role of Google and Facebook as predominant data collectors and commercial reusers of personal information. Data protection and privacy,[150] thus have a much broader remit to 'help prevent consumer harm and protect and enhance consumer welfare'.[151] The implementation of stronger data protection and privacy mechanisms is recommended to encourage further growth in markets flowing from digital platforms by enhancing consumer trust and reducing the types of social and economic harms that can arise from systemic personal information misuse. Stronger data protection and privacy laws would also allow relevant regulators to better monitor and deter 'problematic data practices that result in consumer harm'.[152] Although the DPI is aimed at digital platforms, the data protection considerations are also intended to have an effect across the broader economic sector.[153] However, the ACCC acknowledges that recommendations aimed at the economy more broadly will also need to be supplemented by platform-specific measures, particularly a new enforceable privacy code aimed at major digital platforms.[154]

The ACCC argues that major changes to Australia's information privacy law framework are necessary to combat the market and regulatory failures arising from the market power of dominant digital platforms. The key failures include:

(i) Information asymmetries that undermine a consumer's ability to assess whether services align with their privacy preferences

[149] See [26] and the reference to 'systemic failures'.

[150] As noted above, the ACCC differentiates between data protection and privacy in the DPI as two distinct concepts, similar to the general EU approach.

[151] DPI Final Report, p. 454.

[152] DPI Final Report, p. 448.

[153] DPI Final Report, p. 452.

[154] DPI Final Report, p. 454.

(ii) Bargaining power imbalances that prevent consumers from making genuine choices as to how their personal information is collected, used and shared
(iii) Behavioural biases that work against consumers' ability to select privacy options that better align with their privacy concerns
(iv) A lack of effective deterrence under current consumer protection and privacy laws against certain data practices by digital platforms[155]

These failures require an overhaul of the current information privacy law framework, including amending the object clause of the Privacy Act to include a specific consumer protection element.[156] The ACCC also recommends an update to the Privacy Act's definition of personal information to bring it in line with technological advances and to resolve the confusion arising from the *Telstra* case, as highlighted above. It is important to note, however, that the ACCC did not specifically recommend the adoption of the GDPR definition of personal data but rather called for the further clarification of the technological scope to align it with other 'standards set by overseas regulators'.[157] More broadly, a range of remedial measures were put forward that would increase the ability of Australian consumers to litigate against platforms based on the misuse of personal information. These measures include a new form of private action available through the Privacy Act for breaches of the APPs,[158] yet another call to implement a statutory cause of action for serious invasions of privacy[159] and increased financial penalties for Privacy Act infringements to better align with the ACL.

Given the types of harm identified by the ACCC, it should be no surprise that the DPI recommendations seek to enhance privacy policy requirements as an information disclosure mechanism and to promote stronger forms of consumer control over personal information, particularly centred around friendlier forms of consent acquisition. Stronger notification requirements at the point of collection are recommended. The additional requirements will remove the discretion that data-collecting organisations currently have regarding notice construction and replace it with more standardised forms of notice aimed at reducing the information asymmetry between consumers and collectors.[160] More profoundly, major changes were put forward to strengthen the consent requirements of the Privacy Act so that it better

[155] DPI Final Report, p. 454.

[156] DPI Final Report, p. 477.

[157] DPI Final Report, p. 460.

[158] Currently, s 36(1) and s 40(1)(b) of the Privacy Act require a complainant to make a complaint to the OAIC for determination, which can then be pursued through OAIC investigation at first instance.

[159] This aspect has been a continued bone of contention. It has been recommended by two Australian Law Reform Commission reports and has thus far been ignored by the Australian Government. See Australian Law Reform Commission (2008) For Your Information: Australian Privacy Law and Practice and Australian Law Reform Commission (2014) Serious Invasions of Privacy in the Digital Era.

[160] DPI Final Report, p. 462.

aligns with the GDPR.[161] Consent must specifically be obtained whenever a consumer's personal information is collected, used or disclosed unless an exception applies.[162] The ACCC acknowledges that enhanced requirements are likely to cause 'consent fatigue', which requires a foundational shift in the types of controls provided to consumers in the form of default selections turned to 'off' rather than 'on' as a means of legitimating collection purposes.[163] All of which will significantly impact upon the application of APP6, the disclosure principle, which currently accords far too much leeway for APPs entities in the construction of broad purposes of collection that result in equally broader forms of partially justified disclosure.[164]

The final recommendation of note, as highlighted above, is the development of a new enforceable privacy code for digital platforms. Under Section 26G(2) of the Privacy Act, the OAIC has the power to develop, approve and register enforceable privacy codes. Thus far, the power has again been under-utilised, and only three such codes have been registered.[165] However, during its 2019 budget speech, the Australian government announced plans for a legislated code that would apply to social media and online platforms that trade in personal information.[166] The DPI-proposed code is a consequence of wider governmental intention. It would be developed by the OAIC in consultation with relevant industry bodies and would apply to 'all digital platforms supplying online search, social media, and content aggregation services to Australian consumers and which meet an objective threshold regarding the collection of Australian consumers' personal information'.[167] The proposed code would be enforced by the OAIC, rather than the ACCC, although the latter would be involved from consumer protection and competition law perspectives. The code would provide enhanced protections for consumers relevant to the platform data collection context and would augment DPI general proposals involving heightened privacy policy requirements, consent obligations, information security mechanisms and specific protection for children's data.

The proposed code, and the other recommended changes, would increase the level of hybridity that is currently taking place regarding Australian information privacy law developments, particularly in relation to the GDPR. The DPI makes clear that it does not intend to wholly adopt the component provisions of the GDPR

[161] DPI Final Report, p. 466.

[162] The exceptions recommended are mirrored on the GDPR: 'unless the personal information is necessary for the performance of a contract to which the consumer is a party, is, is required under law, or is otherwise necessary for an overriding public interest reason.'

[163] DPI Final Report, p. 468.

[164] DPI Final Report, p. 465.

[165] OAIC (n.d.) Privacy Codes Register, available at https://www.oaic.gov.au/privacy/privacy-registers/privacy-codes-register/. Accessed 10 September 2021.

[166] DPI Final Report, p. 482. See G. Smith (2019) Government proposes major changes to privacy law, available at https://www.allens.com.au/insights-news/insights/2019/04/government-proposes-major-changes-to-privacy-law/. Accessed 10 September 2021.

[167] DPI Final Report, p. 481.

into Australian law,[168] but the ACCC has, nonetheless, clearly been heavily influenced by the GDPR as part of its overall rationale.[169] In other words, parts of the GDPR are used as a model to supplement and enhance the Privacy Act.[170] This is evident in the new forms of notice collection similar to Articles 13 and 14, the new definition of consent in line with Article 7, a new mechanism for personal information erasure based on Article 17 and the new private right of action similar to Articles 79 and 82. All of these proposed changes indicate the possibility of the Australian government seeking an adequacy ruling from the European Commission that Australian law offers an adequate level of data protection to the GDPR,[171] as a further enhancement of digital economy endeavours. In October 2020, the Commonwealth Attorney-General's Department started its review of the Privacy Act with the release of an issues paper[172] that broadly considers the implementation of the ACCC's key recommendations, including the degree of alignment with the GDPR. The review is expected to be completed in late 2021, with legislative changes to the Act likely to be implemented the following year.

2.8.3 The CDR

Finally, it is also important to briefly cover the CDR, as it exemplifies the ACCC's enhanced focus on consumer privacy protection.

The CDR is a data portability legislative mechanism largely modelled on the Open Banking Scheme in the United Kingdom. It provides CDR consumers with the opportunity to seamlessly access and transfer different types of transaction data and personal information from one service provider to another. The overriding policy purpose of the CDR is to use data portability as a means of enhancing digital economy innovations by reducing the locking in of customers and the corresponding network effects. It is intended to have economy-wide application, but it is being implemented on a sector-by-sector basis, beginning with banking. The DPI did not recommend the application of the CDR to digital platforms because it is not likely to produce a short-term benefit.[173] As things currently stand, the ACCC believes that data portability would have a limited impact on facilitating consumers' switching of services because Google and Facebook have no significant competitors to switch

[168] DPI Final Report, p. 439.

[169] See DPI Final Report 434, where the facilitation of data flow to overseas jurisdictions 'such as the EU' is at the base of broader reforms to the Privacy Act.

[170] Ibid.

[171] DPI Final Report, p. 480.

[172] Attorney-General's Department (2020) Review of the Privacy Act 1988, available at https://www.ag.gov.au/integrity/consultations/review-privacy-act-1988. Accessed 10 September 2021.

[173] DPI Final Report, p. 115.

to. It would therefore not reduce network effects and 'may not have a significant effect on barriers to entry and expansion'.[174]

The CDR has a unique approach to information privacy with the development of a separate set of Privacy Safeguards, which largely mirror the APPs but are prescriptive rather than principle based. In the context of this chapter, the development of the Privacy Safeguards reflects the complex considerations behind data portability that incorporate competition law, consumer protection and information privacy law frameworks. It is this combination that provides the justification for the ACCC to take a stronger role in information privacy law, which has previously been the sole domain of the OAIC.

The ACCC is undertaking a stronger role in the development of the Australian information privacy law framework; its focus on consumer privacy is becoming the dominant regulatory mode for enhancing the Privacy Act and as a basis for further developments in the digital economy. Given that the Australian government has accepted almost all of the ACCC's data protection recommendations in the DPI,[175] it would be prudent to expect a greater alignment of information privacy, consumer protection and competition law imperatives that are intended for the Australian economy as a whole and are specifically enhanced for the major digital platforms. Consequently, future regulatory actions against Google and Facebook are likely to be more commonplace in Australia in the near-term future, especially when compared to the previous era of relative inaction.

The ACCC's recently successful action against Google in the Federal Court is an illuminating example of this new regulatory stance.[176] In that case, the court found that Google's default location data settings, in conjunction with the presentation and substance of privacy policy statements, was conduct capable of being misleading and deceptive conduct under Section 18 of the ACL.[177] The judgement potentially indicates that the ACCC and the ACL is going to be the primary vehicle of future regulatory response against the major platforms rather than the OAIC and the Privacy Act.

[174] DPI Final Report, p. 116.

[175] Australian Treasury (2019) Regulating in the Digital Age: Government Response and Implementation Roadmap for the Digital Platforms Inquiry, available at https://treasury.gov.au/sites/default/files/2019-12/Government-Response-p2019-41708.pdf. Accessed 10 September 2021. Note, however, that there is still tacit scepticism about a statutory cause of action.

[176] *Australian Competition and Consumer Commission v Google LLC (No 2)* [2021] FCA 367.

[177] *Australian Competition and Consumer Commission v Google LLC (No 2)* [2021] FCA 367 at [330]. The Court held that Google also breached s 29(1)(g) of the ACL regarding misleading representations in connection with the supply of goods and services and s 34 of the ACL regarding conduct that was liable to mislead the public.

2.9 Online Advertising

As the existing and past ACCC inquiries suggest, one of the main concerns with regard to issues arising from the digital economy in Australia has been online advertising. While the DPI already dealt with online advertising in connection to competition in media and advertising service markets, the current DASI is directly focussed on digital advertising.[178]

The aim of the DPI with regard to online advertising, as set out in the Terms of Reference, was to consider the impact that digital platforms had on the state of competition in advertising markets. The ACCC was instructed to consider the following three factors:

(i) the extent to which digital platforms are exercising market power in commercial dealings with advertisers
(ii) the impact of digital platforms on advertising markets
(iii) the impact of information asymmetry between digital platforms, advertisers and consumers and the effect on competition in advertising markets.[179]

The ACCC found, as part of the DPI, that Google and Facebook played the most significant role in online advertising, this being the case for the following reasons:

(i) They constituted the most important channels for purchasing and selling the majority of online advertising in Australia.
(ii) They owned over 80 per cent of growth in online advertising in the period of 2016-2018 in Australia and received the majority of online advertising revenue in Australia.[180]

These reasons exist because both Google and Facebook offer various significant benefits, such as targeting specific audiences usually for a lower price than that of traditional forms of advertising.[181]

In the DPI Final Report, the ACCC identified a number of potential competition law and other issues arising in online advertising. (These potential issues then led to the DASI.)[182] In the DPI, the ACCC found that digital platforms with substantial market power had the ability to leverage their power into another market. The lack of transparency, such as the lack of transparency in the pricing of services, could also

[178]The DASI interim report, which will be submitted to the Australian Treasurer, is due by 31 December 2020, with the final report due on 31 August 2021.

[179]DPI Final Report, p. 119.

[180]DPI Final Report, pp. 119-130.

[181]DPI Final Report, pp. 119, 131-132.

[182]Australian Treasury (2019) Regulating in the digital age: Government Response and Implementation Roadmap for the Digital Platforms Inquiry, available at <https://treasury.gov.au/publication/p2019-41708>. Accessed 10 September 2021.

lead to a misuse of market power. Other competition law issues included the bundling of advertising services. The ACCC also identified as issues the significant bargaining power possessed by Facebook and Google and the lack of effective dispute resolution. These issues are attributable to the fact that Facebook and Google are unavoidable platforms in Australia for securing advertising revenue.[183]

Unlike the DPI Final Report, which focussed its analysis on Google and Facebook, the DASI is not just concerned with dominant platforms but also involves analysis of the respective markets as a whole and their specific elements that (can) impact the competitiveness, efficiency and transparency of these markets. Indeed, the DASI added transparency to its focus, reflecting the findings in the DPI. In particular, the DASI investigates 'the competitiveness, efficiency, transparency and effectiveness of markets for the supply of ad tech services and ad agency services'.[184]

The questions the ACCC raises in the DASI are as follows:

(i) whether market participants have enough information (including about pricing, rebates and revenue flows) to make informed choices about the use of ad agency and ad tech services
(ii) competition throughout the ad tech supply chain and in the supply of ad agency services
(iii) the role and use of data in supplying these services, and
(iv) whether competition and efficiency are being affected by supplier behaviour, including vertically integrated suppliers preferencing their own services, or by ad tech services businesses or ad agencies not acting in the best interests of their clients.[185]

The ACCC intends to evaluate the intensity of competition and the efficiency of the markets by investigating:

(i) any concentration of market power in the hands of one or more suppliers of these services,
(ii) the availability of information (including pricing transparency) to advertisers, publishers and other market participants,
(iii) auction and bidding processes used to supply of digital display advertising,
(iv) the impact of mergers and acquisitions on those markets,
(v) supplier behaviour, and

---

[183] DPI Final Report, pp. 119, 133-164.

[184] ACCC (2020) Digital advertising services inquiry Issues Paper ('DASI Issue Paper'), available at https://www.accc.gov.au/focus-areas/inquiries-ongoing/digital-advertising-services-inquiry/issues-paper. Accessed 10 September 2021.

[185] DASI Issues Paper (10 March 2020), p. 2.

(vi) how competition in these markets impacts on competition in the supply of digital display advertising services.[186]

In the DASI interim report, the ACCC mentioned that it would also consider similar inquiries and the findings of other jurisdictions.[187]

2.10 Conclusion

Digital platforms raise complex issues at the intersection of competition, consumer, privacy and data protection laws. The 23 recommendations in the DPI Final Report span each of these areas and have potentially broader implications than digital platform regulations per se. At this juncture, the ACCC appears well placed to deal with competition law issues arising in the digital space. Although the matters the ACCC is entrusted to deal with are broad, the ACCC's powers are limited by the common law approach, where only the court can find infringements. In that regard, it is obvious that there is a lack of competition law cases to make any conclusions regarding the effectiveness of Australian competition law in the digital economy, as it is something that is still due to be tested by the courts.

Competition law issues in the digital world are also likely to have close links with data protection, privacy and/or consumer law issues. As a multi-functional agency, the ACCC would seem to have natural advantages in addressing these issues as compared to a competition authority that deals only with competition law. In particular, as an experienced regulator of both competition and consumer laws, the ACCC can use its powers cumulatively to tackle issues from both areas of law. However, it currently remains uncertain as to whether it is the most appropriate regulator to lead in cases of data protection and privacy issues. Furthermore, an inquiry can, and, with regard to the digital economy, the existing current and past inquiries did, broaden the scope of the areas that the ACCC would usually deal with in order to consider those public interests and laws that are beyond its standard parameters but are essential for a particular inquiry. Although it shows that this is possible in specific inquiries, this does not necessarily mean that it is the best approach from the perspectives of a concentration of powers, the degree of expertise and parties' rights.

The digital economy is, indeed, the new frontier of competition law, presenting both challenges and opportunities for market participants and regulators alike. What is essential regarding the issues or potential issues that the digital economy encompasses is that all areas of law discussed in this chapter (competition, consumer, privacy and data protection laws) and their enforcement are effective as they influence the effectiveness and functioning of the markets, including market and bargaining powers, and competition in those markets.

[186] DASI Issues Paper (10 March 2020), p. 5.

[187] DASI Issues Paper (10 March 2020), p. 2.

Austria

3

Gerhard Fussenegger and Viktoria H. S. E. Robertson

3.1 Introductory Remarks

The digitalisation of markets has led to profound changes in the way goods and services are conceived of, produced, distributed, marketed and bought. The business model of platform markets has taken hold in the digital marketplace, and in many markets, users have become accustomed to being provided 'free' services against the payment of personal data and attention. In addition, the multi-purpose nature of data is coming to play an ever-increasing role in these markets. The advent of the digital era has had a profound impact not only on markets but also on those areas of the law that relate to these markets, among them competition law.

The present report on Austrian competition law and the digital economy addresses the manifold ways in which the Austrian legislator, the Austrian (Supreme) Cartel Court and the Austrian Federal Competition Authority (Bundeswettbewerbsbehörde (BWB)) have dealt with digital markets in the recent past and what developments may lie in store for digital markets in the near future. In order to embed the Austrian experience within a broader picture, these introductory remarks highlight four trends in the application of competition law to digital markets: first, the proliferation of reports to make sense of digital markets from a competition law point of view; second, the interconnection between competition law

Viktoria Robertson contributed the introductory remarks, while Gerhard Fussenegger is the author of the substantive part of this report.

G. Fussenegger (✉)
bpv Huegel, Vienna/Brussels, Belgium
e-mail: Gerhard.Fussenegger@bpv-huegel.com

V. H. S. E. Robertson
Vienna University of Economics and Business and University of Graz, Vienna/Graz, Austria
e-mail: viktoria.robertson@wu.ac.at

B. Kilpatrick et al. (eds.), *Antitrust in Data Driven Markets & Legal Framework for Influencers, Native Advertising and Control over the Use of AI in Marketing*,
LIDC Contributions on Antitrust Law, Intellectual Property and Unfair Competition,
https://doi.org/10.1007/978-3-031-07422-6_3

and other legal areas applying to digital markets; third, the development of new competition law tools for digital markets; and, fourth, the increased need for cooperation among competition authorities where digital markets are concerned.

The first trend that we have been witnessing is the widespread understanding that digital markets raise specific competition issues that require either an adaptation of the current competition law framework or an adapted application of our present rules. Several reports on digital competition law have been published around the globe and have provided both insights into the workings of the digital economy and options for tackling the arising competition issues. They include, to name just a few of these reports, the following:

- Competition Law and Data (Autorité de la concurrence and Bundeskartellamt, May 2016)
- Big Data and Innovation: Key Themes for Competition Policy in Canada (Competition Bureau Canada, February 2018)
- Unlocking Digital Competition: Report of the Digital Competition Expert Panel (Jason Furman et al. for the UK Government, March 2019)
- Competition Policy for the Digital Era (Jacques Crémer, Yves-Alexandre de Montjoye and Heike Schweitzer for the European Commission, April 2019)
- Digital Platforms Inquiry—Final Report (Australian Competition and Consumer Commission, July 2019)
- A New Competition Framework for the Digital Economy—Report by the Commission 'Competition Law 4.0' (Martin Schallbruch, Heike Schweitzer and Achim Wambach for the German Federal Ministry for Economic Affairs and Energy; September 2019)
- Digital Era Competition: A BRICS View (Ioannis Lianos and Alexey Ivanov for the BRICS Competition Law and Policy Centre, September 2019)
- Stigler Committee on Digital Platforms—Final Report (Nolan McCarty, Guy Rolnik, Fiona Scott Morton and Lior Strahilevitz for the University of Chicago Booth School of Business, September 2019)
- Joint Memorandum of the Belgian, Dutch and Luxembourg Competition Authorities on Challenges Faced by Competition Authorities in a Digital World (October 2019)
- Algorithms and Competition (Autorité de la concurrence and Bundeskartellamt, November 2019)
- Online Platforms and Digital Advertising (UK Competition and Markets Authority, July 2020)
- Investigation of Competition in Digital Markets (US House of Representatives, Subcommittee on Antitrust, Commercial and Administrative Law, October 2020)

In June 2020, the Austrian BWB published its own *Proposition Paper on Digitalisation and Competition Law*.[1] On that note, the BWB underlines that

[1]For the English version, see here: BWB, https://www.bwb.gv.at/fileadmin/user_upload/Considerations_on_digitalisation_challenges_in_the_economy.pdf, accessed 26 August 2021.

many of the tools of 'traditional' competition law can directly be applied in the digital economy. It stresses the importance of adequate staffing and resources to enable the authority to gain technical expertise in these markets, in line with the ECN + Directive.[2] The BWB also notes that the Austrian Cartel Act is agnostic as to whether market behaviour occurs online or offline and can therefore adequately deal with both. While Austrian merger control continues to rely on the market dominance criterion, the BWB suggests that the SIEC (significant impediment of effective competition) test, as it is applied under EU merger control, may represent a more useful substantive assessment, especially in digital mergers. In the legislative proposal that is now underway, the Austrian merger control regime will be brought in line with the European SIEC test.[3]

Directly connected to the insights gathered through digital reports on competition law, the second trend is the dawning realisation that competition law is not an island but a part of a broader set of legal rules that apply to digital markets and that, ideally, they should work together in order to provide a digital environment conducive to consumer welfare (understood in a broad sense) and innovation while at the same time respecting fundamental rights that are afforded to every individual in a liberal democratic society. This need is already appreciated by some of the legal reports; e.g., the Stigler Report brings together chapters on antitrust as well as chapters on privacy, the news media and politics. In some jurisdictions, competition authorities have also relied on data protection rules in order to flesh out obligations under competition law.[4]

Following on from the first trend that strives to understand competition issues in digital markets and the second trend that increasingly comes to see competition law as an integral part of a broader legal puzzle, one can observe a third trend, namely competition authorities and legislators that have already moved beyond the stage of gathering intelligence on these markets: both at the European level and at many national levels, new competition tools are being developed through legislative and policy changes to address the particular competition issues and market failures that arise in digital markets. In the European Union, the European Commission is now reviewing its Market Definition Notice of 1997 in an effort to bring it up to date with

[2]Directive 2019/1 of the European Parliament and of the Council to empower the competition authorities of the Member States to be more effective enforcers and to ensure the proper functioning of the internal market (ECN+ Directive), OJ 2019 L 11/3.

[3]At the time of writing, the antitrust amendment act 2021 was still awaiting approval from the first chamber of Parliament; see Kartell- und Wettbewerbsrechtsänderungsgesetz 2021 (KaWeRÄG 2021), 114/ME XXVII. GP (Ministerialentwurf).

[4]See Bundeskartellamt, Case B6-22/16, *Facebook* of 6 February 2019. Facebook's request for a suspensory effect of its appeal was initially granted by the Higher Regional Court Düsseldorf (Case VI-Kart 1/19 (V), of 26 August 2019), but subsequently denied by the German Bundesgerichtshof (Case KVR 69/19, of 23 June 2020). In March 2021, the Düsseldorf Court sent a number of questions to Luxembourg in a request for a preliminary ruling; Case C-252/21, *Facebook v Bundeskartellamt* (pending).

the current economic reality.[5] At the same time, the European Commission has proposed its Digital Markets Act in December 2020, through which it intends to tackle structural issues in digital platforms that act as gatekeepers.[6] In Austria, the draft Austrian Cartel and Competition Law Amendment Act 2021 (Kartell- und Wettbewerbsrechts-Änderungsgesetz 2021—'KaWeRÄG 2021') is set to explicitly incorporate data as a relevant factor for the assessment of dominance and to introduce a new declaratory procedure whereby the BWB and the Federal Cartel Prosecutor can ask the Cartel Court to declare that a company is dominant on a multi-sided digital market.[7] This new instrument at the national level is a first step in order to successfully tackle anti-competitive behaviour in digital markets, but further steps – at the national level, at the European level or at both levels—will need to follow in order to do so successfully.

In 2017, at the same time as the German legislator, Austria introduced an additional transaction value threshold into its national merger regime (§ 9 para 4 Cartel Act 2005). This new threshold requires the notification of mergers where the value of the transaction exceeds €200m, certain (lower) turnover-based thresholds are exceeded and the acquired undertaking is active on the Austrian market to a large extent. While the European Commission does not currently consider such a legislative change,[8] the Austrian and German thresholds mean that mergers in digital markets that raise concerns could be referred to the European Commission from either country (Art. 22 EUMR). It should also be pointed out, however, that contrary to widely held expectations, the new transaction value threshold has so far mainly caught mergers outside of digital markets.[9]

The fourth notable trend relates to the enforcement of competition law in the digital economy: there is an increasing need for competition authorities to cooperate, thanks to the particular competition issues that arise in these markets. This need for collaboration has two dimensions. On the one hand, national competition authorities are called upon to collaborate with each other in order to tackle competition issues that affect several jurisdictions, including competition investigations that span several jurisdictions. The second need for collaboration is directly related to the

[5] On the evaluation, see here: European Commission, 'EU competition law – market definition notice (evaluation)' https://ec.europa.eu/info/law/better-regulation/have-your-say/initiatives/12325-Evaluation-of-the-Commission-Notice-on-market-definition-in-EU-competition-law, accessed 26 August 2021. For the evaluation results, see European Commission, 'Evaluation of the Commission Notice on the definition of relevant market for the purposes of Community competition law of 9 December 1997' SWD(2021) 199 final.

[6] European Commission, Proposal for a Regulation of the European Parliament and of the Council on contestable and fair markets in the digital sector (Digital Markets Act), COM(2020) 842 final.

[7] Kartell- und Wettbewerbsrechtsänderungsgesetz 2021 (KaWeRÄG 2021), 114/ME XXVII. GP (Ministerialentwurf), introducing a new § 28a KartG.

[8] European Commission, Summary of Replies to the Public Consultation on Evaluation of Procedural and Jurisdictional Aspects of EU Merger Control (2017) 4-7 https://ec.europa.eu/competition/consultations/2016_merger_control/summary_of_replies_en.pdf. Accessed 26 August 2021.

[9] See BWB, *Proposition Paper on Digitalisation and Competition Law* (June 2020) 6-7.

second trend noted above, namely that digital markets have made it especially clear that a number of interconnected legal issues—and interconnected market failures—can arise in them relating to competition law, data protection law, democracy, sector-specific regulation and many more. This has led to competition authorities becoming increasingly aware of the need to work together with other authorities—including data protection authorities and sector-specific regulators—in order to comprehensively tackle the challenges of digital markets.[10] This is also recognised by the BWB in its recent *Proposition Paper*.[11]

Regarding the first type of collaboration, Austria is part of the European Competition Network (ECN), which consists of the European Commission and the national competition authorities of all Member States of the European Union. The ongoing collaboration within the ECN represents an important pillar of competition law enforcement, also in digital markets. It can be expected that the implementation of the ECN+ Directive, which was due to be transposed by 4 February 2021 and is currently being implemented by the Austrian Parliament, will add to the efficiency of the ECN. The BWB also has a long-standing tradition of collaboration with the neighbouring German Bundeskartellamt. When the Austrian and German legislators introduced transaction-value-based thresholds into their respective national merger regimes, the BWB and the Bundeskartellamt together issued a Guidance Paper on how they would interpret these new provisions.[12] In 2019, the BWB carried out an investigation into Amazon's competitive behaviour towards retailers active on its Marketplace, in parallel with the German Bundeskartellamt.[13] While several of the competition concerns voiced by the two authorities are now resolved based on the P2B Regulation, which entered into force on 12 July 2020,[14] this case highlighted that competition issues relating to digital platforms are frequently not confined to one jurisdiction.

It is also noteworthy that the Austrian Ministry for the Economy and Digitalisation has recently proposed an amendment to the Austrian Cartel Act

[10]On this, see for instance A. Reyna, Optimizing Public Enforcement in the Digital Single Market Through Cross-Institutional Collaboration (2020), available at https://papers.ssrn.com/sol3/papers.cfm?abstract_id=3529198. Accessed 26 August 2021.

[11]BWB, *Proposition Paper* 11-12.

[12]BKartA and BWB, *Guidance on Transaction Value Thresholds for Mandatory Pre-merger Notification (Section 35 (1a) GWB and Section 9 (4) KartG)* (July 2018) https://www.bundeskartellamt.de/SharedDocs/Publikation/EN/Leitfaden/Leitfaden_Transaktionsschwelle.pdf?__blob=publicationFile&v=2. Accessed 26 August 2021.

[13]For Austria: 'BWB informs: Amazon modifies its terms and conditions' of 17 July 2020, available at https://www.bwb.gv.at/en/news/detail/news/bwb_informs_amazon_modifies_its_terms_and_conditions-1/.Accessed 26 August 2021. For Germany: BKartA, 'Bundeskartellamt obtains far-reaching improvements in the terms of business for sellers on Amazon's online marketplaces' of 17 July 2020, available at https://www.bundeskartellamt.de/SharedDocs/Meldung/EN/Pressemitteilungen/2019/17_07_2019_Amazon.html. Accessed 26 August 2021.

[14]Regulation 2019/1150 of the European Parliament and of the Council on promoting fairness and transparency for business users of online intermediation services (P2B Regulation), OJ 2019 L 186/57.

against the background of digital markets.[15] When kicking off its public debate on that issue, it invited two leading German individuals to join the debate: Achim Wambach, the Chairman of Competition Commission 4.0 and then Chairman of the German Monopoly Commission, as well as Thorsten Käseberg, a leading voice in preparing the tenth German amendment of the German Competition Act, which includes many new approaches to digital platforms and was ultimately passed in January 2021.[16] This highlights that not only competition authorities but also legislators have recognised the need for a converging legal framework for digital platforms.

Concerning the second type of collaboration, the BWB announced in May 2020 that it was joining forces with the national telecommunications regulator—the Austrian Regulatory Authority for Broadcasting and Telecommunications (RTR)—in order to monitor digital platforms.[17] The aim is to allow an insight into competition-related challenges that arise in these platforms. In the future, this might provide the BWB with a basis to start an antitrust investigation. At the European level, the European Data Protection Board (EDPB) has repeatedly offered its expertise on data protection to the European Commission in the case of mergers between digital platforms that may lead to data concentration.[18]

Having noted four recent trends in the application of competition to digital markets, we should bear in mind that digital markets are often highly dynamic, and competition law needs to continuously adapt to them. As the Austrian Report that follows shows, this is what the Austrian competition authority has been doing in its cases as well as in the soft law guidance it provides. The Austrian legislator also sprang into action, adapting the Austrian merger control regime to new market environments in 2017 and proposing a further amendment to the Cartel Act in 2021. While the national advances when it comes to the application of competition law to digital markets are important building blocks that are more easily achieved than change at the European level, as we move forward, it will be vital to ensure that the European Union's competition law framework and that of its 27 Member States remain in sync.

---

[15] Kartell- und Wettbewerbsrechtsänderungsgesetz 2021 (KaWeRÄG 2021), 114/ME XXVII. GP (Ministerialentwurf).

[16] Gesetz zur Änderung des Gesetzes gegen Wettbewerbsbeschränkungen für ein fokussiertes, proaktives und digitales Wettbewerbsrecht 4.0 und anderer wettbewerbsrechtlicher Bestimmungen (GWB-Digitalisierungsgesetz) BGBl I 2021/1.

[17] BWB, 'RTR presents method paper on monitoring digital platforms prepared together with AFCA' of 27 May 2020, available at https://www.bwb.gv.at/en/news/detail/news/rtr_presents_method_paper_on_monitoring_digital_platforms_prepared_together_with_afca/. Accessed 26 August 2021.

[18] European Data Protection Board, 'Statement of the EDPB on the data protection impacts of economic concentration' of August 2018, available at https://edpb.europa.eu/sites/edpb/files/files/file1/edpb_statement_economic_concentration_en.pdf. Accessed 26 August 2021; European Data Protection Board, 'Statement on privacy implications of mergers' of 19 February 2020, available at https://edpb.europa.eu/sites/edpb/files/files/file1/edpb_statement_2020_privacyimplicationsofmergers_en.pdf. Accessed 26 August 2021.

3.2 The Role of Antitrust Authorities Regarding the Digital Economy

3.2.1 Digitalization, a New Field of Law in Austrian Competition Law

The field of digitalisation, e.g. definition of digital markets (including the provision of free services and platforms) or criteria for market dominance (including data collection), is not explicitly mentioned in the current version of the Austrian Cartel Act (*Österreichisches Kartellgesetz 2005*, thereafter 'Cartel Act').[19]

In terms of subject matter, there had been investigations by the BWB concerning Amazon's dual/'gatekeeper' position as a merchant platform and online retailer and, in consequence, Amazon's alleged unfair trade practices against its Austrian retailers. As a consequence, Amazon agreed in a settlement with the BWB to change its terms and conditions in contracts with its retailers. The BWB published a detailed case report (see, in detail, Sect. 3.2.5. below).

Furthermore, the Austrian Regulatory Authority for Broadcasting and Telecommunications (RTR) published in May 2020 its methodology paper 'Monitoring of digital communication platforms and gatekeepers of open Internet access'. On that basis, the RTR is currently developing a monitoring system for digital platforms in order to cover the most important digital communications platforms used in Austria, e.g. voice and messaging apps, voice assistants, operating systems and app stores or browsers. The aim of the monitoring is to provide an overview of the competitive dimension and present an analysis of possible economically damaging developments. A key-element of the RTR will be the monitoring of broad covering ecosystems in which different services and platforms are embedded (e.g. Facebook, which is active in social media, Online advertising, telecommunication (WhatsApp)).

The results of the monitoring process are to be shared between the RTR and BWB on an ongoing basis, thus enabling an initial competitive assessment. Subsequently, indications of competition law violations, in particular suspicions of abuse of market power, can be taken up and thus enable the RTR to initiate investigations.

In June 2020, the BWB published its thesis paper 'Digitalization and Competition' ('Thesenpapier Digitalisierung und Wettbewerbsrecht', 'digitalization paper'). In response to the BWB's draft 'digitalization paper' (from November 2019), the Austrian section of the 'Studienvereinigung Kartellrecht', an association of German-speaking attorneys who are specialising in competition law, summarised the current status quo and recommendations concerning digitalisation and its practical impacts on competition law in Austria ('Stellungnahme der Studienvereinigung Kartellrecht', 10 January 2020, 'Opinion Studienvereinigung').

---

[19] Bundesgesetz gegen Kartelle und andere Wettbewerbsbeschränkungen (Kartellgesetz 2005 – KartG 2005, as amended).

The KaWeRÄG 2021, which is expected to enter into force in autumn 2021,[20] refers, for the first time, to certain aspects of digitalisation (in combination with market dominance). In essence, these cited investigations, initiatives and sources will form the basis of this Austrian report.

3.2.2 Austrian Competition Law and Its General Approach to the Digital Economy

In the BWB's view, the essential question with regard to large technology and Internet undertakings is whether these undertakings gain unjustified advantages over their 'traditional', regional or national competitors due to their 'special standing'. In the authority's view, the 'special standing' of many digital companies and/or the advantage of large networks are based on the collection, processing, linking and transfer of user data.

The 'injustice' can be based on several sources: in the BWB's view, not only antitrust law and fair trading law but also other fields of law may be taken into account when assessing an infringement of antitrust law. Such a broad approach is also backed up by the Cartel Act, wherein 'market dominance' is defined in general terms.[21] Similarly, Section 5 Cartel Act defines abuse of dominance in very general open terms: 'The abuse of a dominant position is prohibited. Such abuse may consist *in particular* of the following: [. . .].'

Therefore, the BWB states that existing competition law rules, which are formulated in an open and neutral manner, *offer certain flexibility for application to new types of situations, even if the historical legislator may not have had this in mind.*[22] The BWB hereby also takes into account § 20 Cartel Act, which states that facts have to be assessed according to their true economic content and not according to their outward appearance. Furthermore, new case law in Austria, and also in the European Union, is to be expected to provide guidance on how to apply existing rules to new digital cases and, more fundamentally, on whether new antitrust issues based on digitisation can be dealt with within the framework of the current Cartel Act. In this context, the BWB considers that a possible inclusion of examples in the law itself (as has been done in Section 18 of the German GWB) or soft law instruments (e.g. Commission notices) might facilitate and speed up investigations

[20]Currently, a more accurate date cannot be specified as the Austrian Federal Council (Bundesrat), the second house of the Austrian Parliament, voted on 15 July 2021 against the draft KaWeRÄG 2021. It is expected that the first house of the Austrian Parliament, the National Council (Nationalrat), which adopted the draft KaWeRÄG 2021 on 8 July 2021, will overrule the Federal Council's refusal to approve it in autumn 2021.

[21]Cf § 4 Cartel Act: "For the purposes of this Federal Act, an undertaking shall be deemed to be dominant if, as a supplier or buyer (1) is not exposed to any or only insignificant competition, or (2) has a superior market position in relation to the other competitors."

[22]Digitalization Report, p. 4.

and proceedings (while the substance of competition/cartel law as such would not be changed).

This correspondens with the view of the Studienvereinigung Kartellrecht, which considers the current open wording of Austrian competition law, especially with regard to market dominance and abuse of dominance, of being sufficient with regard to upcoming digital cases.

Concerning digital markets and dominance, the Studienvereinigung hereby refers to the European Commission's case practice, following which also the provision of free services—a business model that is increasingly encountered especially with regard to digital platforms—can constitute a market (with the consequence that companies may have a dominant position in the provision of free services).[23] Furthermore, the Studienvereinigung stresses the fact that following the Commission's practice, it is acknowledged that, e.g., direct and indirect network effects and related economies of scale, customer usage patterns (e.g. multi-homing or single homing), switching costs and access to data can be taken into account when examining the market dominance of digital undertakings.[24]

Concerning abuse of dominance, the Studienvereinigung states that various proceedings of the EU Commission with regard to undertakings active in digital markets are based on 'traditional' concepts of abuse. For example, the practices investigated by the European Commission in the *Google Android* case can be traditionally classified as abuse based on tying and exclusivity clauses (specifically with regard to the exclusive pre-installation of Google Search by manufacturers of mobile devices).[25] Also in this regard, in the Studienvereinigung's view, the legal standards of the Cartel Act do not have to be changed, at least at this stage. The further development of the decision-making practice of the European Commission, in particular, and the case law of the European Court of Justice (ECJ) should be awaited.[26]

The author agrees with such an approach, as suggested by the BWB and the Studienvereinigung. There seems to be no actual need to amend the respective Cartel Act's legal clauses in Austria, at least in substance. This affects not only the current law concerning market dominance and abuse of dominance but also illegal coordination according to Art. 101 para 1 of the Treaty on the Functioning of the European Union (TFEU) and its Austrian national equivalent, § 1 of the Cartel Act. For example, as concluded in the 'Joint Working Paper – Algorithms and Competition' of the French Autorité de la concurrence and the German Bundeskartellamt of

[23]The Opinion Studienvereinigung hereby refers, inter alia, to the decisions of the European Commission case AT.40099 – Google Android; case AT.39740 – Google Search (Shopping) and, concerning Merger Control, COMP/M.8124 – Microsoft/LinkedIn; COMP/M.7217 – Facebook/WhatsApp; COMP/M.6281 – Microsoft/Skype and GC, case T-79/12 *Cisco Systems and Messagenet/Commission*, ECLI:EU:T:2013:635.

[24]Opinion Studienvereinigung, paras 22 ff.

[25]European Commission, case AT.40099 – *Google Android.*

[26]Opinion Studienvereinigung, paras 25 ff.

November 2019,[27] the contemporary legal framework, in particular Art. 101 TFEU and its accompanying jurisprudence, allows competition authorities to address possible competitive concerns with a certain spectrum of cases involving algorithms. Following this paper, it currently remains open whether algorithms could independently engage in behaviours that resemble explicit forms of collusion.

Only if an evaluation of new cases will prove that the existing Austrian law is not or insufficiently capable of dealing with new questions and challenges of digitalisation would the legislator be obliged to change or supplement the existing law. Therefore, close and prompt monitoring by competition authorities and the legislator must be guaranteed. As an added value for the present time being, the BWB's proposed publication of soft law, e.g. guidelines, which reflect the authority's status quo in dealing with and investigating new digital cases, should be implemented.

However, as mentioned, the KaWeRÄG 2021 now includes specific criteria in assessing market dominance, which refer to the digital market:

§ Section 4 (1) Austrian Cartel Act

> For the purposes of this Act, a dominant undertaking shall be an undertaking which, as a supplier or demand
>
> [..]

2. has a superior market position in relation to the other competitors; in this context, particular account shall be taken of its financial strength, relations with other undertakings, opportunities for access to procurement and sales markets, *the importance of its intermediary services for the access of other undertakings to procurement and sales markets, access to competitively relevant data, the benefits derived from network effects,* and the circumstances restricting market access for other entrepreneurs.

Furthermore, the KaWeRÄG 2021 also edits the legal definition of 'relative market dominance', which—following precedents—exists in existing vertical relationships, *if the customers are dependent on the maintenance of the business relationship in order to avoid serious economic disadvantages.*[28]

Following § 4a KaWeRÄG, even potential customers now can claim 'relevant dominance' towards undertakings that act as an intermediary in a multi-sided digital market. The new § 4a KaWeRÄG reads as follows (new edits of the KaWeRÄG 2021 in *italic*):

> An undertaking which has a superior market position in relation to his customers or suppliers shall also be deemed to have market dominance; such a position exists in particular if the

[27] Available at https://www.bundeskartellamt.de/SharedDocs/Publikation/EN/Berichte/Algorithms_and_Competition_Working-Paper.pdf?__blob=publicationFile&v=5. Accessed 19 September 2021.

[28] *See* Austrian Supreme Cartel Court ("KOG"), case 16 Ok 4/20d, *Büchl/Peugeot*.

latter are dependent on the maintenance of the business relationship in order to avoid serious business disadvantages. *An undertaking which acts as an intermediary in a multi-sided digital market shall also be deemed to have a dominant market position if the customers of his intermediary services are dependent on the establishment of a business relationship in order to avoid serious business disadvantages.*

Lastly, as mentioned in the introductory remarks,[29] § 28a KaWeRÄG introduces a new declaratory procedure whereby the BWB and the Federal Cartel Prosecutor can ask the Cartel Court to declare that a company is dominant on a multi-sided digital market.

It will have to be seen if and, if yes, how these new legal provisions will be applied in practice.

For example, concerning 'relative dominance', it remains to be seen whether also potential customers can claim/prove that they will be *dependent on the maintenance of the business relationship in order to avoid serious economic disadvantages*. Furthermore, the new declaratory procedure of § 28a KaWeRÄG will, if the respective request is successful, only result in a declaratory dominance for the moment of the judgment itself. For the time being, it remains unclear how the (potential) customers of these declared dominant undertakings can claim any rights at a later point in time. Also, the defending rights of the then declared dominant undertakings are not defined in detail yet.

Leaving aside changes in subject matter, in view of the long duration of administrative proceedings and subsequent court proceedings on the one hand and the rapid changes in digitisation on the other, the question arises as to whether there is an actual need now to modify procedural law and/or interim legal protection. For complainants, it will be often essential to obtain a (preliminary) court decision as soon as possible. The digital economy is characterised by strong 'winner takes most' effects, whereby an accumulation of large amounts of data often generates an incontestable competitive advantage within short time and entry barriers for competitors.

Concerning interim relief, such possible modification must be balanced with the predicament of interim legal protection, i.e. the urgency and interest of security on the one side and the rights of defence and the threat of massive interference with the business activities of the company concerned on the other side. Such conflict of interests is especially intensified in the digital sector, where allegations against big tech undertakings are based on complex technical and economic backgrounds, which usually require a detailed and long-lasting investigation, even before deciding on interim measures.

[29] See Sect. 3.1 above.

In its digitalisation paper, the BWB, which is not established as a decision authority (but has the exclusive right[30] to initiate fine proceedings in front of the Cartel Court) suggests completing the already existing provisions on interim relief by the Cartel Court in the Cartel Act (which do require proof of the conditions for prohibition but do not request proof of imminent irreparable damage) by introducing decision deadlines and/or presumption rules and thus making it easier/more practicable for the courts to issue interim injunctions.[31] The Studienvereinigung rejects the BWB's proposed amendments in this regard as, in its view, the (procedural and material) thresholds for interim relief in Austria are already low as compared to other jurisdictions (however, as outlined above, also the low thresholds of interim relief in Austria cannot avoid the possibility that the imposition of interim reliefs may take too long, especially in complex cases).

In recognising that time constrains might be of crucial importance when it comes to cases concerning the digital economy, the Studienvereinigung encourages the strengthening of the legal power of the Cartel Court by also accepting requests for withdrawal in the case of imminent threats and not only in the event of already existing infringements.[32]

In order to facilitate its investigations, the BWB further suggests introducing a reversed burden of proof in the Cartel Act (possibly in combination with instruments of interim relief) if there is a prima facie case of abusive or unfair conduct and if the facts are difficult for an applicant to understand or where investigations of authorities reach their natural or technical limits. In this case, the BWB suggests that it would be up to the companies to explain, using their data, why a certain practice or conduct does not have anti-competitive effects.

Although reasonable from an authority's point of view, the Studienvereinigung doubts whether such reversal of the burden of proof is justified. It hereby mainly refers to the principle of presumption of innocence, as established in EU law and in the precedents of the European Convention on Human Rights (ECHR).[33] Even though introduced, the level of the burden of proof must not be excessive or unrealistic in order to rebut that presumption.[34] Undertakings concerned must have the practicable opportunity to rebut the presumption.

---

[30] Together with the Federal Competition Attorney (Bundeskartellanwalt) as the second official competition authority in Austria.

[31] Digitalization Report, p. 8 ff.

[32] Opinion Studienvereinigung, p. 22.

[33] ECJ, case C-74/14, *Eturas*, ECLI:EU:C:2016:42, para 38; ECHR 14 February 2019, case 5556/10, *SA Capital Oy v Finland*, paras 65 ff.

[34] ECJ, case C-74/14, *Eturas*, ECLI:EU:C:2016:42, para 41.

It also has to be noted that in Austria, contrary to, e.g., the European Union, there is already a legal presumption of market dominance included in the Cartel Act. If the market share thresholds of § 4 (2) and (2a) Cartel Act are triggered,[35] it is on the undertaking or undertakings (if there is an alleged collective dominance) concerned to prove that it is (or they are) not dominant on the respective market.

3.2.3 Austrian Competition Law and Data Protection, Compared to the German Law and Practice

As outlined above, in the BWB's view, the 'special position' of many digital companies and/or the advantage of large networks is based on the collection, processing, linking and transfer of user data. In assessing a respective infringement of antitrust law, the BWB, besides applying antitrust and unfair competition law, hereby generally states that it might base its legal assessment also on other legal sources. It hereby explicitly refers to the handling of and compliance with data protection regulations, such as the General Data Protection Regulation (GDPR). Although these aspects *are not to be dealt with or solved by the competition authorities, they may well be included in the antitrust investigation as factors affecting the market position of a company.*[36]

So while Austria has its own data protection authority,[37] (breach of) data protection as such may also be taken into account by the Austrian competition authorities. It therefore seems to be undoubtful that the BWB might also find that terms and conditions that are referring to data collection result in anti-competitive conduct, especially abuse of dominance.

Concerning the specific combination of data collection and potential abuse of dominance in undertakings' general terms and conditions, no case precedents in Austria exist. Concerning general terms and conditions in general, the BWB's whole settlement with Amazon (see Sect. 3.2.5.) was based on changing Amazon's terms

[35]Following the law (§ 4 (2) Cartel Act), an undertaking is considered of being rebuttable single dominant if it has a market share of (i) at least 30% or (ii) more than 5% and is exposed to competition from a maximum of two undertakings or (iii) has a market share of more than 5% and is one of the four largest undertakings on this market with a combined market share of at least 80%. An undertaking is considered of being rebuttable collective dominant law (§ 4 (2a) Cartel Act), if the undertakings concerned (i) have a combined market share of at least 50 % and consist of maximum three undertakings, or (ii) have a combined market share of at least two thirds and consist of maximum five undertakings.

[36]Digitalization Report, p. 10.

[37]The Austrian data protection authority (Datenschutzbehörde) ensures compliance with data protection in Austria. Austria was one of the first European states to have a data protection authority. The Data Protection Commission was created by the first Data Protection Act, Federal Law Gazette No. 565/1978. The EU Data Protection Directive 95/46/EC was implemented in Austria by the Data Protection Act 2000 (DSG 2000). On 25 May 2018, the Basic Data Protection Regulation (DSGVO) and the revised Data Protection Act (DSG) formed the basis of Austrian data protection law.

and conditions (e.g. by abolishing Amazon's free, irrevocable, unlimited, worldwide right of use of the materials provided by sellers). Some Amazon retailers furthermore complained that in some cases, Amazon requested information on the retailers' purchase price and supplier information. However, based on retailers' feedback, it was unclear whether such information was mandatorily required by Amazon. Amazon argued that the disclosure of the purchase price of marketplace traders was required only in exceptional cases, where it was necessary to examine the authenticity of a product or where there had been a disagreement about the amount of the refund. The BWB acknowledged in general that Amazon might have objectively justified reasons to request such information in exceptional cases. The BWB therefore did not investigate any further into this subject.

In general, there are Supreme Cartel Court precedents, following which general terms and conditions of dominant undertakings can be considered abuse (cf 16Ok46/05, concerning postal deliveries, i.e. a 'traditional sector'): 'The abuse of a dominant position within the meaning of § 5 para. 1 no. 1 Cartel Act may in particular consist in the demand for purchase or sales prices or other terms and conditions which deviate from those which would very probably result from effective competition (16 Ok 1/20p).'

3.2.4 Assessment of Market Power and the Relevance of Network Effects in Digital Markets

So far, no guidelines concerning competition law in digital markets in general and the market power of networks and platforms in particular exist in Austria.

With regard to market power, the BWB refers in its digitalisation paper to the wording of the Cartel Act (§ 4 KartG), which leaves a broad scope of discretion when examining the existence of a dominant position. In the BWB's view, (all) circumstances that restrict the market access of competitors of large technology and Internet firms, e.g. 'network effects, barriers to switching, access to user data, etc.', can be taken into account when assessing dominance. Furthermore, the BWB stresses that already on a very general level, § 20 KartG states that facts are to be assessed according to their true economic content and not according to their outward appearance (economic approach).

Therefore, difficulties in applying the existing antitrust rules to new digital cases are thus likely to result less from the lack of appropriate legal clauses than from the fact that enforcement is not as developed as well as in 'traditional' business segments. In the authority's view, the legal request to provide reasonable evidence for the alleged antitrust infringement in 'digital' cases regularly causes much higher efforts as compared to 'traditional' antitrust infringements.[38]

[38] Digitalization Report, p. 5.

3.2.5 Amazon and the General Definition and Concept of a 'Platform'

As mentioned, there exists no legal definition of (or soft law concerning) platforms in Austria. Some general remarks concerning the authority's approach towards platforms can be derived from the BWB's published case study on its Amazon investigation and settlement.[39]

In the BWB's view, defining platforms as 'one-sided' markets is insufficient with regard to Amazon's online trading platforms; rather, they are to be understood as 'two-sided' or, more general, as 'multi-sided' markets. Whether multi-sided markets should be defined as separate markets for the respective user types or as a single market remains open. However, the authority refers to the fact that in any case, interdependencies between the individual market sides should be considered.

The lower are the prices and the higher are the number and diversity of the products offered on Amazon's website, the higher is the customer's demand. On the other side, based on the increase in the number of customers, the demand of marketplace retailers (4500 Austrian retailers as of 2018) to sell their products via Amazon will increase as well (following the BWB's case study, on average, Austrian retailers that are active on Amazon achieve 64% of their revenues from sales via the Amazon platform).

Adjacent business fields, e.g. Amazon's PRIME subscription, enables paying customers to receive PRIME goods quickly and free of charge, while retailers can distribute their goods as PRIME and store and ship their goods using Amazon's logistics service.

While not finally defining the relevant market for online trading platforms, the BWB preliminarily assumed that the market for online trading platforms includes neither brick-and-mortar shops nor retailer-owned web shops. The latter differ significantly from Amazon's offering in terms of the functionality offered to the retailer (assumption of essential distribution tasks by the platform). In addition, by using a platform, the retailer also acquires the brand trust placed in it, an essential asset compared to an own web shop or a stationary shop. Whether this applies to the same extent as other online trading platforms with related functionality (e.g. eBay, Rakuten, Zalando) could not be conclusively assessed by the BWB on the basis of available data from the survey.

Also in geographical terms, the definition of the relevant market was left open. Criteria that refer to a national market (or at least a separate market for Austria and Germany) were, e.g., the stronger focus of Austrian retailers on Austrian customers and structural factors such as differences in delivery times. On the other side, the similarly strong demand on Amazon.de in Germany and Austria and the fact that Austrian retailers achieve 75% of their Amazon turnover with German customers could indicate that the geographic scope of the market is broader than national.

[39] BWB, *Fallstudie Amazon.de Marktplatz*, 2019, https://www.bwb.gv.at/fileadmin/user_upload/Downloads/standpunkte/BWB_Amazon-Fallbericht_20190717.pdf. Accessed 10 September 2021.

Although not finally defining the relevant market, BWB's retailer survey showed that for a representative sample of a larger Austrian marketplace, retailer Amazon.de has market power.[40] This conclusion is based on, inter alia, the retailer's answers that there are hardly any relevant alternatives to reach their customers, that most retailers would be prepared to accept 5–10% price increases from Amazon, that a large percentage of all retailers sells exclusively on Amazon.de and that other sales channels, such as web shops, stationary sale and other online trading platforms, only contribute to a small share of sales.

In substance, Amazon agreed, in settlement talks with the BWB, to change its terms and conditions in its contracts with retailers, inter alia, concerning the following clauses:

- Termination or suspension of the contract at any time with immediate effect without giving reasons
- Irrevocable, unlimited, worldwide right of use by Amazon of the materials provided by sellers
- Indemnification/compensation of Amazon by dealers only in the event of violations of the law, not in the event of a merely alleged breach of contract
- Amendment of the terms and conditions only after at least 15 days after notification
- No other jurisdiction than Luxembourg
- Extensive disclaimer or indemnity of liability regarding storage in the 'Shipping through Amazon' program

3.2.6 Can There Be an Antitrust 'Market' for Free Services?

The Cartel Act, contrary to Germany.[41] does not explicitly refer to markets for free services. However, it seems to be undisputed that also in Austria, a definition of the relevant market and therefore also market dominance could be based on free services.

For example, in the BWB's Guidance on Transaction Value Thresholds,[42] it is stated that 'despite the absence of monetisation, an activity could conceivably still have market orientation' if, inter alia, 'a service is offered free of charge but is monetized in a different way'.

Furthermore, as mentioned above, the BWB, in its Amazon case study, stated that Amazon is active on 'multi-side' markets, an approach that is also chosen in order to

[40] Case report, p. 8. The BWB hereby followed the European Commission's recommendation that concerning two-sided markets it is essential to focus on the theory of harm and less on a market definition.

[41] Cf, § 18 (2a) GWB.

[42] Prepared and published together with the German Bundeskartellamt, available at https://www.bwb.gv.at/fileadmin/user_upload/Downloads/standpunkte/2018-07_Guidance_Transaction_Value_Thresholds.pdf.

confirm the existence of markets for free services. Lastly, there are precedents in EU law that confirm markets for free services with regard to both antitrust and merger control purposes.[43]

3.2.7 Online Advertising

So far, the BWB has not initiated sector enquiries concerning online advertising. In general, traditional media undertakings (publishing houses, free TV channels) and also traditional advertising undertakings claim that online advertising more and more suppresses traditional advertising in print titles or free TV. Following such comments, the previous approach defining separate product markets, e.g. for advertisements in newspapers, magazines, classified ads or free TV, would be arguably too narrow.

Therefore, a possible focus of a sector enquiry could target advertising customers and request information on whether, from their point of view, separate forms of advertising (including print and online advertising) are viewed as substitutes for each other or whether, from the point of view of advertisers, there is at least a competitive pressure from online advertising on traditional forms of advertising, such as print or TV advertising. Media undertakings could be asked whether their print advertising gets more and more substituted by online advertising on their web presence.

Based on feedback from the market received, one could ask whether the market definition of advertising markets must be eventually broadened in consequence of the significant impact of online advertising on the total advertising market and its differing advertising segments.

3.2.8 Best-Price Clauses and Most-Favoured-Nation Clauses

In 2015, the BWB settled its investigations on Booking.com and Expedia. Unlike the later quashed decision of the German Bundeskartellamt, the BWB accepted that Booking.com and Expedia operate so-called narrow most-favoured-nation (MFN) clauses (or best-price clauses) on their hotel booking platforms. That is, booking platforms were allowed to oblige hotels in Austria not to offer better prices or conditions on their own Internet sites. The narrow MFN clauses were considered to be justified in order to prevent hotels from 'free riding', which might occur if potential guests find hotels via the booking platform and then book the rooms at better prices on the hotel's own website.

[43] See European Commission, AT.40099 – Google Android; AT.39740 – Google Search (Shopping); or – concerning merger control - COMP/M.8124 – Microsoft/LinkedIn; COMP/M.7217 – Face-book/WhatsApp.

On the other side, Booking.com and Expedia offered and agreed not to oblige hotels on broad MFN clauses, which would prohibit the latter to offer better prices through other sales channels, including phone calls, other hotel portals, etc.

The Austrian Supreme Cartel Court stated in its grocery 'Spar' decision (and therefore concerning a traditional, non-digital economy) that most-favoured-nation clauses, e.g. a customer obliging its supplier to dispose of its goods to other customers not at lower prices, generally infringe Article 101(1) TFEU (the Supreme Court did not refer to Vertical Block Exemption Regulation 330/2010, which covers such clauses).[44] In the author's view, it seems to be undoubtful that such approach can be also applied to dominant undertakings and/or content providers, also with regard to digital markets. However, also dominant undertakings may use MFNs if justified for objective reasons (such as the prevention of free riding in the above-mentioned booking platform cases).

3.2.9 Network Cooperation in the Telecommunication Sector

So far (and besides case-related investigations on filed merger transactions), network cooperation as such has not been made subject to a sector enquiry of the BWB.

In general, the responsible authority is RTR, which is established by Austrian law. Its core mandate is to promote competition in the broadcasting, telecommunications and postal markets as well as to achieve the goals set out in the KommAustria Act and the Telecommunications Act. RTR also serves as an administrative agency, providing support to the Austrian Communications Authority (KommAustria), the Telekom-Control-Kommission (TKK) and the Post-Control-Kommission (PCK). Furthermore, the RTR offers alternative dispute resolution services, which can also be provided via its officially recognised consumer arbitration centres.

In its role as the administrative agency of the TKK, RTR is active in matters including the regulation of competition, managing broadcasting frequencies, spectrum awards, approving providers' terms of business, net neutrality and the electronic signature system. RTR's activities on behalf of the PCK specifically involve postal service points, providers' terms of business, regulation of competition and legal supervision.

However, as clearly defined by precedents and the law, any approval of the tariff by the RTR does not necessarily mean that the tariff is admissible under competition law (cf. 16Ok13/08 h, 16 Ok 4/11 and also § 2 (4) Telecommunication Act (TKG)). Therefore, e.g., the BWB and also undertakings in private enforcement are entitled to initiate proceedings against undertakings that rely on approved tariffs by the RTR.

[44] KOG, case 16 Ok 2/15b.

3.2.10 Abuse Without Dominance?

In Austrian competition law enforcement, there is a broad range of legal sources, following which undertakings can be pursuit for unilateral behaviour without necessarily having a dominant market position (based on market shares).[45]

In its above-mentioned Amazon investigations, the BWB did not finally define the relevant market. Nevertheless, the authority—'regardless of the market definition ultimately chosen'—came to the conclusion that Amazon has market power towards its Amazon retailers. In a footnote, the authority hereby refers to the Commission. When analysing multifaceted markets, the Commission apparently[46] advises focusing less on a precise market definition but rather primarily on a plausible damage theory.

Furthermore, pursuant to Section 4(3) of the Cartel Act, an undertaking is also deemed to be dominant if—regardless of its market shares—it has a predominant position in relation to its customers or suppliers. This is the case, in particular, when customers and suppliers are obliged to maintain business relations with the dominant company in order to avoid serious economic disadvantages.[47]

In addition to the cartel law regulations concerning abuse of a market-dominant position, the Austrian Act on Local Supply (Federal Law on Improvement of Local Supply and Competitive Conditions, hereinafter 'Act on Local Supply')[48] includes provisions on 'good business behaviour' (*Kaufmännisches Wohlverhalten*). According to the general clause in Section 1 (1) of the Act on Local Supply, business practices between any undertakings (thus, despite the title of the Act, not only undertakings involved in local supply) can be interdicted (by the Cartel Court) insofar as they are likely to threaten performance-related competition.

Pursuant to Section 1 (2) of the Act on Local Supply, such behaviour is particularly the *offering or requesting, granting or accepting of money or other benefits, also of discounts or special conditions, among suppliers and retailers that are not objectively justified, especially if the additional benefits are not offset by appropriate compensation*. This provision was created primarily to counteract the 'tapping' of suppliers by market-powerful companies on the demand side. It can be seen as a

---

[45] Concerning Austria's approach to dominance and abuse, and also alternative legal concept, where dominance is not necessarily requested for pursuing an abuse, reference can be made to the national Austrian LIDC report, G. Fussenegger, F. Schuhmacher and R. Tahedl, Abuse of Dominant Position and Globalization & Protection and Disclosure of Trade Secrets and Know-How, In LIDC Contributions on Antitrust Law, Intellectual Property and Unfair Competition, Springer 2017, pp 27-44, by Mag.

[46] In its source, the BWB refers not to the Commission's practice but to a publication, i.e., S. Wismer, and A. Rasek, Market definition in multi-sided markets, OECD 2017, pp. 4 f., available at http://www.oecd.org/officialdocuments/publicdisplaydocumentpdf/?cote=DAF/COMP/WD(2017)33/FINAL&docLanguage=En.

[47] See, e.g., KOG, case 4 Ob 214/97t, Filmverleihgesellschaft I; Austrian Cartel Court, case 25 Kt 48, 49/99, Leiner/Kika/Michelfeit.

[48] Nahversorgungsgesetz, NahVersG, BGBl. (Federal Law Gazette) No. 392/1977 as last amended by BGBl. No. 50/2012.

supplement to the cartel law prohibition of abuse and does not require the undertaking concerned to hold a market-dominant position (the Austrian Supreme Cartel Court left open whether at least a certain amount of market power was required.)[49]

Section 2 of the Act on Local Supply provides for a ban on discrimination similar to Section 5 (1) No. 3 of the Cartel Act (*disadvantaging of contractual partners in the competition by applying dissimilar conditions to equivalent transactions*—which is corresponding to Article 102 TFEU), whereby market dominance, again, is not required: *(1) Who, in spite of the same conditions prevailing as a supplier grants or offers different conditions to authorized resellers without objective justification, can be claimed for cease-and-desist. (2) In the same way also a reseller can be claimed for requesting or accepting objectively unjustified conditions from suppliers.* In this context, 'condition' actually stands for 'price' (including the granting of discounts); thus, judicial practice so far has focused on unjustified price discrimination.

Another notable provision is Section 4 (1) of the Act on Local Supply, which refers to the abuse of a market-dominant position by refusal to deal: according to this provision, undertakings that usually deliver to final sellers may be obliged to conclude a contract if non-supply would threaten local supply or would significantly affect the competitiveness of the final seller on the market of the type of goods not supplied.[50] The Austrian Supreme Court recently stated that if the behaviour of a market-dominant undertaking is objectively justified (and therefore does not constitute abuse of dominance), this behaviour must also be justified under the Act on Local Supply (16Ok1/18k).

The Austrian Act Against Unfair Competition (Federal Law Against Unfair Competition),[51] constituting substantial regulation of the Austrian law on unfair practices, does not include any provision that would expressly refer to the exercise of market power or abuse of a market-dominant position, like the ban on abuse does in the Cartel Act. However, as can be seen from case law, the abusive exercise of market power may well (under certain circumstances) constitute a violation except of this law since various behaviours of market-powerful companies, such as the tapping of suppliers, discriminative practices and refusal to supply, and certain exclusivity obligations have been repeatedly considered unfair business practices within the meaning of the general clause of Section 1 of the Act Against Unfair Competition or Section 1a of the Act Against Unfair Competition (aggressive commercial practices). Furthermore, an infringement of the prohibition of the abuse of a market-dominant position stipulated in Section 5 Cartel Act can also be considered as 'other unfair practice' within the sense of Section 1 (1) No. 1 Act Against Unfair Competition. Pursuant to Section 2 (1) No. 7 of the Competition Act (Federal Law on the

[49] See KOG, case16 Ok 3/08, Sägerundholz.

[50] See also Section 3 Act on Local Supply (which corresponds to Section 6 Cartel Act), which provides that proceedings pursuant to Sections 1 and 2 Act on Local Supply may not be taken as a reason by the defendant to exclude the undertaking affected (by a conduct as defined by those provisions) from further supplying or demanding on reasonable conditions.

[51] Bundesgesetz gegen den unlauteren Wettbewerb 1984 - UWG, BGBl. (Federal Law Gazette) No. 448/1984 as last amended by BGBl. I No. 49/2015.

Establishment of a Federal Competition Authority),[52] the Austrian Federal Competition Authority[53] (Bundeswettbewerbsbehörde (BWB)) is entitled to claim injunctive relief under the Act Against Unfair Competition.

Lastly, the Austrian Telecommunications Act[54] includes provisions relating to 'undertakings with significant market power'.[55] These articles state that under specific conditions, such undertakings can be obliged by the regulatory authority to stop certain practices as there are, in particular (cf. Section 43 (2) Austrian Telecommunications Act), excessive pricing, hampering of the entry of new market participants, predatory pricing to eliminate competition, inappropriately preferring certain end users or unjustifiable bundling of services.

3.2.11 Consumer Protection

Until the beginning of 2020, the BWB was the responsible independent Austrian authority for Cooperation of EU Authorities in Consumer Protection (Consumer Authorities Cooperation Act—VBKG). Within this cooperation, the BWB was entitled to remedy intra-EU (cross-border) infringements of certain relevant consumer protection rules that harm collective interests (i.e., interests of a large number of consumers).

At the request of a requesting authority, the BWB had to take all necessary enforcement measures to bring about a cessation or prohibition of the intra-Community infringement without delay or, at the request of the requesting authority, had to provide all relevant information necessary to establish whether an intra-Community infringement has occurred or whether there are reasonable grounds to suspect that such an infringement might occur. The BWB hereby could file civil court claims against undertakings for an injunction against a suspected intra-Community infringement. In addition, the Federal Competition Authority could obtain a cease-and-desist order.

However, as of 17 January 2020, the new Regulation 2017/2394 on cooperation between national authorities responsible for the enforcement of consumer protection law came into force. Since the national law, which grants these cooperation rights to the BWB, exclusively refers to the old EU Regulation 2006/2004, the BWB had lost all its authority to act in these matters as of 17 January 2020.

[52] Bundesgesetz über die Einrichtung einer Bundeswettbewerbsbehörde (Wettbewerbsgesetz - WettbG), BGBl. (Federal Law Gazette) I No. 62/2002 as last amended by BGBl. I No. 129/2013.

[53] Bundeswettbewerbsbehörde (BWB).

[54] Telekommunikationsgesetz 2003 - TKG, BGBl. (Federal Law Gazette) I No. 70/2003 as last amended by BGBl I No. 44/2014.

[55] Pursuant to Section 35 (1) of the Telecommunications Act, an undertaking is considered to have a significant market power if such undertaking, either alone or together with other undertakings, holds such a strong economic position that it has the capability of acting, to a considerable extent, independently of its competitors, its customers and ultimately the consumers.

4 Belgium

Friso Bostoen

4.1 Introduction

A recent book titled *Digital Competition Law in Europe* reviewed the work of the European Commission (EC) and a number of Western European national competition authorities (NCAs), namely the UK, French, German, Dutch and Belgian ones. The chapter on Belgium was by far the shortest, with the authors explaining with regard to its NCA:

> The 'BMA-ABC' is not visibly active in respect of issues surrounding digital competition. . . . To date, its track record has been limited to a single cartel procedure. The BMA-ABC started an investigation in 2016 under both the cartel prohibition and the prohibition of abuse of a dominant position into the most popular real estate site in Belgium, 'Immoweb'.[1]

This statement is not necessarily wrong if one interprets 'digital competition' very restrictively. Indeed, *Immoweb* can be considered the only case concerning *online platforms*. However, these days, all markets and forms of competition are to some extent digital. Then again, reviewing every single case by the Belgian Competition Authority (BCA) is not feasible or useful either. This work tries to strike a balance

In accordance with the ASCOLA Transparency and Disclosure Declaration, the author has nothing to disclose. This article was updated until August 2021.

[1] M. Wiggers, R. Struijlaart and J. Dibbits, Digital Competition Law in Europe: A Concise Guide, Kluwer 2019, Chapter 7.

F. Bostoen (✉)
Consumer Competition Market, KU Leuven, Leuven, Belgium

Research Foundation – Flanders, Brussels, Belgium
e-mail: friso.bostoen@kuleuven.be

B. Kilpatrick et al. (eds.), *Antitrust in Data Driven Markets & Legal Framework for Influencers, Native Advertising and Control over the Use of AI in Marketing*, LIDC Contributions on Antitrust Law, Intellectual Property and Unfair Competition, https://doi.org/10.1007/978-3-031-07422-6_4

between an overly restrictive and an overly broad view of case law in digital markets.[2]

At the same time, it must be acknowledged that the BCA has not taken up the kind of digital cases that would make international news headlines, in contrast to some of the neighbouring NCAs, such as the Dutch Authority for Consumers & Markets (ACM) and the German Bundeskartellamt. Nevertheless, a focus on the work of the BCA is useful, both nationally, to evaluate its work, and internationally, to learn from legal assessments, the value of which is independent of the 'size' of the case.

Therefore, this chapter takes a deep dive into the work of the BCA in digital markets. I take a broader view of 'digital' than some by examining cases where the market is eminently digital (e.g. search engines) or where the conduct is digital, even in a market not traditionally considered digital (e.g. lotteries). Cases regarding more traditional digital markets, such as telecommunications and media, are functionally included but do not constitute this work's focus. Finally, I focus on *conduct* cases (restrictive agreements and abuse of dominance), but merger cases are examined when useful to inform broader points (e.g. regarding market definition and power).

This work is structured as follows. Section 4.2 reviews institutional aspects of the BCA and how they (can) inform its work in digital markets. Section 4.3 then moves on to a substantive assessment. It first deals with the preliminary questions of market definition and power (Sect. 4.3.1) and then examines selected issues of anti-competitive conduct (Sect. 4.3.2). Those issues are selected based on the BCA's work so far and include self-preferencing, use of data, most favoured nation clauses (MFN clauses/MFNs) and new (*ex ante*) regulation. Section 4.4 concludes.

4.2 Institutional Aspects: From Advocacy to Case Law in Digital Markets

To understand the work of the BCA, one needs to understand its structure.[3] In contrast to the EC, the BCA has a dual structure: it consists of an investigative service, the Investigation and Prosecution Service ('Auditoraat' or 'Auditorat') and a decision-making body, the Competition College ('Mededingingscollege' or 'Collège de la Concurrence'). Depending on the case in question, the investigative service can adopt a decision itself or has to present its case to the decision-making body.[4] An important role is also played by the president, who is part of the board and

[2] This view is also taken in R. Alderweireldt, J. Froidmont, A. Lepièce and D. Van Liedekerke, Belgium, In: D. Mândrescu (ed.), EU Competition Law and the Digital Economy, Eleven International Publishing 2020, pp. 105–130. The author acknowledges reliance on this work, in particular for case selection.

[3] See Wetboek Economisch Recht (hereafter: Code of Economic Law or CEL), arts IV.16–29.

[4] The Competition College decides on infringement cases that are not settled or closed by the Investigation and Prosecution Service as well as non-simplified merger control procedures.

presides over the Competition College. Since the BCA became an autonomous institution in 2013, Jacques Steenbergen has fulfilled this role.[5]

Advocacy is a crucial part of the president's mandate. This is reflected in the law, which charges the president with contributing to national and international competition policy[6] and also gives them the power to informally opine on questions and disputes regarding the application of competition rules.[7] This power has been used in the digital sphere. In 2011, the Belgian economy minister urged the BCA to investigate a potential abuse of dominance by Apple, which required iPad users to subscribe to newspapers via iTunes.[8] An informal arrangement was reached with Apple, under which publishers were free to offer iPad users subscriptions outside of Apple's distribution channel.[9]

Interestingly, the obligation to use Apple's distribution channels and, in particular, its App Store rages on to this day. The EC is investigating 'the mandatory use of Apple's own proprietary in-app purchase system and restrictions on the ability of developers to inform iPhone and iPad users of alternative cheaper purchasing possibilities outside of apps'.[10] The ACM is investigating whether Apple violated 'the prohibition of abuse of dominance, for example, by giving preferential treatment to its own apps'.[11]

These continued investigations highlight the strengths and weaknesses of informal arrangements. One strength is the speed of intervention, which is especially important in digital markets, where a dominant position can be attained quickly (through anti-competitive conduct), after which it can be difficult to restore the pre-existing competitive situation. Further, they allow a small competition authority to punch above its weight. On the other hand, informal arrangements offer less legal certainty, given the lack of precedent. And partially settled issues may return later, requiring a decision rather than an opinion.

In a Joint Memorandum with the ACM and the Luxembourgish Conseil de la Concurrence, the BCA argues for an extension of guidance procedures. The Benelux

[5] Before that, from 2007 to 2013, he already served as director general of the directorate general for competition in the Belgian ministry of economic affairs.

[6] Jacques Steenbergen has done so not only through active cooperation with national (e.g. parliament) and supra-/international (e.g. ECN, ICN, OECD) institutions, but also by being a frequent speaker at conferences. See BCA, Annual Report (2019), pp. 28–30.

[7] Code of Economic Law, art IV.19, §1, 5.

[8] Euractiv, Belgium to launch antitrust probe on Apple's iPad (18 January 2011), available at https://www.euractiv.com/section/competition/news/belgium-to-launch-antitrust-probe-on-apple-s-ipad/ last accessed 10 February 2021.

[9] BCA, Annual Report (2011), p. 12. At present, this arrangement appears formalized in Article 3.1.3(a) "Reader" Apps of the App Store Review Guidelines, available at https://developer.apple.com/app-store/review/guidelines/ last accessed 10 February 2021.

[10] EC, Commission opens investigations into Apple's App Store rules (16 June 2020) IP/20/1073.

[11] ACM, ACM launches investigation into abuse of dominance by Apple in its App Store (11 April 2019), available at https://www.acm.nl/en/publications/acm-launches-investigation-abuse-dominance-apple-its-app-store last accessed 10 February 2021.

NCAs note that '[t]he digital economy and other fast-moving markets confront us with the challenge of having a real impact on market behaviour within a time period that meets the legitimate expectations of stakeholders'.[12] Therefore:

> We propose to examine the possibility of developing an approach that would allow the European Commission and Member State competition authorities to sidestep the infringement route in a much less formal and fast track commitment procedure, e.g. as a development of the practice under article 10 of Regulation 1/2003 or in line with the Notice on informal guidance.[13]

In 2020, the BCA adopted a new communication regarding the president's informal opinions.[14] Among others, an opinion can only be requested on novel legal questions of societal importance on which there are no pending procedures. Questions on competition law in the digital economy often fulfil the first limb of that test. The opinions are published on the BCA's website, which promotes legal certainty.

However, it remains to be seen whether the BCA will take up more digital cases in the future. Its 2020 Guidance did mark the digital economy as one of its enforcement priorities, referring to the Joint Memorandum discussed above.[15] Of course, the complexity of (certain) digital cases implies they require sufficient resources. On this point, the BCA's president remarked laconically that the authority's 'lack of resources is not limited to digital cases'.[16] Moreover, given the nature of the products and services involved, digital cases often do not have a local nexus. Therefore, an EU rather than a national solution may often be preferable.

The above reasons for not pursuing digital cases were reflected in the *Dialo/ Google* decision (2012), in which the BCA's investigative service chose not to examine the complaint of a classified ad website that it was discriminated against in Google's search results.[17] One reason for not taking up the case was the fact that Google's ranking algorithms are global, and the BCA is thus not the best-placed

---

[12] Joint memorandum of the Belgian, Dutch and Luxembourg competition authorities on challenges faced by competition authorities in a digital world, p. 4, available at https://www.belgiancompetition.be/en/about-us/publications/joint-memorandum-belgian-dutch-and-luxembourg-competition-authorities last accessed 10 February 2021 (hereafter 'Joint Memorandum of the Benelux NCAs').

[13] Ibid, p. 5 (the Memorandum reads "allowsthe"—an apparent typo that has been corrected here).

[14] Mededeling van de Belgische Mededingingsautoriteit betreffende de informele zienswijzen van de voorzitter van de Belgische Mededingingsautoriteit, Official Gazette 2020, N. 143, p. 37397.

[15] Het prioriteitenbeleid van de Belgische Mededingingsautoriteit voor 2020, available at https://www.bma-abc.be/nl/over-ons/publicaties/nota-prioriteitenbeleid-2020 last accessed 10 February 2021 (hereafter 'Prioriteitenbeleid Belgische Mededingingsautoriteit 2020'). Such a priority was absent most of the preceding years, see the review of 2014–19 in R. Alderweireldt, J. Froidmont, A. Lepièce and D. Van Liedekerke, Belgium, In: D. Mândrescu (ed.), EU Competition Law and the Digital Economy, Eleven International Publishing 2020, p. 109.

[16] OECD, Abuse of dominance in digital markets – contribution from Belgium (2020) DAF/COMP/GF/WD(2020)2, p. 3.

[17] BCA, Decision 2012-P/K-07-AUD, *SPRL Dialo/SA Google Belgium* (17 April 2012).

authority to remedy any problem. Moreover, an investigation would require significant resources, and the EC had already started investigating similar conduct in 2010.[18]

These reasons against taking up digital cases continue to apply. It is an open question whether the updated enforcement priorities change the BCA's calculus.

4.3 Substantive Assessments in Digital Markets

Now that I have surveyed the more institutional aspects of the BCA's approach to digital markets, the next sections examine the BCA's more substantive assessments. Section 4.3.1 investigates the preliminary steps of defining markets and establishing market power. Section 4.3.2 delves into questions of conduct, in particular restrictive agreements and abuse of dominance.

4.3.1 Market Definition and Power in Digital Markets

One important feature of digital markets is the presence of online platforms. Those platforms can be defined as intermediaries connecting two or more user groups (the 'sides' of the multi-sided market).[19] A useful distinction can be made between *matching* platforms, which connect two user groups (buyers and sellers) for the purpose of a transaction, and *advertising* platforms, which connect two user groups so that one can serve the other ads. While Amazon Marketplace and the App Store are examples of the former, Google Search and Facebook illustrate the latter. To maximize participation, platforms maintain an uneven price structure, which often means that one of the user groups does not pay any price. Consumers using Google or Facebook, for example, receive the service for 'free'. 'Free' requires quotation marks because these users do provide consideration, in particular their data (on the basis of which ads are targeted) and their attention (to view the ads they are served)—two valuable commodities.

When it comes to market definition in the case of online platforms, various questions arise, but two stand out: how *many* markets should be defined, and can a market for *free* services even be defined? Let us look at how the BCA has answered each question while putting those answers in their broader context.

Firstly, on the question of how *many* markets need to be defined, the research of Filistrucchi et al. has been defining. They held that in the case of advertising platforms, two (interrelated) markets need to be defined, and in the case of matching

[18] EC, Commission probes allegations of antitrust violations by Google (30 November 2010) IP/10/1624.

[19] See J.-C. Rochet and J. Tirole, Platform competition in two-sided markets, Journal of the European Economic Association 2003(4), pp. 990–1029 for the foundational article.

platforms, only one market should be defined.[20] This rule of thumb has also been followed by NCAs and the EC.[21] The EC, for example, has defined a single market for matching platforms, such as app stores,[22] while suggesting at least two markets for advertising platforms, such as search engines.[23]

The BCA, too, has defined markets in line with this distinction. In *Immoweb* (2016), the BCA's investigative service assessed MFN clauses maintained by a real estate intermediation website (see, further, Sect. 4.3.2.3 below). The investigative service explicitly qualified the real estate website as a 'matching platform' operating in a multi-sided market, with sellers on one side and buyers on the other.[24] Consequently, it defined a single market for 'online real estate portals'.[25] In cases regarding advertising-funded media, by contrast, the BCA defined two markets: one for the delivery of content (e.g. television) and another for the sale of advertising.[26]

Secondly, there is the question of *free* services—can they constitute the basis of a market? The question has a more fundamental aspect (can there even be a market for non-priced services?) and a methodological aspect (how do you go about defining such markets?).

Importantly, most online platforms only offer a free service to *one* user group (e.g. social media users) while charging the other (e.g. advertisers). In the case of matching platforms—where a single market is defined—the need to define a non-priced market does not arise; in the case of advertising platforms—with two markets, one of which has no price—it does. By analogy, with its cases in media sectors, the BCA would define a market on the non-paying side, e.g. for general

[20] L. Filistrucchi, D. Geradin, E. van Damme and P. Affeldt, Market definition in two-sided markets: theory and practice, Journal of Competition Law and Economics 2014(2), pp. 293–339.

[21] See e.g. Bundeskartellamt, The market power of platforms and networks (Working Paper 2016), pp. 25–32.

[22] *Google Android* (case AT.40099) Commission Decision, where the EC acknowledges the market for Android app stores is multisided (para 638), serving both OEMs and developers—as well as consumers, but since the abuse was not consumer-facing, this side was largely left out of focus. The EC considered substitutability from these different perspectives (in particular paras 284–294), but ended up defining a single market for 'Android app stores'.

[23] *Google Search (Shopping)* (case AT.39740) Commission Decision, para 159, where the EC acknowledges that "[g]eneral search services and online search advertising constitute the two sides of a general search engine platform" (para 159), but then focuses only on the market for general search services (in which the abuse took place).

[24] BCA, Decision ABC-2016-I/O-31-AUD, *Immoweb* (7 November 2016), paras 9–21.

[25] The BCA relied on Bundeskartellamt, Decision B6-39/15, *Immowelt/Immonet* (20 April 2015). The investigative service also recognized the advertising functionality of online real estate portals (and tested its substitutability with other means of advertising), but found the matching function predominant.

[26] See recently BCA, Decision BMA-2019-C/C-16, *Telenet Group BVBA/De Vijver Media NV* (13 May 2019). The BCA also defined other (e.g. upstream) markets, but those are less important for our purposes.

social networks or horizontal search services. The mistaken idea that there cannot be a market for non-priced services has been properly dispelled across the EU.[27]

However, defining non-priced markets comes with methodological difficulties. Obviously, a small but significant non-transitory increase in price (SSNIP) test is not workable when the price of the product in question is zero: a 5% increase would not change the price and thus demand. Yet while suppliers of zero-priced services do not compete on price, they do compete on other dimensions, such as quality. Based on this idea, competition authorities have relied on the alternative small but significant non-transitory decrease in quality (SSNDQ) test, which seeks to measure shifts in customer demand in response to a quality degradation rather than a price increase.[28] Recognizing that precise quantitative measurements are all the more difficult in case of quality degradations, however, authorities have mostly relied on more qualitative evidence of substitutability, such as consumer surveys, internal business documents and views from the industry.[29] Going forward, the BCA's president recognizes that we may need 'a different kind of economics' and, in particular, more behavioural economics, which may be useful to define non-price markets.[30] Given such challenges, the BCA has also called for updated guidelines.[31]

Finally, the BCA is well aware of the indicators of market power in digital markets, although it did not have many opportunities to apply that knowledge in practice yet. The *traditional* indicators of market power include market share and barriers to entry, such as economies of scale. When it comes to market share in digital markets, the BCA has estimated market share based on the number of daily visitors to the platform rather than more classic measures, such as sales volume/value.[32] Regarding barriers to entry, the BCA has identified direct and indirect network effects (and the winner-takes-most dynamic they lead to) as well as access to data as particularly important in digital markets.[33] The dynamic, fast-moving

[27]The EC regularly defines markets for non-priced services such as messaging apps, see e.g. *Facebook/WhatsApp* (case M.7217) Commission Decision, paras 20–34. In Germany, legislators left no room for doubt when they amended their Competition Act to state that free services can constitute a market, see Act against Restraints of Competition, §18(2a) ("The assumption of a market shall not be invalidated by the fact that a good or service is provided free of charge.").

[28]See e.g. *Google Android* (case AT.40099) Commission Decision, paras 284–322 and 483–566.

[29]On this difficulty, see J. Crémer, Y.-A. de Montjoye and H. Schweitzer, Competition policy for the digital era (Special Advisers' Report 2019), p. 45 ("it is unclear how this test could be made operational in practice without a precise measurement of quality . . . and without a way to quantify the effects of the quality degradation on the firm's revenues in order to determine whether such a degradation would be profitable").

[30]J. Steenbergen, On competition policy and online markets, European Competition and Regulatory Law Review 2019(1), VII.

[31]Joint Memorandum of the Benelux NCAs, p. 6.

[32]BCA, Decision ABC-2016-I/O-31-AUD, *Immoweb* (7 November 2016), paras 22–27.

[33]See *inter alia* Prioriteitenbeleid Belgische Mededingingsautoriteit 2020; Joint Memorandum of the Benelux NCAs; and BCA, Decision BMA-2019-C/C-16, *Telenet Group BVBA/De Vijver Media NV* (13 May 2019) for an application, in particular regarding the role of data.

nature of digital market power *can* make market power transient, but that process should be assessed on a case-by-case basis—not assumed.

4.3.2 Restrictive Agreements and Abuse of Dominance in Digital Markets

Given the relative paucity of cases originating in digital markets handled by the BCA, it is impossible to give a comprehensive account of how the authority *has* assessed—let alone *will* assess—the various kinds of potentially anti-competitive conduct under the provisions relating to restrictive agreements (Article IV.1 of the Code of Economic Law (CEL)/101 of the Treaty on the Functioning of the European Union (TFEU)) and abuse of dominance (Article IV.2 CEL/102 TFEU). Still, by combining the limited decisional practice with policy documents and past experience, an incomplete yet informative picture emerges. In what follows, I survey the most relevant and interesting developments, including self-preferencing, the role of data (protection), MFNs and regulatory changes that may help tackle the challenges of digital markets.

4.3.2.1 Self-Preferencing by Gatekeepers

One idea that is gaining traction is that certain online platforms, in particular the so-called GAFAM (Google, Amazon, Facebook, Apple, Microsoft), serve as 'gatekeepers' to the online ecosystems they manage.[34] An app developer, for example, has to go through the App Store to reach iPhone users; a business wanting to be found online needs to show up on Google. At the same time, these platforms compete with the businesses on their platform (suppliers), e.g. by providing their own apps or specific search services (verticals). This can lead to a conflict of interest whereby the platform has an incentive to preference its own products over those of suppliers, e.g. by algorithmically ranking them higher than competing offers. Such conduct is at the heart of the EC's *Google Shopping* decision and the ongoing *Amazon Marketplace* investigation, both under Article 102 of the Treaty on the Functioning of the European Union (TFEU).[35]

The BCA has stated that it 'do[es] not consider self-preferencing to be a new theory of harm'.[36] Indeed, self-preferencing is only a specific application of vertical

[34] See Joint Memorandum of the Benelux NCAs, pp. 5–6.

[35] *Google Search (Shopping)* (case AT.39740) Commission Decision and EC, Commission sends Statement of Objections to Amazon for the use of non-public independent seller data and opens second investigation into its e-commerce business practices (10 November 2020) IP/20/2077.

[36] OECD, Abuse of dominance in digital markets—contribution from Belgium (2020) DAF/COMP/GF/WD(2020)2, p. 3, describing how it "imposed e.g. remedies with regard to self-preferencing in respect of data access on a cable network for the operator and third party customers in a merger case" (presumably referring to BCA, Decision BMA-2019-C/C-16, *Telenet Group BVBA/De Vijver Media NV* (13 May 2019), see discussion under next section).

foreclosure that raises rivals' costs.[37] The mechanism does not differ much, for example, from retailers producing private-label goods and distributing them in competition with goods from suppliers, which it may then have an incentive to discriminate against (e.g. through shelf placement). The BCA has studied this exact mechanism in the supermarket sector, noting its potential pro- and anti-competitive effects.[38] One important difference, however, is that—at least in Belgium—supermarkets are unlikely to be dominant, given the existence of at least three national chains and some smaller but significant players.[39] Some of the aforementioned platforms are likely dominant in their respective markets, which may alter the balance between the pro- and anti-competitive effects of self-preferencing.

4.3.2.2 Access to and Use of Data

In addition to self-preferencing (through ranking algorithms and otherwise), access to and use of data are a growing concern. As mentioned in the previous section, the BCA recognized data as a potential source of market power in digital markets. However, they can also be a source of anti-competitive conduct and in particular abuse of market power, both of an exclusionary and an exploitative nature.[40]

Firstly, data can be used in an exclusionary manner. For example, an online marketplace may rely on the data it gathers on its sellers' products and sales to then produce and sell competing products. This is a form of self-preferencing, which is part of the aforementioned Amazon investigation.[41] Once more, an analogy can be made with supermarkets, which also have access to large amounts of data on their retailers, which can be used to out-compete them.[42]

In *The Great Circle* (2019), the BCA had to rule on a case concerning access to data.[43] The Great Circle sought access to raw meteorological data from the Royal Meteorological Institute, which it needed to operate its maritime navigation software SQUID, but was refused. That refusal to supply was allegedly discriminatory, given that the Institute did furnish the Great Circle's competitors with such data. However, the BCA rejected the request for interim measures, given that the Great Circle could

[37] See T. Krattenmaker and S. Salop, Anticompetitive exclusion: raising rivals' costs to achieve power over price, The Yale Law Journal 1986(2), pp. 209–293.

[38] BCA, Prijsniveau in supermarkten (Study 2012), pp. 88–91.

[39] Moreover, the online interface offered by online platforms may offer more effective and certainly subtler opportunities for exclusion than those of supermarkets.

[40] Great reliance on data and the algorithmic processing thereof can also have important consequences for the assessment of restrictive agreements. For example, it may make markets more transparent, which can increase the risk of collusion. However, given the lack of cases dealt with by the BCA in this area, the focus is on abuse of dominance.

[41] EC, Commission sends Statement of Objections to Amazon for the use of non-public independent seller data and opens second investigation into its e-commerce business practices (10 November 2020) IP/20/2077.

[42] BCA, Prijsniveau in supermarkten (Study 2012), pp. 88–91.

[43] BCA, Decision ABC-2019-V/M-10, *SPRL The Great Circle/l'Institut Royal Météorologique* (15 February 2019).

not establish a *prima facie* case. In particular, it was unclear whether the Institute had a dominant position on the relevant market for the commercial supply of digital meteorological information. The data were not perceived to be *unique*—a measure that other NCAs also rely on to decide questions of data access.[44]

In the earlier *Nationale Loterij* case (2015), the issue was not access to but rather the use of data.[45] The National Lottery, which held a statutory monopoly for public lotteries, increasingly had to compete against undertakings in the related market for sports betting. In doing so, however, it relied on the advantages of its statutory monopoly. In particular, it used consumer data gathered through its monopoly activity (including e-mail addresses) to promote its competitive activity. In addition, it requested data on competitors from the distributors of its lottery tickets. The BCA's investigative service found that the cross-use of data may have distorted competition in the sports betting market and thus constituted an abuse of dominant position.

Secondly, data-related practices can be exploitative. Of primary importance here is the *Facebook* decision by the Bundeskartellamt, in which it found that Facebook abused its dominant position by combining the data gathered on users via its main social network with data collected via its other properties (Instagram and WhatsApp) and other websites (*inter alia*, those that integrate its 'like' or 'share' application programming interfaces).[46] However, the BCA appears unlikely to follow its example. In the first place, exploitative cases are significantly more difficult to bring than exclusionary cases, even more so when they relate to data rather than price.[47] The recent amendment of Belgian competition law (see Sect. 4.3.2.4 below) may, however, change this. In the second place, data exploitation cases raise questions regarding the interaction of competition and data protection law. The EC has held that 'privacy-related concerns flowing from the increased concentration of data . . . do not fall within the scope of the EU competition law rules but within the scope of the EU data protection rules'.[48] While the Bundeskartellamt views things differently, the BCA appears more likely to follow the EC's line.

Of course, that does not mean the BCA is oblivious to questions of data protection. The primary example is its *Telenet/De Vijver Media* decision, in which the BCA approved—with commitments—a merger between a content provider and a distributor. One concern was that Telenet would use the viewing data it gathered as a

[44] See in particular Autorité de la concurrence and Bundeskartellamt, Competition law and data (Joint Report 2016), pp. 18 (data can meet the conditions of the refusal to supply doctrine if it is "truly unique and . . . there is no possibility for the competitor to obtain the data that it needs to perform its services.").

[45] BCA, Decision BMA-2015-P/K-27-AUD, *Stanleybet Belgium NV et al./Nationale Loterij NV* (22 September 2015).

[46] Bundeskartellamt, Decision B6-22/16, *Facebook* (6 February 2019). Note that the decision is still under appeal.

[47] For a discussion that includes the *Facebook* decision, see Friso Bostoen, Online platforms and pricing: adapting abuse of dominance assessments to the economic reality of free products, Computer Law & Security Review 2019, pp. 273–80.

[48] *Facebook/WhatsApp* (case M.7217) Commission Decision, para 164.

content distributor to improve its offering as a content provider, to the detriment of competing providers. Telenet relied on the General Data Protection Regulation (GDPR) to argue against a potential commitment to share viewing data with competing content providers.[49] Interestingly, the BCA considered this not just an issue of data access but also one of self-preferencing, depending on whether the data were gathered via Telenet's own channels or whether they were gathered via other channels in its role as distributor.[50] To address both concerns, commitments were imposed under which Telenet was forced to give other content providers access to data *and* was limited in the way it could use data gathered via the channels it distributed.[51] Furthermore, the BCA has cooperated closely with the Belgian telecommunications regulator in the past[52] and could extend such cooperation to the Belgian data protection authority.

4.3.2.3 Most Favoured Nation Clauses

The most prominent efforts of the BCA in digital markets relate to MFNs implemented by online platforms.[53] In 2016, the BCA participated in a monitoring exercise of MFNs in the hotel sector and also dealt with a case on MFNs in the real estate sector.

Together with the EC and nine other NCAs, the BCA was involved in a monitoring exercise in the hotel sector,[54] the results of which were published in 2017.[55] NCAs had been investigating MFNs agreed between hoteliers and online platforms since 2010. The general approach was to prohibit *wide* MFNs, which oblige hotels to offer their rooms on platforms such as Booking.com at the lowest prices (and best availability) relative to *all* other sales channels. By contrast, *narrow* MFNs, which allow hotels to offer lower room prices on other platforms and offline but prevent them from publishing lower prices on their own websites, were generally

[49] BCA, Decision BMA-2019-C/C-16, *Telenet Group BVBA/De Vijver Media NV* (13 May 2019), pp. 207–210. It argued viewing data constitutes personal data, and that sharing such data would infringe various GDPR provisions.

[50] Ibid, pp. 228-229, referencing CFI, case T-201/04, *Microsoft v Commission*, EU:T:2007:289 and *Google Search (Shopping)* (case AT.39740) Commission Decision.

[51] BCA, case MEDE-C/C-19/0006, *Liberty Global / De Vijver Media* – Voorstel Verbintenissen (8 mei 2019), Section F.

[52] Based on Koninklijk besluit betreffende de samenwerking tussen het Belgisch Instituut voor Postdiensten en Telecommunicatie enerzijds, en de Belgische Mededingingsautoriteit anderzijds, Official Gazette 2014, N. 198, p. 53403.

[53] On this topic, see F. Bostoen, Most favoured nation clauses: towards an assessment framework under EU competition law, European Competition and Regulatory Law Review 2017(3), pp. 223–236.

[54] BCA, Resultaten van de monitoring in de hotelsector (6 April 2017), available at https://www.belgiancompetition.be/en/node/5779 last accessed 10 February 2021.

[55] EC, Monitoring exercise carried out in the online hotel booking sector by EU competition authorities (Report 2017), available at https://ec.europa.eu/competition/ecn/hotel_monitoring_report_en.pdf last accessed 10 February 2021.

allowed.[56] The heads of the European Competition Network (ECN)—who commissioned the report—held that the enforcement efforts were 'go[ing] in the right direction' but decided to keep the online hotel booking sector under review.[57]

Also in 2016, the BCA's investigative service intervened in the real estate sector.[58] Immoweb, the main real estate platform, included MFN clauses in its contracts with software developers, who in turn intermediated between the Immoweb and real estate agencies by listing their offers. According to these MFNs, the developers had to ensure that the financial conditions of their offers on Immoweb were equally or more favourable than those accorded to other partners. These clauses restricted competition between real estate platforms: given that real estate offers could not be more favourable than those posted on Immoweb, competing platforms had no incentive to implement lower commission fees (as those would not affect the final price). The BCA suggested that the MFNs infringed Article 101 TFEU and—given Immoweb's apparent dominance—also Article 102 TFEU (and Belgian equivalents). While a preliminary assessment thus showed anti-competitive/illegal conduct, the case was closed when Immoweb committed to withdraw the MFNs and not reintroduce them for a period of 5 years.

In 2018, the Belgian regulatory landscape regarding MFNs shifted again. This time, not the BCA but the Belgian parliament intervened, adopting a law regarding 'the freedom of tariffs by hoteliers in contracts concluded with platform operators for online reservation'.[59] As its name implies, this law prohibits platform operators from restricting in any way—e.g. through MFNs—the price-setting freedom of hotel owners (when the property is located in Belgium).[60] However, parliament went further than most NCAs by prohibiting not only *wide* but also *narrow* MFNs, albeit only in *one* sector.

This interaction between competition law enforcement and *ex ante* regulation is also increasingly observed at the EU level. The recent EC proposal for a Digital Markets Act (DMA) is the most prominent example of this dynamic.[61] This proposal, too, includes a provision on MFNs. In particular, it obliges certain platforms (gatekeepers) to 'allow business users to offer the same products or services to end users through third party online intermediation services at prices or conditions that

[56] Among others, narrow MFNs prevent free-riding by hotels. The Bundeskartellamt is an exception, as it also prohibited *narrow* MFNs in the case of Booking.com. It justified this diverging position by pointing to the specific nature of the German market and the failure of Booking.com to substantiate its efficiency claims, see OECD, Hearing on across platform parity agreements – note by Germany (2015) DAF/COMP/WD(2015)56, pp. 9–10.

[57] EC, Outcome of the meeting of ECN DGs (17 February 2017), available at http://ec.europa.eu/competition/antitrust/ECN_meeting_outcome_17022017.pdf last accessed 10 February 2021.

[58] BCA, Decision ABC-2016-I/O-31-AUD, *Immoweb* (7 November 2016), paras 9–21.

[59] Wet betreffende de tariefvrijheid van exploitanten van toeristische logies in de contracten afgesloten met platform-operators voor online reservatie, Official Gazette 2018, N. 188, p. 62710.

[60] Ibid, arts 3–6.

[61] EC, Proposal for a Regulation of the European Parliament and of the Council on contestable and fair markets in the digital sector (Digital Markets Act) COM(2020)842.

are different from those offered through the online intermediation services of the gatekeeper'.[62] Given its reference to 'third party online intermediation services', this ban would be limited to *narrow* MFNs.

4.3.2.4 Abuse of Dominance Without Dominance—and Even Without Abuse

A growing number of policymakers and scholars are concerned that the current provisions on abuse of dominance are insufficient to remedy harms in the digital economy. Some rely on institutional reasons, zooming in on the speed of intervention, and argue for the increased use of interim measures, among others.[63] Unlike the EC, it must be noted that the BCA has repeatedly used its powers to impose interim measures.[64] Others call for substantive reform, arguing that the requirements of (proving) market power and abuse lead to under-enforcement. In Belgium, policy has been shifting on both fronts.

Firstly and most importantly, in 2020, the Belgian law on abuse of economic dependence entered into force.[65] This law is stricter than the EU's abuse of dominance provision, as allowed under Article 3(2) of Regulation 1/2003.[66] In particular, the law does not require *absolute* market power but only 'economic dependence', which is—or should be—easier to prove. Economic dependence is a form of *relative* market power, defined as:

> A position of subjugation of one undertaking to another—characterized by the absence of a reasonable equivalent alternative, available within a reasonable period, and under reasonable conditions and costs—that allow the latter undertaking to impose conditions that it could not have obtained under normal market circumstances.[67]

Undertakings are prohibited from abusing such positions of economic dependence when they affect competition on (a substantial part of) the relevant Belgian

[62] Ibid, art 5(b). The DMA proposal includes a number of other obligations/prohibitions, including on app stores. Unlike the MFN prohibition, those are not self-executing but are "susceptible of being further specified" (art 6).

[63] For the Belgian provisions on interim measures, see Code of Economic Law, arts IV.26, §3, 12° and IV.71–73.

[64] See T. De Meese and E. Wijckmans, Voorlopige maatregelen door de Belgische Mededingingsautoriteit: stand van zaken na 5 jaar *prima facie*-vaststellingen onder Boek IV van het Wetboek economisch recht, Tijdschrift voor Belgisch Handelsrecht 2018(8), pp. 799–817.

[65] Wet houdende wijziging van het Wetboek van Economisch Recht met betrekking tot misbruiken van economische afhankelijkheid, onrechtmatige bedingen en oneerlijke marktpraktijken tussen ondernemingen ['Law amending the Code of Economic Law with regard to abuses of economic dependence, abusive clauses and unfair commercial practices between undertakings'], Official Gazette 2019, N. 115, p. 50066; Koninklijk besluit tot wijziging van de boeken I en IV van het Wetboek van economisch recht met betrekking tot misbruiken van economische afhankelijkheid, Official Gazette 2020, N. 228, p. 59654.

[66] Council Regulation 1/2003 of 16 December 2002 on the implementation of the rules on competition laid down in Articles 81 and 82 of the Treaty, OJ 2003 L 1, p. 1.

[67] Code of Economic Law, art I.6, 12bis.

market.[68] Given its recent entry into force, the new provision remains largely untested. However, a similar French provision—which served as inspiration for the Belgian one—has been used effectively in the digital economy.[69] Based on the said provision, a French court ordered Amazon to modify its clauses with third-party sellers on its Marketplace.[70] Even though the Belgian provision was not adopted with the digital economy in mind, it could thus play a role there.

Once you have done away with the (absolute) market power requirement, why stop there? Why not do away with the abuse requirement as well? A proposal in that direction has indeed been launched, although it is obviously more complex than these tongue-in-cheek questions suggest. In the aforementioned Joint Memorandum, the Benelux NCAs support '[t]he introduction of an ex ante instrument providing for imposed remedies *without the establishment of an infringement*'.[71] This proposal was put forward by the Dutch Secretary of State for Economic Affairs[72] and previously endorsed by the ACM.[73]

The driving concern behind the initiative is—once more—the slow nature of *ex post* enforcement, which becomes particularly problematic in fast-moving digital markets. The envisaged instrument would allow NCAs to impose non-punitive, behavioural remedies, such as platform access, data portability, data sharing and non-discriminatory ranking.[74] The instrument would closely follow the interpretation of dominance and abuse in the context of Article 102 TFEU, but NCAs would not actually have to find an infringement. Rather, the mechanism would be one akin to 'the powers that the CMA has to impose remedies following market studies and the powers of the Member States' telecom authorities to impose remedies on companies with significant market power'.[75]

In 2020, the EC launched a proposal for a 'new competition tool' that aligns with the requests of the Benelux NCAs. Under that tool, however, the EC would be the

[68] Ibid, art IV.2/1. See further J. Blockx, Belgian prohibition of abuse of economic dependence enters into force, Journal of European Competition Law & Practice 2021 (advance access).

[69] Code de Commerce, art L420–2.

[70] Tribunal de commerce de Paris, case 2017050625, *Ministre de L'Economie et des Finances/ Amazon* (2 September 2019). For a discussion, see Friso Bostoen, Abuse of relative dominance in the platform economy: a French court finds Amazon's contracts with third-party sellers significantly imbalanced, CoRe Blog (12 November 2019), available at https://www.lexxion.eu/en/coreblogpost/amazon-case-france/ last accessed 10 February 2021.

[71] Joint Memorandum of the Benelux NCAs, p. 5 (own emphasis).

[72] Secretary of State for Economic Affairs and Climate, Letter to the President of the Second Chamber (17 May 2019), available at https://www.tweedekamer.nl/kamerstukken/brieven_regering/detail?id=2019Z09936&did=2019D20357.

[73] ACM, Extension of enforcement toolkit to increase effectiveness in dealing with competition problems in the digital economy (6 August 2019), available at https://www.acm.nl/sites/default/files/documents/ex-ante-tool_0.pdf.

[74] Joint Memorandum of the Benelux NCAs, p. 5–6 ('Rather than broad-stroked regulation, these remedies should be proportionate and tailored to specific situations.').

[75] Ibid, p. 5.

main enforcer, with more of a cooperative role reserved for NCAs.[76] In the end, that initiative was abandoned in favour of the aforementioned DMA, which gives the EC an even more central role.[77] Given the current focus on the DMA, a regulatory initiative in line with the Benelux NCAs' demands appears unlikely at this point, unless their respective national policymakers come up with an equivalent.

4.4 Conclusion

In conclusion, the track record of the BCA in digital markets is fairly limited, at least if one only examines its case law. When one considers the authority's broader contributions to competition policy, a richer picture emerges. That picture shows a modern, cooperative NCA with in-depth knowledge of digital competition but—it must be acknowledged—fairly few opportunities to put that knowledge into practice. One can make educated guesses about the BCA's apparent reticence to intervene in digital markets. Resources are certainly a part of the equation, together with a—in my view, correct—belief that some digital issues are best dealt with at the EU level.

However, its new enforcement priorities, which include the digital economy, may preclude a more active role. This should be welcomed, given that we know from experience that NCAs can have a meaningful impact on competition in the digital sphere. This is illustrated by the Bundeskartellamt's *Amazon* case, in response to which Amazon amended its terms for Marketplace sellers *globally*.[78] Moreover, the work of NCAs, when limited to its national territory, can serve as a regulatory experiment that—when successful—can be replicated or extended by other NCAs or even the EC. Building on acquired knowledge, the BCA can certainly play a role here, be it through more informal guidance or more formal interventions.

[76] See EC, New Competition Tool (Inception Impact Assessment) Ares(2020)2877634.

[77] EC, Proposal for a Regulation of the European Parliament and of the Council on contestable and fair markets in the digital sector (Digital Markets Act) COM(2020)842.

[78] Bundeskartellamt, case B2-88/18, *Amazon* (17 July 2019).

France

5

Lauren Mechri, Nizar Lajnef, Jean-Julien Lemonnier,
Liliana Eskenazi, Daniel Green, and Laurence Charvoz

5.1 Tools and Types of Regulatory Intervention to Address Market Power in Digital Markets

5.1.1 The General Competence of the Competition Authority

The French Competition Authority (FCA) has a general competence to ensure that economic operators comply with competition rules, regardless of the economic sector (i.e. the rules apply to all production, distribution and service activities).[1]

[1] See Article L.410-1 of the French Commercial Code.

L. Mechri
Fieldfisher, Paris, France
e-mail: lauren.mechri@fieldfisher.com

N. Lajnef (✉)
UGGC Avocats, Paris, France
e-mail: n.lajnef@uggc.com

J.-J. Lemonnier
Stephenson Harwood LLP, Paris, France
e-mail: jj.lemonnier@shlegal.com

L. Eskenazi
Fréget Glaser and Associés AARPI, Paris, France
e-mail: liliana.eskenazi@freget-glaser.fr

D. Green
Linklaters LLP, Paris, France
e-mail: daniel.green@linklaters.com

L. Charvoz
Avocat, Angers, Paris, France
e-mail: cabinet@avocat-charvoz.eu

B. Kilpatrick et al. (eds.), *Antitrust in Data Driven Markets & Legal Framework for Influencers, Native Advertising and Control over the Use of AI in Marketing*, LIDC Contributions on Antitrust Law, Intellectual Property and Unfair Competition,
https://doi.org/10.1007/978-3-031-07422-6_5

In regulated sectors, the FCA and sectorial regulators have concurrent powers to promote competition for the benefit of businesses and consumers. While the FCA's action is in principle limited to *ex post* intervention on the market, it increasingly relies on its consultative powers to undertake a "review of the competitive situation" of the relevant market and issues "opinions" that can eventually identify failures or even companies and practices that are likely to be found anti-competitive.[2,3]

The FCA has recently enhanced its capacity to better address competition law issues specific to digital markets. A department dedicated to the digital economy was created within the FCA in January 2020, which focuses on new issues linked to the development of digital technology, such as the collection and exploitation of data by online platforms.[4] The FCA intends to use information technology (IT) tools, such as *web scraping* or searches on the The Onion Router (TOR) network,[5] to gather information on data or algorithms on the Internet to detect potential infringements of competition law.

There are close links between the French electronic communications regulator (Autorité de Régulation des Communications Électroniques et des Postes (ARCEP)) and FCA. Indeed, ARCEP may refer a case to the FCA when it suspects that one or more operators are infringing competition rules[6] or are under specific regulatory circumstances;[7] on the other hand, the FCA is obliged to request an opinion from ARCEP on any matter falling within the latter's jurisdiction.[8]

In addition, ARCEP's contribution to the regulation of "structuring digital platforms"—which is based on the main principles of regulation in the electronic

[2] See Article L.462-4 of the French Commercial Code: the FCA can issue an opinion *ex officio* on any questions relating to competition.

[3] The FCA's conclusions can encourage market players to file complaints before the FCA against companies identified in the opinions. For instance, following the publication of opinion No. 18-A-03 on the online advertising sector, Criteo announced that it had filed a complaint against Facebook for practices relating to the exploitation of user data. Available at: https://www.criteo.com/news/press-releases/2019/10/criteo-files-fca-complaint-against-facebook/. Accessed 28 September 2021.

[4] Press release of the French Competition Authority of 9 January 2020, available at https://www.autoritedelaconcurrence.fr/fr/communiques-de-presse/lautorite-cree-un-service-de-leconomie-numerique. Accessed 28 September 2021.

[5] The TOR (short for The Onion Router) network is a secure and encrypted network that ensures the privacy of data and communications on the web.

[6] See Article L.36-10 of the French Postal and Electronic Communications Code.

[7] See Article L.37-1 of the French Postal and Electronic Communications Code: ARCEP consults the FCA when it wishes to impose specific ex ante obligations on operators that exercise a "significant influence" on the relevant market(s). For example, in decision No. 2017-1568 of 21 December 2017, ARCEP first consulted the FCA, which issued Opinion No. 17-A-14 of 25 October 2017.

[8] See Article R.463-9 of the French Commercial Code: ARCEP will be granted two months to issue an opinion. For example, in decision No. 20-D-02 of 23 January 2020 relating to practices implemented by Orange in the electronic communications sector, the FCA first consulted ARCEP which issued an Opinion No. 2019-1385 of 30 September 2019.

communications sector—shows the significant role played by sectoral regulators in the competitive functioning of markets.

5.1.2 Sectoral Regulators and Public Authorities Can Also Intervene in the Field of Digital Platforms

Sectoral regulators (e.g. media and telecoms) generally intervene on an *ex ante basis*, although they may have *ex post* intervention powers (e.g. through dispute settlement procedures) in sectors where specific regulation (often derived from general competition law) applies.

The following examples illustrate the role of sectoral regulators and public authorities in relation to networks and platforms:

- ARCEP is responsible for regulating the electronic communications, postal and press distribution sectors. It has broad powers to ensure a high degree of competition in the market. It defines the regulations that apply to all or some of the operators, enacts "soft law" such as guidelines or recommendations and investigates and eventually imposes sanctions in case of violations of sectoral regulations. Its responsibilities include protecting net neutrality.
- The DGCCRF (which is, among other things, the investigative arm of the Ministry of Economy and Finance) pursues broader objectives than the application of competition law, such as general interest and consumer protection. For example, following an investigation by the DGCCRF, Apple was fined EUR 25 million for an update of its operating system that limited the performance of aging devices. The company also had to display a statement on its website for a month.[9]
- The Minister of Economy and Finance also plays an important role: it acts in the general interest and has the power to summon a company before the Commercial Court of Paris for cases of unfair commercial practices. For example, the Minister filed a suit against Apple and Google in March 2018 on the ground that several provisions of their app store template contracts were likely to cause significant harm to both developers and the economy in general.[10]

[9]DGCCRF, press release of 7 February 2020, available at: https://www.economie.gouv.fr/files/files/directions_services/dgccrf/presse/communique/2020/CP-Ralentissement-fonctionnement-iPhone200207.pdf. Accessed 28 September 2021.

[10]Ministry of Economy, press release of 14 March 2018, available at: https://www.economie.gouv.fr/files/files/directions_services/dgccrf/presse/communique/2018/cp-google-apple.pdf. Accessed 28 September 2021.

Since August 2020, the FCA has also benefited from the expertise and resources of the new Digital Regulation Expertise Unit (PEReN),[11] an administrative unit attached to the Ministry of Economy, whose mission is to support the government departments involved in the regulation of digital platforms, in particular by providing them with technical assistance in data analysis. The FCA was the driving force behind the creation of this new service.

Finally, on 31 January 2019, the DGCCRF and CNIL (Commission Nationale de l'Informatique et des Libertés) signed a cooperation protocol whose objective is to strengthen the collaboration of these two entities through the possibility of conducting joint controls and to adapt to the new digital challenges, according to an analysis conducted jointly in the field of consumer law and the General Data Protection Regulation (GDPR).

The work of the FCA, sectoral regulators and public authorities is therefore multifaceted, and they are quite well equipped to handle the issues of market power of platforms in France from the different angles of their respective fields of competence.

5.2 Competence of Antitrust Authorities in the Field of Personal Data Protection in the Case of Social Networks and Other Digital Platforms

In France, the FCA does not have any specific competences in the field of personal data protection. The independent administrative authority in charge of enforcing compliance with data protection regulations is the Commission Nationale Informatique et Libertés (CNIL).

The CNIL is therefore at the forefront of data-related issues. However, this has not prevented the FCA from being active and from developing a rich decisional practice with regard to the collection, use and sharing of data.

The CNIL has been very active in recent years in sanctioning unlawful behaviour by Google and Amazon for non-compliance with cookie-related legislation after finding that the search engine Google Search and the website Amazon.fr were storing cookies on the terminals of users residing in France even before any action was taken by them without users being properly informed and being able to consent to the deposit of cookies.[12] The CNIL also sanctioned Facebook in April 2017 following Facebook's refusal to provide users of its website with information considered essential for the processing of their personal data, such as the identity

[11] Decree No. 2020-1102 of 31 August 2020 creating a service with national competence called "Pôle d'expertise de la régulation numérique" (PEReN), Official Journal of the French Republic No. 214 of 2 September 2020.

[12] CNIL, Deliberation No. SAN-2020-012 of 7 December 2020 concerning the companies GOOGLE LLC and GOOGLE IRELAND LIMITED; CNIL, Deliberation No. SAN-2020-013 of 7 December 2020 concerning the company AMAZON EUROPE CORE.

of the data controller, the purposes of the processing or information on data transfers outside the European Union.[13]

Although issues *strictly* relating to the protection of personal data do not fall within its jurisdiction, the FCA has looked into the collection and use of data by digital players, as well as into the competition issues raised by algorithms. The FCA published, in 2016, a joint study with the Bundeskartellamt on data and the challenges connected with them in the enforcement of competition law in order to better understand the issues raised by the increased collection, processing and commercial use of data in digital markets.[14] The FCA is aware of the critical importance of data for online platforms, particularly in the context of its merger control analysis. In February 2018, the FCA examined for the first time a merger between two online platforms: the Axel Springer group (which owns the real estate portal seloger.com) and the Concept Multimédia company (which owns the real estate portal logic-immo.fr). The FCA then took a close look at the data access issue and, in particular, the risk of foreclosure of competitors in relation to the acquirer's possession of significant volumes of data.[15]

It is therefore worth considering the allocation of competences between the CNIL, the FCA and other authorities in the area of data: while the debate seems to focus on personal data, what about other types of data? Can we consider that the CNIL's natural competence is limited to personal data and that "other" types of data could therefore call for a specific intervention and/or regulation insofar as they are likely to raise their own competition concerns? Such a debate leads to the question of which data fall into the category of "personal data" and which do not (for example individual energy consumption, data produced by connected cars).

Competition law and data protection issues must therefore be resolved separately, although it cannot be assumed that there is no link between the two. This is also in line with the desired cooperation between independent administrative authorities in France. In this regard, the FCA specified in its opinion No. 18-A-03 that "varying compliance with rules on personal data protection could strengthen or weaken the positions of stakeholders",[16] i.e. that data protection rules have an impact on competition, which is likely to be taken into account in the FCA's analysis. The FCA could, for example, rely on the findings of the CNIL in terms of personal data protection to assess the practices of an operator from the point of view of competition rules.

[13] CNIL, Deliberation No. SAN-2017-006 of 27 April 2017.

[14] FCA and Bundeskartellamt, Competition Law and Data, Joint Study, 10 May 2016. Available at https://www.bundeskartellamt.de/SharedDocs/Publikation/DE/Berichte/Big%20Data%20Papier.pdf?__blob=publicationFile&v=2. Accessed 28 September 2021.

[15] Decision No. 18-DCC-18 of 1 February 2018 on the acquisition of sole control of Concept Multimedia by the Axel Springer group, paras. 470-477.

[16] Opinion No. 18-A-03 of 6 March 2018 on data processing in the online advertising sector, page 112.

5.3 Determining "Market Power" in Digital Markets

5.3.1 The Historical Absence of Guidelines to Assess the Market Power of Platforms

The market power of platforms is not a new topic in the antitrust sphere. Online businesses, because of their peculiarities, have been the subject of heated academic debates in the past 20 years, including in France.[17]

Unlike the Bundeskartellamt, which published a working paper in June 2016 dedicated to the market power of platforms and networks,[18] the FCA has developed its understanding of this issue through decisions and opinions concerning digital players and issued as part of its consultative and litigation prerogatives, as well as in merger control.[19]

For example, in the decision accepting the commitments offered by Booking.com in the online hotel reservation sector, the FCA relied on several factors usually mentioned by the economic literature. In assessing the market power of the booking platform, it considered (i) market shares, (ii) the multi-sidedness of the relevant market, (iii) the existence of strong network effects (and the risk of a "snowball effect") and (iv) the existence of economies of scale.[20]

In its recent interim measures decision in the *Google/Publishers* case, the FCA also took into account the experience effects and Google's superior capacity to generate revenue and invest in comparison with that of most of its competitors.[21]

In fact, over the last 12 months, several French regulators and public authorities published contributions on networks and platforms in which they discuss, in

[17] For example, Rochet J. and J. Tirole (2003), Platform competition in two-sided markets; Rochet J. and J. Tirole (2006), Two-Sided Markets: a Progress Report, Rand Journal of Economics 37(3), pp. 645-667; Wauthy X. (2008), Competition and regulation in platform markets: an introduction, Reflets et perspectives de la vie économique 2008/1 (Tome XLVII), pp. 39-54; Maxwell W. and T. Pénard T, Quelle régulation pour les plateformes numériques en Europe ?, Annales des Mines - Réalités industrielles 2016/3 (August 2016), p. 42-46.

[18] Bundeskartellamt, The Market Power of Platforms and Networks, June 2016.

[19] See, for example, Opinions No. 10-A-29 of 14 December 2020 and No. 18-A-03 of 6 March 2018 on online advertising, the joint study by the Authority and the Bundeskartellamt on data and its implications for the application of competition law published on 10 May 2016, or Decision No. 18-DCC-18 of 1 February 2018 on the acquisition of sole control of Concept Multimédia by the Axel Springer group.

[20] Decision No. 15-D-06 of 21 April 2015 regarding practices implemented by Booking.com B.V., Booking.com France SAS and Booking.com Customer Service France SAS in the online hotel booking sector, paras. 112 and 113.

[21] Decision No. 20-MC-01 of 9 April 2020 on requests for interim measures by the Syndicat des éditeurs de la presse magazine, the Alliance de la presse d'information générale and others and Agence France-Presse, para. 171.

particular, the notion of "structuring digital platforms".[22] In France, like in the European Union and beyond, the debate has indeed shifted to the notion of "structuring platforms" to capture new expressions of the market power of digital platforms. Market power becomes a *trigger* for intervention by public authorities instead of a *legal standard* to demonstrate an abuse of dominant position.

5.3.2 Market Power as a Criterion to Define a "Structuring Digital Platform"

The FCA and ARCEP have made proposals relying on several criteria and/or factors to identify the "structuring platforms" and the practices that raise competition concerns and are specific to these players. These proposals are similar to the criteria set out in Article 3 of the draft Regulation proposed by the European Commission on 15 December 2020 on contestable and fair markets in the digital sector (which is part of the "Digital Markets Act") for the identification of "gatekeepers".

5.3.2.1 Contribution of the FCA[23]

The Authority proposes a general definition of "structuring platforms", which may be supplemented by guidelines on the three steps of the definition.

A structuring digital platform could be defined as follows:

i. a company that provides online intermediation services for exchanging, buying or selling goods, content or services, and
ii. which holds structuring market power:
    a. because of its size, financial capacity,[24] user community and/or the data it holds,
    b. enabling it to control access to or significantly affect the functioning of the market(s) in which it operates,
iii. with regard to its competitors, users and/or third party companies that depend on access to the services it offers for their own economic activity.

The FCA introduces the notion of "structuring market power", which refers to the strategic nature of the platforms' behaviour on the dominated market and other markets. This notion goes beyond the criteria usually followed to assess market power (for example, network effects, economies of scale...) by taking into account

[22] See, for example, the contributions published by the French Senate in October 2019, by the Commission des affaires numériques of the Assemblée Nationale in June 2020 or by ARCEP in September 2020.

[23] Contribution of the *Autorité de la concurrence* to the debate on competition policy and digital challenges, published on 21 February 2020.

[24] Webinar with Isabelle de Silva, Nicolas Petit and Alexandre de Steel on 11 June 2020: the FCA President mentioned the financial capacity (i.e. the possibility of using the market to raise resources) and level of profitability of the activity as factors that can distinguish the GAFAs.

the gatekeeper role in the access to some markets and the regulator role on the functioning of other markets. The concepts of "entrenched and durable position" and "core platform service"' set out by the European Commission in the draft Regulation are similar to the criteria set out by the FCA (structuring market power) to identify structuring platforms.

According to the Report submitted by the *Commission des Affaires Economiques* before the *Assemblée Nationale* on 24 June 2020, the criteria for identifying structuring digital platforms communicated by the FCA are the following: "access to the data needed to enter and/or grow on the market, the existence of a dominant position in one or more neighbouring markets, vertical integration and activity in neighbouring markets, financial capacity, access to financial resources, the existence of massive network effects, the existence of multi-sided markets, the degree of data portability and interoperability, the ability of the operator to define the market rules itself or its ability to place the regulator in a strong position of information asymmetry".

The FCA appears to have relied on these criteria in its recent decisions concerning Google and Apple: the FCA considered that market players depended on Google and Apple and drew the consequences of the circumstance that they were unavoidable to access the relevant markets or their functioning.

5.3.2.2 ARCEP Contribution

ARCEP proposes a general definition of "structuring digital platforms", which is partly based on the criteria used by the Commission to identify operators with significant individual power in the electronic communications sector.[25]

It relies on a body of "primary" and "secondary" evidence to determine if an actor falls within the definition of "structuring digital platforms":

i. Main evidence:
   - The platform is unavoidable either because it has "bottleneck power" or because of its downstream positioning, which entails a form of economic dependency on other actors.
   - The platform has a certain size, defined in terms of the number of unique users of the platform.
   - The platform is integrated into an ecosystem controlled by the group to which it belongs, and this ecosystem enables a leverage effect from one sector to another.
ii. Secondary evidence:
   - The platform is a key gateway to a range of digital content ("gatekeeper").
   - It has access to significant volumes of qualitative data.
   - Market shares of the platform.
   - Financial value of the company that owns the platform.

[25] ARCEP, Plateformes numériques structurantes – Eléments de réflexion relatifs à leur caractérisation, December 2019.

The proposed approach is different from a classic dominance assessment in that it proposes a "definition essay with an operational scope". The criteria laid down by ARCEP overlap with those set out by the European Commission in the draft Regulation of 15 December 2020, in particular the "secondary" factors of Article 3, paragraph 6 (for example, the financial value of the company, the number of users, the size, activities and position of the core platform services' providers). ARCEP had responded to the European Commission's consultation on the Digital Markets Act.[26]

ARCEP published another document in September 2020 in which it shared its thoughts on the "Remedies to issues raised by digital structuring platforms".[27]

5.4 The Competition Law Definition of a "Market" for Free Services

The question of the definition of a market for free services, like the nature of two-sided markets, is not a new subject in the antitrust sphere, whether in the French, European or international level. These subjects, which are inextricably linked, have been the subject of an abundant economic and legal literature.[28]

In the same way as the European Commission now accepts a market for free services, for example among users of search engines, the FCA has since 2010 developed a decisional practice that shows that it is possible to define a market even in the case of free provision of goods or services. As regards free print media, the FCA considers that "it is relevant to draw a distinction between free print media with editorial or informative content and free advertising print media".[29] It also considered in a recent case relating to Google that the market for general search services is likely to constitute a relevant market, in particular because "the decision to provide a service without direct financial consideration is the result of a choice by undertakings, and that there is nothing to prevent an undertaking from providing a service without receiving payment from Internet users in return".[30]

[26] ARCEP, Arcep's contribution to the public consultation on the DSA Package and the New Competition Tool—An assymetric ex ante regulation of structuring digital platforms for open digital infrastructures, September 2020.

[27] ARCEP, Remèdes aux problèmes posés par les plateformes numériques structurantes. Available at https://www.arcep.fr/uploads/tx_gspublication/plateformes-numeriques-structurantes-remedes_reflexion_sept2020.pdf. Accessed 28 September 2021.

[28] See for example E. Malavolti and F. Marty, La gratuité peut-elle avoir des effets anticoncurrentiels, OFCE, January 2013.

[29] See for example decision No. 18-DCC-18 of 1 February 2018 relating to the acquisition of sole control of the company Concept Multimedia by the Axel Springer Group, para. 109.

[30] Decision No. 20-MC-01 of 9 April 2020 on requests for interim measures by the Syndicat des éditeurs de la presse magazine, the Alliance de la presse d'information générale and others and Agence France-Presse, paras. 137 and seq.

Finally, in the audiovisual sector, the Authority seemed to accept in its Opinion No. 19-A-04 a distinction between free television and pay television markets,[31] in accordance with the practice of the European Commission, which recognises that "the distribution of pay television and free television constitute distinct product markets".[32] However, it did not make such a distinction in the context of the decision to authorise the Salto joint venture.[33] To date, its decisional practice in the audiovisual sector therefore differs from its decisional practice on online markets, where it considers that the fact that a service is free of charge does not prevent the delimitation of a relevant market within the meaning of competition law.

5.5 Sector Enquiries on Online Advertising by Antitrust Authorities

The FCA has noted the growth and importance of the online advertising market in recent years and has published two opinions on this subject, in 2010[34] and 2018.[35]

In its 2010 opinion, the FCA carried out a general study of the market, at the end of which it concluded that there was a market for online advertising, distinct from the market for offline advertising.[36] In this opinion, the FCA further delineated the online advertising market, distinguishing between *search* advertising which operates via search engines and *display* advertising, *i.e.* ads that appear via blocks, banners and wrappers integrated into the content of a website.[37]

In its 2018 opinion, the FCA focused on the more specific issue of data exploitation in the advertising sector, noting the key value of data for players in this market. It noted the strong overall growth of the online advertising sector while emphasising that this growth was mainly captured by a few stakeholders such as Google and Facebook. In this respect, the FCA highlighted the significant competitive advantages of these two players over their competitors, namely (i) the management

[31] Opinion No. 19-A-04 of 21 February 2019 on a request from the Committee on Cultural Affairs and Education of the French National Assembly (Assemblée Nationale) for an opinion on the audiovisual sector, para. 188.

[32] Decision of the European Commission of 18 July 2007, case M.4504 SFR/Télé 2 France, para. 45.

[33] Decision No. 19-DCC-157 of 12 August 2019 on the creation of a joint-venture by France Télévisions, TF1 and Métropole Télévision. The decision has been appealed (available at: https://www.autoritedelaconcurrence.fr/en/decision-de-controle-des-concentrations/regarding-creation-joint-venture-france-televisions-tf1-and). The referral decision by the European Commission had adopted the free/pay distinction; see the decision of the European Commission of 18 March 2019, case M.9040 France Télévisions/Métropole Télévisions/TF1/JV, para. 27.

[34] Opinion No. 10-A-29 of 14 December 2010 on the competitive functioning of online advertising.

[35] Opinion No. 18-A-03 of 6 March 2018 on data processing in the online advertising sector.

[36] Opinion No. 10-A-29 of 14 December 2010 on the competitive functioning of online advertising, para. 136.

[37] Ibid, para.160.

of very large, diversified volumes of data that are frequently updated and (ii) the presence of these two players on both the publishing and technical intermediation levels.

The FCA therefore concluded that the online advertising market has a fragile competitive equilibrium.[38] It should also be noted that, following the 2018 opinion, several contentious investigations were opened by the FCA in the online advertising sector, several of which are expected to be concluded in 2021.[39]

5.6 Regulators' Treatment of Data Collection: The FCA's Approach to the Use of General Terms and Conditions and Data Collection

5.6.1 The Absence of a "Facebook Case" in France

Unlike in Germany, where the Bundeskartellamt found that Facebook had breached competition rules due to violations of data collection rules (a decision that was subsequently suspended on appeal by the Düsseldorf Court pending a decision on the merits and overturned by the German Federal Court of Justice on 23 June 2020), there is no similar precedent in France regarding digital players.

The Authority clarified in its joint study with the Bundeskartellamt on data and competition (2016) the conditions under which data processing could constitute a competition concern:

- first, privacy policies could be considered from a competition standpoint whenever these policies are liable to affect competition, notably when they are implemented by a dominant undertaking for which data serves as a main input of its products or services; and
- second, a close connection between the dominance of the company, its data collection processes and competition on the relevant markets could justify the consideration of privacy policies and regulations in competition proceedings.[40]

[38] Opinion No. 18-A-03 of 6 March 2018 on data processing in the online advertising sector, see the summary.

[39] Press release of the French Competition Authority on the priorities for 2021, available at https://www.autoritedelaconcurrence.fr/fr/communiques-de-presse/apres-une-activite-tres-soutenue-en-2020-lautorite-de-la-concurrence-annonce. Accessed 28 September 2021.

[40] Bundeskartellamt and French Competition Authority, Joint Study on Competition Law and Data, 10 May 2016, pp. 23-24.

5.6.2 The FCA's Decisional Practice on the Use of Certain Data by Dominant Operators Outside Digital Networks and Platforms

In Opinion No. 10-A-13,[41] the FCA specified the restrictions of competition that could result from the cross-usage of customer data sets, for instance when the operator is active on several distinct or related markets. The FCA takes into account (i) the conditions under which the data set was constituted, (ii) the replicability of the data set by competitors under reasonable conditions and (iii) the likelihood that the use of the data set would result in a significant competitive advantage.

Based on this opinion, the FCA adopted several important decisions, in particular in the energy sector. For instance, in an interim measures decision No. 14-MC-02, the FCA ordered GDF Suez, the former incumbent gas supplier, to disclose parts of its database relating to consumers whose gas tariffs are regulated. The FCA found that (i) GDF Suez used the database of customers on regulated tariffs to offer them deals on gas and electricity in the non-regulated market open to competition, (ii) the advantages gained through this behaviour cannot be replicated under reasonable conditions since no database exists that would allow competitors to precisely locate gas consumers and know their consumption level in order to propose them offers that are better suited to their profile and (iii) there was a high risk of market pre-emption.

5.6.3 The FCA's Decisional Practice on the Terms of Use of Platforms and Networks

Since 2010, the FCA has, in decisions related to sanctions, interim measures and commitments, specified on several occasions the conditions for the legality of the establishment and application of the terms of use of Google's services in the advertising market. It particularly insists on the predictability and non-discriminatory application of the terms of use.

In the *Navx* case,[42] the FCA considered that the AdWords content policy concerning roadside bypass devices was not implemented in a transparent and objective manner and had led to several differences in treatment between advertisers. As such, the content policy was likely to violate the prohibition of abuses of dominant position and led Google to offer commitments to make the operation of its AdWords service more transparent and predictable for users.

More recently, in the *Gibmedia* case,[43] the FCA ultimately sanctioned Google for not having complied with the commitments made in 2010. First, it considered that

[41] Opinion No. 10-A-13 of 14 June 2010 on the cross-usage of customer datasets.

[42] Decision No. 10-MC-01 of 30 June 2010 on the request for interim measures submitted by Navx; Decision No. 10-D-30 of 28 October 2010 on practices implemented in the online advertising sector.

[43] Decision No 15-D-13 of 9 September 2015 on the request for interim measures by the company Gibmedia (rejected); Decision No 19-D-26 of 19 December 2019 on practices implemented in the online search advertising sector.

Google Ads operating rules applicable to advertisers were established and applied in non-objective, non-transparent and discriminatory conditions. Second, the opacity and lack of objectivity of these rules would make it very difficult for advertisers to apply them, while Google would have full discretion to change its interpretation of the rules in a way that is difficult to predict. This allows Google to apply its operating rules in a discriminatory or inconsistent manner.

The FCA (in a decision that is currently under appeal) imposed the following sanctions:

- A EUR 150-million penalty
- A series of injunctions to clarify the terms of use of its service
- The introduction of mandatory annual training for relevant employees; and
- The communication of reports detailing the measures taken by Google and the complaints received from users

It results from the above that the FCA has extensive powers to require the operators concerned to modify the terms of use of their services.

Finally, in the recent No. 21-D-07 of 17 March 2021 decision, a request for interim measures against Apple was filed before the FCA, which led it to assess "the freedom of the dominant undertaking to define access and maintenance rules of the digital platform". The FCA applies its traditional decisional practice to so-called "structuring" platforms", stating that "[t]he principle of commercial freedom also applies to so-called 'structuring' platforms, which have considerable market power in the market in which they are primarily active, but also in neighbouring markets, because of their status as gatekeepers, provided that the rules governing access or maintenance of economic operators to or on these platforms do not have the purpose or effect of restricting competition".[44]

It specifies that the dominant operator is "free to define rules providing additional protection for Internet users compared with what is otherwise required by the regulations" but must not "direct the economic model of the economic operators listed on its platform, limit their freedom of enterprise, and influence the quality and diversity of the offer open to Internet users" by means of anti-competitive practices.[45]

[44]Decision No. 21-D-07 of 17 March 2021 on a request for interim measures filed by Interactive Advertising Bureau France, Mobile Marketing Association France, Union Des Entreprises de Conseil and Achat Media, and Syndicat des Régies Internet associations in the sector of mobile applications advertising on iOS, para. 136.

[45]Decision No. 21-D-07 of 17 March 2021 on a request for interim measures filed by Interactive Advertising Bureau France, Mobile Marketing Association France, Union Des Entreprises de Conseil and Achat Media, and Syndicat des Régies Internet associations in the sector of mobile applications advertising on iOS, paras 137 and 138.

5.7 Parity Clauses, Market Power and the Special Liability of Dominant Firms in the Case of Digital Platforms

The most-favoured-nation clause is a clause stipulating that the supplier undertakes to pass on automatically to a customer any more favourable conditions that another customer would subsequently obtain. As a result, the customer automatically obtains the most favourable tariffs granted by the supplier to other customers, leading to a kind of non-discrimination obligation for the supplier. Most-favoured-customer clauses are known as "parity clauses" in the hotel sector.

The "competitiveness clause", or "competing offer clause" or "English clause", is one that is "requiring the buyer to report any better offer and allowing him only to accept such an offer when the supplier does not match it"[46] or a clause by which a supplier commits towards a customer to match its price to the lowest price charged by any competitor at a given time,[47] which implies that the supplier is informed of the prices charged by its competitors and hence, in practice, by the buyer.

The English clause can be viewed either under competition law, where it is assessed on the basis of its anti-competitive effects (Sect. 5.7.1), or under "restrictive practices" law, where it is prohibited as such (*per se*) (Sect. 5.7.2).

5.7.1 Effect-Based Analysis from the Perspective of Competition Law

The English clause constitutes a vertical agreement between the supplier and the distributor in that it results from a vertical contract between two parties located at a different level of the production and distribution chain. The European Commission considers that it "can be expected to have the same effect as a single branding obligation, especially when the buyer has to reveal who makes the better offer".[48]

It can also be an abuse of dominance when it is imposed unilaterally.

The competitiveness clause may also produce horizontal effects when distributors agree to generalise its use to all their customers. Indeed, if each of the distributors in a market applies this clause, any price reduction implemented by one of them will result in the alignment of competitors on this lowered price, resulting in a neutralisation of the effect of such price reduction.[49]

Competitiveness clauses can also increase market transparency: the application of the clause can give the customer who benefits from it valuable information on the price level supported by its competitors.

---

[46] European Commission, Guidelines on Vertical Restraints, OJ 2010 C 130, p. 1-46, para. 129.

[47] Lamy Droit économique, para. 760.

[48] European Commission, Guidelines on Vertical Restraints, OJ 2010 C 130, p. 1-46, para 129.

[49] Competition Commission, opinion of 25 June 1981 on the competitive situation in the fertiliser production and marketing sector, BOCCRF 12 Dec. 1981.

In an opinion of 2003, the *Conseil de la concurrence*, the predecessor of the FCA, gave its opinion on the offer to contract made by a lead insurer to insurers wishing to join a pool. This offer included a "most favoured insurer clause, under which the most favourable investment conditions obtained by one or more participating co-insurer(s) apply to the whole co-insurance".[50]

The *Conseil* considered that "the fact that an insurer accepts the conditions to enter into a contract offered by the lead insurer constitutes an agreement that falls within the scope of Article L. 420-1 of the French Commercial Code. By extending to all insurers' offers the conditions obtained by only one of them, the clause substitutes the conditions obtained by a company to the assessment that each company in the grouping makes of its costs, its selling price and, more generally, its offer, when applying to join the grouping. In doing so, the clause implements a practice likely to restrict competition on the market by favouring an artificial increase in prices to the detriment of their free determination by market forces."[51]

More recently, the FCA ruled on the lawfulness of parity clauses included in the referencing contracts of online hotel booking centres.

The online booking platforms Booking.com, Hotel Reservation Services (HRS) and Expedia had, from 2012 to 2015, imposed clauses granting the automatic benefit to the said platforms of the advantages in terms of rate, number of nights and conditions of the offer (booking conditions, inclusion or not of breakfast, etc.) granted by the hotels to the competing platforms, as well as on all other distribution channels (online and offline), among which the hotel's own distribution channels (website, telephone, email, at the hotel desk, etc.).

As a result, the offer made by the platforms in question was necessarily the most advantageous for the consumer, even if the latter went directly to the hotel.

The FCA found that Booking.com had a market share above 30% in France and was considered to be dominant on the European market.[52] It also found that the parity clauses were likely to stifle competition between online travel agencies (OTAs): "parity clauses break the link that usually exists between the prices charged by an operator (in this case the amount of commission required by Booking.com's from the hotel) and the level of demand for its services (in this case the number of bookings made on Booking.com), reducing price competition between platforms";[53] "these clauses also entail risks of exclusion of small and new entrant platforms"; "the

[50] *Conseil de la concurrence*, Opinion No. 03-A-19 of 17 November 2003 relating to a request from the Fédération française des courtiers d'assurances et de réassurances concerning the conditions for negotiating co-insurance contracts for industrial risks.

[51] *Conseil de la concurrence*, Opinion No. 03-A-19 of 17 November 2003 relating to a request from the Fédération française des courtiers d'assurances et de réassurances concerning the conditions for negotiating co-insurance contracts for industrial risks.

[52] Decision No. 15-D-06 of 21 April 2015 regarding practices implemented by Booking.com B.V., Booking.com France SAS and Booking.com Customer Service France SAS in the online hotel booking sector, para. 101.

[53] Paragraph 115 of the decision.

parity clause prevents competitors [. . .] from attracting internet users by offering lower commissions".[54]

In order to address the restrictions on competition identified, the FCA, in coordination with the European Commission and the Italian and Swedish competition authorities, made binding the commitments offered by Booking.com to modify the parity clause on prices and to remove any clause imposing parity obligations in terms of room availability or commercial conditions, in relation to not only competing platforms but also hotels' direct offline sales channels and part of their online sales channels.

The FCA disjoined the practices implemented by HRS and Expedia, and, in a separate decision, rejected the referral on the grounds that other national competition authorities had dealt with the same facts and, for the other complaints raised, for lack of evidence.[55]

5.7.2 The *Per Se* Approach Under the "Restrictive Practices" Law

The "most-favoured-customer clause" may be considered from the point of view of the provisions specifically prohibiting it as such on the one hand (Sect. 5.7.2.1), or from the point of view of the provisions on significant imbalance in the rights and obligations of the parties on the other hand (Sect. 5.7.2.2).

5.7.2.1 The Prohibition of Most-Favoured-Customer Clauses as Such

The French Commercial Code includes provisions specifically prohibiting the most-favoured-customer clause in the following terms: "Shall be void any clauses or contracts providing for any person engaged in production, distribution or service activities the possibility of benefiting: (b) Automatically from more favourable conditions granted to competing undertakings by the co-contracting party".[56]

This article therefore prohibits an economic operator from requiring its partner to grant it the same advantages as those of its competitors.

This sanction was introduced by the Law on the Modernisation of Economy in 2008[57] in the former Article L.442-6 II. d. of the French Commercial Code, now Article L.442-3, to accompany the abolition of the ban on discrimination introduced by the same law. The most-favoured-customer clause indeed produces a similar effect by requiring the contracting party to treat its competing customers equally.

This text has also been applied in the online hotel booking sector, firstly before the *Commission d'examen des pratiques commerciales* (CEPC) in 2013.

[54]Paragraph 123 of the decision.

[55]Decision No. 19-D-23 of 10 December 2019 regarding practices implemented in the online hotel booking sector.

[56]Article L. 442-3 of the Commercial Code.

[57]Law No. 2008-776 of 4 August 2008 on the modernisation of Economy, Official Journal of the French Republic No. 181 of 5 August 2008.

Organisations representing the hospitality sector questioned the CEPC about the validity—with respect to Article L.442-6 II d. of the French Commercial Code mentioned above—of clauses by which the hotel undertakes to provide the hotel booking website with rates that are equal to or more advantageous than the rates and conditions available from the hotel (directly or via its website or its call centres), and also with any competitor of the central reservation site (parity of rates and conditions), and that the rooms available for reservation on the website concerned are at least as favourable as those provided to any competitor, this commitment sometimes including the hotel itself (parity of availability).

The CEPC considered the conditions referred to in this provision of the French Commercial Code to be related to price or not, so that all parity clauses, whether they relate to tariffs, availability or other conditions, are null and void as soon as they provide for an automatic alignment with the more favourable conditions granted to competitors.[58]

Secondly, in 2016, the Paris Commercial Court applied this provision to the parity clauses applied by Booking.com, at the initiative of the Minister of Economy.[59]

After recalling that the FCA's decision of 21 April 2015 making the commitments proposed by Booking.com binding did not entail a finding of anti-competitive practice or an acknowledgment by Booking.com of the unlawfulness of its practices, that the legal basis of the action is not the same and that the purpose of the Minister's action is to have past clauses sanctioned and declared void and to impose a civil fine, the court considered that the challenged clauses, insofar as they allow Booking.com to ensure that it automatically obtains more favourable conditions than those applied through other distribution channels and to Booking.com's competitors, infringed the provisions of Article L.442-6 II d of the French Commercial Code. It ordered Booking.com to put an end to these practices and declared the clauses null and void.[60]

The Paris Court of Appeal also ruled on the conditions of application of the former Article L.442-6 II d of the French Commercial Code in the online hotel booking sector, on the initiative of the Minister of Economy.[61]

The action was brought against Expedia, Hotels.com and other online travel agencies operating hotel search engines. The investigation carried out by the DGCCRF revealed that the contracts included price and non-price parity clauses, whereby the subject hotel undertook not only to grant the customer the most favourable rates and other conditions otherwise granted (directly or indirectly) by

[58] Commission d'examen des pratques commerciales (CEPC), Opinion No. 13-10 on the commercial relations of hoteliers with the companies operating the main hotel booking sites.

[59] Paris Commercial Court, 29 November 2016, RG n°2014027403.

[60] Paris Commercial Court, 29 November 2016, RG n°2014027403.

[61] Paris Court of Appeal, 21 June 2017, RG n°15/18784.

the hotel to others but also to grant him rates and conditions at least 25% or EUR 20 lower than the lowest rate.[62]

The Minister requested the nullity of these clauses.

The court found that the companies benefited not only from the lowest rate charged by the hotel on its own website or in direct sales but also from any additional benefits, such as upgrades, free breakfast, more flexible cancellation conditions and loyalty benefits.

The court considered that the clause exceeded the scope of Article L.442-6 II d in that it extended to the most favourable conditions granted by the hotelier itself, which is not covered by Article L.442-6 II d, which only covers the conditions granted to competing companies. The clause is therefore contrary to Article L.442-6 II d, but only insofar as it aims to align it with the best conditions granted to third-party competitors.[63]

5.7.2.2 Prohibition of Most-Favoured-Customer Clauses Creating a Significant Imbalance in the Rights and Obligations of the Parties

The imbalance in the rights and obligations of the parties also makes it possible to sanction the most-favoured-client clauses.

Under Article L. 442-6 I 2°, now L.442-1, of the French Commercial Code, it is provided: "I. – Any producer, trader, or service provider shall be held liable and obliged to remedy the damage caused in the course of business negotiation, the conclusion or the execution of a contract for the following actions: /. . ./ 2) To submit or attempt to submit a trading partner to obligations creating a significant imbalance in the rights and obligations of the parties."

In the context of the nullity action brought by the Minister of the Economy before the Paris Commercial Court, which gave rise to the above-mentioned judgment in 2016, the court assessed the lawfulness of the parity clauses applied by Booking.com with regard to the prohibition of significant imbalance in the rights and obligations of the parties.[64]

The clauses in Booking.com's general terms and conditions of service enabled it to ensure that the hotel owner automatically obtained the more favourable terms and conditions granted by Booking.com, not only to other distribution channels and competitors of Booking.com, which contravene the provisions of the former Article L442-6 II of the French Commercial Code, but also to those implemented by the hotel itself, which the Minister deems constituting a violation of Article L.442-6 I 2 of the French Commercial Code—now Article L.442-1 of the French Commercial Code—which deals with a significant imbalance in the rights and obligations of the parties.

---

[62] Paris Court of Appeal, 21 June 2017, RG n°15/18784.

[63] Paris Court of Appeal, 21 June 2017, RG n°15/18784.

[64] Paris Commercial Court, 29 November 2016, RG n°2014027403.

The court considered that the parity clauses are likely to fall within the scope of Article L442-6 I 2 of the French Commercial Code as the significant imbalance is characterised by a lack of negotiation insofar as the price parity and availability clauses are systematically included in the same terms in all the contracts signed between Booking.com and the hoteliers, at the cost of excessive contractual concessions. In addition, the burden was on Booking.com to prove that other clauses in the contract compensated for the significant imbalance that resulted from the practice, which it did not do in this case.

Despite the fact that the FCA accepted the parity clause applied by Booking.com as part of its commitments, the court found that the effects of the clause "show that the obligations of the parties are significantly imbalanced and that the clause, taking into account the general balance of the contract, is contrary to Article L. 442-6 I 2 of the French Commercial Code".[65]

The court ordered Booking.com to put an end to the practices in question and declared the clauses null and void.

In 2017, the Paris Commercial Court was again called upon to rule on parity clauses, this time on the initiative of a hotel restaurant against the central booking system Relais & Châteaux.

The contract included a price parity clause "whereby the member undertakes to advertise on the chain's website the best available rate for an equivalent length of stay, even if obtained through a direct booking".[66] The hotel restaurant requested that the clause be declared null and void under Article L. 442-6-2 of the French Commercial Code "on the grounds that it has no counterpart and creates a discretionary power that is a source of significant imbalance between the parties".[67] Noting the absence of consideration, the court declared the clause void.[68]

In the above-mentioned judgment of 17 June 2017, the Paris Court of Appeal was also called upon to rule on the lawfulness of parity clauses in light of the provisions of the French Commercial Code on a significant imbalance in the rights and obligations of the parties.[69]

It considered that the rate parity and availability clauses contained in the contracts between hoteliers and companies of the Expedia group characterised a significant imbalance in the rights and obligations of the parties. It fined the companies of the Expedia group EUR one million and ordered them to put an end to the practice of mentioning the clauses in their contracts signed with the platform's member hotels.

However, this judgment was overturned on this point by the Commercial Chamber of the French Supreme Court (*Cour de Cassation*) in 2020.[70] According to the Commercial Chamber, "In order to hold that the so-called 'last available room'

[65] Paris Commercial Court, 29 November 2016, RG n°2014027403.

[66] Paris Commercial Court, 28 March 2017, No. 2015000203.

[67] Paris Commercial Court, 28 March 2017, No. 2015000203.

[68] Paris Commercial Court, 28 March 2017, No. 2015000203.

[69] Paris Court of Appeal, 21 June 2017, n°15/18784.

[70] French Supreme Court, Commercial Chamber, 8 July 2020, n°17-31.536.

clause, correlated with the price, non-price and promotional parity clause, leads to a significant imbalance between the rights and obligations of the parties, the judgment held that this clause obliges the hotelier who has rooms still available to sell them through the companies of the Expedia group. By ruling as such, in spite of the fact that the clauses relating to the last available room only imposed on hoteliers to allow the reservation of this room through the channel of the companies of the Expedia group under the conditions provided for other channels, the Court of Appeal, which disregarded the will expressed by the parties in the disputed clauses, violated the aforementioned text."[71]

5.7.2.3 Sectoral Ban *Per Se*

In 2015, the "Macron law"[72] introduced the prohibition of any parity clause, restricted or extended, in the Tourism Code (Subsection 2: Relations between hoteliers and online booking platforms).

This text imposes the use of the agency contractual model and prohibits clauses limiting the hotelier's pricing freedom in the following terms: "the hotelier retains the freedom to grant the client any discount or pricing advantage of any kind, any clause to the contrary being deemed unwritten".[73] The text is of international public order nature: "This subsection applies regardless of the place of establishment of the online reservation platform, provided that the rental is made to a hotel established in France."[74] It is also immediately applicable, including to current contracts: "Contracts between hoteliers and online reservation platforms concluded before the publication of Law No. 2015-990 of 6 August 2015 for growth, activity and equal economic opportunity cease to have effect as soon as the same law comes into force."[75]

The fact that a legal representative of the online reservation platform operates without a contract concluded in accordance with the requirements of Article L.311-5-1 is punishable by a fine of EUR 30,000, which may be increased to EUR 150,000 if it is a legal person.

On 9 February 2017, the FCA published an assessment of the commitments made by Booking.com.[76] It found that more hoteliers were differentiating their rates across online hotel booking platforms. This price differentiation was an initial development in the sector, but there were no other external signs of increased competition between

[71] French Supreme Court, Commercial Chamber, 8 July 2020, n°17-31.536.

[72] Law No. 2015-990 of 6 August 2015 for growth, activity and equal economic opportunities, Official Journal of the French Republic No. 181 of 7 August 2015.

[73] Article L. 311-5-1 of the French Tourism Code.

[74] Article L. 311-5-4 of the French Tourism Code.

[75] Article L. 311-5-4 of the French Tourism Code.

[76] Press release of the French Competition Authority, available at https://www.autoritedelaconcurrence.fr/fr/communiques-de-presse/9-fevrier-2017-plateformes-de-reservation-hoteliere. Accessed 28 September 2021.

online travel agencies (market share, quality of the offer, changes in commission rates).

On 20 June 2018, an information report, submitted on behalf of the Senate's *Commission des affaires économiques* by the "tourism" working group, has highlighted, through hearings and written contributions from hoteliers, the generally satisfactory nature of the normative developments in terms of parity clauses linking hoteliers to hotel booking platforms. Indeed, it is recorded that "The hotel unions emphasise the difference between Europe and France in the share of direct bookings in the total hotel booking market in order to consider that the ban on the price parity clause has had a beneficial effect (62.8% compared with 55.1%). (...) According to the data provided (...) by the national group of independents, after the adoption of the ban on the price parity clause, almost 59% of hotels offered lower rates directly than through Booking.com, compared to 31% before."[77] According to the report, the FCA's action and the Macron Law have "encouraged hoteliers to better control their commercial policy", "probably contributed to a relative stabilisation of the rate of intermediation fees" and "enabled hoteliers and platforms to improve their relationship".[78]

5.8 Comparison Portals, the Role of Competition Authorities in Consumer Protection and Other Intervention Tools

5.8.1 Competition Law and the Role of the FCA Regarding Comparison Portals

The FCA is seeking to address new issues related to the development of digital platforms and online comparison portals that offer intermediation services.

The FCA controls and sanctions infringements from the traditional angle of competition law: cartels by assessing the effects of an agreement or abuses of a dominant position via exploitation and exclusionary abuses (gatekeeping role)[79] or the control of concentrations of all economic players by evaluating the effects of a merger or acquisition.

---

[77] Information report on behalf of the *Comité des affaires économiques* by the "Tourism" working group, on tourist accommodation and digital technology, by Viviane ARTIGALAS and Patricia MORHET-RICHAUD, Senators.

[78] Information report on behalf of the *Comité des affaires économiques* by the "Tourism" working group, on tourist accommodation and digital technology, by Viviane ARTIGALAS and Patricia MORHET-RICHAUD, Senators.

[79] Decision No. 21-D-04 of 24 February 2021 regarding practices implemented in the publishing and sales of professional software sector.

The FCA has, however, demonstrated its ability to deal with new behaviours by using innovative reasoning or by applying well-established solutions to new services, such as online advertising services (*Google/Gibmedia* decision[80]).

However, no specific competence has been attributed to the FCA outside its traditional scope of action with regard to comparison portals.

The FCA has nevertheless had the opportunity to mention price comparison websites in its Opinion No. 12-A-20 of 18 September 2012 on the competitive operation of e-commerce. In this opinion, the FCA indicated that by allowing better price comparison, price comparison sites can reinforce the competitive pressure exerted by e-commerce. The FCA then identifies the characteristics of price comparison sites, the way they operate and the competitive risks they present and discusses a move towards the self-regulation of these comparison sites as well as avenues for improvement. The competitive risks were qualified by the FCA as being limited. Finally, the FCA emphasised that the concerns raised by these sites were more related to consumer law than to competition law.

5.8.2 Consumer Law as a Tool for Understanding the Practices of Comparison Portals

The French Consumer Code provides a definition of online platform operators,[81] which has made it possible to introduce a general information obligation for these operators that varies according to the nature of the activity of the platform concerned.

This article covers, inter alia, “any online platform operator whose activity consists of providing information enabling the comparison of prices and characteristics of goods and services offered by professionals, the information provided to consumers concerning the elements of this comparison and what is advertising within the meaning of Article 20 of Law No. 2004-575 of 21 June 2004 on confidence in the digital economy”.

The regulatory part of the French Consumer Code specifies the nature of online goods and service comparison and stipulates that “the provision of online information allowing the comparison of prices and characteristics of goods and services is understood to mean the activity of sites comparing goods and services and allowing, where appropriate, access to the sites for the sale of these goods or the provision of these services”.[82]

This also applies to online marketplaces that offer, as their main activity, the comparison of goods or services, whether sold by themselves or by third parties, and

---

[80] Decision No. 19-D-26 of 19 December 2019 regarding practices implemented in the sector of online search advertising sector; press release available at: https://www.autoritedelaconcurrence.fr/en/press-release/autorite-de-la-concurrence-hands-down-eu150m-fine-abuse-dominant-position. Accessed 28 September 2021.

[81] See Article L. 111-7 of the French Consumer Code.

[82] See Article D. 111-10 of the French Consumer Code.

to any person who uses the terms "comparator" or "comparison" for his or her commercial activity by electronic means.

Comparison portals are therefore subject to a certain number of obligations (compulsory information about the site relating to its ranking criteria, contractual or capital links between the comparison site and the professionals listed, remuneration of the site by the professionals listed, etc.).[83]

Failure to comply with these obligations may result in an administrative fine of up to EUR 75,000 for a natural person and EUR 375,000 for a legal person.[84]

The DGCCRF is competent to sanction these practices. Nevertheless, it is worth emphasising the particularity of this authority, which also has powers in the area of competition and which may, if it finds anti-competitive practices during its investigations, refer the investigation file to the FCA.

5.8.3 Civil Law and the Role of the Courts

In civil matters, actions may be brought against these comparison portals before the ordinary courts on the ground of unfair competition, disparagement or restrictive practices and discriminatory practices.

On the basis of restrictive practices, the Minister of Economy and Finance, as the guarantor of economic public order, or any interested party may summon the offender before the courts, which will impose a penalty. Amazon was fined EUR 4 million in September 2019 by the Paris Commercial Court for unbalanced contractual clauses.[85]

The French Civil Code defines the standard-form contract. Article 1240 of the French Civil Code is the legal basis for unfair competition actions, which includes parasitism (use of another person's reputation), disorganisation (of a company and/or its working methods by hacking into its website and/or disclosure of its he know-how, and/or misappropriation of a competitor's customer or supplier lists, electronic couponing allowing the systematic solicitation of competitors' customers).

[83] See Articles D. 111-11 to D.111-13 of the French Consumer Code.

[84] See Article L.131-4 of the French Consumer Code.

[85] Paris Commercial Court, 2 September 2019, n°2017050625.

5.9 Abuse and Treatment of Non-dominant Business Practices

5.9.1 The Concept of Abuse of Economic Dependency

Abuse of economic dependency is a specific offence under French law that allows the FCA to sanction abuses without having to demonstrate a dominant position, pursuant to Article L.420-2 §2 of the French Commercial Code.[86] Unlike abuse of a dominant position, which examines the position of an actor on the market, abuse of economic dependence focuses solely on the bilateral relationship between a company in a strong position and its commercial partner.[87]

To do so, three cumulative conditions must be met: (i) the existence of a situation of economic dependency, (ii) abusive exploitation of this situation and (iii) an actual or potential effect on the functioning of the structure of competition on the market. The most difficult condition to satisfy is often the demonstration of economic dependency and, in particular, the impossibility for the dependent undertaking to find a technically and commercially equivalent solution.

This offence, which was initially introduced to protect suppliers against large retailers, was very rarely applied until recently because of the strict conditions required by case law. The abuse of economic dependency is, however, nowadays the subject of renewed interest against Google, Apple, Facebook and Amazon (GAFA), for which market definition and dominance are not easy to establish.

In its study "Competition and E-commerce" published in May 2020, the FCA indeed stated that the abuse of economic dependence may be particularly relevant to assessing unfair trading conditions in the field of online platforms and marketplaces.[88]

For example, the FCA recently used the concept of abuse of economic dependency in two sanction decisions against Apple and Google:

- On 16 March 2020, the FCA fined Apple and its two main wholesalers, Tech Data and Ingram Micro, EUR 1.24 billion – of which EUR 1.1 billion for Apple.[89] It considered *inter alia* that Apple had abused the economic dependency of its premium resellers, imposing on Apple a fine of EUR 218 million for this practice alone. These premium resellers, dedicated exclusively to Apple products and with no possibility of leaving the Apple universe, had been subjected to supply delays, discriminatory treatment and non-transparent remuneration conditions in relation to the Apple Stores and the Apple website.

---

[86] Which prohibits "where it is liable to affect the functioning or structure of competition, the abuse by an undertaking or group of undertakings of the state of economic dependence of a customer or supplier on it".

[87] T. Tombal, Economic Dependence and Data Access, International Review of Intellectual Property and Competition Law, 51(1), 2020, p. 75.

[88] FCA, Study Concurrence et commerce en ligne, May 2020, paragraph 150 ».

[89] Decision No. 20-D-04 of 16 March 2020 regarding practices implemented in the Apple products distribution sector.

- On 9 April 2020, the FCA once again applied the concept of abuse of economic dependency in its interim measures decision against Google in the context of the entry into force of the law of 24 July 2019 on the neighbouring rights of press publishers.[90] Faced with the obligation to pay press publishers for excerpts of articles, photographs, infographics and videos, Google unilaterally decided that it would no longer display this content within its various services (Google Search, Google News and Discover), unless the publishers gave it permission to do so free of charge.
  The FCA found that Google's practices were likely to constitute an abuse of a dominant position on the general search services market by imposing unfair trading conditions on publishers and news agencies, and an abuse of economic dependency since search engines account for between 26% and 90% of the traffic redirected to the pages of publishers and news agencies who have no choice but to comply with Google's display policy without financial compensation.
- Finally, the FCA very recently examined a complaint of abuse of economic dependency in the professional software sector in a decision of 24 February 2021 in which it considered that the situation of economic dependency on a supplier was not supported by sufficient evidence by the distributor.[91]

5.9.2 Restrictive Competition Practices

Restrictive practices are also a specific feature of French competition law, allowing certain contractual clauses or abusive behaviour of economic actors vis-à-vis their partners to be sanctioned, regardless of their effect on the market. Here again, the demonstration of a dominant position is not necessary.

The provisions of the French Commercial Code qualify as unlawful practices the fact of (i) obtaining an advantage without consideration or one that is clearly disproportionate to the consideration,[92] (ii) subjecting the other party to obligations that create a significant imbalance between the parties,[93] (iii) applying penalties that are disproportionate with regard to the non-performance of contractual commitments[94] or (iii) abruptly terminating established commercial relations.[95]

---

[90] Decision No. 20-MC-01 of 9 April 2020 on requests for interim measures by the Syndicat des éditeurs de la presse magazine, the Alliance de la presse d'information générale and others and Agence France-Presse.

[91] Decision No. 21-D-04 of 24 February 2021 regarding practices implemented in the publishing and sales of professional software sector.

[92] Article L.442-1, I, 1° of the French Commercial Code.

[93] Article L.442-1, I, 2° of the French Commercial Code.

[94] Article L.442-1, I, 2° of the French Commercial Code.

[95] Article L.442-1, II of the French Commercial Code.

Several recent cases illustrate the strict application of these provisions. After Booking[96] and Expedia,[97] it was Amazon that, on 2 September 2019, was fined EUR 4 million by the Paris Commercial Court for a significant imbalance in its relations with third-party sellers on its platform.[98]

The court held that several clauses presented a manifest imbalance, in particular the one allowing Amazon to impose a modification to the contract unilaterally and without any prior notice or the one allowing Amazon to suspend or terminate the contract with immediate effect at its own discretion by simple notification.

Finally, it should be noted that French consumer law also contains certain provisions that make it possible to penalise unfair terms (and in particular significant imbalance) in the context of commercial relations between professionals and consumers.[99]

[96] Paris Commercial Court, 29 November 2016, n°2014027403.

[97] Paris Commercial Court, 21 June 2017, n°15/18784.

[98] Paris Commercial Court, 2 September 2019, n°2017050625.

[99] Article L.212-1 of the French Consumer Code.

Germany

6

Mark-Oliver Mackenrodt

6.1 The Collection of User and Device-Related Data

In February 2019, the Bundeskartellamt[1] qualified the data processing policy of *Facebook* as abuse of dominance according to German competition law, § 19 I, Gesetz gegen Wettbewerbsbeschränkungen (GWB).[2] *Facebook* appealed this decision to the OLG Düsseldorf (Oberlandesgericht, Higher Regional Court). A court decision on the merits in the main proceedings has—as of September 2021—not yet been taken. However, there have been several preliminary rulings regarding the immediate enforceability of the Bundeskartellamt's decision. In case of serious doubts regarding the legality of the Bundeskartellamt's decision, the court may order the appeal to have a suspensive effect (§ 65 III sentence 1 No. 2 GWB). In a preliminary ruling, the OLG Düsseldorf[3] had initially suspended the decision of the Bundeskartellamt. This interim decision by the OLG Düsseldorf was in turn reversed by the Bundesgerichtshof (BGH, Supreme Court) in June 2020.[4] The BGH stated that it does not have serious doubts as to the existence of abuse of market power by

[1] Bundeskartellamt, Decision of 6 February 2019 – B6-22/16.

[2] Gesetz gegen Wettbewerbsbeschränkungen, German Law against Restraints of Competition.

[3] Higher Regional Court Düsseldorf, Judgement of 26 August 2019 – VI-Kart 1/19 (V), Beck-RS 2019, 188837 – *Facebook* (interim decision).

[4] BGH, Judgment of 23 June 2020 – KVR 69/19 – *Facebook* (interim decision). For a more detailed analysis see M.-O. Mackenrodt and K. Wiedemann, Zur kartellrechtlichen Bewertung der Datenverarbeitung durch *Facebook* und ihrer normativen Kohärenz mit dem Datenschutzrecht und dem Datenschuldrecht, ZUM 2021, pp. 89–103.

M.-O. Mackenrodt (✉)
Technical University Munich, TUM School of Management, Munich, Germany
e-mail: mackenrodt@tum.de

B. Kilpatrick et al. (eds.), *Antitrust in Data Driven Markets & Legal Framework for Influencers, Native Advertising and Control over the Use of AI in Marketing*, LIDC Contributions on Antitrust Law, Intellectual Property and Unfair Competition,
https://doi.org/10.1007/978-3-031-07422-6_6

Facebook[5] and confirmed the immediate enforceability of the Bundeskartellamt's decision. As a next step, the *Facebook* case is to be decided in the main proceedings by the OLG Düsseldorf. This decision may then again be appealed to the BGH. However, in March 2021, the OLG Düsseldorf has made a referral to the Court of Justice of the European Union (CJEU).[6]

As to substance, the Bundeskartellamt has prohibited *Facebook* from combining user data and device-related data collected on the *Facebook* website on the one hand and the so-called off-*Facebook* data on the other hand if there is no sufficient consent in the sense of the General Data Protection Regulation (GDPR). "Off *Facebook* data" are data collected by other *Facebook*-owned services, like Instagram and WhatsApp, or data that have been collected from third-party websites and made available to *Facebook*. The Bundeskartellamt argued that *Facebook*'s general terms and conditions, which provided for such a data policy, were in contravention of the GDPR[7] and that imposing these terms and conditions upon *Facebook* users constituted exploitative abuse of dominance.[8]

The OLG Düsseldorf[9] had serious doubts that this constituted abuse by exploitation. But the BGH, by contrast, held that there were no serious doubts that *Facebook*'s data policy constitutes an abuse of dominance.[10] In its unusually long decision, the BGH establishes its own line of argument, which—as compared to the reasoning of the Bundeskartellamt—is more competition oriented and does not derive the exploitative abuse from a mere violation of the GDPR. At the same time, the BGH points out that, still, the Bundeskartellamt's decision and findings are at their core in accordance with the BGH's more specific line of argument and that the BGH's reasoning does neither change the nature of the Bundeskartellamt's decision nor restrict *Facebook* in its legal defense.[11]

The BGH notes that users can only use *Facebook* if *Facebook* is left free to combine off-*Facebook* data with personal data that are generated using *Facebook* and to process these combined data.[12] The data policy allows an extended highly personalized user experience. However, the BGH holds that this extension of the service is imposed upon users by *Facebook* and that users are forced to use the social network with this extended highly personalized experience based on the additional processing of off-*Facebook* data.[13] Users may not have wanted that these off-*Facebook* data are made available to or used by *Facebook*. The BGH argues

[5]BGH, Judgment of 23 June 2020 – KVR 69/19 – *Facebook* (interim decision).

[6]Higher Regional Court Düsseldorf, 24 March 2021, press release No. 9/2021.

[7]Bundeskartellamt, Decision of 6 February 2019 – B6-22/16, paras 573 ff.

[8]Bundeskartellamt, Decision of 6 February 2019 – B6-22/16, paras 525 ff.

[9]The OLG's main concerns are summarized at BGH, Judgment of 23 June 2020 – KVR 69/19, para 9 – *Facebook* (interim decision).

[10]BGH, Judgment of 23 June 2020 – KVR 69/19, para 53 – *Facebook* (interim decision).

[11]BGH, Judgment of 23 June 2020 – KVR 69/19, para 131 – *Facebook* (interim decision).

[12]BGH, Judgment of 23 June 2020 – KVR 69/19, para 53 – *Facebook* (interim decision).

[13]BGH, Judgment of 23 June 2020 – KVR 69/19, para 58 – *Facebook* (interim decision).

that users are deprived of their free choice to use the social network in a way that is limited to the personalized experience which is based on the personal data which are generated only by their use of the social network and not based on off-Facebook data.[14]

In its competitive assessment the BGH finds that the imposition of the unwanted extended services results in negative competitive effects in two respects:[15] in the vertical relationship customers are exploited[16] and, in addition, in the horizontal dimension[17] competition is impeded on the market for social networks and potential competitors are excluded. It is important to note that, by contrast, the Bundeskartellamt's theory of harm was centered around the question of whether abuse by exploitation could be inferred from the imposition of invalid terms and conditions and from a violation of the GDPR.[18] The BGH, instead, refers to the freedom of choice of the users and, in addition, to the creation of an impediment[19] to competition by exclusionary conduct.[20] Thereby, the BGH's reasoning is more competition oriented and minimizes any interference with the data protection regime. As a general feature of the business model of social networks, data are used for placing targeted advertisements with Internet users in order to finance the social network. Therefore, a dominant social network that—possibly without the consent required by the GDPR—collects and processes more and more data gains an even higher advantage over potential competitors. The BGH notes that a huge collection of data by a dominant social network raises entry barriers for potential competitors,[21] which on their part due to *Facebook*'s extensive data processing practice become less likely to acquire advertising contracts, which are required for the amortization of a competing social network. Furthermore, the BGH found that a restraint of competition on the market for online advertising could not be excluded.[22] A detailed analysis of the online advertising market was not required because the potential restraints of competition do not necessarily need to occur on the same market where market power has been established.

---

[14] BGH, Judgment of 23 June 2020 – KVR 69/19, para 58 – *Facebook* (interim decision).

[15] BGH, Judgment of 23 June 2020 – KVR 69/19, para 64 – *Facebook* (interim decision).

[16] A discussion of the exploitative abuse and in particular of the disputed causality requirement can be found at BGH, Judgment of 23 June 2020 – KVR 69/19, para 66 ff. – *Facebook* (interim decision). See also M.-O. Mackenrodt and K. Wiedemann, Zur kartellrechtlichen Bewertung der Datenverarbeitung durch *Facebook* und ihrer normativen Kohärenz mit dem Datenschutzrecht und dem Datenschuldrecht, ZUM 2021, pp. 89, 92 ff.

[17] A more detailed discussion of an impediment to competition can be found at BGH, Judgment of 23 June 2020 – KVR 69/19, para 92 ff. – *Facebook* (interim decision).

[18] Bundeskartellamt, Decision of 6 February 2019 – B6-22/16, para 522 ff.

[19] BGH, Judgment of 23 June 2020 – KVR 69/19, para 92 ff. – *Facebook* (interim decision).

[20] BGH, Judgment of 23 June 2020 – KVR 69/19, para 64 – *Facebook* (interim decision).

[21] BGH, Judgment of 23 June 2020 – KVR 69/19, para 93, 94 – *Facebook* (interim decision).

[22] BGH, Judgment of 23 June 2020 – KVR 69/19, para 96 – *Facebook* (interim decision).

In addition to identifying a potential restraint of competition, finding an abuse according to § 19 GWB requires a balancing of the interests involved.[23] In the particular case, the interests to be weighed include freedom of competition as the main objective of competition law, the fundamental right of informational self-determination, which forms the constitutional basis of the data protection regime, and also interests like economic freedom and the fact that for many users access to a social network is a way of participating in social life.[24] It is in this balancing process that data protection law comes in as one factor, among others. The BGH states that the consent requirement of the GDPR represents one aspect of the constitutional right of self-determination.[25] And in a quite detailed discussion of the GDPR, the BGH finds that *Facebook* actually does not meet the requirements of Article 6 GDPR.[26] The BGH maintains that within the balancing process, the fact that a behavior is illegal can be one and possibly a decisive and sufficient factor to attribute less weight to the respective interest of the party that is in contravention of the law.[27] This is because competition law does not intend to support a behavior that is disapproved by the legal order. At the same time, the BGH points out that the illegality of a contractual condition or of a behavior, like data collection, is not a necessary requirement for finding an abuse.[28]

This holding illustrates the relationship between competition law and data protection law in the BGH's line of reasoning: the BGH refers to privacy considerations as one factor within the balancing of interests, and finding a GDPR violation is not a necessary requirement for a competition law infringement. By contrast, in the Bundeskartellamt's line of argument, abuse is strongly related to the infringement of the GDPR. Consequently, the BGH expressly holds that no doubt exists when it comes to the competence of the Bundeskartellamt in this case and that the Bundeskartellamt's competence is not excluded only because the application of competition law in this case takes data protection issues into account, just as competition law, in other cases, takes into account other fields of European or national law.[29]

6.2 Determination of Market Power in Digital Markets

Section 18 I GWB contains a legal definition of market power. It states that a supplier or purchaser has market power if he has no competitors, is not exposed to significant competition or has a paramount market position in relation to his

[23] BGH, Judgment of 23 June 2020 – KVR 69/19, para 98 – *Facebook* (interim decision).

[24] BGH, Judgment of 23 June 2020 – KVR 69/19, paras 97 ff. – *Facebook* (interim decision).

[25] BGH, Judgment of 23 June 2020 – KVR 69/19, para 106 – *Facebook* (interim decision).

[26] BGH, Judgment of 23 June 2020 – KVR 69/19, paras 97 ff. – *Facebook* (interim decision).

[27] BGH, Judgment of 23 June 2020 – KVR 69/19, para 99 – *Facebook* (interim decision).

[28] BGH, Judgment of 23 June 2020 – KVR 69/19, para 99 – *Facebook* (interim decision).

[29] BGH, Judgment of 23 June 2020 – KVR 69/19, para 126 – *Facebook* (interim decision).

competitors. Section 18 III GWB enumerates factors particularly to be taken into account when assessing market power. These criteria are also of value when determining market power in digital markets. However, digital markets exhibit economic features that are distinct from traditional markets and are of high relevance for the determination of market power. The Bundeskartellamt has in 2016 published a working paper on the determination of market power in digital platform markets and networks.[30] In addition, drawing from the Bundeskartellamt's case practice, the ninth amendment to the GWB in 2017 has introduced an explicit list of factors in § 18 IIIa GWB, which are particularly relevant for the determination of market power, especially with regard to digital and multisided platform markets or networks. The Bundeskartellamt has already applied the new provision in merger cases and abuse proceedings, for example for assessing the market position of a ticketing platform.[31] The criteria in § 18 IIIa GWB are of equal rank to those in § 18 III GWB, and both catalogues are not exhaustive.[32] These catalogues do not extend the definition of market power in § 18 I GWB.[33] Rather, when determining the market position, the overall picture of the competitive forces has to be assessed.[34]

6.2.1 Market Shares as Starting Point

Market shares are the starting point for the assessment of the market position of an undertaking in traditional markets and also in digital markets.[35] In digital markets, however, market shares are of lower significance as an indicator of market power. For example, in the merger of two real estate platforms, the Bundeskartellamt considered a joint market share of the merger partners of over 75% not to be objectionable as such.[36] In a merger of two digital dating platforms, the Bundeskartellamt has not raised objections even though the (turnover-based) joint market share of the merging parties rose to 45–50%[37] and the nearest competitor held only 30–35% of the market.[38] The methodology of determining market shares

[30] Bundeskartellamt, Working Paper, Marktmacht von Plattformen und Netzwerken, 2016.

[31] Bundeskartellamt, Decision of 4 December 2017 – B6-132/14-2, paras 163 ff.; Bundeskartellamt, Decision of 23 November 2017 – B6-35/17, paras 154 ff.

[32] M.-O. Mackenrodt, Germany, In: Mandrescu (ed.), EU Competition Law and the Digital Economy: Protecting Free and Fair Competition in an Age of Technological (R)evolution, Dongress Publications 2020, pp. 251, 260.

[33] C. Grave, Marktbeherrschung bei mehrseitigen Märkten und Netzwerken, In: Kersting, Podszun (eds.), Die 9. GWB-Novelle, C.H. Beck 2017, pp. 17–44, para 7.

[34] Legislative reasons, Official Gazette 18/10207, p. 49.

[35] See for example Bundeskartellamt, Decision of 8 September 2015 – B6-126/14, para 154.

[36] Bundeskartellamt, Case Report of 25 June 2015 – B6-39/15, p. 3.

[37] It has to be kept in mind that § 18 IV GWB establishes a presumption for dominance if the market share is above 40%.

[38] Bundeskartellamt, Decision of 22 October 2015 – B6-57/15, para 138.

in digital markets often requires a modification as opposed to traditional markets. For example, in the case of services provided free of charge, the turnover cannot be used to determine market shares. As alternatives to turnover figures, the Bundeskartellamt has, for example, used the number of search queries with regard to search engines,[39] and with regard to dating platforms,[40] it referred to the number of registered members and individual visitors.

6.2.2 Competitive Pressure Through Innovation, § 18 IIIa No. 5 GWB

If digital markets are very dynamic, market shares may change in a short time and are therefore only a weak indicator of market power. Therefore, a high degree of innovation in a particular market can indicate that a market position is contestable and can be an argument against assuming market power irrespective of high market shares. Accordingly, § 18 IIIa no. 5 GWB lists innovation-driven competitive pressure as a factor for determining market power. However, the legislative reasons expressly note that it is not a sufficient argument against assuming market power if contestability by innovation has only been put forward in an abstract and vague way.[41] Quite similarly, the Bundeskartellamt rejects a general "internet defense" and states that an assumption of market power cannot be ruled out simply by generally pointing to the dynamic character of the Internet.[42] Rather—as the Bundeskartellamt continues—an individual case examination is required, and concrete indications of sufficiently probable, innovation-driven, or disruptive changes within the respective forecast period have to be ascertained.

In a merger case between two online dating platforms, the Bundeskartellamt[43] has—even before the introduction of § 18 IIIa no. 5 GWB—investigated the dynamic character of the market: it found that there was a concrete dynamic competitive pressure in the market, which would counteract a possible market power of the merged online dating platforms. Mobile dating apps such as the smartphone app Tinder had only been launched about 3 years prior to the merger, had experienced a strong growth in user number, and offered novel functionalities such as the localization of persons and accessibility at any time. Therefore, the concern regarding market power was lower, and the merger was approved.

[39] Bundeskartellamt, Decision of 8 September 2015 – B6-126/14, para 154.

[40] Bundeskartellamt, Decision of 22 October 2015 – B6-57/15, para 132.

[41] Legislative reasons, Official Gazette 18/10207, p. 51.

[42] Bundeskartellamt, Decision of 22 October 2015 – B6-57/15, para 176.

[43] Bundeskartellamt, Decision of 22 October 2015 – B6-57/15, paras 177 ff.

6.2.3 Network Effects, § 18 IIIa No. 1 GWB

In digital markets, network effects are—as spelled out in § 18 IIIa No. 1 GWB—a significant factor for assessing the market position of an undertaking. Even before the introduction of this provision, the Bundeskartellamt has argued that network effects constitute an entry barrier to multisided markets and reinforce market power.[44] This is because successful entry into a multisided market requires building up a critical mass of users simultaneously on both sides of the market in order to match the network effects of the incumbents. In the case of a ticketing platform, the Bundeskartellamt has held that indirect network effects create a considerable barrier to market entry.[45] On the other hand, in a merger between dating platforms, the Bundeskartellamt assumed that in this particular case, entry barriers did not give rise to market power despite the existence of network effects:[46] the Bundeskartellamt argued that in that specific case—as the example of the Tinder app had shown—market entry and the development of a critical mass and the achievement of a certain level of awareness remain possible even with lower investment costs.

6.2.4 Parallel Use of Networks, § 18 IIIa No. 2 GWB

According to § 18 IIIa no. 2 GWB, the parallel use of several platforms (multihoming) and the cost of switching between platforms represent factors for determining market power.

A widespread parallel use and low costs for switching between platforms reduce the impact of network effects and the associated barriers to market entry. Consequently, the tendency toward concentration and toward tipping of the market is also reduced.[47] Therefore, it argues against an assumption of market power if multihoming on both sides of a platform is common and if switching cost is low.

In the case of a merger between two real estate platforms, the Bundeskartellamt determined that there was a high degree of multihoming on both sides of the platform and considered this sufficient to counteract the tipping of the platform market.[48] Reviewing a merger of dating platforms, the Bundeskartellamt also found that multihoming sufficiently restricted the emergence of market power as 72% of users were using multiple dating platforms at the same time.[49] By contrast, in the case of ticketing platforms, the Bundeskartellamt found that on both sides of the

[44] Bundeskartellamt, Decision of 22 October 2015 – B6-57/15, para 142.

[45] Bundeskartellamt, Decision of 4 December 2017 – B6-132/14-2, paras 165 f.; Bundeskartellamt, Decision of 23 November 2017 – B6-35/17, para 162.

[46] Bundeskartellamt, Decision of 22 October 2015 – B6-57/15, paras 184 ff.

[47] Legislative reasons, Official Gazette 18/10207, pp. 50 f.; Bundeskartellamt, Decision of 22 October 2015 – B6-57/15, para 151.

[48] Bundeskartellamt, Case Report of 25 June 2015 – B6-39/15, p. 4.

[49] Bundeskartellamt, Decision of 22 October 2015 – B6-57/15, paras 153 f.

platform, multihoming was only possible to a limited extent.[50] Event organizers on one side of the platform and advance booking outlets on the other side rarely used competing platforms. In this case, the low degree of multihoming resulted from the fact that the ticketing platform had concluded exclusivity agreements with event organizers, and ticket quotas that were allotted to competing ticketing systems were not comparable in size.[51] Overall, advance booking outlets had used competing platforms only in a complementary or sequential manner.[52] This was not considered to be sufficient to restrict market power.

6.2.5 Economies of Scale, § 18 IIIa No. 3 GWB

A further factor to be considered when assessing market power is economies of scale (§ 18 IIIa no. 3 GWB). When assessing the market power of ticketing platforms, the Bundeskartellamt noted that economies of scale existed due to the high fixed costs of establishing a database and due to low variable costs.[53] As a consequence, this barrier to market entry was found to strengthen the market power of the leading platform and further reduce the ability of potential competitors to catch up.

6.2.6 Access to Data, § 18 IIIa No. 4 GWB

Pursuant to § 18 IIIa no. 4 GWB for determining market power in digital markets, another factor is access to data. According to the legislative reasoning, not only the quantity of data is important but also the actual ability to analyze and process huge amounts of data.[54]

In its practice related to ticketing platforms, the Bundeskartellamt has considered the collection and use of customer data as a factor for market power.[55] In the case of third-party event organizers, the Bundeskartellamt noted that they did not receive the data of their own customers who had booked with them through the dominant ticketing platform.[56] Therefore, third-party event organizers could not build up their own customer database and could not receive data about their customers

[50] Bundeskartellamt, Decision of 23 November 2017 – B6-35/17, paras 177 ff.; Bundeskartellamt, Decision of 4 December 2017 – B6-132/14-2, paras 170, 180.

[51] Bundeskartellamt, Decision of 23 November 2017 – B6-35/17, paras 178 ff.; Bundeskartellamt, Decision of 4 December 2017 – B6-132/14-2, paras 187, 191.

[52] Bundeskartellamt, Decision of 23 November 2017 – B6-35/17, para 186.

[53] Bundeskartellamt, Decision of 23 November 2017 – B6-35/17, paras 188 f.; Bundeskartellamt, Decision of 4 December 2017 – B6-132/14-2, para 194.

[54] Legislative reasons, Official Gazette 18/10207, p. 51.

[55] Bundeskartellamt, Decision of 23 November 2017 – B6-35/17, paras 190 ff.; Bundeskartellamt, Decision of 4 December 2017 – B6-132/14-2, paras 195 ff.

[56] Bundeskartellamt, Decision of 23 November 2017 – B6-35/17, para 195; Bundeskartellamt, Decision of 4 December 2017 – B6-132/14-2, para 200.

when they wanted to switch to a competing ticket distribution system.[57] By contrast, the ticketing platform had passed on data to other companies like *Facebook*.

6.2.7 Relative Market Power and Superior Market Power

In German competition law, an intervention against abusive unilateral conduct is already possible below the threshold of dominance if there is relative market power or superior market power. These concepts are discussed in Sect. 6.10.

6.2.8 Intermediation Power

In 2021, the tenth amendment to the GWB has entered into force.[58] The new § 18 IIIb GWB introduces the concept of intermediation power into German competition law. This new paragraph provides that when determining the market position of a company that acts as an intermediary on a multisided platform, the importance of the intermediary services for access to supply and sales markets should be a relevant factor. Intermediation power constitutes a third form of market power distinct from the demand-side and supply-side market power.

6.2.9 Undertakings with Paramount Significance for Competition Across Markets

In addition, in the course of the tenth amendment to the GWB in 2021, the new § 19a I GWB has established the concept of undertakings with paramount significance to competition across markets. This concept applies to undertakings that are active to a significant extent in platform or network markets. When assessing their market position, the following factors are particularly important: the dominant position in one or more markets, financial strength and access to resources, vertical integration and activities in related markets, access to data, and their importance to third parties' access to supply and sales markets. The new provision would provide for a two-step procedure: in the first step, the Bundeskartellamt may declare by order that a particular company holds such a market position. In the second step, the Bundeskartellamt could prohibit certain conduct of such a company. This includes—under certain circumstances—self-preferencing, impeding competitors, certain abuses by using data, making data portability more difficult, and the retention

[57] Bundeskartellamt, Decision of 23 November 2017 – B6-35/17, para 195; Bundeskartellamt, Decision of 4 December 2017 – B6-132/14-2, para 200.

[58] Federal Government, draft, legislative reasons, GWB-Digitalisierungsgesetz, Official Gazette 19/23492, 19 October 2020.

of information from other companies. The addressee of the norm carries the burden of proof to show that its conduct is objectively justified.

6.3 Platform Markets

As a general starting point, the term "platform" refers to two-sided or multisided markets that act as intermediaries between different groups of users.[59] The German competition law rules that were introduced into the GWB in 2017 to better address digital markets use the terms "multi-sided markets" and "networks" in § 18 IIIa GWB and not the term "platform". However, according to the legislative reasons, the terms "platform" and "multi-sided markets" are used as synonyms.[60]

The legislative reasons also point out that the term "platform markets" is not used in a uniform sense in the economic literature.[61] The Bundeskartellamt discusses different economic concepts of "platforms" in a working paper.[62] The Bundeskartellamt, when defining the notion of a "platform", considers, as a starting point, that such a definition serves the objective of delineating certain constellations that pose similar analytical problems in competition law, in particular with regard to determining market power, delineating markets, and devising theories of harm.[63] Therefore, the Bundeskartellamt combines several elements of different economic approaches in its definition of platforms.[64] In its decision in *Google VG Media*, the Bundeskartellamt defines a platform for competition law purposes:[65] It considers it a defining element that a platform acts as an intermediary and enables direct interaction between different user groups between which indirect network effects exist. Further, it is regarded as a characteristic that the platform is no longer involved in the interaction between these groups.

Indirect network effects occur if the utility of one user group is affected by the number of users in a different group who are also using the platform.[66] In the Bundeskartellamt's approach, the term "indirect network effects" is used in a broad sense.[67] It encompasses unilateral network effects and also negative network effects. For example, reading an additional advertisement can be of little or even

[59] Bundeskartellamt, Working Paper, Marktmacht von Plattformen und Netzwerken, 2016, p. 8.

[60] Legislative reasons, Official Gazette 18/10207, p. 47.

[61] Legislative reasons, Official Gazette 18/10207, p. 47.

[62] Bundeskartellamt, Working Paper, Marktmacht von Plattformen und Netzwerken, 2016, p. 8 ff.

[63] Bundeskartellamt, Decision of 8 September – B6-126/14, para 121, footnote 137; Bundeskartellamt, Working Paper, Marktmacht von Plattformen und Netzwerken, 2016, p. 14.

[64] Bundeskartellamt, Working Paper, Marktmacht von Plattformen und Netzwerken, 2016, p. 8 ff.

[65] Bundeskartellamt, Decision of 8 September – B6-126/14, para 121. See also Bundeskartellamt, Working Paper, Marktmacht von Plattformen und Netzwerken, 2016, p. 14.

[66] M. Armstrong, Competition in Two-Sided Markets, RAND Journal of Economics 2006 (37), p. 668.

[67] A discussion of different concepts of network effects can be found at Bundeskartellamt, Working Paper, Marktmacht von Plattformen und Netzwerken, 2016, p. 9 f.

negative additional utility for the reader,[68] for example, in terms of nuisance or time loss. As an example, where indirect network effects might be negative for one group, the Bundeskartellamt has considered the Google search engine as a platform with advertisers on one side of the market and Internet users on a different side.[69]

It should be noted that the "interaction" that is enabled between different user groups by a platform does not necessarily need to be a transaction in the sense of purchase,[70] but it rather refers to bringing together the user groups. Whether—as a result of the interaction—there is an additional transaction in the sense of a purchase depends on the type of the platform. For example, a selling platform like Amazon brings together potential purchasers and traders, which might result in a purchase or not. And a search engine platform enables the interaction of advertisers with Internet users who perceive the advertisements. But an actual purchase of the advertised product—in most cases from a third party—is not required for assuming a platform market.

In its practice, the Bundeskartellamt distinguishes between different categories of platforms depending on the purpose of the interaction which the platform enables between different user groups.[71]

One category is formed by matching platforms and transaction platforms. A matching platform seeks to achieve the best possible matchmaking between users who are from different heterogeneous groups. It is not required that this interaction results in an actual transaction, like a purchase. For instance, the Bundeskartellamt has qualified dating platforms as matchmaking platforms.[72] Transaction platforms are a subgroup of matching platforms where the interaction of the users is geared toward completing a purchase transaction. For example, the Bundeskartellamt has qualified real estate portals[73] and comparison portals[74] where contracts could be concluded as transaction platforms. Matching and transaction platforms usually exhibit bidirectional indirect network effects between the different user groups,[75] and the existence of the second market side is a requirement for the platform to function.[76]

The second category of platforms in the Bundeskartellamt's case practice is called "attention platforms" or "advertising platforms." On such platforms, offering a

---

[68] J.-C. Rochet and J. Tirole, Two-sided Markets: a progress report, RAND Journal of Economics 2006 (37), pp. 645, 652.

[69] Bundeskartellamt, Decision of 8 September – B6-126/14.

[70] J.-C. Rochet and J. Tirole, Two-sided Markets: a progress report, RAND Journal of Economics 2006 (37), pp. 645, 652.

[71] An overview of platform categories can be found at Bundeskartellamt, Working Paper, Marktmacht von Plattformen und Netzwerken, 2016, p. 19 ff.

[72] BKartA, 22.10.2015, B6-57/15 – *OCPE/EliteMedianet.*

[73] BKartA, 25.06.2015, B6-39/15 – *Immonet/Immowelt.*

[74] BKartA, 05.08.2015, B8-76/15 – *ProSiebenSat.1/Verivox.*

[75] Bundeskartellamt, Working Paper, Marktmacht von Plattformen und Netzwerken, 2016, p. 20.

[76] Bundeskartellamt, Case Report of 5 August 2015 – B8-76/15, p. 2; Bundeskartellamt, Case Report of 25 June 2015, B6-39/15, p. 2.

service to one group of users does not necessarily require the existence of another market side.[77] For example, offering the service of a search engine or offering the consumption of music or a video clip would not necessarily require the existence of a second market side in relation to advertising providers whose advertising is presented to the Internet users. Rather, adding the second market side of advertising for financing the other market side is a strategic decision within the business model.[78] On attention platforms, the indirect network effects are often only unilateral in character.[79] For example, indirect network effects operate in favor of advertisers that have a higher willingness to pay if their advertisements are communicated to a higher number of Internet users. By contrast, Internet users do often not derive a higher utility from a platform because it shows them more advertising. The Bundeskartellamt has qualified a search engine as an attention platform.[80]

For the purpose of market definition, the Bundeskartellamt has, as a rule, considered the two market sides of a matching or transaction platform to form a single market,[81] while it qualified the sides of an attention platform as two separate relevant markets.[82] However, the market definition and the question of whether the two sides of a platform form a single market will be decided on a case-by-case basis.

6.4 "Markets" for Free Services

It is a common feature of digital business models that consumers do not have to make direct monetary payments in return for receiving digital goods or services. Particularly in multisided platform markets, such services are financed by payments of users or businesses on the other side of the market. Such a pricing scheme can be motivated by the need to build up a large user base on one side of the market in order to generate high network effects. Network effects can be of different intensity on different sides of the market. In such a case, to be successful, the platform may have to apply a different strategy depending on the market side.

For example, search services are seemingly offered to consumers for free. However, search engines like Google collect consumers' data and garner high profits in the advertising markets.[83] Also, users do not have to pay a direct fee for using a

[77] Bundeskartellamt, Case Report of 25 June 2015 – B6-39/15, p. 2.

[78] Bundeskartellamt, Decision of 8 September 2015 – B6-126/14, paras 124 ff.; Bundeskartellamt, Case Report of 5 August 2015 – B8-76/15, p. 2; Bundeskartellamt, Case Report of 25 June 2015 – B6-39/15, p. 2; Bundeskartellamt, Decision of 22 October 2015 – B6-57/15, para 80.

[79] Bundeskartellamt, Working Paper, Marktmacht von Plattformen und Netzwerken, 2016, p. 25.

[80] Bundeskartellamt, Decision of 8 September 2015 – B6-126/1.

[81] Bundeskartellamt, Case Report of 5 August 2015 – B8-76/15, p. 3; Bundeskartellamt, Case Report of 25 June 2015 – B6-39/15, p. 3.

[82] Bundeskartellamt, Decision of 8 September 2015 – B6-126/14, paras 124 ff.

[83] For example, Google enjoys a share of 90% of the 7.3-billion-pound search advertising market in the United Kingdom. See Competition and Market Authority UK, Online platforms and digital advertising, market study final report, 2020, para 18.

hotel booking platform, but hotels have to pay 15% of their revenues from bookings that have been made through the platform.[84] However, economically, such fees very often have to be indirectly borne by the consumers. Quite similarly, for consumers using app store platforms like the Apple App Store or Google Play Store is free of charge and also most apps are offered for free. However, app businesses that want to offer their app on app stores have to pay a fixed fee and, in addition, a 30% fee for certain revenues they make from their app.[85]

The fact that services are seemingly offered for free to consumers is not an obstacle to defining a relevant market with regard to these services when applying competition law. In Germany, this is explicitly spelled out in § 18 IIa GWB. The provision introduced into the law in 2017 applies not only to multisided digital markets but also to traditional markets.[86] However, this does not mean that every nonpaid relationship constitutes a market. The absence of payment according to the wording of the provision simply does not preclude a competition law analysis. Assuming the existence of a market requires that the free service is offered for economic reasons or is part of a strategy that is at least indirectly profit oriented.[87] For example, the provision of a private scholarship may not qualify as a market relationship.[88]

Prior to the introduction of § 18 IIa GWB, the courts had occasionally refused to recognize the market character of a relationship that is free of charge.[89] The clarification in § 18 IIa GWB simplifies the enforcement of the law, especially in the digital economy. For example, the assumption of a market does not depend on the discussion as to whether, in the absence of monetary consideration, another form of consideration exists, for example, in the form of provision of data or attention to advertising.

The Bundeskartellamt had—even before the introduction of the legal provision—already assumed that a relationship that is free of charge could qualify as a market relationship, for example, in the case of a merger between two dating platforms[90] and a merger between two real estate platforms.[91] In the dating platform case, the Bundeskartellamt has even expressly included the use of a dating platform which

[84]Booking.com applies a standard rate of 15%. See for example the Swedish case against Svea Hovratt, 9 May 2019, PMT 7779-18 – booking. An English translation of the decision is reprinted at IIC 2019, pp. 1167, 1172. See also M.-O. Mackenrodt, Price and condition parity clauses in contracts between hotel booking platforms and hotels, IIC 2019, pp. 1131, 1132.

[85]The 30% fee is at issue in the investigations of the European Commission against Apple, see European Commission, Press Release IP/20/1073 of 16 June 2020, p. 1.

[86]Legislative reasons, Official Gazette 18/10207, p. 48. See also R. Podszun, Unentgeltliche Leistungen, In: Kersting, Podszun (eds.), Die 9. GWB-Novelle, C.H. Beck 2017, pp. 1–16, paras 29 ff.

[87]Legislative reasons, Official Gazette 18/10207, p. 48.

[88]Legislative reasons, Official Gazette 18/10207, p. 48.

[89]Higher Regional Court Düsseldorf, Judgement of 9 January 2015 – VI-Kart 1/14 (V), para 43.

[90]Bundeskartellamt, Decision of 22 October 2015 – B6-57/15, para 79.

[91]Bundeskartellamt, Case Report of 25 June 2015 – B6-39/15, p. 3.

was financed by advertising and which was therefore free of charge for users into the same relevant product market as the use of a dating agency service for which users had to pay a fee.[92]

The Bundeskartellamt has applied the new § 18 IIa GWB to a platform that provided services for event ticketing. The relationship between the ticketing platform and advance booking agencies was qualified as a market, even though the service was free of charge.[93] Also in the *Facebook* case, the social network services were considered to be a market even though there was no monetary payment by private users.[94] The Bundeskartellamt pointed out that the services provided to private social network users is closely connected with the activities of the platform on the market side of advertising and noted that the overarching business purpose of the platform activities is the same irrespective of the market side toward which a service is directed.[95] In any case, the Bundeskartellamt argued that a relationship of exchange with private users could be found because *Facebook*'s general terms and conditions constitute a contract with private users and private users are giving away their data.[96]

6.5 Online Advertising

In 2018, the Bundeskartellamt[97] opened a sector inquiry on online advertising, which has not yet been completed.[98] Online advertising is regarded as a complex, highly technology-driven sector with substantial economic significance. Some market actors claim that a few powerful online advertisers have established "walled gardens", which restrain competition. The sector inquiry in particular seeks to create a better understanding of the role of the different technical services, like data collection and processing, audience targeting, benchmarking, and auctioning of advertising spots.[99]

[92] Bundeskartellamt, Decision of 22 October 2015 – B6-57/15, paras 71, 81 ff.

[93] Bundeskartellamt, Decision of 4 December 2017 – B6-132/14-2, para 144; Bundeskartellamt, Decision of 23 November 2017 – B6-35/17, para 123.

[94] Bundeskartellamt, Decision of 6 February 2019 – B6-22/16, paras 239 ff.; BGH, Judgment of 23 June 2020 – KVR 69/19, paras 27 ff. – *Facebook* (interim decision).

[95] Bundeskartellamt, Decision of 6 February 2019 – B6-22/16, para 241.

[96] Bundeskartellamt, Decision of 6 February 2019 – B6-22/16, para 244.

[97] Bundeskartellamt, Press Release of 1 February 2018.

[98] The United Kingdom's CMA has meanwhile completed a similar sector inquiry; see Competition and Market Authority UK, Online platforms and digital advertising, market study final report, 2020.

[99] Current issues in online advertising were sketched out by the Bundeskartellamt in a short working paper, see Bundeskartellamt, online advertising, paper series, 2018, p. 7 f.

6.6 General Terms and Conditions with Regard to Data Collection

In its original decision dealing with the data processing policy of *Facebook*, the Bundeskartellamt's[100] main line of reasoning was focused on abuse by exploitation through the imposition of invalid general terms and conditions on private users.[101] This line of argument was largely rooted in a finding that *Facebook* used invalid general terms and conditions and violated the GDPR.

The German *Facebook* decision was based on the German prohibition of abuse in § 19 I GWB. For exclusionary abuses it is uncontested that under German law a fairly lenient causation requirement applies.[102] By contrast, for exploitative abuses in the form of imposition of unfair or invalid terms and conditions, the causation requirement is contentious under German[103] law.[104] The Bundeskartellamt held that only a lenient, so-called normative causation is required.[105] This means that the dominant position has to contribute to an amplification of the negative competitive effects of the conduct.[106] By contrast, the OLG Düsseldorf in the interim proceedings demanded that there be strict causation between the dominant position and the conduct.[107] This means that the imposition of the terms and conditions was possible only because of the dominance.[108] The OLG Düsseldorf argued that there was no such strict causation because users would accept general terms and conditions without reading them anyway. Therefore, the unilateral imposition of general terms and conditions would not be an expression of the market power.[109]

The OLG Düsseldorf[110] criticized several other dogmatic aspects of the Bundeskartellamt's reasoning, but the BGH[111] held that there are no serious doubts

[100] Bundeskartellamt, Decision of 6 February 2019 – B6-22/16, paras 169 ff.

[101] See already above 1.

[102] BGH, Judgment of 23 June 2020 – KVR 69/19, para 78 – *Facebook* (interim decision).

[103] For a discussion of the causation requirement in Article 102 TFEU see A. Fuchs, In: U. Immenga and E.-J. Mestmäcker (eds), EU Wettbewerbsrecht, C.H. Beck 2019, Art. 102 TFEU, paras 135 ff.

[104] See for example M.-O. Mackenrodt and K. Wiedemann, Zur kartellrechtlichen Bewertung der Datenverarbeitung durch *Facebook* und ihrer normativen Kohärenz mit dem Datenschutzrecht und dem Datenschuldrecht, ZUM 2021, pp. 89, 94.

[105] Bundeskartellamt, Decision of 6 February 2019 – B6-22/16, para 873.

[106] A. Fuchs, In: U. Immenga and E.-J. Mestmäcker (eds), GWB, C.H.Beck 2019, § 19 GWB, para 215.

[107] Higher Regional Court Düsseldorf, Judgement of 26 August 2019 – VI-Kart 1/19 (V), Beck-RS 2019, 188837, paras 64 ff. – *Facebook* (interim decision).

[108] A. Fuchs, In: U. Immenga and E.-J. Mestmäcker (eds), GWB, C.H. Beck 2019, § 19 GWB, para 215.

[109] Higher Regional Court Düsseldorf, Judgement of 26 August 2019 – VI-Kart 1/19 (V), Beck-RS 2019, 188837, para 77 – *Facebook* (interim decision).

[110] Higher Regional Court Düsseldorf, Judgement of 26 August 2019 – VI-Kart 1/19 (V), Beck-RS 2019, 188837 – *Facebook* (interim decision).

[111] BGH, Judgment of 23 June 2020 – KVR 69/19 – *Facebook* (interim decision).

as to the validity of the Bundeskartellamt's decision. However, as discussed above in Sect. 6.1, the BGH has chosen a line of reasoning in which a possible illegality of the general terms and conditions or an actual violation of the GDPR is no more a necessary condition for finding an abuse.[112] Rather, the BGH has additionally focused on abuse by exclusion and on treating data as a parameter of competition. In any case, the BGH rejected the OLG Düsseldorf's quite strict view regarding the causation requirement in this case because the BGH found that the conduct of *Facebook* was a combination of abuse by exploitation and abuse of exclusion.[113]

In addition, the new wording of § 19 GWB after its amendment in 2021 also suggests that a fairly low degree of causation is sufficient for finding an abuse.[114] The role of data as a factor in determining market power is explicitly mentioned in § 18 IIIa no. 4 GWB (see Sect. 6.2), and this argument has also been used in competition law decisions.[115]

6.7 Most Favored Nation Clauses and Cooperation Between Content Providers

Most favored nation clauses are also referred to as best-price clauses or parity clauses. They can take many different shapes depending on their drafting. In digital markets, competition authorities have scrutinized most favored nation clauses that state that the seller is not allowed to sell its goods in competing distribution channels at a lower retail price or at better conditions.[116] Different forms of such parity clauses have been dealt with more widely on a Member State level and are discussed in Sect. 6.11.

As one of the few decisions at the European level, a commitment decision of the European Commission deals with parity clauses of Amazon with regard to e-books.[117] Parity clauses related not only to price but also to other aspects through which competing distributors could differentiate themselves from Amazon, for example, by developing innovative alternative distribution models. The Commission pointed to the special responsibility of a dominant undertaking[118] and uttered concerns that these parity clauses would impede incentives for innovations or

[112] BGH, Judgment of 23 June 2020 – KVR 69/19, para 99 – *Facebook* (interim decision).

[113] BGH, Judgment of 23 June 2020 – KVR 69/19, para 77 – *Facebook* (interim decision).

[114] Federal Government, draft, legislative reasons, GWB-Digitalisierungsgesetz, Official Gazette 19/23492, 19 October 2020, p. 11.

[115] See for example BGH, Judgment of 23 June 2020 – KVR 69/19, para 43 – *Facebook* (interim decision).

[116] See for example T. Hoeren, Germany, In: Kilpatrick et al. (eds), Antitrust analysis of online sales platforms & copyright limitations and exceptions, Springer C.H. Beck 2017, pp. 157–185; M.-O. Mackenrodt, Price and conditions parity clauses in contract between Hotel booking platforms and hotels, IIC 2019 (50), p. 1131; F. Bostoen, Most favoured nation clauses, CoRe 2017, pp. 223, 224.

[117] European Commission, Decision of 4 May 2017 – AT.40153.

[118] European Commission, Decision of 4 May 2017 – AT.40153, para 70.

incentives to support innovations with regard to e-book distribution models that might be more appealing to consumers, like different book formats, bundles with print books, establishment of book clubs, etc.[119] Amazon's commitments[120] included an obligation not to enforce the parity clauses that required publishers to offer Amazon similar nonprice and price conditions as those offered to its competitors. Also, Amazon would not be allowed to use clauses that require publishers to inform the company about certain terms and conditions of its competitors, particularly regarding new business models, promotions, or prices.

6.8 Cooperation Between Network Operators

Telekom Deutschland and EWE are two leading Internet network companies in certain regions in northern Germany, and they wanted to cooperate regarding the rollout of a fast fiber-optic Internet network in the country. They concluded a cooperation agreement, which was reviewed by the Bundeskartellamt, and they established a joint venture, which required a parallel merger review by the Bundeskartellamt. After negotiations with the parties, the Bundeskartellamt issued a consent decree[121] that contains binding obligations for the parties of the cooperation agreement, and the Bundeskartellamt subsequently also approved the merger.[122]

The Bundeskartellamt had several concerns about the possible negative effects on competition of the cooperation between the network operators.[123] The commitments that were declared binding in December 2019 were meant to mitigate or eliminate these concerns. At the same time, the commitments aimed at accelerating the cost-intensive rollout of high-speed Internet.

The commitments concerning the rollout of high-speed Internet contain an obligation to connect 300,000 households or company locations to the fast Internet until the end of 2023.[124] As a safeguard, noncompliance with this obligation would lead to the termination of the joint venture.[125] Also, a certain percentage of the rollout has to be accomplished in rural areas and areas with a low coverage of cable infrastructure.[126] Such commitment addressed the concern that the cooperation agreement would create an incentive to shift the rollout efforts away from rural areas to urban areas, which already have a high level of infrastructure.[127] Because of the cooperation agreement, the two leading companies would not need to fear

[119] European Commission, Decision of 4 May 2017 – AT.40153, para 79.

[120] European Commission, Press Release IP/17/1223 of 4 May 2017.

[121] Bundeskartellamt, Commitment of 4 December 2019 – B7-21/18.

[122] Bundeskartellamt, Decision of 30 December 2019 – B7-21/18.

[123] See Bundeskartellamt, Commitment of 4 December 2019 – B7-21/18, paras 32 ff.

[124] Bundeskartellamt, Commitment of 4 December 2019 – B7-21/18, paras 110 ff.

[125] Bundeskartellamt, Commitment of 4 December 2019 – B7-21/18, paras 121 f.

[126] Bundeskartellamt, Commitment of 4 December 2019 – B7-21/18, para 131.

[127] Bundeskartellamt, Commitment of 4 December 2019 – B7-21/18, para 132.

competition from each other anymore within the area covered by the cooperation. Therefore, they might be incentivized to direct their rollout efforts away from rural areas and toward more lucrative urban areas where they fear that other companies could become market leaders. Thus, the cooperation agreement would dampen the rollout efforts of the cooperating parties within the area affected by the cooperation.[128] By establishing certain rollout objectives, the commitment seeks to mitigate such negative side effects of the cooperation agreement.

Further, the parties to the cooperation agreement committed to participating in future public tenders independently and assured that their joint venture would not participate in these public tenders.[129] This commitment, however, was criticized because the cooperating enterprises might still pursue a less aggressive bidding strategy even if they are not bidding jointly.[130]

Another category of commitments relates to access:[131] the cooperating parties are obliged to grant nondiscriminatory access to competitors. This includes an agreement that third-party telecommunication enterprises can buy services from the joint venture under the same conditions as those for the cooperating parties.[132] Accordingly, this access regime seeks to achieve an "equivalence of inputs," which is stricter than the "equivalence of outputs" principle.[133] Together with certain marketing objectives, the access commitment aims at promoting competition for end customers as third parties would be enabled to offer the same service to end consumers, like the cooperating parties.

In sum, cooperation agreements regarding the rollout of an advanced technology infrastructure can accelerate the dissemination of new technology. However, there is a danger of negative effects on competition in particular when dominant companies are involved. Therefore, such cooperation agreements should be limited to regions where absent a cooperation, a rollout of the new technology infrastructure would be impossible.[134] In addition, carefully drafted commitments with effective sanctioning mechanisms should serve as a safeguard.

6.9 Comparison Portals and Consumer Protection

Comparison portals can provide guidance to users and help inform the decisions of consumers. If users are put in a better position to take well-informed decisions, competition and consumer protection are strengthened. However, users of comparison portals rely on the neutrality of results, on which they base their decision. In

[128] Monopolies Commission, Twenty-third Biennial Report, 2020, para 460.

[129] Bundeskartellamt, Commitment of 4 December 2019 – B7-21/18, para 132.

[130] Monopolies Commission, Twenty-third Biennial Report, 2020, para 464.

[131] Bundeskartellamt, Commitment of 4 December 2019 – B7-21/18, paras 136 ff.

[132] Bundeskartellamt, Commitment of 4 December 2019 – B7-21/18, para 137.

[133] Monopolies Commission, Twenty-third Biennial Report, 2020, para 463.

[134] Bundeskartellamt, Commitment of 4 December 2019 – B7-21/18, para 5.

2019, the BKartA has published a sector inquiry on comparison portals[135] that are active in the sectors of energy, telecommunications, loans, insurance, hotels, and flights. The sector inquiry analyzes how consumer interests are accounted for by the comparison portals, in particular with regard to the issues of cooperation between comparison portals, market coverage of the portals, preselection of offers, ranking of results, and consumer reviews.

In all sectors, the inquiry found frequent cooperation between different comparison portals, for example, by using the same databases, but this was often not made transparent to consumers.[136] As a consequence, users of multiple comparison portals could erroneously interpret similar results by different portals as an independent confirmation of the results which they have received from a seemingly competing comparison portal.

Regarding the market coverage of the comparison, the inquiry showed that in some sectors, the comparison portals reflected only a small portion of the markets—as low as 25%—and that some large suppliers were not included in the comparison. The reasons for a low market coverage include the fact that some portals excluded from the comparison those suppliers that were not willing to pay a provision fee to the portals or some portals required the inclusion of exclusivity agreements from the suppliers.[137] In many cases, the consumers were not sufficiently informed about the market coverage of the comparison.

With respect to the preselection of what offers would be displayed in the list of results, the criteria were often not made sufficiently transparent to consumers.[138] Many portals listed offers in "position 0," which is displayed prior to the ranking, and from this position 0, about 25% of users selected an offer.[139] In some sectors like energy and telecommunications, the inclusion in this section was only based on a minimum qualitative assessment and depended on whether the company had paid a fee to the portal.[140] Consumers would, in many cases, not be informed if a contractual arrangement was the reason for including an offer in position 0.[141]

In particular in the hotel sector, the ranking was influenced by the provision which hotels had paid to the comparison portal.[142] In some instances, the ranking could be perceived as recommendations to users, and the criteria for the ranking were not sufficiently made transparent.[143] User ratings in the respective branches were—

[135] Bundeskartellamt, Sektoruntersuchung Vergleichsportale, 2019.

[136] Bundeskartellamt, Sektoruntersuchung Vergleichsportale, 2019, p. 55.

[137] Bundeskartellamt, Sektoruntersuchung Vergleichsportale, 2019, p. 65.

[138] Bundeskartellamt, Sektoruntersuchung Vergleichsportale, 2019, pp. 3, 70.

[139] Bundeskartellamt, Sektoruntersuchung Vergleichsportale, 2019, p. 78.

[140] Bundeskartellamt, Sektoruntersuchung Vergleichsportale, 2019, p. 75.

[141] Bundeskartellamt, Sektoruntersuchung Vergleichsportale, 2019, p. 78.

[142] Bundeskartellamt, Sektoruntersuchung Vergleichsportale, 2019, pp. 87 ff., 97.

[143] Bundeskartellamt, Sektoruntersuchung Vergleichsportale, 2019, p. 98.

with some exceptions in the hotel sector—only allowed by users who had actually made a transaction through the portal.[144]

The German Bundeskartellamt traditionally did not dispose of competences in the field of consumer protection. Following the ninth amendment of the GWB in 2017, the preexisting competence of the Bundeskartellamt to conduct sector inquiries was extended to the field of consumer protection.[145] Under the new provisions, the Bundeskartellamt has carried out or is carrying out several sector inquiries, in particular with regard to user ratings,[146] smart TVs,[147] comparison portals,[148] and messenger services. But the introduction of further enforcement competences in the field of consumer protection for the Bundeskartellamt was rejected by the lawmaker. Several consumer protection issues addressed in the sector inquiry on comparison portals are covered by the German unfair competition law (UWG).[149] However, the German UWG relies on private enforcement, which might not be sufficiently effective in some fields and might need to be cautiously complemented by public enforcement by the Bundeskartellamt.[150]

6.10 Abuse Proceedings Below the Threshold of Dominance

The Bundeskartellamt and the German courts can, according to § 20 GWB, apply the German provisions on the prohibition of unilateral conduct (§ 19 GWB) already below the threshold of market power, namely, in cases of relative market power in a vertical relationship (§ 20 I, II GWB) or in cases of superior market power in a horizontal relationship (§ 20 III GWB). This national regulation is stricter than Article 102 Treaty on the Functioning of the European Union (TFEU) because the threshold for intervention is lower. But according to Article 3 (1) sentence 2 of

[144] Bundeskartellamt, Sektoruntersuchung Vergleichsportale, 2019, p. 116. User ratings were the subject of a separate sector inquiry, see Bundeskartellamt, Sektoruntersuchung Nutzerbewertungen, vorläufige Ergebnisse, 2020.

[145] See § 32e V GWB.

[146] Bundeskartellamt, Sektoruntersuchung Nutzerbewertungen, Press Release of 23 May 2019; Bundeskartellamt, Sektoruntersuchung Nutzerbewertungen, vorläufige Ergebnisse, Press Release of 18 May 2020.

[147] Bundeskartellamt, Sektoruntersuchung Smart TVs, 2020.

[148] Bundeskartellamt, Sektoruntersuchung Vergleichsportale, 2019.

[149] See for example C. Alexander, Aktuelle lauterkeitsrechtliche Problemfelder von Online-Vergleichsportalen, WuW 2018, p. 765; F. Henning-Bodewig, Vergleichsportale, UWG und Befugnisse des Bundeskartellamts, WRP 2019, p. 537.

[150] For a general discussion of a possible public enforcement regime with competences for the Bundeskartellamt see R. Podszun, C. Busche and F. Henning-Bodewig, Behördliche Durchsetzung des Verbraucherrechts?, 2018.

Regulation 1/2003,[151] the Member States are free to adopt and apply stricter national laws in their territory regarding unilateral conduct.

6.10.1 Relative Market Power Within a Vertical Relationship

Section 20 I GWB applies to enterprises that are suppliers or purchasers of goods or services. Relative market power requires that these companies depend on another undertaking (the holder of relative market power) in such a way that sufficient and reasonable possibilities of switching to other undertakings do not exist.[152] In such a scenario, the company with relative market power is prohibited from unfairly impeding another undertaking or treating it differently without objective justification. In many cases in the digital sector in Germany, § 20 I GWB has been applied with regard to the relationship between platforms and independent companies that offer their products or services on the platform. For example, wide parity clauses on the hotel booking platform HRS have been found to be illegal according to Article 101 TFEU and also § 20 I GWB.[153]

The tenth amendment to the GWB in 2021 has introduced several changes to § 20 I GWB, which are of particular importance to digital markets: the applicability of the newly adapted § 20 I GWB is—unlike before the amendment—not anymore limited to small and medium-sized enterprises (SMEs). Rather, its applicability has been extended to all undertakings and associations irrespective of their size. In addition, the new § 20 I sentence 2 GWB is now explicitly applicable to multisided platforms on whose intermediation services companies are dependent as regards their access to sales and procurement markets. This dependency has to be of such a character that sufficient and reasonable alternative possibilities do not exist. Further, the new § 20 Ia GWB clarifies that relative market power can also be based on the fact that one company is dependent on access to data controlled by another company. The new § 20 Ia GWB also spells out that a refusal of access to data may constitute an unfair impediment even if there is no commerce opened for such data yet. These newly introduced amendments to § 20 GWB reflect the concept of intermediation power and could apply even in favor of a bigger company that is selling its goods on a superdominant trading platform.

[151] Regulation 1/2003 of 16 December 2002 on the implementation of the rules on competition laid down in Articles 81 and 82 of the Treaty, OJ 2003 L 1 p. 1.

[152] Examples for such a dependency can for example be found with: U. Loewenheim, In: U. Loewenheim, C. Kersting, H. J. Meyer-Lindemann (eds), Kartellrecht, C.H. Beck 2020, § 20, paras 20 ff.

[153] Higher Regional Court Düsseldorf, Judgement of 9 January 2015 – VI-Kart 1/14 (V), para 193; Bundeskartellamt, Decision of 20 December 2013 – B9-66/10, paras 235 ff.

6.10.2 Superior Market Power Within a Horizontal Relationship

Section 20 III GWB deals with the superior market power of an undertaking in relation to small and medium-sized enterprises that are its competitors. Accordingly, superior market power refers to the relation between an undertaking and a competing small or medium enterprise. The concept of superior market power requires an imbalance of power which allows the superior undertaking a leeway for unilateral actions which is not sufficiently controlled by competition.[154] The threshold for assuming superior market power is lower than for establishing market power in the sense of Article 102 TFEU.[155] An undertaking with superior market power is prohibited from unfairly impeding such competitors.

The newly created § 20 IIIa GWB is also only applicable to small and medium-sized competitors. It defines a special scenario of abusive conduct on multisided platform markets or networks. On such markets, an undertaking with superior market power is prohibited from impeding the independent attainment of distinctive positive network effects by competitors if this creates a serious risk that competition on the merits is restricted to a not inconsiderable extent. In sum, new amendments of § 20 GWB might be of high significance for securing competition in digital markets.

6.11 Parity Clauses

Best-price clauses or parity clauses that are often used by digital platforms in different sectors forbid sellers to offer their goods or services at lower prices or at better conditions in other distribution channels than on the platform.[156] These distribution channels can, for example, be the seller's own website, competing platforms, or even offline distribution or advertising channels. Depending on the drafting of the contract, best-price clauses can take different forms, which may influence their legal assessment.

While only a few cases have addressed parity clauses on the EU level (see Sect. 6.7), quite many cases have been opened on the Member State level, and parity clauses have partly received a divergent treatment in different Member States.[157] In the digital sector, the Bundeskartellamt and the courts in Germany have repeatedly

[154] U. Loewenheim, In: U. Loewenheim, C. Kersting, H. J. Meyer-Lindemann (eds), Kartellrecht, C.H.Beck 2020, § 20, para 57.

[155] U. Loewenheim, In: U. Loewenheim, C. Kersting, H. J. Meyer-Lindemann (eds), Kartellrecht, C.H.Beck 2020, § 20, para 57.

[156] See for example T. Hoeren, Germany, In: Kilpatrick et al. (eds), Antitrust analysis of online sales platforms & copyright limitations and exceptions, Springer C.H. Beck 2017, pp. 157–185; M.-O. Mackenrodt, Price and conditions parity clauses in contract between Hotel booking platforms and hotels, IIC 2019 (50), p. 1131.

[157] For a comparison of the Swedish and the German hotel booking case see M.-O. Mackenrodt, Price and conditions parity clauses in contract between Hotel booking platforms and hotels, IIC 2019 (50), p. 1131.

classified agreements containing various forms of parity clauses as restrictions of competition by effect, and a justification has in most cases been denied.

In 2013, the Bundeskartellamt had opened proceedings against the Amazon sales platform. The parity clauses used by Amazon essentially prohibited independent third-party traders from offering the goods they offered via the Amazon platform at lower prices on other Internet platforms or in their own online shops. The Bundeskartellamt considered these price parity clauses to be restrictions of competition by object in horizontal trade cooperation, which could not be justified by efficiencies.[158] After Amazon had committed to abstaining from using such parity clauses, the Bundeskartellamt issued a commitment decision and discontinued the proceedings.

The Bundeskartellamt had also objected to the use of parity clauses on a comparison portal for electricity and gas supply contracts (Verivox) where consumers could also conclude contracts with suppliers. The Bundeskartellamt pointed to the positive effects of comparison portals on competition but also uttered concerns that parity clauses would restrict the freedom of suppliers to set prices and hinder competition between platforms.[159] After Verivox had abandoned the use of these best-price clauses, the Bundeskartellamt discontinued the proceedings.

In 2018, the German Bundeskartellamt[160] and the Austrian Bundeswettbewerbsbehörde[161] initiated abuse proceedings with regard to the general terms and conditions that Amazon used in its contracts with third-party traders. The competition authorities had concerns that Amazon was a gatekeeper that controlled the access of third-party traders to their customers and that Amazon was abusing its position by forcing disadvantageous contract clauses upon third-party traders. After negotiations on a wide range of clauses that may have constituted an abuse of market power, Amazon changed its clauses, and the competition authorities ended the investigation. In particular, Amazon dropped a quality parity clause which required traders to use presentation material for their products which is of the highest quality which the particular trader is using in any other of its distribution channels.

Formal decisions on best-price clauses with an extensive discussion of the relevant substantive issues have been taken by the Bundeskartellamt and the courts with regard to hotel booking platforms. In the German cases regarding hotel booking platforms, clauses have been referred to as "narrow" best-price[162] clauses if they prohibited the hotel owner from offering his rooms at lower prices on his own website but allowed him to offer lower prices on competing digital platforms. In

---

[158] Bundeskartellamt, Case Report of 9 December 2013 – B6-46/12.

[159] Bundeskartellamt, Press Release of 3 June 2015; https://www.bwb.gv.at/fileadmin/user_upload/Downloads/standpunkte/BWB_Amazon-Fallbericht_20190717.pdf.

[160] Bundeskartellamt, Case Report of 17 July 2019 – B2-88/18; Bundeskartellamt, Press Release of 29 November 2018 – B2-88/18; Bundeskartellamt, Press Release of 17 July 2019 – B2-88/18.

[161] Bundeswettbewerbsbehörde, Case Report amazon.de Marktplatz, 2019.

[162] The cases also dealt with parity clauses referring to other attributes of the offer than the price, like quality and contractual conditions.

contrast, "wide" best-price clauses, which were, for example, one of the issues in the HRS case, prohibited offering hotel rooms at lower prices on competing platforms. It should, however, be noted that—depending on the drafting of the contract—there are not only two categories but also a wide range of best-price or parity clauses. For example, in the Czech case against the hotel booking platform the "wide" best-price clauses at issue were even wider than the "wide" best-price clauses that were prohibited by the authorities and the courts in Germany. In the Czech case, hotel owners were prohibited from offering better prices and conditions not only on their website and in other online distribution channels but also in offline distribution channels.[163] Such an extremely "wide" best-price clause, which also refers to offline scenarios, contains an even stronger limitation of the economic freedom of hotel owners.

Wide parity clauses of the hotel booking platform HRS were, in 2013, found by the Bundeskartellamt to be infringing Article 101 TFEU.[164] This decision was upheld by the OLG Düsseldorf[165] and is final. It was found that wide parity clauses restrict the economic freedom of hotel owners and have the effect of restraining competition on the market for booking platforms and on the market for hotel rooms. Regulation 330/2010[166] was not applicable because in this case, the 30% market share threshold was exceeded. However, the Bundeskartellamt and the OLG Düsseldorf pointed to several contentious issues with regard to the application of Regulation 330/2010 to parity clauses.[167] An individual exemption according to Article 101 (3) TFEU on the ground of alleged efficiencies was denied.[168]

Wide parity clauses have also been at issue in a private competition law case that was decided by the OLG Düsseldorf (Expedia).[169] In this case, the defendant, the hotel booking platform Expedia, held a market share of below 30%. The OLG Düsseldorf upheld the wide parity clauses because due to the low market share of Expedia, Regulation 330/2010 would, at any rate, apply.[170] It was therefore left open in this case the question of whether there was a restriction of competition or an individual justification.

---

[163] UOHS (Czech competition authority), Press release of 11 November 2019.

[164] Bundeskartellamt, Decision of 20 December 2013 – B9-66/10.

[165] Higher Regional Court Düsseldorf, Judgement of 9 January 2015 – VI-Kart 1/14 (V).

[166] Vertical Block Exemption Regulation 330/2010 of 20 April 2010 on the application of Article 101(3) of the Treaty on the Functioning of the European Union to categories of vertical agreements and concerted practices, OJ 2010 L 102 p. 1.

[167] Bundeskartellamt, Decision of 20 December 2013 – B9-66/10, paras 181 ff.; Higher Regional Court Düsseldorf, Judgement of 9 January 2015 – VI-Kart 1/14 (V), paras 164 ff.

[168] Regional Court Düsseldorf, Judgement of 9 January 2015 – VI-Kart 1/14 (V), para 172; Bundeskartellamt, Decision of 20 December 2013 – B9-66/10, para 195.

[169] Higher Regional Court Düsseldorf, Judgement of 4 December 2017 – VI-U (Kart) 5/17. This decision upheld the first instance decision of the District Court Köln, Judgement of 16 February – 88 O (Kart) 17/16.

[170] Higher Regional Court Düsseldorf, Judgement of 4 December 2017 – VI-U (Kart) 5/17, para 54.

Narrow best-price clauses of the hotel booking platform were declared to be invalid by the Bundeskartellamt in 2015.[171] The decision was initially reversed by the OLG Düsseldorf[172] but upheld by the BGH in 2021.[173] Unlike the OLG Düsseldorf, the BGH rejected the application of the ancillary restraint doctrine, an unwritten exemption to Article 101 (1) TFEU.[174] The BGH pointed out that the elimination of best-price clauses in 2016 had not harmed Booking.com's market success, and had even strengthened its market position. Accordingly, the narrow best-price clauses were held to be not objectively necessary for offering the platform booking services, as required by the ancillary restraints doctrine. The restriction of competition regarding the offering of hotel rooms on the hotel's own website constitutes an infringement of Article 101 (1) TFEU,[175] which is not justified by efficiencies according to Article 101 (3) TFEU.[176] The BGH held that narrow best-price clauses were not a precondition for efficiency gains generated by the platform service[177] and that investigations had not shown that free riding severely threatened the functioning of the platform service.[178]

Unlike Germany, several Member States have introduced special legislation that deals with best-price clauses.[179] At the EU level, Article 10 (1) of the platform to business Regulation 2019/1150[180] addresses best-prices clauses. It requires an intermediation platform that uses best-price clauses in its terms and conditions to spell out the main economic, commercial, or legal considerations for these restrictions. The reasons need to be easily available to the public. Regulation 2019/1150 has applied since July 12, 2020, in the Member States. However, a mere transparency approach may not be sufficient to address the potential harm from best-price clauses. In any case, Article 10 (2) Regulation 2019/1150 explicitly states that it does not exclude other legal limitations on the use of best-price clauses that may derive from other legal acts of the European Union or from laws of the Member States. According to Art. 5 1. (d) of the new Vertical Block Exemption Regulation 2022/720 across-platform retail parity clauses are not exempted while other types of parity obligations can benefit from the exemption.

[171] Bundeskartellamt, Decision of 22 December 2015 – B9-121/13.

[172] Higher Regional Court Düsseldorf, Judgement of 4 June 2019, VI-Kart 2/16 (V).

[173] BGH, Judgment of 18 June 2021 – KVR 54/20 – booking.

[174] BGH, Judgment of 18 June 2021 – KVR 54/20, para 23 – booking.

[175] BGH, Judgment of 18 June 2021 – KVR 54/20, para – booking.

[176] BGH, Judgment of 18 June 2021 – KVR 54/20, para 54 – booking.

[177] BGH, Judgment of 18 June 2021 – KVR 54/20, para 59 – booking.

[178] BGH, Judgment of 18 June 2021 – KVR 54/20, para 69 – booking.

[179] For example, Austria has included a prohibition of best-price clauses in the hotel sector into the blacklist of illegal clauses of its unfair competition law; see No 32 in the blacklist of UWG.

[180] Regulation 2019/1150 of 20 June 2019 on promoting fairness and transparency for business users of online intermediation services, OJ 2019 L 186 p. 57.

Hungary

7

Judit Firniksz and Péter Mezei

7.1 Introduction

The digital era presents National Competition Authorities (NCAs)[1] with new challenges since the impact of the fourth industrial revolution reaches all markets and affects market behaviour. Datafication, digitalisation, data-driven markets, online platforms, digital economy, the Internet of Things—even this short list of examples highlights that a new set of terminology requires interpretation and the underlying new market phenomena require new assessment methods. The digitalisation of the value chain is reshaping processes from the development of products through production, logistics and sales systems to their acquisition by end users. Product digitalisation is creating new markets and is fundamentally changing data-driven business models, which are simultaneously generating new organisational structures and working methods. There are industries where consumers pay with their personal data for 'free' services, and consumers' data serve as the fuel for economic growth. These parallel processes result in new

Members of the Working Group (in alphabetical order): Virág Balogh, Bálint Bassola, Balázs Bence, Borbála Dömötörfy, Krisztina Grimm, Maxim Járdai, Anikó Keller, Márton Kocsis, Ádám Liber, Ákos Réger and Ildikó Sopov.

[1] National competition authorities.

J. Firniksz (✉)
Firniksz Law Firm, Budapest, Hungary

Pázmány Péter Catholic University, Budapest, Hungary
e-mail: judit.firniksz@ecommercelawyer.eu

P. Mezei
Hegymegi-Barakonyi & Partner Baker & McKenzie Law Firm, Budapest, Hungary
e-mail: peter.mezei@bakermckenzie.com

B. Kilpatrick et al. (eds.), *Antitrust in Data Driven Markets & Legal Framework for Influencers, Native Advertising and Control over the Use of AI in Marketing*, LIDC Contributions on Antitrust Law, Intellectual Property and Unfair Competition, https://doi.org/10.1007/978-3-031-07422-6_7

channels and context in the communication with buyers, consumers, customers, clients, users and employees.[2] At the same time, traditional roles have been expanded since consumers are present on the demand side not only by virtue of their purchasing decisions but also because they create new content, share information, rank products and discuss their performed or planned consumer choices in all available public fora.[3]

In the changing economic environment, the Hungarian Competition Authority (Gazdasági Versenyhivatal (GVH)) has a crucial role in articulating the legal requirements. Since the GVH is among those NCAs that have competencies both in antitrust and in consumer protection, it is a key issue of the efficiency of enforcement to equilibrate the enforcement tools in order to understand the underlying market developments and, if necessary, to find the adequate form of intervention.

7.2 The Role of Data in the Digital Markets: Data Protection Implications in Competition Law

Like other competent authorities in the European Union (e.g. Germany, France) and elsewhere (e.g. United Kingdom, United States of America), the GVH has made certain efforts to assess the behaviour of social network operators, taking into account data protection requirements in the EU. Firstly, the GVH generally recognises the data protection angle of 'consumer welfare' because consumers consider the privacy aspects of online products as a significant product characteristic. In the *Google Allo* case,[4] the GVH investigated whether consumers received the information necessary to make an informed decision on Google's data processing activity. In its decision, the GVH considered that data processing is an essential aspect of Google's products and that the consumers' informational self-determination right has not only privacy-related but also market- and competition-related aspects. Accordingly, the GVH acknowledged that the non-transparent nature of data processing may be a relevant factor in consumers' transactional decisions. Secondly, in the *Facebook* case,[5] the GVH stated that the business model of large online providers (such as Facebook and Google) involves user data monetisation, and it recognised that such data have a monetary value when used for advertising or providing access to advertising. However, the GVH has not yet

[2] A. Preta, A. and M. Maggiolino (Eds.), Data Driven Economy: Market Trends and Policy Perspectives (2018), available from: http://www.itmedia-consulting.com/DOCUMENTI/datadrivensummary.pdf. Accessed 10 August 2020.

[3] J. Firniksz and B. Dömötörfy, Information Exchange Going Digital – Challenges to Hungarian Competition Law Enforcement. YARS, 12(19), pp. 133–133.

[4] Decision of the GVH no. Vj/88-71/2016, GVH Press Release, available at https://www.gvh.hu/en/press_room/press_releases/press_releases_2018/competition_proceeding_against_google_is_closed. Accessed 9 September 2021.

[5] Decision of the GVH no. Vj/85-189/2016.

explicitly concluded that large social networks gain advantages over competitors by collecting and merging data from several applications.

Regarding the data privacy angle of the use of social networks and the related data collection in Hungary, pursuant to the Fundamental Law of Hungary, the enforcement of data protection rules is the exclusive responsibility of the National Authority for Data Protection and Freedom of Information (Nemzeti Adatvédelmi és Információszabadság Hatóság (NAIH)). However, since data protection may be regarded as a part of consumer welfare and a data-related practice might also constitute an unfair commercial practice (UCP) against consumers, the GVH adopted a policy to intervene and enforce unfair competition rules (see Sect. 4 below). The GVH has not yet applied the antitrust rules to the data processing activities performed by social networks and similar online providers.

It is worth mentioning that, unlike in certain Member States of the European Union (e.g. *Facebook Bundeskartellamt* case[6]), there is no precedent in Hungary relating to the use of inadmissible general terms and conditions regarding data collection.

7.3 Platforms

7.3.1 Definition of the Term 'Platform'

In its approach to assessing what constitutes a platform, the GVH refers to the definition laid down by the Guidance on the Implementation of UCP Directive[7] and indicates as the key function of platforms their capacity to provide infrastructure and enable interactions between suppliers and users for the provision of goods, services, digital content and information online. Further, the GVH emphasises the diversity of business models applied by online platforms, ranging from allowing users to search for information supplied by third parties to facilitating, often against remuneration, contractual transactions between third-party traders and consumers or advertising and selling, in their own name, different kinds of products and services, including digital content.[8]

7.3.2 Market Power of Networks and Platforms

Although the GVH has been proactive in the field of digital markets in the last couple of years, it has focused predominantly on the enforcement of consumer protection norms. Other major competition-law-enforcing jurisdictions (such as the European

[6] Bundeskartellamt Beschluss no. B6-22/16 Facebook.

[7] Guidance on the Implementation/Application of Directive 2005/29/EC on Unfair Commercial Practices, section 5.2.1.

[8] Facebook case, para 38–39.

Union and Germany) have heavily relied on regulations concerning abuse of dominance and, thus, assessed the relevant issues of digital markets even in the context of business-to-business (B2B)[9] relationships. In contrast, the GVH's practice is quite sporadic in this regard, and it almost entirely lacks dominance cases of this kind. Nevertheless, its reliance on consumer protection norms allowed the GVH to tackle similar issues (e.g. the way platforms use users' data in the context of the *Facebook* case) and to reflect on them from a business-to-consumer (B2C)[10] perspective. Given the fact that consumer protection procedures are significantly shorter because they do not require market analysis, they enable the GVH to intervene in the underlying anti-competitive practices in a timely and effective manner. However, although the GVH issued multiple guidelines and other soft law instruments concerning digital markets, none of them addressed specifically market power. Consequently, there is somewhat of a vacuum when it comes to the definition of market power in relation to digital markets since there is no developed case law in this field yet, given that market analyses focusing on abuse of dominance are largely absent.

As regards market power, the GVH's practice and case law in merger control and a recent decision on vertical restrictions can also be considered relevant. These decisions suggest that the GVH's practice acknowledges that access to data and the increasing potential of online platforms may be considered significant factors for market power in digital markets, and thus in future cases, competition law enforcement should consider these factors as well.

7.3.2.1 Effect of Online Presence on Offline Market Power

In the *Mömax/Kika* case,[11] where the Mömax group was about to take over Kika stores, the GVH stated that the online and offline sales of home furnishings belonged to one and the same product market. In the *Tesco/MediaMarkt* case,[12] where MediaMarkt was to take over from Tesco the sale of electronic consumer devices by establishing a shop-in-shop construction, the GVH did not provide a definitive answer to the issue regarding the separation of online and offline markets, stating that the merger would not result in a significant lessening of competition in any way. A conclusion may be drawn that the GVH's practice acknowledges the existence of competition pressure from online channels to offline stores, but the level of such pressure and the underlying market circumstances are to be examined on a case-by-case basis.

---

[9] Business-to-business.

[10] Business-to-consumer.

[11] Decision of the GVH no. Vj/24-60/2019, paras 43–46.

[12] Decision of the GVH no. Vj/34-68/2018, para 55.

7.3.2.2 The Role of Data in Mergers Between Players of Digital Markets

In the *Netrisk/Biztosítás.hu*[13] case concerning the acquisition of Biztosítás.hu,[14] a Hungarian online insurance broker, by a major competitor, Netrisk.hu, the GVH assessed in detail the effects that the takeover of Biztosítás.hu's database by Netrisk could have in the market. Citing the European Commission's *Apple/Shazam EU* case,[15] the GVH stated that the main criteria in the assessment of data are composition, quantity, speed and overall value, concluding that the most valuable data are the ones that are hard to obtain and, at the same time, have a significant potential to indicate consumer behaviour and habits. In this regard, the GVH's examination addressed three potential competition issues concerning the merger of the databases, focusing on (i) whether Netrisk would become capable of providing services that it previously was unable to provide, (ii) whether Netrisk would be able to provide custom-made offers of higher quality and (iii) whether the data are essential for the operation of neighbouring markets. Finally, the GVH concluded that the merger of the databases did not raise any competition concerns because the data stored by them are largely similar, and due to privacy regulations, such data can be used only in a narrow scope and in aggregated form; therefore, it did not provide a genuine competitive edge.

In the *eMAG/Extreme Digital* case,[16] Dante International Kft., operating the eMAG.hu online store, intended to acquire control over Extreme Digital Zrt., operating the edigital.hu online store and 16 retail outlets in Hungary. The GVH noted that Hungarian consumers generally attach a large degree of importance to the ability to visit the physical versions of online stores; therefore, the combination of eMAG's digital capabilities with Extreme Digital's network of offline stores will follow the market-wide trend of multi-channel distribution, which most likely will result in fostering innovation. As in the *Netrisk/Biztosítás.hu* case, the GVH, concluded that the merger of databases was unlikely to lessen competition since the data are not essential for competitors, they can easily be reproduced and their use was limited by data privacy regulations.[17]

7.3.2.3 'New' Merger Regime and Ancillary Restrictions in the Context of Start-Ups

In January 2017, a new merger notification regime was introduced in Hungary, one aim of which was to allow the GVH to exercise control over some lower-value transactions (below the relevant thresholds) that otherwise could cause significant changes in the market structure and alter the position of market players. Examples include the acquisition of start-ups or newly established players on the digital markets with zero or low net revenue but with a high growth, development and

[13] Decision of the GVH no. Vj/12-248/2019.
[14] Netrisk/Biztosítás.hu case, paras 74–79.
[15] Case M.8788 - Apple/Shazam.
[16] Decision of the GVH no. Vj/14-175/2019.
[17] eMAG/Extreme Digital case, para 74.

innovation potential. GVH Order no. 11/2017[18] made it clear that the new regime's open intention was to catch the acquisitions of start-ups as well. The relevant criteria in the assessment of a start-up are, in the GVH's view, its potential for growth, the know-how, innovation and technology, and goodwill that are the basis of its value, and how easily the relevant technology is substitutable.[19]

Another relevant issue in mergers concerning start-ups is the ancillary restrictions applied by investors acquiring the business to limit the founders' or inventors' opportunities to re-utilise the original idea that has already constituted the base value of a start-up. Previously, the GVH—in line with the existing EU approach on ancillary restrictions—accepted non-compete obligations for a limited time period (generally 2 or 3 years), to be applied to founders and investors completely leaving the firm. This is because the Hungarian Competition Authority (HCA) acknowledged that the exit of an inventor often means that also professional knowledge (necessary to use or develop the original idea) would abandon the original business and that it may entail the devaluation of the financial investment made by other investors by unfairly generating competition to the original business. Now, in its recent *CodeCool*[20] case, the GVH extended the above approach also to innovators who remain with the firm as minority owners; thus, they can also be imposed a non-compete requirement during the period of their minority ownership and for an additional two years after the future termination of such ownership.

7.3.2.4 Issues Raised in the *NetPincér* Case[21]

In the *NetPincér* case (see Sect. 5.2.2 below), the GVH assessed both sides of the relevant two-side market, i.e. consumers and restaurants. For the consumer side of the market, the GVH concluded that consumers have some degree of choice and, therefore, would probably switch in case of a 5–10% price increase (small but significant and non-transitory increase in price (SSNIP) test); however, it also noted that 'comfort services' (i.e. the lack of need for registration or the simplicity of the ordering process) are among the most important aspects for consumers. Nevertheless, the GVH could not definitively state whether direct orders from restaurants exert competitive pressure on NetPincér. Further, the GVH clearly laid down that direct orders could not substitute, for restaurants, the engagement on NetPincér's platform since the platform presents a marketing opportunity and reaching similar audiences would require significant marketing investments.[22] Although the *NetPincér* case concerned the assessment of vertical agreements, it is apparent from the presented market data that NetPincér could have been considered a dominant player in the market of online food delivery platforms between 2012 and

[18] GVH Order no. 11/2017. (XII. 20.) on the notification form and related fill-in guidance to be applied for the notification of concentrations under Section 24 of the Competition Act.

[19] GVH Order no. 11/2017, point V.9.2.

[20] Decision of the GVH no. Vj/19-60/2019.

[21] Decision of the GVH no. Vj/89-236/2015.

[22] NetPincér case, paras 69–70.

2016 (its market share did not go below 70%). The GVH noted that the barriers to entry are high since although establishing an online presence does not present an overly heavy financial burden, building a reputation and achieving optimal capacity utilisation that would allow profitable long-term market presence would require significant resources and investments. The GVH also highlighted the relevance of existing network effects as a key factor for the assessment of these online platforms, referring to the fact that, for restaurants, only platforms that have a high number of consumers are attractive and, for consumers, only platforms that have a high number of restaurants are attractive.[23]

7.4 Market for Free Services

Generally speaking, services offered for 'free', i.e. at a monetary price of zero, are not new to the NCAs' practice. Therefore, to interpret the concept of 'market relationship for free services', considerable attention must be paid to the following facts. On the one hand, some NCAs have already delivered opinions regarding 'free services' that can be used as a starting point. On the other hand, some NCAs have competencies in other areas of law, i.e. in the area of consumer protection law. Therefore, investigations may be launched and notices/documents may be issued regarding 'free services' also in such areas, and the lessons drawn in this field can still be relevant inputs to antitrust analyses.

The GVH is among those NCAs that are competent for both competition and consumer protection. Regarding zero price markets, in general, the GVH stated in its 2018 note to the Organisation for Economic Co-operation and Development (OECD) that the 'GVH is not precluded from addressing any competition or consumer protection issue related to quality in zero price markets, as the general rules of consumer protection/competition law apply to investigations concerning these markets as well'.[24] The GVH also noted in accordance with its Digital CP Strategy[25] that since it 'believes that consumer protection enforcement can ensure a faster, more active invention and enforcement in a timely manner, the GVH has taken steps in this field'.[26]

Some of these steps concern data-related questions, but as the GVH also noted in the *Google Allo* case, the competition authority investigates unfair commercial practices, but it does not assess the information flow between the parties from a

[23] NetPincér case, paras 76–78.

[24] Quality considerations in the zero-price economy – Note by Hungary, available at https://one.oecd.org/document/DAF/COMP/WD(2018)131/en/pdf, point 5. Accessed 9 September 2021.

[25] Hungarian Competition Authority - Mid-Term Digital Consumer Protection Strategy, 2018 https://www.gvh.hu/pfile/file?path=/en/gvh/strategy/GVH_Stategia_Digitalis_fogyved_startegia_2018_09_27_a&inline=true, point 13.

[26] Quality considerations in the zero-price economy – Note by Hungary, point 6.

data protection point of view. Notwithstanding this, the commercial practice can be unfair even if there is no breach of data protection regulations (see Sect. 2 above).[27]

In the GVH's approach, the consumer decisions regarding non-monetary, 'paid-with-data' transactions qualify as transactional decisions with market significance. Consequently, the relevant B2C commercial communications were assessed as commercial practices concerning essential features of the services, and the GVH concluded that consumers might make a different choice if they are well informed. Thus, consumers and businesses are involved in transactions regarding 'free services', and the GVH's approach and case law appear to outline a 'market relationship for free services'.

7.4.1 Case Law

In its decisions prior to the emergence of digital markets,[28] the GVH had already established that monetary payment was not the only consideration to be made by the consumer: other circumstances, among others, limiting the scope of future decisions might serve as a consideration of equal value as well.[29] The GVH went even further when saying that monetary compensation does not belong to the conceptual elements of a transactional decision, and the consumers' decision to provide their personal data can be a transactional decision in itself.[30]

The GVH reinforced its earlier assessment in the *Google Allo* case when it assessed the consumer decision related to the downloading, installation and use of the Allo chat client and its new functions as a transactional decision.[31] While partly terminating the case and partly closing it with commitments,[32] the GVH also added digital-market-related elements when it investigated whether the consumer received adequate information about the data processing activity in order to reach well-informed decisions. The information received by the consumer during the installation of the application qualified as a commercial practice.[33] Furthermore, the GVH established that the products and services of the digital world 'can be described with demand and supply patterns', that market participants are profit oriented but at least seek to work in a rational way and that they do not offer their products and services

---

[27] Google Allo case, para 245.

[28] To conducts until 31 August 2018 the GVH applied the Competition Act, and from 1 September 2018, it applies the UCP Act. The latter ensures compliance with the UCP Directive.

[29] GVH press release, available at https://www.gvh.hu/en/press_room/press_releases/press_releases_2008/5022_en_broadband_paid_narrow-band_provided.html. Accessed 10 August 2020.

[30] Decision of the GVH no. Vj/93-34/2011.

[31] Google Allo case, para 214.

[32] The commitments had been fulfilled as the GVH established it in the Google Allo follow-up case (Decision of the GVH no. Vj/18-14/2019).

[33] Google Allo case, para 207.

without remuneration.[34] The primary consideration provided by the consumer is information that has economic value for Google,[35] and the consumer has to be aware of this fact and should not to be 'mistaken [. . .] about the possibilities and nature of its own – market-significant - transactional decision'.[36]

In addition, in the *Facebook* case, specific elements were added to the assessment of 'free services'. The GVH established that Facebook breached the prohibition of unfair commercial practices by actively misleading the consumers regarding the price of its services by misleadingly claiming that the services are free to its users.[37] The GVH here repeatedly noted that the consideration provided by consumers in the form of information is of serious market value. Further, the GVH emphasised that consumers' sharing their user data held a significant weight, which cannot be disregarded either in the calculation of the price of the service or in the relevant B2C commercial communication.[38] Finally, the GVH added that 'if the average prudent consumer knew that his/her free registration and use of Facebook [. . .] represents significant market value, he/she could make different transactional decisions with regard to the use of Facebook'.[39]

7.4.2 Free Services in the Context of DCTs[40]

Regarding the use of zero-price offers, faster and easier decision-making of consumers might be supported by digital comparison tools (DCTs) that could foster competition. Whether DCTs are zero-price services or not is also a relevant factor. Generally, they are not promoted as 'free', but at the same time the use of consumer data is part of some of the business models. Therefore, undertakings offering these services must be aware of the prohibitions regarding unfair commercial practices, as was noted by the GVH in its DCT Market Study.[41]

[34] Google Allo case, para 245.

[35] Google Allo case, para 220.

[36] Google Allo case, para 245.

[37] UCP Act, Section 3 (1) and (6); Section 3 (1) point c).

[38] Facebook case, para 256.

[39] Facebook case, para 274.

[40] Digital Comparison Tools.

[41] GVH Press Release, available at https://www.gvh.hu/en/press_room/press_releases/press-releases-2020/how-can-digital-comparison-tools-affect-consumer-decision-making. Accessed 9 September 2021.

7.5 Enforcement Measures in the Field of Digital Markets and Online Services

The focus of enforcement measures in respect of digital markets and online services depends mostly on the specifics of the case and the issues the concerned NCA faces and/or market characteristics. To understand new market phenomena, NCAs are entitled to perform certain kinds of market surveys (sectoral inquiries or market studies) and address anti-competitive and unfair market practices experienced during such market surveys.

Firstly, surveys focused on digital markets might be useful to gain an insight into issues being subject to potential future investigations even beyond national boundaries and to assess such issues within the framework of international comparison. Secondly, surveys offer unique opportunities to map the relationships and interactions among the numerous entities operating on these markets, such as platform providers, retailers, advertising agencies, marketing agencies, sales experts, data-mining and data-analyst companies, data scientists, software firms developing algorithms, etc. Thirdly, surveys can serve as a basis for a forward-looking approach by understanding the market players' views on the current forms of cooperation and potential alternative solutions. Finally, surveys may allow a more comprehensive understanding of the demand side through opinions of consumer organisations and cooperation with agencies working in the given industries. As a result, surveys may provide relevant inputs to provide a picture of the actual status of these markets as accurately as possible and offer the opportunity to take the most effective steps and/or determine the most effective recommendations.

In Hungary, the GVH's measures taken so far have had a clear focus on the online hotel booking segment and the diversity of DCTs.[42] In recent years, it has actively used a range of tools at its disposal to assess underlying market developments. For example, the GVH may initiate competition supervision proceedings in connection with the application of a DCT that constitutes an unfair commercial practice. In several cases, the GVH has initiated proceedings against companies that use DCTs as well. Secondly, the GVH may launch a sector inquiry into a given sector in order to identify and assess its market developments, as it did for example between 2013 and 2016 in the online accommodation booking market, which is characterised by the application of DCTs. Thirdly, the GVH may also carry out market studies to assess and analyse certain market practices and their impact on competition and,

[42]The concept of DCT is defined in the DCT Market Study in accordance with the concept outlined by the multi-stakeholder group also referred to in Section 5.2.7 of the Guidance on the Implementation of UCP Directive. Thus, also for the purposes of this document, the term DCT includes all digital content and applications developed to be used by consumers primarily to compare products and services online, irrespective of the device used or the parameter(s) on which the comparison is based (e.g. price, quality, user reviews). To the extent that operators of search engines, travel or ticket booking sites, e-commerce platforms acting as a marketplace for several traders develop functions or applications dedicated to the comparison of products and services, these functions or applications are also covered by the term DCT.

most importantly, on consumers—such as the market study it conducted between 2019 and 2020 inquiring about the effects of DCTs on consumer decision-making.

7.5.1 GVH Proceedings Against DCTs Constituting Unfair Commercial Practices

Pursuant to the UCP Act,[43] the Hungarian regulation implementing the UCP Directive,[44] the GVH may initiate competition supervisory proceedings in the event of conduct that is likely to violate the prohibition of unfair commercial practices and falls within its competence, meaning the underlying commercial practice can substantially affect competition on the market. As for the DCTs that are available online nationwide, or across the European Union or globally, there is a high probability of substantial effects on competition. Consequently, if a market practice related to the application of a DCT falls under the scope of the UCP Act, the GVH's competence extends to the examination of the related commercial practices. For this, the UCP Act requires the operator of the DCT to be a 'business entity' and for its activity relating to the operation of the DCT to be a 'B2C commercial practice', conducted in the territory of Hungary.[45]

7.5.1.1 DCT as Commercial Practice

In terms of the UCP Act, in accordance with the UCP Directive's approach,[46] various DCTs can be considered a 'B2C commercial practice' and their operators a 'business entity'. On the one hand, DCTs that are operated by a given trader in connection with its own goods clearly qualify as 'B2C commercial practices'. In addition, comparison websites as well as applications and product comparison functions of search engines and for which third-party operators receive some sort

---

[43] Act XLVII of 2008 on the Prohibition of Unfair Business-to-Consumer Commercial Practices.

[44] Directive 2005/29/EC of the European Parliament and of the Council of 11 May 2005 concerning unfair business-to-consumer commercial practices in the internal market and amending Council Directive 84/450/EEC, Directives 97/7/EC, 98/27/EC and 2002/65/EC of the European Parliament and of the Council and Regulation (EC) No 2006/2004 of the European Parliament and of the Council.

[45] Pursuant to the UCP Act, 'consumer' means 'any natural person who is acting for purposes which are outside his/her independent profession or economic activity' (UCP Act, Section 2., point a)), while a 'business entity' means 'any person who, in commercial practices, is acting for purposes relating to his/her independent profession or economic activity' (UCP Act, Section 2., point b)). Further, according to the UCP Act, 'B2C commercial practices' means 'any act, omission, course of conduct or representation, commercial communication including advertising and marketing, by a person for a business entity or in the name and on behalf of a business entity, directly connected with the promotion, sale or supply of a product to consumers' (UCP Act, Section 2.., point d).

[46] Comparison Tools - *Report from the Multi-Stakeholder Dialogue* - Providing consumers with transparent and reliable information, Report presented at the European Consumer Summit 18–19 March 2013. Available at https://ec.europa.eu/info/sites/info/files/consumer-summit-2013-msdct-report_en_0.pdf. Accessed 9 September 2021.

of consideration (including also advertising and sponsorship) from manufacturers or distributors for the display or comparison of goods also belong to this category.[47] The latter category also covers DCTs that offer direct opportunities to establish a contract; i.e., upon selecting the compared or displayed product, the DCT does not redirect the user to the website of the third party that produces or distributes the goods but allows them to purchase the product directly. Examples include the online platforms and the product comparison functions of the multi-merchant e-commerce platforms and online marketplaces (Booking.com, Szállás.hu, Amazon, eBay, etc.). Finally, for the sake of completeness, 'paid for DCTs' that allow consumers to use the comparison tools as a service provided in exchange for payment may also constitute a 'B2C commercial practice'.

7.5.1.2 GVH Case Law Concerning Online Platforms

In order to ensure the lawful operation of online platforms, the GVH launched several investigations into companies that apply DCTs.[48] In these proceedings, the GVH assessed mainly misleading statements concerning insurance products and online booking services (statements on the adequacy of the indications on prices and costs, the favourable nature of the offer or the extent of the savings, market leadership, the scope of the comparison, the actuality of information, the allegation of a best-guaranteed price or a free cancellation option). In several cases, the proceedings were closed by an infringement decision or by commitments. In addition to the assessment of the misleading nature of information, in its latest decision of April 2020, the GVH has also examined commercial practices because of the inconsistent presentation of the preferred means of payment that violates the requirement of professional diligence and their aggressive, urging nature that may put consumers under psychological pressure caused by scarcity bias. In order to confirm the 'Fear of Missing Out' effect that the aggressive commercial practices have on the transactional decision, the GVH referred to both its public-survey-supported DCT

[47] The DCTs that are imposed by law, cannot, in our view, be considered a 'B2C commercial practice'. In Hungary, the operation of a comparison portal is currently required by law for the monetary and credit market and telecommunications services. Concerning financial services, the Hungarian National Bank (Magyar Nemzeti Bank) has prepared financial product search and comparison programs and calculators, while for the telecommunications services, the joint information page of the NRA and the Commissioner for Media and Communications provides information on fixed and mobile tariffs, data on the costs of roaming (using a mobile phone abroad) and information enabling a comparison between bundled service providers (landline, Internet and television together).

[48] The Netrisk I. case (Decision of the GVH no. Vj/43/2008/19), the Netrisk II. case (Decision of the GVH no. Vj/75-84/2013), the Biztosítás.hu case (Decision of the GVH no. Vj/19-81/2010), the Szállás.hu case (Decision of the GVH no. Vj/112-61/2013), the Allegro Group/Hotel Outlet case (Decision of the GVH no. Vj/113-103/2013), the Agoda case (Decision of the GVH no. Vj/93-37/2014), the Airbnb case (Decision of the GVH no. Vj/89-90/2016), and the Booking.com I. case (Decision of the GVH no. Vj/70-59/2014) and Booking.com II. case (Decision of the GVH no. Vj/17-110/2018) are to be mentioned as leading cases.

Market Study (see Sect. 5.3 below) and the findings of behavioural economics and neuromarketing.[49]

As it is apparent from the above cases, in recent years, the GVH has been very active in the field of consumer protection concerning market practices involving DCTs, and this is clearly reflected also in the increasing amounts of fines in consumer protection cases.[50]

7.5.2 Sector Inquiries

If the change in prices or other circumstances indicates that competition may be restricted or distorted in the market of any particular industry, the GVH may initiate a sector inquiry in order to study and analyse market developments. The GVH communicates the initiation of the sector inquiry by way of a notice. In the framework of a sector inquiry, the GVH may request entities to provide information. Failure to comply with the request, filing it belatedly or supplying it with false or untrue data may result in an administrative penalty of up to 1% of the net turnover for the previous financial year.[51] The GVH prepares a report on the outcome of the sector inquiry, which it publishes on its website. Before the adoption of the report, the GVH provides an opportunity for companies operating in the affected sector to present their opinion. If, based on the outcome of the sector inquiry, a market disturbance exists and such disturbance cannot be remedied in whole or in part by means of competition supervisory proceedings, the GVH informs the competent legislative or law enforcement body thereof and, if necessary, initiates the enactment or amendment of legislation. Furthermore, the GVH may publish non-binding recommendations for market participants on best practices for maintaining and promoting fair and effective competition, to facilitate the supply of information to customers, and to the market behaviour recommended to be followed. In addition, one of the most effective tools at the GVH's disposal is that it can initiate proceedings in respect of individual cases. If, based on the results of a sector inquiry, it considers that the eventual competition concerns justify it, it can also initiate competition supervisory proceedings in order to properly address the concerns. The GVH may not impose a substantive competition supervisory fine during the sector

[49] GVH Press release is available at https://www.gvh.hu/en/press_room/press_releases/press-releases-2020/gigantic-fine-imposed-on-booking.com-by-the-gvh. Accessed 9 September 2021.

[50] As compared to the previous five years, in 2019 the fines imposed in both the cartel and the consumer protection cases were significant, and the sum of the consumer protection related fines (HUF 4,891,162,500 – approx. 14,162,504 EUR) materially exceeded that of the cartel fines (HUF 3,389,910,300 – approx. 9,815,585 EUR) (calculated on the official daily exchange rate published by the Central Bank of Hungary [Magyar Nemzeti Bank, MNB] on 10 August 2020, when EUR 1 equalled HUF 345.36). The GVH's report to the Hungarian Parliament, 2019 – available at https://gvh.hu/pfile/file?path=/gvh/orszaggyulesi_beszamolok/gvh_ogy_pb_2019&inline=true, pp. 22–23. Accessed 9 September 2021.

[51] Section 43/D of the Competition Act.

inquiry; however, it may impose penalties in the competition supervisory proceedings initiated following the sector inquiry for the identified unlawful market conduct.

7.5.2.1 The Outcomes of the Online Accommodation Booking Sector Inquiry[52]

On 29 July 2013, the GVH launched the Online Accommodation Booking Sector Inquiry on the online accommodation booking market of the tourism sector, where the market players typically operate DCTs as well. The sector inquiry covered the period 2010–2015, and the GVH published its final report on the Online Accommodation Booking Sector Inquiry on 8 June 2016.[53]

As a result of the Online Accommodation Booking Sector Inquiry, the GVH established that, through the appearance of online travel agencies, the market has become more transparent as several accommodation offers can be compared on the website of an online travel agency. The increased transparency also has led to heightened inter-brand price competition among the entities offering accommodation. Simultaneously, the capacity utilisation of the accommodations has increased. However, the GVH also established that there was no substantive intra-brand price competition among the online travel agencies in the market, which, according to the GVH, was caused by the use of the price parity clauses stipulated by agreements between the online travel agencies. The GVH took the view that the general use of the parity clauses could restrict competition in the market by standardising prices and increasing entry barriers. At the same time, as in the case law of many European countries, the GVH considered it acceptable for the online travel agencies to introduce the so-called narrow price parity clauses instead of the wide price parity clauses.[54]

7.5.2.2 The GVH Approach and Case Law on Price Parity Clauses

In Hungary, as in many other Member States, a stricter approach applies to the best price or most favoured nation (MFN)[55] clauses, also known as price parity clauses, if they are fully applied (wide parity clauses).[56] They are considered to violate the general prohibition on the restriction of competition regulated by Section 11 of Act

---

[52] Hungarian Competition Authority - Report on the Sector Inquiry on the Online Accomodation Booking Market - May 2016.

[53] Final Report on the Online Accommodation Booking Sector Inquiry published by the GVH on 8 June 2016 is available at https://www.gvh.hu/pfile/file?path=/en/resolutions/sectoral_inquiries_market_analyses/sectoral_inquiries/Agazati_vizsgalat_online_szallashelyfoglalas_piacan_vegleges_jelentes_2016_06_08_a&inline=true. Accessed 9 September 2021.

[54] Online Accomodation Booking Sector Inquiry, pp. 6–7.

[55] Most-favoured-nation.

[56] 'Fully applied' or 'wide' price parity means that the obligor of the price parity clause (e.g. an accommodation provider) is not allowed to offer/advertise either itself (on its own website) or through a third party (e.g. a third-party accommodation booking platform intended to utilise) more favourable prices or conditions concerning its product/service than those offered towards the party

LVII of 1996 on the Prohibition of Unfair and Restrictive Market Practices (Competition Act) and Article 101 of the Treaty on the Functioning of the European Union (TFEU). Such restrictions may be exempted from the prohibition by a block exemption or an individual exemption, provided that the related conditions are met. Under Hungarian competition law, however, narrow parity clauses may be considered acceptable[57] since these agreements do not expressly restrict competition by object (rather, they may do so based on effect), in particular due to their ability to counteract the so-called free-riding phenomenon.[58] This approach is reflected both in the GVH's position in the Online Accommodation Booking Sector Inquiry and in recent case law.[59]

As a conclusion of the Online Accommodation Booking Sector Inquiry, the GVH found that there was no substantive intra-band price competition among the online travel agencies in the market; therefore, the prices of the accommodations were roughly the same in every sales channel. According to the GVH, the lack of intra-brand competition was rooted in the use of the price parity clauses, i.e. contractual clauses in which the accommodation providers bound themselves not to offer their rooms on other distribution channels for prices lower than those offered to the given online travel agencies. Accommodation providers were able to comply with this clause by standardising the prices of the rooms on all distribution channels during the concerned contractual period. This has led to the elimination of intra-brand competition and, simultaneously, to the elimination of price competition between online travel agencies offering the same rooms. The wide price parity was included in contracts with the international and major domestic travel agencies. The GVH, however, concluded that a wide price parity clause even in one agreement established by a significant online travel agency may result in the standardisation of room prices on all distribution channels on the whole market.[60]

According to the findings of the Online Accommodation Booking Sector Inquiry, price parity is more likely to have a negative effect on accommodation providers

---

entitled by the price parity clause (e.g. another accommodation booking platform with whom the accommodation provider already concluded a contract).

[57] 'Narrow' price parity means that the obligor of the price parity clause (e.g. an accommodation provider) itself is not allowed to offer/advertise (on its own website) more favourable prices or conditions concerning its product/service than those offered towards the party entitled by the price parity clause (e.g. another accommodation booking platform with whom the accommodation provider already concluded a contract), but it is allowed to offer/advertise more favorable prices or conditions concerning its product/service through a third party (e.g. a third-party accommodation booking platform that the obligor intends to utilise).

[58] Platforms usually invest heavily in the design and development of the online interface, as well as, while aiming at the widest possible consumer base, allocate significant amounts to advertising, marketing and other support functions (consulting, warranty, etc.). The return on these investments may be jeopardised, and the incentive for further investments may be reduced, if partners, leveraging the platform's services, make these products more affordable for consumers through their own channels (the so-called 'free-rider' phenomenon) (NetPincér case, para 95).

[59] NetPincér case.

[60] Online Accomodation Booking Sector Inquiry, pp. 6–7.

since they would only be able to maximise their profits if they could apply favourable prices on their own distribution channels and on platforms applying lower commission rates. According to the GVH, the reduction of searching costs could serve as an advantage of price parity for consumers, but the search is performed online; therefore, there is no extra cost beyond the time spent on the search. Consequently, the GVH concluded that the full use of parity clauses may restrict competition by standardising market prices and increasing barriers to entry. This may give rise to competition concerns also in markets with a similar level of concentration as in Hungary. The GVH summarises that the application of wide parity clauses does not result in significant efficiency benefits and does not enhance consumer welfare to a degree that could justify the total restriction of intra-brand competition.[61]

The approach followed also by the other Member States, and finally adopted and confirmed by the GVH as an outcome of the Online Accommodation Booking Sector Inquiry, typically indicates that travel agencies may introduce the so-called narrow price parity clauses instead of using the wide price parity clauses. Based on the narrow price parity, a travel agency expects the accommodation providers not to grant less favourable conditions to it than those conditions displayed on their own website. The parity obligation in this latter case does not apply to other booking channels and other accommodations. The GVH considered that, based on the market conditions and taking into consideration the danger and adverse effects of the free-rider phenomenon, the use of narrow price parity clauses may be a suitable solution to the current market issues as it may pave the way for accommodation providers to be able to freely apply different pricing strategies on each sales channel. Further, travel agencies can use their income from commissions to organise promotion activities and participate in price competition.[62]

Following the publication of the draft report of the Online Accommodation Booking Sector Inquiry at the end of 2015, the largest Hungarian-owned market participant in the domestic online travel agency market, Szállás.hu, informed the GVH that it would switch to the use of narrow parity. Later, narrow parity appeared in the practice of all major market participants.[63]

In the framework of the European Competition Network, the GVH also participates in a monitoring working group, comprising 11 competition authorities of the Member States[64] and the Swiss competition authority, coordinated by the European Commission and aimed at the observation of the changes on the online accommodation booking market and the effect of introducing the narrow MFN

[61] Online Accomodation Booking Sector Inquiry, p. 7.

[62] Online Accomodation Booking Sector Inquiry, p. 7.

[63] Online Accomodation Booking Sector Inquiry, page 7. Booking.com has previously switched to this model on a global level; the press release is available at https://news.booking.com/bookingcom-to-amend-parity-provisions-throughout-europe/. Accessed 10 August 2020.

[64] Austria, Belgium, Czech Republic, Germany, France, Hungary, Ireland, Italy, Netherlands, Sweden, United Kingdom.

clause. The working group monitors the development of the market processes, and in this regard, the GVH has stated that it is ready to intervene if the market developments are unlikely to strengthen competition.[65] It is also noteworthy that cooperation between national authorities, in particular the cooperation of the European Commission's Consumer Protection Cooperation Network, can also be particularly effective and useful in addressing market failures in this field. This is demonstrated, for example, by the joint action of the European Commission and national consumer protection authorities leading to global change concerning an aspect of Booking.com's practice that raised consumer protection concerns.[66]

Finally, it is important to mention that the GVH further elaborated its approach on narrow price parity clauses in its decision of 17 April 2018 in the *NetPincér* case to the extent that it found that in the online food ordering market, the narrow MFN clause, which covered all the obligated party's own distribution channels, including both its online channel and offline (telephone, flyer) sales, constituted a restrictive agreement by effect. As a result of a commitment following the HCA decision, the operator of the Netpincér platform removed the relevant clause regarding offline sales. Consequently, based on current practice, the GVH primarily considers acceptable a narrow price parity clause that is related only to own online offers/advertising, while narrow clauses extended to offline (telephone, flyer) sales, and wide MFN clauses are qualified as anti-competitive.

7.5.3 Market Studies

To perform its statutory responsibilities more effectively and efficiently, the GVH may conduct market studies. By collecting publicly available information and voluntarily provided data, with the involvement of third-party experts or consultants, market studies serve as tools to survey and assess the operation of certain markets, trends, market practices and the impact these may have on competition and consumers. The GVH publishes on its website a summary report on the facts and findings that the market study revealed and indicates the necessary further measures.[67] Within the GVH's toolbox, the market study is a fast and flexible measure with a relatively wide scope. As compared to the sector inquiry, undertakings are not obliged to cooperate with the GVH; the answers are provided on a voluntary basis. Although the GVH is not entitled to impose a procedural fine for failure to respond voluntarily to an inquiry, the contacted undertakings are generally interested in cooperating with the GVH, given that the GVH may interpret the refusal to cooperate as an attempt to conceal anti-competitive practices, and this may entail the launching of competition supervisory proceedings.

[65] Online Accomodation Booking Sector Inquiry, pp. 7–8.

[66] Press release by the European Commission is available at https://ec.europa.eu/commission/presscorner/detail/en/ip_19_6812. Accessed 10 August 2020.

[67] Section 43/C of the Competition Act.

7.5.3.1 Findings of the DCT Market Study[68]

The GVH has already expressed in its Digital CP Strategy issued in September 2018 that it has identified an increased presence of DCTs and, therefore, planned a market study focusing on DCTs.[69] Accordingly, the GVH launched the DCT Market Study on 27 March 2019 to explore consumer protection specific market developments relating to DCTs and their effects on consumers' decision-making process. It is interesting to note that this was the first time that a market study was initiated by the GVH to examine consumer protection aspects. The paper prepared in the framework of the DCT Market Study[70] was published by the GVH on 12 March 2020.

The primary aim of the DCT Market Study was to assess the factors that influence the decisions of consumers and undertakings appearing on the platforms of DCTs and their experience in using them. Further, the study attempted to identify the extent to which market participants operating DCTs are aware of consumer protection expectations concerning the platforms and whether they operate accordingly.[71] The scope of the DCT Market Study included the operation of platforms specialised in the comparison of retail, accommodation and travel, financial and info-communication products and services.[72]

In the GVH's view, the use of comparison tools can generally assist consumers to reach more efficient, informed decisions, and it can significantly contribute to transparency, and thus, it may also strengthen competition in the market. In this context, the GVH considers it important that the consumer receives adequate information when using DCTs. To meet this requirement, among others, the compared product characteristics must be real, the consumer must be in a position to easily understand the criteria according to which the comparison or ranking is set up and the DCTs must contain objective, impartial information in a user-friendly manner.[73]

The GVH concluded that DCTs as digital services are characterised by the specificities of digital markets—they are mostly bilateral or multilateral markets that have significant network effects, as well as being characterised by economies of scale and scope and by increasing returns to scale.[74] Therefore, according to the GVH, entry barriers may be high in the market of DCTs, and monopolistic or oligopolistic market structures may easily emerge—where unilateral consumer preference appears (tipping effect) and competition may be significantly distorted

---

[68] Hungarian Competition Authority – Paper Summarizing the Results of the Market Study on Exploring the Impacts of Digital Comparison Tools on Consumer Decisions – 2020.

[69] The GVH's Digital CP Strategy is available at https://www.gvh.hu/pfile/file?path=/gvh/strategia/GVH_Stategia_Digitalis_fogyved_startegia_2018_09_27, point 29. Accessed 9 September 2021.

[70] Hungarian Competition Authority – Paper Summarizing the Results of the Market Study on Exploring the Impacts of Digital Comparison Tools on Consumer Decisions – 2020.

[71] DCT Market Study, pt 27.

[72] DCT Market Study, pt 29.

[73] DCT Market Study, pts 1, 7 and 30.

[74] DCT Market Study, pt 39.

or reduced. The GVH considers that, under these circumstances, competition authorities have very limited means at their disposal to prevent these adverse effects on the market structure. At the same time, different types of DCTs operating in different sectors (retail, accommodation booking, etc.) may be exposed to these adverse effects to varying degrees.[75]

The GVH also found that the revenue models applied by DCTs may distort competition, the commissions paid to the DCTs may be built in the prices by the retailers. Thus, the retailers' prices may grow as a consequence of their use of the DCTs. In addition, comparability can be reduced, for example, if the products are highlighted in return for additional financial benefits or if the comparison of complex products is solely based on a single criterion. The GVH considers that given the high degree of complexity of some products, DCTs may present an offer that does not reflect reality in the absence of detailed information.[76]

The GVH expressed its concern that comparability between products is in many cases incomplete, differences between product characteristics are either difficult to identify or not identifiable at all (e.g. the differences in product definitions) and information is not sufficiently accurate and detailed. In addition, changes made in the services cannot be properly monitored—there is no possibility to display discounts and promotions, meaning that DCTs can also have a strong market and competition-distorting effect.[77]

In the context of the DCT Market Study, the GVH collected the most common DCT behaviours that mislead or misinform consumers. The list of examples includes providing incomplete information on prices and additional costs (for example, shipping and handling costs are not included in the price), lack of information on the sponsored nature of the offers, manipulation of the ranking of the products presented, withholding or misrepresenting certain relevant details of product features (for example, displaying a particular technical feature for some products but not for others), representing only a limited range of the presented products or retailers on the DCT's site and displaying products that are no longer available. The GVH found it problematic that some consumers are not aware of these problems, so they are unlikely to make an informed consumer decision.[78]

The GVH acknowledged that DCT platforms applied a business model in which the costs of the set-up and operation were basically financed by commissions and advertising fees paid by online stores and retailers. At the same time, the GVH expressed its concern that business-driven DCTs may have an obvious interest in manipulating the ranking of products if they are financially encouraged to do so by retailers. Retailers must generate profits to cover the commissions and fees; thus, in the long run, these are going to be incorporated into the consumer prices of the products. The GVH emphasised its concern that these costs would be built into the

[75] DCT Market Study, pt 40.

[76] DCT Market Study, pt 3.

[77] DCT Market Study, pt 4.

[78] DCT Market Study, pts 63–64.

prices and, thus, some DCTs would not improve consumer welfare; on the contrary, the outcome may even be detrimental to consumers. The GVH also pointed out that if a DCT could obtain market power, it would enable it to take advantage of this by demanding increased commissions and fees from retailers, which would mean in the long run a further price increase for consumers.[79]

In order to eliminate the above competition concerns, the GVH formulated a series of recommendations and expectations related to DCTs:

(i) As for the list of comparison results and rankings, clear and perceptible information is required on the relevant ranking criteria, and concerning highlighted products, clear and perceptible reference must be made to the fact that such highlighting was the result of a paid-for service.[80]
(ii) Solely real consumer experiences, opinions and evaluations can be published as user reviews.[81]
(iii) If an evaluation system is applied, comprehensible and perceptible information is required on the criteria on which it is based, how these are observed and on the methodology of the calculation.[82]
(iv) If the complex, customisable nature of the product/service requires, the consumers' attention have to be drawn to the fact that to find the most appropriate offer it is not enough to ponder the particular aspects of the comparison, but further steps are necessary to understand also other elements.[83]
(v) It must be clear whether the DCT is independent or it is operated or sponsored (either directly or indirectly) by any retailer.[84]
(vi) The identity and contact details of the operator of the DCT and the liability for transactions should be indicated in an easily comprehensible and clear manner.[85]
(vii) The consumer must be informed about the number, identity, contact details and location of the retailers presented by the website.[86]
(viii) The identical, actual, up-to-date prices and conditions must be displayed both on the interface of the DCT and on the websites and webshops of the partners available through it.[87]

[79] DCT Market Study, pts 65–67.
[80] DCT Market Study, pt 152.
[81] DCT Market Study, pt 155.
[82] DCT Market Study, pt 160.
[83] DCT Market Study, pt 164.
[84] DCT Market Study, pt 168.
[85] DCT Market Study, pt 171.
[86] DCT Market Study, pt 176.
[87] DCT Market Study, pt 181.

(ix) Clear and transparent information must be provided on how consumer data will be treated, particularly on potential profiling and on the use of targeted advertising to fund the platform.[88]

(x) The design and usability of the platforms must ensure transparency, easy navigation, simple and comprehensible wording; complex legal or technological concepts must be avoided; and personalised searches or detailed filters, as well as multi-level ranking must be available.[89]

(xi) Enhanced caution is required when claiming market leadership, as the DCTs must be able to demonstrate the criteria and substantiation of their claims throughout the entire period, for all market participants and for the entire market.[90]

(xii) Only such certificates on reliability can be displayed that are provided according to a well-defined set of criteria from an actually existing organization, ensuring that the consumer can easily find out their content.[91]

(xiii) Finally, when displaying educational articles, it must be clearly indicated whether the content is an independent content drawn up by the operator of the DCT or the DCT has received remuneration for its publication.[92]

7.5.4 The GVH's Case Law on the Responsibility of Dominant Companies in Digital Markets

Practically, there are no abuse of dominance cases in the GVH's practice in respect of digital markets and online platforms, and a kind of market analysis that focuses on these cases are practically missing as regards the Hungarian market.

In the Online Accommodation Booking Sector Inquiry, the GVH clearly stated that 'MFN clauses are considered vertical agreements'[93] and further discussed the pros and cons of different types of MFN clauses as vertical agreements in the context of Section 11 of the Competition Act and Article 101 of the TFEU (see Sect. 5.2.2 above). It also briefly noted that if an MFN clause is applied by a dominant market player, it can create barriers to market entry and/or expansion or can exclude players already present in the market, thereby stabilising or strengthening an existing dominant position or an already high market share.[94]

The GVH examined MFN clauses in the *NetPincér* case conducted in the market of online food delivery. Although the *NetPincér* case concerned the assessment of vertical agreements, with a market share of above 70% during the entire investigated

[88] DCT Market Study, pts 184 and 187.

[89] DCT Market Study, pts 194–195.

[90] DCT Market Study, pts 200–201.

[91] DCT Market Study, pt 203.

[92] DCT Market Study, pt 204.

[93] Online Accomodation Booking Sector Inquiry, point 5.2.4., p. 79.

[94] Online Accomodation Booking Sector Inquiry, point 5.3.9., pp. 84–85.

period, that is, between 2012 and 2016 (and with competitors holding much smaller market shares between 1% and 14%), Viala Kft., the operator of the NetPincér platform was found to be having a strong market position. In this regard, the GVH noted that building a reputation and achieving optimal capacity utilisation that would allow a profitable long-term market presence on the online food delivery market would require significant resources and investments. The GVH also highlighted the relevance of existing network effects as a key factor for the assessment of online platforms.[95] In GVH's view, these additional circumstances also supported the market power of Viala Kft. However, as this was a vertical agreement case, the GVH did not focus on the abuse of dominance angle and, thus, did not discuss the special responsibility of a dominant undertaking operating an online platform.

Notwithstanding the above, it can be stated that the GVH's practice acknowledges that the increasing potential of online platforms and the data they collect may be considered significant factors for market power on the digital markets, and thus, in future cases, competition law enforcement is expected to take into account these factors as well.

7.5.5 Other Potential Legal Consequences of the Unlawful Use of DCTs

If the application of a DCT constitutes a misleading commercial practice, it may have other possible legal consequences besides competition law.

On the one hand, in accordance with Hungarian criminal law provisions, any person who, before the public at large or in respect of an essential feature of any good of substantial quantity or value, states false facts or true facts in a deceptive way or provides deceptive information for the purpose of promotion commits the crime of misleading consumers, which is punishable by up to 3 years' imprisonment.[96] Pursuant to the Criminal Code, 'public at large' covers also the cases where a crime is committed by means of publication on an electronic communication network. Consequently, misleading commercial practices through the DCT's online platforms may satisfy this criminal law criterion.[97] Misleading the consumers may also constitute fraud, considering that the consumer protection type offense may be established in a combination of offences, together with the crime against property.[98] However, it is also worth noting that criminal sanctions concerning the misleading of consumers should be mentioned primarily as a theoretical possibility

[95] NetPincér case, para 76–78.

[96] Act C of 2012 on the Criminal Code (Criminal Code), Section 417, paras (2)–(4).

[97] Criminal Code, Section 459, para (1), pt 22.

[98] See for example: Békés Ádám: A fogyasztók érdekeit sértő bűncselekmények - In: Hollán Miklós, Barabás A. Tünde (szerk.) - A negyedik magyar büntetőkódex: régi és újabb vitakérdések. p. 384 - Budapest: MTA Társadalomtudományi Kutatóközpont, 2017. pp. 247–258 (https://jog.tk.mta.hu/uploads/files/17_BekesA.pdf) (*in Hungarian*). Accessed 9 September 2021.

only since a relatively small number of cases have been registered [99] and the detailed conditions are not known. Consequently, no stable and consistently applied practice in this field can be outlined.

On the other hand, according to Hungarian civil law provisions, a person who acted under a misapprehension regarding any material circumstance at the time a contract is concluded shall be entitled to contest his/her contract statement if his/her mistake was caused or could have been recognised by the other party. In this regard, the mistake shall be considered to impact a material circumstance if the party, being aware of such a mistake, would not have concluded the contract or would have done so only on fundamentally different contractual terms. Further, a party that has been persuaded to conclude a contract by misrepresentation or fraud by the other party shall be entitled to contest the contractual obligations arising as a consequence of such misrepresentation. The latter also applies if the misrepresentation was committed by a third person and the other party had or should have had knowledge of such conduct.[100] Consequently, the use of misleading conduct by a DCT may provide the consumer the opportunity to contest a transaction based on the use of a DCT if his/her mistake was caused by the undertaking operating the DCT or the distributor of the compared product, regardless of whether the act was intentional or not. The consumer is entitled to enforce this right within 1 year of the conclusion of the contract and, as a result of a successful claim, if the transaction becomes invalid as of the date on which it was concluded.[101] Moreover, if the GVH finds that the application of the DCT constituted an unfair commercial practice, the contract concluded on the DCT platform or through it may be considered illegal and therefore null and void.[102] Finally, Hungarian civil procedure law allows, under certain conditions, consumers to bring a class action (in order to enforce their claims arising from a consumer contract, in which case the court may decide uniformly on the claims of a large number (at least ten) of consumers.[103] Within 3 years following the infringement, the GVH may, in the public interest, bring a civil action on behalf of consumers if such infringement resulted in a grievance that affected a wide range of consumers.[104]

[99] According to the Criminal Statistics System (Bűnügyi Statisztikai Rendszer) operated by the Ministry for Home Affairs, there have been a total of 15 registered cases in the last two years. The data are available on the website: https://bsr.bm.hu/. Accessed 9 September 2021.

[100] Act V of 2013 on the Civil Code (Civil Code), Section 6:91, para (1) and (3).

[101] Civil Code, Section 6:89 para (1) and (3).

[102] Civil Code, Section 6:95.

[103] Act CXXX of 2016 on the Code of Civil Procedure, Part Eight (Actions for the collective enforcement of claims), Chapter XLIII (Class actions), Sections 580–591.

[104] Competition Act, Section 85/A., para (1) and (3).

7.5.6 The Issue of Interim Measures in the Digital Sphere

It may be observed that the GVH has been rather reluctant to use its power to impose interim measures (injunctive relief) in the digital sphere.[105] Due to the dynamically changing and fast-moving nature of the online field, in certain cases, the reluctance to use this enforcement tool might result in scenarios in which consumers and/or competitors can quickly suffer irreparable harm. For example, a small digital firm is attempting to provide online services (e.g. specialised online search services) that are in rivalry with those of a dominant market player. In such a case, however, the small company may, as a result of the exclusionary conduct of the dominant competitor, disappear from the market by the time the GVH establishes the abuse of dominance at stake.

7.6 Latest Hungarian Developments and Case Law in the Telecommunication Sector

7.6.1 Hungarian Approach Regarding Broadband Competition

In Hungary, the market conditions—including the level of competition—in the telecommunication and, more broadly, the electronic communication markets are regulated by the National Regulatory Authority for Media and Infocommunications (Nemzeti Média– és Hírközlési Hatóság (NMHH)), and the GVH also participates in controlling and supervising the operation of these markets by means of ex post competition law interventions. The NMHH, in its two Broadband Market Resolutions with respect to mass broadband markets,[106] established a list of telecommunication service providers with significant market power (SMPs) and a detailed set of obligations that require SMPs to provide access to their infrastructure so that competitors without infrastructure could also enter the market. The NMHH has a broad spectrum of powers; e.g., it can initiate a market analysis at any point if it deems that certain SMPs could distort competition on the relevant market, and it can prescribe a variety of commitments in order to remedy such potential issues. The adoption of the Broadband Market Resolutions was the first occasion that the NMHH applied the bottom-up pure long-run incremental cost (LRIC) calculation

[105] In recent years there were only two cases in which interim measures have been imposed: one in 2017 (Orangeways et Al. case - Decision of the GVH no. Vj/100-98/2016) and another one in the first half of 2020 (Global AQA Pty Ltd. / Globoport Média Holding Kft./Glenwood Media Kft. case – Investigation of the GVH no.Vj/3/2020).

[106] Decisions of the President of the National Media and Infocommunications Authority no. PC/ 17915-64/2017. and PC/17920-64/2017. - 'Ensuring the wholesale provision of fixed local and central access to broadband networks' (dated 14 December 2017).

methodology recommended by the European Commission[107] to calculate the regulatory prices at which SMPs are required to provide access to competitors.

The GVH has also dealt with the state of competition on electronic communication markets and, in particular, the question of competition between broadband networks.[108] Recently, this issue has come up mainly in connection with the GVH clearance adopted in the *DIGI/Invitel* merger case,[109] with abuse of dominance cases against Telekom and UPC and with the ongoing cartel investigations into the *Telekom/Telenor Broadband Sharing* cases.[110]

In the *DIGI/Invitel* case, both participants in the underlying merger were important and active players on the Hungarian market of wired telecommunications; i.e., they provided wired, telephone, Internet and wired broadcasting (TV) and other services for hundreds of thousands of subscribers. Regarding broadband networks, the GVH stated that the relevant vertical issues in case of mergers in the electronic communication markets are primarily regulated by the NMHH; therefore, the GVH would only act in a secondary manner.[111] However, given the extensive networks of both undertakings, the horizontal effects of the relevant transaction would have resulted in the significant lessening of competition in certain geographical areas.[112] To remedy these issues, DIGI offered commitments, according to which DIGI had to sell a part of its broadband networks in order to provide competitors access to the affected geographical markets.[113] Interestingly, instead of requiring DIGI to provide access to its infrastructure, the GVH decided to require DIGI to actually sell the relevant networks. Presumably, this competition-strengthening approach in respect of integrated info-communication services was based on the actual state of the Hungarian market. The *Vodafone/UPC* case[114] resulted in a second integrated operator in the Hungarian market, besides Magyar Telekom, and this market development will most likely lead to better services and foster innovation.

The *DIGI/Invitel* case clearly showed the main difference between the approaches of the European Commission and the rest of the major European competition law jurisdictions and that of the GVH. While the European Commission (and the NMHH) generally requires network operators to allow access to their networks in

---

[107] Calculation on the basis of forward-looking long-run incremental costs (LRIC), see Commission Recommendation of 7 May 2009 on the Regulatory Treatment of Fixed and Mobile Termination Rates in the EU (2009/396/EC).

[108] For earlier case-law concerning infocommunications markets see the UPC/RubiCom case (Decision of the GVH no. Vj/80-98/2011).

[109] Decision of the GVH no. Vj/42-272/2018.

[110] Telekom/Telenor Broadband Sharing I. case - Investigation of the GVH no. Vj/18/2015, Telekom/Telenor Broadband Sharing II. case, Investigation of the GVH no. Vj/1/2018.

[111] DIGI/Invitel case, para 139.

[112] DIGI/Invitel case, paras 134–136.

[113] DIGI/Invitel case, paras 158–165.

[114] CASE M.8864 – Vodafone/Certain Liberty Global assets. This transaction also affected the Hungarian market since it resulted in the merger of Vodafone's Hungarian branch (primarily mobile services) with UPC (retail fixed telephony, retail fixed Internet access and retail TV services).

order to ensure the possibility of market entry of competitors, based on the commitments mentioned in the *DIGI/Invitel* case, it seems that the GVH encourages network providers to individually build/acquire/operate parallel networks that compete with each other rather than to cooperate and share the networks or jointly exploit new network technologies.

It is also worth noting that the GVH investigated Telekom for abuse of dominance allegations in the *Telekom Flip*[115] case. The GVH noticed that Telekom made available its Flip service—which was introduced in 2017 as a triple-play service package at a discounted price (in addition to the already existing triple-play service under the Telekom brand)—only in territories where it faced serious competitive pressure. In some municipalities, where the necessary technical requirements were met (i.e. the necessary network was available), Telekom did not make Flip available; thus, it forced consumers to buy traditional product packages with a significantly higher subscription fee. During the proceedings, Telekom offered a commitment to introduce a new three-play service package at a price that is higher than Flip's but the range of available services is wide and could provide advantages for consumers, making it worth for them to pay a higher price.[116]

In the *UPC Monor* case,[117] the GVH investigated UPC for a somewhat similar allegation. The GVH launched an investigation based on a large number of consumer complaints claiming that in numerical district 29 of UPC (Monor and its surroundings), the quality of the Internet service provided on ADSL technology was lower (the upload and download speeds were slower), while the price of the Internet subscription was higher, as compared to UPC's cable footprint in the country. UPC was the incumbent telecommunication service provider in Monor and its neighbouring villages. The proceedings were closed with commitments offered by UPC.

Another angle of the broadband competition issue is expected to be decided when the GVH closes its horizontal cooperation investigations into the first network sharing agreements in Hungary between Telekom and Telenor. In December 2014, Telekom and Telenor entered into a cooperation agreement on the sharing of mobile frequencies and for the joint development and operation of their 4G networks in the 800 MHz spectrum band, which is made applicable to the whole territory of Hungary, except Budapest. While NMHH approved the spectrum sharing, the GVH launched a horizontal cooperation investigation into the arrangement in February 2015 to determine whether it conforms to or conflicts with Hungarian and EU competition laws.[118] In January 2018, the GVH launched another

[115] Decision of the GVH no. Vj/56-244/2017.

[116] GVH press release available at https://www.gvh.hu/en/press_room/press_releases/press_releases_2019/magyar_telekom_launches_a_new_low_priced_tv_intern. Accessed 9 September 2021.

[117] Decision of the GVH no. Vj/15-147/2015.

[118] GVH press release available at https://www.gvh.hu/en/press_room/press_releases/archive/press_releases_2015/gvh_investigating_cooperation_between_telekom_and_. Accessed 28 February 2022.

investigation and held dawn raids at the premises of Telekom and Telenor based on a suspicion that the parties may have coordinated their behaviours in respect of tenders published by the NMHH in 2014 in order to obtain rights to use the spectrum required to provide broadband services and that their agreement may have been aimed at sharing the relevant market and coordinating their bids to fulfil tender conditions in advance.[119] Both proceedings are still pending.

Articles 76 and 79 of Directive 2018/1972 of the European Parliament and of the Council of 11 December 2018 establishing the European Electronic Communications Code also encourage and regulate the conditions of co-investment and other forms of cooperation between providers of electronic communication networks and services. This approach provides some new considerations to the complex dilemma of service-based competition versus infrastructure-based competition.

7.6.2 Ensuring Effective Consumer Protection

Traditionally, the GVH also puts significant emphasis on the enforcement of consumer protection norms regarding mobile service providers, especially concerning pricing and discount practices, as well as market leadership statements. In 2019, the GVH adopted infringement decisions in respect of all three market players: Telenor, Vodafone and Telekom.

In the *Telenor* case,[120] the GVH imposed a record fine of approx. EUR 5.45 million on Telenor for misleading communications since it advertised mobile devices as free and failed to inform customers that if they exercised the option to purchase a device, the subscription fee would be higher than if they had only concluded a subscription contract with the same contractual terms without purchasing a device.[121]

In the *Vodafone* case,[122] the GVH established that Vodafone falsely claimed to hold the position of a market leader with its slogans ‘Europe’s largest 4G partner network’ and ‘Europe’s largest 4G network’.[123]

[119] GVH press release available at https://www.gvh.hu/en/press_room/press_releases/press_releases_2018/the_gvh_held_unannounced_inspections_at_the_premis. Accessed 28 February 2022.

[120] Decision of the GVH no. Vj/13-66/2018.

[121] GVH press release available at https://gvh.hu/en/press_room/press_releases/press_releases_2019/fine-of-over-eur-5.45-m-imposed-by-the-hungarian-competition-authority-on-telenor-for-its-misleading-commercial-practices. Accessed 9 September 2021.

[122] Decision of the GVH no. Vj/76-142/2016.

[123] GVH press release available at https://gvh.hu/en/press_room/press_releases/press_releases_2019/fine-amounting-to-a-total-of-over-eur-3-m-imposed-on-vodafone-for-repeated-infringements. Accessed 9 September 2021.

In the *Telekom iPhone* case,[124] the GVH examined the legality of a comparative advertisement regarding the price of an iPhone 7. The GVH established that Telekom misled consumers by projecting an image that the device was much cheaper than the same device offered by Vodafone; however, taking into calculation the subscription fees, the overall sum to be paid was nearly identical.

[124] Decision of the GVH no. Vj/4-50/2018.

Norway

8

Eirik Østerud, Ronny Gjendemsjø, and Ingrid Halvorsen Barlund

8.1 Introduction[1]

This report presents and discusses the Norwegian competition authorities' role in the digital and data-driven economy. By the Norwegian competition authorities, we mean the Norwegian Competition Authority (NCA) (*Konkurransetilsynet*), the Norwegian Competition Appeals Committee (*Konkurranseklagenemnda*) and the Norwegian government, in particular the Ministry of Trade, Industry and Fisheries and its Department of Competition Policy. We also include the European Free Trade Association (EFTA) Surveillance Authority (ESA).

In the report, we take a relatively broad view of the notion of the digital and data-driven economy. Our report is therefore not limited to markets for commercial data and digital services. We will also consider how digital data and communication affect Norwegian competition policy in markets for traditional goods or services.

The formal enforcement of Norwegian competition law is an important, although not an exclusive, role of the Norwegian competition authorities. We will, of course, present and discuss Norwegian enforcement practice in our digitalised economy. We

[1] The authors have also written a report on Norwegian competition law and policy and the digital economy for the FIDE XXIX Congress. See I. H. Barlund, R. Gjendemsjø and E. Østerud, Norway, in: D. Mândrescu (ed. Board), EU Competition Law and the Digital Economy: Protecting Free and Fair Competition in an Age of Technological (R)evolution, Eleven International Publishing 2020, pp. 399–414.

E. Østerud
University of Oslo, Oslo, Norway
e-mail: Eirik.Osterud@jus.uio.no

R. Gjendemsjø · I. H. Barlund (✉)
University of Bergen, Bergen, Norway
e-mail: Ronny.Gjendemsjo@uib.no; Ingrid.Halvorsen@uib.no

B. Kilpatrick et al. (eds.), *Antitrust in Data Driven Markets & Legal Framework for Influencers, Native Advertising and Control over the Use of AI in Marketing*, LIDC Contributions on Antitrust Law, Intellectual Property and Unfair Competition, https://doi.org/10.1007/978-3-031-07422-6_8

also aim to consider the competition authorities' priorities, public statements and views on specific sectors and various public measures related to digital and data-driven sectors.

Norwegian competition policy is implemented within the Norwegian legal context. Norwegian substantive antitrust law consists of the rules of the Norwegian Competition Act[2] and the rules of the European Economic Area Agreement (EEA) between the EFTA states and the EU member states.[3] The substantive antitrust prohibitions in the Norwegian Competition Act and the EEA Agreement correspond to Articles 101 and 102 of the Treaty on the Functioning of the European Union (TFEU) and shall, in principle, be interpreted and applied equivalently. The Competition Act (2004) thus contains prohibitions on anti-competitive agreements/concerted practices (Section 10) and abuse of dominance (Section 11). Under the EEA Agreement, Articles 53 and 54 correspond to Articles 101 and 102 TFEU.

With regard to public enforcement, the NCA investigates and sanctions infringements of Sections 10 and 11 of the Competition Act and also of Articles 53 and 54 EEA where there is a cross-border dimension. Administrative appeals against the NCA's infringement decisions are handled by the Competition Appeals Committee and subsequently by the Norwegian courts of law. The competence to enforce Articles 53 and 54 EEA is further divided between ESA and the European Commission, pursuant to Article 56 EEA.

The digital economy has not led to amendments of the Norwegian Competition Act. In the past, the enforcement of Norwegian competition law in digital service markets has also been relatively limited. We do, however, observe that digital developments and communication affect the competition authorities' enforcement priorities as well as competition analyses of markets for tangible goods and services. In 2020, digital platforms and marketplaces have also been high on the NCA's policy agenda.

The Norwegian telecom sector has been a priority for the NCA and has led to active enforcement, especially under the merger control rules. As regards abuse of dominance, the NCA and the Competition Appeals Committee's findings against the Norwegian telecom incumbent Telenor are worth mentioning, in addition to ESA's recent decision against Telenor. As regards anti-competitive coordination, there have to date not been any infringement decisions in the digital sector. In September 2020, however, the NCA issued a statement of objections to four publishing houses and their jointly owned database company. The NCA's preliminary assessment was that the undertakings had exchanged competitively sensitive information concerning,

---

[2] Act of March 5, 2004, No 12 on competition between undertakings and control of concentrations.

[3] All the EU Member States are members of the European Economic Area (EEA), in addition to three of the EFTA States, i.e., Norway, Iceland and Liechtenstein. The EEA Agreement allows the EEA/EFTA States to participate in the internal market of the EU where free movement of goods, capital, services and people creates the world's largest common market for trade. Note that this is not the case for Switzerland, which is not a party to the EEA Agreement, but only a member of EFTA.

inter alia, future book prices and the timing of book releases through their joint database company. A preliminary fine totalling NOK 502 million was calculated.

Notably, in September 2020, the NCA, in collaboration with the other Nordic competition authorities in Denmark, Finland, Iceland and Sweden, issued a joint report on 'Digital platforms and the potential changes to competition law at the European level – The view from the Nordic competition authorities'.[4] The report discusses opportunities and challenges created by digital platforms and also makes some recommendations and policy suggestions from a Nordic perspective.

Our Norwegian report is structured as follows: in Sect. 8.2, we provide an overview and a general perspective on the Norwegian competition authorities' role and policies in the digital economy. In Sect. 8.3, we consider certain specific issues and topics in more detail. From a Norwegian competition law and policy perspective, we will here discuss market definition and market power in the digital economy, telecom and digital infrastructure, data protection and privacy, digital platforms, online advertising, content providers and comparison portals.

8.2 Norwegian Competition Law and Policy in the Digital Economy: A General Perspective

The emergence and development of the digital and data-driven economy have not led to amendments of, or proposals to amend, general, substantive Norwegian competition law. The 'pillars' of substantive Norwegian competition law (the prohibitions on anti-competitive agreements and abuse of dominance and the standard of intervention against concentrations) are harmonised with EU competition law.

Due to certain particular features of Norwegian competition law, some concerns raised in the discourse on the digital economy may be less urgent. Pursuant to the Norwegian Competition Act, the NCA is competent to intervene against concentrations below the general turnover thresholds for mandatory notification (NOK 1 billion/NOK 100 million) and also against acquisitions of non-controlling shareholdings. The NCA is thus equipped with *ex ante* enforcement tools to handle potential competition concerns caused by, *inter alia*, 'killer acquisitions' of start-ups and cross-shareholdings, also in the digital sector. In 2019, the NCA notably intervened against Sector Alarm's acquisition of 49.99% of the shares in Nokas (non-controlling interest) and against Sector Alarm's acquisition of Nokas Small Systems (below the general turnover thresholds) in markets for alarm systems.[5] In 2020, the NCA ordered a notification of an acquisition of a majority shareholding in

[4] Digital platforms and the potential changes to competition law at the European level – The view from the Nordic competition authorities (2020), available at https://konkurransetilsynet.no/nordic-competition-authorities-release-joint-memorandum-on-digital-platforms-and-the-future-of-european-policy/?lang=en. Accessed 14 October 2020.

[5] V2019-17, Sector Alarm – Nokas. (Conditional approval).

an online marketplace for used cars, which was below the general turnover-based notification threshold.[6] On 30 September 2020, the NCA arrived at a preliminary conclusion that the concentration would result in a significant impediment to competition in the online market for the sale of used cars.

Under section 14 of the Competition Act, the Ministry of Trade, Industry and Fisheries may, if necessary to promote competition in the markets, issue regulations intervening against terms of business, agreements or actions that restrict or are liable to restrict competition contrary to the purpose of the Act. In 2009, the Ministry enacted a regulation that requires Internet portals that offer residential property advertisements to provide general access to the advertisement service to everyone on non-discriminatory conditions.[7] The NCA held the view that access to Internet portals had become a necessary marketing channel to reach most potential buyers. The Internet portals' practice of refusing advertising access to anyone other than real estate agents constituted a significant entry barrier, which in effect forced sellers to use estate agents, thereby reducing competition.[8]

The application of section 14 is generally restricted by the EEA Competition Act[9] Section 7 second paragraph, according to which the application of Norwegian competition law may not lead to the prohibition of agreements, decisions by associations of undertakings or concerted practices that may affect trade between Member States, but which do not restrict competition within the meaning of Article 53 (1) EEA, or which fulfil the conditions of Article 51(3) EEA or which are covered by a block exemption. The competition authorities are not precluded, however, from applying stricter national laws prohibiting or sanctioning unilateral conduct engaged in by undertakings. The provision corresponds to Regulation 1/2003 Article 3(2). In principle, a section 14 regulation could thus be used to intervene against unilateral conduct by non-dominant undertakings in the digital economy.

Under Section 3 of the Competition Act, certain sectors may, through the adoption of regulations, be exempted from the Competition Act. In 2018, the Norwegian government proposed retaining the regulation exempting coordination of the sale of books from the prohibition on anti-competitive agreements in Section 10 of the Competition Act.[10] However, the regulation does not and cannot exempt the application of Article 53 EEA insofar as trade between EEA states is affected, which may be the case for eBooks.

With regard to other specific sectors affected by digitalisation, it is noteworthy that in 2017, a government-appointed committee submitted a report on the 'sharing economy' (NOU 2017:4 *Delingsøkonomien*). One of the proposals considered was

[6]V2020-16, Schibsted – Nettbil.

[7]Regulation 16 April 2004 No 631 on access to residential property advertisements.

[8]The NCA's note to the OECD Hearing on Digital Economy 2012.

[9]Act 5 March 2004 No 11 on the implementation and control of the competition rules of the EEA Agreement.

[10]Regulation 19 December 2014 No 1716 on exemption from Section 10 of the Competition Act for cooperation on the sale of books.

the deregulation of the taxi industry in order to enhance competition and technological innovation made possible by digital communication. After a political process, amendments to Act No 45 of 21 June 2002 on professional transport by motor vehicle and vessel entered into force on 1 July 2020, thereby partly liberalising the taxi sector.

In relation to the food and groceries sector, in 2018, the Norwegian parliament ordered the government to consider policies to promote competition, new market entry and innovation, including a possible sector-specific prohibition on price discrimination for suppliers. The Norwegian grocery sector is characterised by high concentration, substantial entry barriers and the absence of international grocery chains. E-commerce/home delivery also has a very limited market share. To date, a sector-specific price discrimination rule in the food/groceries sector has not been put to the Norwegian parliament.

With regard to competition policy targeting the digital sector, in January 2020, the Ministry for the first time instructed the NCA to give particular priority to monitoring the digital economy. The Ministry pointed out that digitalisation contributes to significant technological development and innovation but may also lead to market dominance and potential collusion. The NCA should therefore ensure that digitalisation contributes to increased competition in the interest of consumers and the business community. These instructions could potentially represent a significant policy shift as the NCA has previously expressed the view that it should deal with global digital competition issues in a uniform manner and in close cooperation with the European Commission.[11] The NCA has also expressed scepticism about using antitrust law as a means for furthering data protection, with reference to the German Bundeskartellamt's decision against Facebook.[12]

The NCA has, however, also voiced competitive concerns about digitalisation. It has opposed the establishment of a publicly funded Internet web portal in the grocery sector that would give consumers access to real-time prices for Norwegian groceries. The NCA has also opposed the exemption from Section 10 on coordination of the sale of books, *inter alia* based on concern that the exemption could reduce the number of publications of eBooks. The NCA has publicly warned against the competitive risks posed by digital coordination technologies and AI.[13] The NCA has further monitored the use of MFN clauses in the online hotel booking sector.[14]

The NCA has been relatively active in its merger control enforcement in markets linked to the digital economy. Thus, the NCA has intervened against several

[11] Interview with the Director of the NCA, 7 February 2019, available at https://www.tek.no/nyheter/nyhet/i/vQ8EqX/tyske-myndigheter-ut-mot-facebook-applauderes-av-norske-kollegaer. Accessed 14 October 2020.

[12] "Digital regulering", Dagens Næringsliv 8 April 2019, available at https://konkurransetilsynet.no/kronikk-digital-regulering/. Accessed 14 October 2020.

[13] The Norwegian Competition Authority (2018), available at https://konkurransetilsynet.no/kronikk-er-kunstig-intelligens-fremtidens-kartell/. Accessed 14 October 2020.

[14] The Norwegian Competition Authority (2019), available at https://konkurransetilsynet.no/kronikk-hotellbooking-pa-nett-er-til-fordel-for-hotellgjestene/. Accessed 14 October 2020.

concentrations in the telecom sector[15] and in the information technology (IT) service sector.[16] The impact of the digitalisation of information and digital communication is regularly considered in the assessment of markets for traditional goods or services. In 2008, for example, telephone number searches on the Internet were not regarded as a sufficient substitute for telephone number directory services by short message services (SMSs).[17] And in 2012, Google and Facebook were not considered to be part of the same relevant advertising market as local newspapers.[18] With regard to competition between payment systems, the NCA considered potential competition from Internet giants: Apple Pay, Google Pay, Facebook Pay and Amazon Pay.[19] In some cases, the NCA has also intervened against concentrations based on coordinated effect theories, whereby market transparency was found to be enhanced by price announcements on the Internet.[20]

As to abuse of dominance in the digital economy, NCA fined telecom incumbent Telenor NOK 788 million for having changed the network access prices it charged to a potential network competitor.[21] The alleged infringement was based on the theory that the access prices reduced a potential competitor's incentives to develop its own national communication grid. The NCA's decision was upheld by the Competition Appeals Committee and is currently pending in the Norwegian courts. ESA recently also found Telenor guilty of abusive conduct in the form of margin squeeze in the mobile broadband service market.[22] ESA is also investigating airline operator Widerøe for refusing to supply competitors with its satellite-based approach system receivers, which are used to operate local routes in Norway.

The Norwegian competition authorities' enforcement of anti-competitive agreements in the digital economy has been limited. In April 2019, ESA announced that it had discontinued its investigation into whether Finance Norway, Bits, BankID, DNB and Nordea had engaged in agreements, decisions or concerted practices aimed at preventing a potential competitor from entering the e-payment market in Norway. As mentioned, in September 2020, the NCA issued a statement of objections to four publishing houses and their jointly owned database company. The NCA's preliminary assessment was that the undertakings had exchanged competitively sensitive information concerning, *inter alia*, future book prices and the timing of book releases through their joint database company.

---

[15] V2005-6 Telenor – Tiscali, V2012-8 Telenor – LOS, V2015-1 Telia – Tele2/NwN.

[16] V2019-23, Tieto – Evry.

[17] V2008-22 Opplysningen Mobil AS – Aspiro Søk AS.

[18] V2012-11 A-pressen AS – Mecom Europe AS.

[19] V2018-18 Vipps AS – BankAxept AS/BankID Norge AS.

[20] V2006-490 Falck – Viking, V2015-29 St1 Nordic OY – Smart Fuel.

[21] V2018-20 Telenor.

[22] 'ESA fines Telenor EUR 112 million for anticompetitive practices', Press release, 2020, 39.

The joint report from September 2020 by the Nordic competition authorities on digital platforms[23] is also a sign of the increased interest in these sectors of the economy on the NCA's part.

8.3 Norwegian Digital Competition Policy: Specific Issues

8.3.1 Introduction

In this section, we discuss some specific issues relating to Norwegian competition policy towards the digital economy in more detail. We will therefore consider market definition and market power (Sect. 8.3.2), telecom and digital infrastructure (Sect. 8.3.3), data protection and privacy (Sect. 8.3.4), digital platforms (Sect. 8.3.5), online advertising (Sect. 8.3.6), content providers and comparison portals (Sect. 8.3.7).

8.3.2 Market Definition and Market Power in the Digital Economy

The Norwegian competition authorities have not formally addressed the question of relevant antitrust markets for free services. Norwegian, substantive competition law is to a significant extent harmonised with EU/EEA competition law. For guidance, the Norwegian competition authorities generally also look to the European Commission's policy on various issues. If a comparable case were to emerge, this means that the Norwegian competition authorities would consider the European Commission's practice concerning market definition and free services.

When defining relevant markets, the Norwegian competition authorities generally apply the guidance provided by ESA and the European Commission's Notice on the definition of a relevant market. In concrete cases, the NCA and ESA will generally also look to the European Commission's decisional practice for guidance as to how various markets have been delineated in relation to their product and geographical dimensions.

There are a few enforcement decisions by the Norwegian competition authorities on digital service markets in the narrow sense. Moreover, the NCA rarely enforces competition rules in markets that are wider in geographical scope than Norway. The NCA's stated policy in response to global tech giants has been that national competition authorities in the EU/EEA should act in unison under the direction of the European Commission.[24]

[23] Digital platforms and the potential changes to competition law at the European level – The view from the Nordic competition authorities (2020), available at https://konkurransetilsynet.no/nordic-competition-authorities-release-joint-memorandum-on-digital-platforms-and-the-future-of-european-policy/?lang=en. Accessed 14 October 2020.

[24] Interview with the Director of the NCA, 7 February 2019, available at https://www.tek.no/artikler/facebook-palagt-store-begrensninger/457500. Accessed 14 October 2020.

The issue of whether an antitrust market can exist for free services is closely related to competition policy in relation to two-sided markets, where an undertaking serves two different customer groups that provide each other with network benefits. By lowering the price on one side of the market, or even providing a service for free, the 'platform' may increase its prices and profits on the paying side.

The NCA intervened against concentration in a two-sided market in connection with a media merger in 2012. In V2012-11 A-pressen AS – Mecom Europe AS, the NCA identified significant anti-competitive effects in the markets for printed news in one county and for advertising in printed newspapers in a Norwegian town. The NCA defined separate markets for the different customer groups (readers and advertisers) and addressed the two-sidedness of the service in its competitive assessment by discussing whether network effects could offset the impact of reduced competition in the individual markets separately. The merger was given conditional clearance.

Looking beyond Norwegian competition policy, 'free' services may be offered at zero prices but are not necessarily cost-free. If an intermediary, for example a digital platform, is 'paid' in the form of personal data, the platform can exploit the data to sell targeted advertising. When defining zero-price markets, the traditional SSNIP test may require analytical adaptations or modifications, for example concentrating on the quality dimensions of the service rather than price.

With regard to the issue of market power, the Norwegian competition authorities have not developed a particular policy or issued specific guidelines for digital networks or platforms. Nor have the Norwegian competition authorities yet finally decided on any cases against digital networks or platforms, and they have thus not considered issues such as network effects, economies of scale, access to data or innovation potential in these kinds of digital markets.

On a more general note, it is worth mentioning the NCA's gradually increasing reliance on other analytical tools than structural market factors to measure market power, especially in merger cases. Over the last decade, inspired by US economic theory and UK merger practice, the NCA has increasingly based its decisions—albeit predominantly in horizontal cases concerning unilateral effects and differentiated products—on customer surveys, analyses of the closeness of competition, diversion ratios and price pressure indexes.

8.3.3 Telecom and Digital Infrastructure

The responsibility for ensuring effective competition in the telecom market in Norway is shared between the competition authorities and the National Communications Authority. The agencies' respective competences are mostly complementary and do not overlap. The Communications Authority primarily regulates the market *ex ante* through general regulations and by imposing obligations on undertakings, while the NCA carries out control of mergers and enforces the prohibitions on abuse of dominance and anti-competitive agreements. The Communications Authority may impose certain obligations on undertakings found to have a dominant position in the telecom sector, and it has done so in relation to Telenor. The obligations imposed on Telenor consisted of offering service providers and mobile virtual network operators (MVNOs) access to their network on non-discriminatory terms and an obligation not to engage in a margin squeeze. The case concerning Telenor's alleged abuse of a dominant position, which we comment on below, illustrates that the obligations imposed by the Communications Authority do not prevent the NCA from enforcing the prohibition on abuse of dominance and that their competences are complementary. It is also worth mentioning that the Communications Authority has incentivised the rollout of a third mobile phone network through its regulation of network owners' pricing of voice call termination. Voice call termination is the transfer of a call from another network to the operator's own network.

Promoting competition in the telecom sector has been an enforcement priority for the Norwegian competition authorities. This is seen, for instance, in controls of concentration in the telecom sector, in which the NCA has intervened several times.[25] A common denominator in all these concentration cases is that the concentrations were in principle found to have anti-competitive effects. Under the Norwegian Competition Act, however, undertakings can propose modifying commitments in order to avoid harm caused to competition in markets otherwise affected by the concentration. In the cases referred to in this report, the NCA authorised all the concentrations after accepting the proposed modifying commitments. Note that, as under EU law, the modifying commitments are subject to a proportionality test, which means that the commitments must not go further than is necessary to address the significant competitive restraints caused by the concentration.

As part of their enforcement focus on the telecom sector, the Norwegian competition authorities have investigated several cases against Telenor, Norway's leading telecom company. Two of these investigations led to Telenor being found guilty of abusing its dominant position by using pricing structures that distort competition. In the first decision, adopted by the Norwegian NCA in 2018, Telenor was found to have prevented the establishment of a new third mobile network operator in Norway. In the second decision, recently adopted by ESA in June 2020, Telenor was found to have hindered competition by squeezing the margins of operators providing stand-alone mobile broadband services to Norwegian customers.[26]

More specifically, the first case concerned whether, pursuant to Section 11 of the Norwegian Competition Act and Article 54 EEA, both of which mirror Article 102 TFEU, Telenor abused its dominant position between 2010 and 2014 by preventing the establishment of a third mobile network in Norway. During this period, Telenor and Telia were the only two mobile operators with a mobile network infrastructure that covered all of Norway. When a third actor, Network Norway, tried to establish a third national mobile network, Telenor arguably prevented this from

[25] V2005-6 Telenor – Tiscali, V2012-8 Telenor – LOS, V2015-1 Telia – Tele2/NwN.

[26] Last updated 12 August 2020.

happening by changing its pricing structure for Network Norway, which rented access to Telenor's network.

Access to Telenor's infrastructure was a precondition for Network Norway being able to establish a network of its own. The new terms from 2010 included a fixed monthly 'SIM fee' that Network Norway had to pay Telenor for each of Network Norway's subscribers. The SIM fee was independent of Network Norway's actual data traffic passing through Telenor's network. The new terms allegedly reduced Network Norway's incentives to continue work on establishing its own network. Network Norway was incentivised to merely continue as a service provider in the downstream market. Telenor thus arguably hindered more competition in the upstream market by preventing the establishment of a third network. In 2018, the NCA fined Telenor NOK 788 million for this behaviour, pursuant to the Norwegian Competition Act Section 29, seen in conjunction with Article 6 of the EEA Competition Act.[27] The NCA's decision was upheld by the Competition Appeals Committee as well as by the Norwegian Court of Appeal.

It is interesting to note in connection with this case that the changed pricing structure was actually found to increase competition in the downstream market, for example by providing Network Norway with more customers. However, the improvements in the downstream market were not found to outweigh the negative impact that the new terms had on competition in the upstream market, which discouraged the establishment of a third mobile network. The Competition Appeals Committee found, in particular, that the changed pricing structure was not reconcilable with the fundamental responsibility of a dominant operator; i.e., Telenor must not disturb competition in a market. Telenor was therefore found to be in breach of Section 11 of the Norwegian Competition Act and of Article 54 EEA since a cross-border dimension was found to be present.

In the second case, the formal investigations opened by ESA back in 2014 concerned possible abuse of a dominant position contrary to Article 54 EEA in the market for mobile broadband services for Norwegian users. ESA found that Telenor charged higher prices for wholesale access to its network than the prices offered to its own purchasers of mobile broadband services. This type of margin squeeze made it impossible for stand-alone mobile broadband service providers, which were dependent on access to Telenor's mobile network, to operate in the fast-evolving market without selling these types of services at a loss. Telenor had increased wholesale prices for network access during a period of critical growth between 2008 and 2012 when demand for mobile data increased dramatically. Note that the case concerned universal serial bus (USB)-enabled mobile broadband services, i.e. mobile broadband accessed on large-screen devices such as tablets and laptops and not phones.

[27] V2018-20 Telenor.

ESA issued a historically high fine of EUR 112 million, declaring that Telenor had abused its dominant position pursuant to Article 54 EEA.[28] Telenor appealed the decision to the EFTA Court who dismissed the application for annulment.

8.3.4 Data Protection and Privacy

In Norway, the enforcement of the GDPR and the protection of privacy more generally are tasks assigned to the Norwegian Data Protection Authority. The NCA has not focused on big data issues in its enforcement or policy, at least not in a manner that is publicly known. Instead, the NCA has chosen to follow the EU Commission's lead on these issues. There is no direct overlap between the competences of the NCA and the Data Protection Authority. Instead, their respective competences are complementary. This means that for the NCA to be able to interfere due to concerns about the collection of data, the conduct would have to amount to an anti-competitive agreement or an abuse of dominant position, or in the case of a merger, it would have to give rise to competitive concerns due to the collection of data.

As regards the use of general terms and conditions for data collection, this is primarily a matter for the Data Protection Authority or the consumer protection authorities.

8.3.5 Digital Platforms and Online Advertising

As regards competition policy relating to digital platforms, the NCA has advocated the deregulation of the taxi sector, thereby enabling the emergence of new business models that use digital platforms to connect drivers and customers. In relation to platforms for booking travel and hotels, and issues concerning MFN clauses etc., the NCA's policy is to await recommendations from the European Commission.[29]

As to the Norwegian competition authorities' preferred policy tools, it is worth mentioning that through an amendment of the Norwegian Competition Act in 2013, the NCA became competent to issue commitment decisions pursuant to Section 12 third paragraph, which is inspired by Article 9 Regulation 1/2003. The NCA has used this competence once, in a case concerning the distribution of beer to Norwegian pubs.[30]

Where it is found necessary in order to promote competition in a market, Article 14 of the Norwegian Competition Act also empowers the Ministry of Trade, Industry

[28] Press release 2020, 39.

[29] The Norwegian Competition Authority (2019), available at https://konkurransetilsynet.no/kronikk-hotellbooking-pa-nett-er-til-fordel-for-hotellgjestene/. Accessed 14 October 2020.

[30] V2020-20, Ringnes.

and Fisheries to take regulatory action against certain types of anti-competitive behaviour.[31]

The provision is enforced by the NCA and has rarely been used, although there is a seminal example of a regulatory intervention by the Ministry from 2009, when it was found necessary to change the practice for the online advertising of real estate. Online advertising platforms, such as the Norwegian *Finn.no*, had up until then only allowed professional operators, such as licensed real estate agents, to advertise their properties online. As online advertising became a more and more common way of reaching potential buyers, the NCA found that refusing private individuals to sell their own houses and apartments through these channels could distort competition. In practice, this meant that private individuals were forced to use professional sellers in order to reach the online advertising market. This, in turn, allowed professional sellers to charge higher prices since they faced less competition.

By including, *inter alia*, a non-discrimination clause, the Ministry prohibited this exclusionary practice, enabling private individuals to sell their houses and apartments online.[32] Today, the platform allows users to have their own profile and offers not just online advertising for the sale and rental of properties but also a range of other products and services in different markets. This development shows that Article 14 of the Norwegian Competition Act was a useful tool for the Norwegian authorities when dealing with online anti-competitive practice. The regulatory intervention has enabled new online advertising markets to evolve, and a broader range of online services and products is now offered to consumers at a better price. At the same time, it has led to *Finn.no* becoming the largest online advertising platform for housing (and most likely for other products and services as well) in Norway. As mentioned earlier in this report, with regard to competition policy targeting the digital sector, the Ministry has instructed the NCA to give particular priority to monitoring the digital economy. This case seems to fit quite well with the Ministry's message that digitalisation contributes to significant technological development and innovation but may also lead to market dominance, thereby necessitating close monitoring.

In 2020, cases concerning digital platforms and online advertising have been a priority for the NCA. The NCA ordered a notification of an acquisition of a majority shareholding in an online marketplace for used cars, which was below the general turnover-based notification threshold.[33] On 30 September 2020, the NCA reached a preliminary conclusion that the concentration would result in a significant impediment to competition in an online market for the sale of used cars. In September 2020, the NCA also issued a statement of objections to four publishing houses and their jointly owned database company. The NCA's preliminary view was that the

[31] See Sect. 8.2 above.

[32] Act on access to online advertising of real estate by the Norwegian Ministry of Trade, Industry and Fisheries, 2009, available at https://lovdata.no/dokument/SF/forskrift/2009-09-09-1169?q=forskrift%20om%20tilgang%20til%20boligannonsering. Accessed 14 October 2020.

[33] V2020-16, Schibsted – Nettbil.

undertakings had exchanged competitively sensitive information concerning, *inter alia*, future book prices and the timing of book releases through their joint database company, which supplies online advertising and marketplace services. A preliminary fine was calculated to be a total of NOK 502 million.

Finally, it should be noted that the Nordic competition authorities' joint report on 'Digital platforms'[34] of September 2020 identifies four challenges that digital platforms pose in relation to achieving effective competition: (1) 'Tipping scenarios', where a market may 'tip' in favour of one single platform and leave no room for competing platforms; (2) 'Gatekeeper scenarios', where a platform acts as a regulator and unilaterally set the rules for accessing the ecosystem that it creates; (3) 'Leveraging scenarios', where digital platforms integrate with upstream or downstream markets; and (4) 'The dynamic and fast-moving nature of digital markets', which makes predictions and forward-looking competition analyses uncertain. The report further adopts a balanced and knowledge-based approach to various competition policy proposals targeting digital platforms.

8.3.6 Comparison Portals and Sector Inquiries

As we briefly mentioned in section A, the Norwegian Consumer Agency developed an Internet web portal that was intended to show real-time prices of groceries. This was done on behalf of the government, and its purpose was, among other things, to increase consumer information about prices. The NCA was highly critical of the portal because it would lead to increased market transparency in an oligopolistic market, increasing the risk of collusion.[35] In this case, neither the NCA nor the Consumer Agency had the power of decision. It was the government that had the final say and hence could resolve the disagreement. In the end, the development of the portal was terminated in May 2018.

The case of the grocery Internet portal illustrates that the NCA is mainly concerned with comparison portals from an efficiency or welfare point of view and not from a consumer protection point of view. Unlike in some other jurisdictions, the Norwegian Competition Act does not contain any provisions relating to fair or unfair competition, but only provisions relating to efficient competition. This prevents the NCA from interfering for consumer protection reasons as such, unless the consumer interest at risk is consumer welfare.

---

[34] Digital platforms and the potential changes to competition law at the European level – The view from the Nordic competition authorities, 2020, available at https://konkurransetilsynet.no/nordic-competition-authorities-release-joint-memorandum-on-digital-platforms-and-the-future-of-european-policy/?lang=en. Accessed 14 October 2020.

[35] The Norwegian Competition Authority's consultation submission of 30 November 2011 concerning NOU 2011:4, available at https://konkurransetilsynet.no/decisions/731-horingssvar-rapport-fra-matk/. Accessed 14 October 2020.

8.3.7 Abuse Proceedings in the Absence of Dominance

The Norwegian Competition Act does not contain a general provision allowing for abuse proceedings against non-dominant undertakings. However, pursuant to Section 14 of the Act, the government may adopt market-specific regulations if that is considered necessary to promote competition in a market. Such regulations may include the possibility of intervening against non-dominant undertakings' unilateral conduct. Introducing such regulations for the digital economy has not been on the agenda, however. Instead, the debate in Norway about a regulation allowing for intervention against non-dominant undertakings' unilateral conduct has been limited to the grocery market.

9 Switzerland

Johana Cau

9.1 Introductory Remarks on the Approach of the Swiss Competition Authorities in the Digital Economy

This contribution discusses the role of the Swiss competition authorities, *i.e.* the Competition Commission (Comco) and its Secretariat, in the digital economy.

It is the responsibility of the Competition Commission to prevent the harmful consequences of cartels and other restrictions on competition, including in the context of digitalisation. This encompasses the fight against illegal agreements, the monitoring of abuses committed by companies in a dominant position and merger control. The prevention of abusive pricing is, however, within the competence of the Price Supervisor.[1]

The increasing digitisation of the economy has not yet led to amendments to the fundamental principles of Swiss competition law. The wording of the relevant provisions of the Swiss Cartel Act[2] (hereafter "CartA") and of the federal law on the supervision of prices is broad enough to allow for an adaptation of the Swiss authorities' practice to react to practices deriving from a world that is becoming more and more digitalised. It is noteworthy that the Swiss federal government issued a report in January 2017[3] that exposes the existing framework conditions for the

[1] Federal law on the Supervision of Prices, RS 942.20. Available (in French) at: https://www.fedlex.admin.ch/eli/cc/1986/895_895_895/fr.

[2] RS 251. Available at: https://www.fedlex.admin.ch/eli/cc/1996/546_546_546/en.

[3] Report of the Federal Council on the main framework conditions for the digital economy, 11 January 2017, available (in French) at https://www.newsd.admin.ch/newsd/message/attachments/46894.pdf. Accessed 10 August 2021.

J. Cau (✉)
Lenz & Staehelin, Geneva, Switzerland
e-mail: johana.cau@lenzstaehelin.com

B. Kilpatrick et al. (eds.), *Antitrust in Data Driven Markets & Legal Framework for Influencers, Native Advertising and Control over the Use of AI in Marketing*, LIDC Contributions on Antitrust Law, Intellectual Property and Unfair Competition, https://doi.org/10.1007/978-3-031-07422-6_9

development of the digital economy in Switzerland and underlines the tools of the CartA that can be directly applied to the digital economy.

Nevertheless, the increasing digitalisation of the economy poses new challenges for Swiss competition policy. Two phenomena are notable: the attention given to data and the increasing diffusion of platform markets. While the digital economy has not directly led to amendments to the CartA, the analysis of the facts relevant to competition must be adapted, particularly with regard to the delimitation of the market, such as online platforms and marketplaces, or the examination of possible abuses of a dominant position or merger control. The Competition Commission stated that errors of appreciation could lead to regulations that hinder competition, instead of allowing different players to compete on equal terms: *a fight with equal arms is more useful to competition than the blind application of old rules in the face of new forms of economy. It is therefore necessary to take a critical look at existing regulations.*[4]

In 2014, the Secretariat of the Competition Commission created an internal working group named *Digitization* with the aim of maintaining an up-to-date basis for the evaluation of digital markets.[5] This working group supports the Competition Commission in its decision-making and follows international developments and case law in this field. In addition, in 2017, the Competition Commission declared digitisation to be one of its priority topics, thereby signalling the special attention it intends to devote to this area.[6] It is *aware of the particularities of the digital markets, observing its current developments and taking the potential for innovation in this sector particularly into account in its decision-making practice.*[7] In addition to this organisational and material reorientation, the Competition Commission conducted a series of investigations and merger control cases in which digitalisation played an important role. To name a few, the use of broad price parity clauses on targeting online reservation platforms was prohibited by the Competition Commission in 2015.[8] As regards telecommunication, the Competition Commission approved the planned merger between Sunrise and Liberty Global without remedies in 2019,[9] after refusing the one between Sunrise and Orange in 2010.[10]

Finally, it will be necessary to consider whether an adaptation of merger control criteria would be relevant, considering mergers or acquisitions of Internet platforms. Despite the trend in the convergence of Swiss and European competition law,

[4] Competition Commission annual report, 2016, pp. 24 and 28.

[5] Competition Commission annual report, 2020, p. 31.

[6] Competition Commission annual report, 2020, p. 31.

[7] Competition Commission annual report, 2017, p. 4.

[8] Decision of the Competition Commission "*Online-Buchungsplattformen für Hotels*", LPC 2016/1, p. 67ff. See also Sect. 9.2.6 below.

[9] Decision of the Competition Commission "*Sunrise/Liberty Global*", LPC 2020/2, p. 841. See also Sect. 9.2.4 below.

[10] Decision of the Competition Commission "*France Télécom SA/Sunrise Communications AG*", LPC 2010/3, pp. 507ff. See also Sect. 9.2.4 below.

notable differences persist as Switzerland's merger review is more permissive than the one in the European Union. The Swiss government explained, in the report on the legal framework of the digital economy published in 2017, that it may be necessary and useful to adapt the merger notification criteria, in particular introducing the SIEC (*significant impediment to effective competition*) test, so that the authorities can examine the mergers or acquisitions of young Internet platforms that could possibly impact competition.[11]

9.2 Swiss Digital Competition Policy: Specific Issues

In this section, we consider specific issues relating to Swiss competition policy towards the digital economy in more detail. We will discuss hereafter data protection and data collection (Sect. 9.2.1), market definition and market power (Sect. 9.2.2), online advertising (Sect. 9.2.3), telecom and digital infrastructure (Sect. 9.2.4), abuse of a dominant position (Sect. 9.2.5) and price parity clauses (Sect. 9.2.6).

9.2.1 Data Protection and Data Collection

Swiss data protection and competition laws have different purposes. The goal of data protection is to ensure the rights of individuals who are the subjects of data processing. The goal of competition law is to prohibit firms from engaging in conduct that will harm competition. Data protection plays a part in competition law only when, by being disrespected, it has or presents the risk of distorting the competitive process.[12]

In Switzerland, the enforcement of the Federal Act on Data Protection[13] and the protection of privacy are assigned to the Federal Data Protection and Information Commissioner. The Competition Commission hence follows the lead of the European Commission on these issues: having complementary competences, the Competition Commission will only interfere due to concerns about anti-competitive conduct or agreements or abuse of dominant position. In the case of a merger, the collection of data would have to give rise to competitive concerns.

There are no case precedents in Switzerland pertaining to the specific combination of data collection and potential abuse of a dominant position in an undertaking's general terms and conditions. The Secretariat investigated the use of data by companies with monopoly positions at the cantonal level, such as companies in

[11]Report of the Federal Council on the main framework conditions for the digital economy, 11 January 2017, pp. 8ff, available (in French) at: https://www.newsd.admin.ch/newsd/message/attachments/46894.pdf. Accessed 10 August 2021.

[12]Report of the Federal Council on the main framework conditions for the digital economy, 11 January 2017, pp. 162ff, available (in French) at: https://www.newsd.admin.ch/newsd/message/attachments/46894.pdf. Accessed 10 August 2021.

[13]RS 235.1. Available at: https://www.fedlex.admin.ch/eli/cc/1993/1945_1945_1945/en.

the field of electricity. Even if the Comco did not open any in-depth investigation on this topic, it is worthy to mention that, in January 2021, the Swiss Federal Office of Energy fined four employees of the energy company Groupe E for infringing the Law on Electricity Supply by using customer data to advertise the company's new online store.[14]

The debate concerning the interface between competition and data protection issues is hence in its early stages. Nevertheless, the Competition Commission expressed itself on the challenges related to big data and the use of data, opening new possibilities when it comes to pricing. The Competition Commission explained that *personalized prices, such as individual discounts, can be offered on the basis of a large amount of data specific to clients. The peaks and declines of demand can be recognized more easily and faster thanks to data collected at different moments.*[15] The Competition Commission then specifies that such practices can potentially deploy hindering effects or exploiting effects on competitors. Thus, new questions arise that the Competition Commission finds difficult to answer with a *final judgement from the point of view of cartel law, which is currently not possible due to a lack of experience.*[16]

9.2.2 Market Definition and Market Power

In the digital economy, the simple approach aimed at determining the markets on which an undertaking is effectively active, especially platforms, can be complex. The specificities of platforms, be they two sided or multi-sided, can influence the delimitation of the market, which usually constitutes the starting point of competition analysis. In such a context, it can be extremely complicated to properly understand the competitive relationships linking the different actors. For example, on the Android platform, at least five parties interacted in addition to Google: the product suppliers, the smartphone manufacturers, the application developers, the telecommunication companies and the users. Once the multi-sided market is identified, one must evaluate whether only one market covering the entirety of the multi-sided market should be defined or if it must be distributed over two or more markets linked between them.[17]

Furthermore, digital platforms and online services adopt specific features that differ from traditional offers: given the fact that many digital goods are "free", the

[14]RS 734.7. Available (in French) at: https://suissetec.ch/files/PDFs/Kommunikation/Medienmitteilungen/Franz/210129_CP_suissetec_Nouveau%20succ%C3%A8s%20dans%20la%20lutte%20contre%20les%20distorsions%20de%20concurrence.pdf. Accessed 20 September 2021.

[15]Competition Commission annual report, 2016, p. 26.

[16]Competition Commission annual report, 2016, p. 27.

[17]Report of the Federal Council on the main framework conditions for the digital economy, 11 January 2017, pp. 162ff, available (in French) at: https://www.newsd.admin.ch/newsd/message/attachments/46894.pdf. Accessed 10 August 2021.

possibility of analysing prices disappears. Users on online markets no longer pay their use with money but rather pay it with their interest or with data pertaining to them and their behaviours when it comes to research and shopping, which makes it possible for the platform to perform targeted analysis of shopping behaviours and to differentiate pricing. Hence, in the practice of the Competition Commission, the services offered "for free" by a platform are usually considered an economic activity.[18] Indeed, the fact that a requesting party does not pay for the use of a platform is not a valid reason not to consider them to be a part of a commercial relationship and not to take them into consideration when defining the relevant market, first, because that side of the market often pays a price, albeit an implicit one, for example in the form of data pertaining to their usage of the platform or the provision of personal data. Second, the fact that one side of the market does not have to pay precisely reflects a given price model, which is optimal for the platform.[19]

In principle, the relevant test for market definition in Switzerland is the substitutability of products and services[20] and, in particular, the cross-price elasticity and small but significant and non-transitory increase in price (SSNIP) test.[21] While it is often used to delimit relevant markets based on prices, it cannot be used as such for two-sided markets since one of the user groups does not pay a (monetary) price for the use of the platform, and the platforms usually apply different prices for different user groups. Finally, the use of the SSNIP test is made even more complex because the determination of prices for each side depends on how the prices are determined by the other side, and therefore it cannot be considered separately for both parties.[22]

When it comes to the market share of online undertakings, turnovers do not always disclose much information on their respective positions in the market. There are no specific guidelines in Switzerland to assess the market power of a digital platform or social media network. In this author's view, the criteria to determine the position of an undertaking on the digital economy market are, on a case-by-case basis, qualitative competition, the ability of an undertaking to control access to services online and to charge high fees or the vulnerability of its position in the market in relation to innovation cycles.

Finally, rules pertaining to interventions on merger controls in the European Union show significant discrepancies. For example, in Germany, an additional

[18]Report of the Federal Council on the main framework conditions for the digital economy, 11 January 2017, p. 160, available (in French) at: https://www.newsd.admin.ch/newsd/message/attachments/46894.pdf. Accessed 10 August 2021.

[19]Report of the Federal Council on the main framework conditions for the digital economy, 11 January 2017, pp. 160ff, available (in French) at: https://www.newsd.admin.ch/newsd/message/attachments/46894.pdf. Accessed 10 August 2021.

[20]Pursuant to Article 11 (3) lit. a of the Ordinance on the Control of Concentrations of Undertakings.

[21]Decision of the Swiss Federal Supreme Court of 29 June 2012, case 2C_484/2010, para 9.2.3.5.

[22]Report of the Federal Council on the main framework conditions for the digital economy, 11 January 2017, pp. 160ff, available (in French) at: https://www.newsd.admin.ch/newsd/message/attachments/46894.pdf. Accessed 10 August 2021.

intervention threshold has already been instituted for mergers exceeding a certain transactional value, allowing the competition authorities to examine mergers between online platforms despite their respective turnovers.[23]

In this context, the annual report of the Competition Commission in 2016 notably stated that the turnover-based thresholds in merger control could lead to a situation wherein mergers are not controlled, even though, in relation to customer data, a dominant position exists.[24] Following this statement from the Competition Commission, the Swiss government explained, in the report on the legal framework of the digital economy published in 2017, that it may be necessary and useful to adopt merger notification criteria so that authorities can examine mergers or acquisitions of young Internet platforms that could possibly impact competition; however, there are no plans to modify notification thresholds related to turnover. The introduction of the SIEC test would allow the Competition Commission to take into account potential efficiency gains during platform concentrations while broadening the discretion of the Competition Commission.[25]

9.2.3 Online Advertising

As of today, the number of decisions issued by the Competition Commission in the field of online advertising and comparison portals is relatively scarce. The decisions described hereafter were taken in specific merger control cases and do not result from sector-specific investigations.

The first significant decision pertaining to online advertising dates back to 2015. In this matter, the Competition Commission authorised, without charges nor conditions, a co-enterprise between Swisscom, the SSR and Ringier whose purpose was to commercialise and place advertisements in different media (the Internet, television, print and radio).[26] The investigation focused on televised advertising aimed at specific target groups that the co-enterprise had planned to introduce in Switzerland.

Notwithstanding the fact that the joint venture would create one of the strongest actors in advertising, the Competition Commission ruled that it would not lead to a dominant position that would put at risk effective competition in the 2 or 3 following years. Other than the actual competition, the investigation specifically took into

[23] More information in the "*Guidance on Transaction Value Thresholds for Mandatory Pre-merger Notification (Section 35 (1a) GWB and Section 9 (4) KartG)*" report, available at: https://www.bundeskartellamt.de/SharedDocs/Publikation/EN/Leitfaden/Leitfaden_Transaktionsschwelle.pdf?__blob=publicationFile&v=2. Accessed 20 September 2021.

[24] Competition Commission annual report, 2016, p. 28.

[25] Report of the Federal Council on the main framework conditions for the digital economy, 11 January 2017, pp. 8 and 10, available (in French) at: https://www.newsd.admin.ch/newsd/message/attachments/46894.pdf. Accessed 10 August 2021.

[26] Decision of the Competition Commission "*Schweizerische Radio and Television Company/Swisscom AG/Ringier AG*", LPC 2016/1, pp. 299ff.

account the following dynamic competitive relationships: the substitutability with other advertisement channels and the possibility of televised advertisement aimed at specific target groups; the strong growth of over-the-top services such as Zatoo, Wilmaa or Teleboy; and the uncertainties pertaining to the association effects or synergies emanating from targeted advertising.

In the field of media and advertising, the Competition Commission assessed two mergers: Tamedia/Tradono Switzerland[27] and Tamedia/Neo Advertising,[28] through which Tamedia wanted to gain exclusive control over Tradono Switzerland and Neo Advertising SA, respectively. Tradono Switzerland operates a digital marketplace for small ads, which can only be used by means of a small ad application. Neo Advertising SA is active in the field of outdoor advertising (out-of-home advertising). The Competition Commission has given its approval after having conducted a preliminary investigation on these projects.

It is interesting to note that the Tamedia/Tradono Switzerland merger was approved notwithstanding the high market share of Tamedia, thanks to the dynamic evolution of online advertisement, market share volatility, low barriers to entry and Tradono's marginal market share. Such considerations were not required for the Tamedia/Neo Advertising merger, which was authorised on the following grounds: the low market share of Neo Advertising and the fact that both companies, while both active in the advertising business, were not competitors (indeed, Tamedia is active in the print and Internet advertisement market, but Neo Advertising is active in outdoor advertising).

In the context of the acquisition of Websheep GmbH by Swisscom Directories SA in 2019, the Competition Commission investigated the situation from the point of view of competition law regarding comparison platforms and did not notice the elimination of effective competition.[29] The ground for the approval was the fact that the resulting increase of Swisscom Directories' market share was insignificant.

Finally, more recently, in December 2019, the Competition Commission opened a preliminary investigation into the case of Swisscom Directories due to potential abuse of a dominant position on the directories and digital marketing market.[30] Suspicions are related to data withholding from third parties and the linkage of various services. The investigation was ongoing in 2020,[31] and no decision has been issued by the Competition Commission yet.

---

[27] Decision of the Competition Commission "*Tamedia/Tradono Switzerland*", LPC 2017/4, pp. 579ff.

[28] Decision of the Competition Commission "*Tamedia AG/Neo Advertising AG*", LPC 2017/3, pp. 505ff.

[29] Decision of the Competition Commission "*Swisscom Directories/Websheep*", LPC 2019/2, pp. 535ff.

[30] LPC 2020/1, p. 66.

[31] LPC 2021/1, p. 73.

9.2.4 Telecom and Digital Infrastructure

The enforcement of competition law in Switzerland in the telecom sector happened on multiple instances with merger controls.

In 2019, the Competition Commission approved the planned merger between Sunrise and Liberty Global without remedies. With the takeover of UPC and its cable network infrastructure, Sunrise would become the second-largest telecommunications company in Switzerland, which would be able to offer fixed network, broadband Internet and mobile telephony services, as well as digital television on its own infrastructure in Switzerland, like the market leader Swisscom. The Competition Commission concluded that there would not be any collective dominance with Swisscom and that coordination between the two companies was unlikely because the parties to the mergers and Swisscom are differently positioned on the market.[32]

In 2010, the Competition Commission prohibited the planned concentration between Orange (France Télécom) and Sunrise, which would have reduced the number of competitors from three to two on the mobile communication market.[33] According to the Competition Commission, the merger between Orange and Sunrise would have created a collectively dominant position with Swisscom in the mobile telephony market, which is likely to end effective competition without new entrants on the market being able to exercise a disciplining power. The new entity born from the merger and the third actor on the market, Swisscom, would not have offered effective competition to each other: it would have been more advantageous for both undertakings to maintain a high price range rather than to attack each other, hoping to get their market shares. However, since UPC is mainly active in the cable network market, the current merger project would not have resulted in a three-to-two merger in the mobile communication market.[34]

Regarding access to competition in the field of network infrastructures, the Competition Commission examined in 2016 the behaviour of network providers in the field of peering (direct interconnection agreements between networks) as well as the communication between such providers.[35] Furthermore, the Commission, in 2015, opened an investigation against Naxoo SA, which is suspected of holding a dominant position on the wired network of the city of Geneva, for abusing its position vis-à-vis building owners, third-party system providers and end users.[36] Reportedly, Naxoo imposed unfair conditions of trade through its exclusive contracts to connect buildings to the cable television network, which consequently limited both market opportunities and technical development. Whenever it

---

[32] Decision of the Competition Commission "*Sunrise/Liberty Global*", LPC 2020/2, p. 841.

[33] Decision of the Competition Commission "*France Télécom SA/Sunrise Communications AG*", LPC 2010/3, pp. 507ff.

[34] Decision of the Competition Commission "*Sunrise/Liberty Global*", LPC 2020/2, pp. 759ff.

[35] LPC 2017/1, pp. 73ff.

[36] LPC 2016/1, pp. 54ff and LPC 2017/1, p. 62.

physically connects buildings to the cable television network, Naxoo took advantage of the fact that proprietors viewed its services as indispensable by taking control over the usage of the buildings' wiring and, in this way, preventing proprietors from installing third-party systems using the same coaxial cable socket. In doing so, Naxoo ensured its sole ability to use the indoor installations required to distribute the coaxial signal to end users. Thus, by being prevented from setting up competing parallel systems, proprietors were impeded by Naxoo's business practice to freely use their buildings' wiring. This has also limited market opportunities for third parties providing, for instance, satellite-based systems and has hampered technological development. Finally, end consumers have been prevented from accessing other telecommunication services that are complementary to or in competition with the cable network, in particular satellite broadcasting services. The Competition Commission fined Naxoo SA in December 2017.[37] In 2021, the Federal Administrative Court upheld, in principle, the sanction imposed by the Competition Commission.[38]

9.2.5 Abuse Proceedings in the Absence of a Dominant Position

As of now, the prohibition of unilateral anti-competitive conduct applies only to dominant companies, pursuant to Article 7 CartA and similarly to Article 102 of the Treaty on the Functioning of the European Union (TFEU).

However, in its spring session in 2021, the Swiss Parliament strengthened the concept of relative market power. With the new revised CartA, which will enter into force on 1 January 2022,[39] the rules on abuse of a dominant position will be clearly extended to companies with relative market power.[40] A company has relative market power if another company depends on it as it does not have sufficient and reasonable possibilities of switching to other companies. Individual consumers and public authorities that do not qualify as undertakings cannot benefit from the protection against abuse of relative market power.[41] The application of the concept of relative market power will potentially lead to a large number of currently non-dominant companies being newly deemed as having relative market power. The legal

[37] Decision of the Competition Commission "*Supermédia*", LPC 2018/3, pp. 508ff.

[38] Decision B-2798/2018 of the Federal Administrative Court. An appeal to the Swiss Federal Supreme Court is pending.

[39] See Press release of 17 September 2021 "Indirect counter-proposal to the Fair Prices Initiative enacted", available at https://www.admin.ch/gov/en/start/documentation/media-releases.msg-id-85152.html. Accessed 20 September 2021.

[40] Following the "*Fair-price initiative*" (19.037). See, in more detail (in French), at: https://www.parlament.ch/fr/ratsbetrieb/amtliches-bulletin/amtliches-bulletin-die-verhandlungen?SubjectId=48671. Accessed 11 August 2021.

[41] See Press release of 17 September 2021 "Indirect counter-proposal to the Fair Prices Initiative enacted", available at https://www.admin.ch/gov/en/start/documentation/media-releases.msg-id-85152.html. Accessed 20 September 2021.

consequence of the abuse of relative market power is a prohibition of the conduct in question and, potentially, a claim for damages.

New types of abuse will also be introduced: companies with relative market power (as well as dominant companies) are prohibited from restricting the ability of customers to purchase goods or services offered in Switzerland and abroad at local prices and conditions.[42]

The concept of relative market power is inspired by the provisions of German competition law; however, unlike German law, the amendment does not explicitly include relative dominance for intermediation on platforms (e.g. Amazon) and data. Furthermore, it is still uncertain how the concept will be used in Switzerland, compared to the German practice. Due to the need to interpret the new rules, the Competition Commission and the Swiss civil courts are expected to develop their own practice when applying the new law. Therefore, there will be considerable legal uncertainty until enough leading cases have been litigated.

9.2.6 Price Parity Clauses

With the emergence of Internet platforms, some contractual agreements, such as price parity clauses, became more important. Price parity clauses aim to ensure that the products presented on the platform cannot be offered at a better price through another commercial channel. They can have distinct reaches. The "tight" clauses simply forbid the provider from offering better conditions himself. The "broad" ones widen this prohibition to other platforms.

Whether a price parity clause is harmful in practice must always be established on a case-by-case basis. In 2015, the Competition Commission—after an investigation targeting online reservation platforms Booking.com, Expedia and HRS from 2012 onwards—prohibited the use of broad price parity clauses.[43] However, it did not act against tight clauses that prohibit hotels from offering more advantageous prices on their own websites. According to the Competition Commission, those are not problematic because they do not exclude competition between platforms and aim at avoiding providers from taking advantage of the services of a commercial platform (avoidance of free riding).[44] The Competition Commission decided to observe the development of new clauses in the market, reserving its right to intervene again if deemed necessary.[45]

Nevertheless, the Price Supervisor opened an investigation in 2017 against the online booking platform Booking.com after finding indications of abuse in the rates

[42] See the new Art. 7 par. 2 let. g CartA.

[43] Decision of the Competition Commission "*Online-Buchungsplattformen für Hotels*", LDC 2016/1, p. 67ff.

[44] Decision of the Competition Commission "*Online-Buchungsplattformen für Hotels*", LDC 2016/1, p. 76 pt. 67.

[45] Competition Commission annual report, 2015, p. 5.

paid by hotel owners to the platform.[46] Furthermore, the Swiss Parliament has recently approved a motion prohibiting the unfair contracts of online booking platforms, introducing a general ban on all clauses pertaining to price parity in the hotel industry, although the Federal government considered that the main problem of the broad clauses had been solved and that competition was working between the various platforms.[47] This should lead to an amendment of the Swiss Unfair Competition Act to prohibit all price parity clauses in the future.[48]

9.3 Conclusion

In an overall assessment, Swiss competition law can effectively deal with various anti-competitive practices related to the digitalisation of the economy. Existing regulations are drafted broadly enough to take into account the characteristics of this new economy on a case-by-case basis. This is highlighted by numerous decisions issued by the Competition Commission over the last decade regarding, for example, online sales platforms or merger control and abuse of dominant positions in the field of telecommunications and network infrastructures. That being said, the Swiss legislator still addresses some of the new challenges deriving from the digital economy, notably with a recent motion to introduce a general ban on all clauses pertaining to prices in the hotel industry.

Finally, Swiss and foreign companies must follow the implementation of the recent amendment to the CartA strengthening the concept of relative market power and closely monitor how it will be enforced, also in view of the challenges posed by digitalisation. While the German legislator already adapted the German competition law for digital regulation and although German case law may provide useful indications in this regard, the scope of the revised CartA on relative market power and its interpretation by the Competition Commission and Swiss courts are difficult to predict at this stage.

---

[46] See the Price Supervisor Newsletter Nr. 4/17 of 12 September 2017, p. 3. Available at: https://blog.preisueberwacher.ch/file.axd?file=/NL_PDF/Newsletter_04_17_f.pdf. Accessed 20 September 2021. The investigation is still ongoing.

[47] Motion "*Bischof*" to prohibit the unfair contracts of online booking platforms, which are hurting the hotel industry of 30 September 2016 (16.3902) approved on 18 September 2017. Consultation for amendments to the Law on unfair competition was opened in November 2020. See in more details (in French) at: https://www.parlament.ch/fr/ratsbetrieb/suche-curia-vista/geschaeft?AffairId=20163902. Accessed 20 September 2021.

[48] RS 241, available (in French) at: https://www.fedlex.admin.ch/eli/cc/1988/223_223_223/fr.

United Kingdom

10

Sir Christopher Bellamy

10.1 Introduction

This chapter analyses how the United Kingdom has, and is proposing to, use competition law and regulation to create holistic solutions in tackling fast-moving competition issues in relation to digital networks, online platforms and most favoured nation clauses (**MFNs**).

The UK government is presently considering the introduction of a new legislative regime designed to regulate businesses that have attained a position of strategic market power within the digital economy. Two key reports have driven UK policy thinking in this area. These are:

- Unlocking Digital Competition, March 2019 (*the Furman Report*),[1] by an expert panel led by Harvard economist and former White House economic adviser Jason Furman; and
- Competition and Markets Authority's Online Platforms and Digital Advertising Market Study final report, July 2020 (*the Digital Report*)

These reports are considered, in detail, below. Following on from these reports, there have been a number of further developments of significance. A Digital Markets Taskforce was created in March 2020 to provide advice to inform the government's

[1] Digital Competition Expert Panel, Unlocking digital Competition, Report of the Digital Competition Expert Panel ('Furman Report') (March 2019), available at https://assets.publishing.service.gov.uk/government/uploads/system/uploads/attachment_data/file/785547/unlocking_digital_competition_furman_review_web.pdf. Accessed 31 August 2021.

S. C. Bellamy (✉)
Monckton, London, UK
e-mail: cbellamy@monckton.com

B. Kilpatrick et al. (eds.), *Antitrust in Data Driven Markets & Legal Framework for Influencers, Native Advertising and Control over the Use of AI in Marketing*, LIDC Contributions on Antitrust Law, Intellectual Property and Unfair Competition,
https://doi.org/10.1007/978-3-031-07422-6_10

decisions on how to promote competition in digital markets to the benefit of consumers and businesses.[2] It is led by the Competition and Markets Authority (**CMA**), with support from the telecoms regulator Ofcom and the Information Commissioners Office (**ICO**) and has also had substantial input from the Financial Conduct Authority (**FCA**). The Taskforce produced proposals for the new regulatory regime for digital markets in December 2020.

Subsequently, a Digital Markets Unit (**DMU**) was established within the framework of the CMA in April 2021. As explained below, for the time being, the DMU is tasked with engaging in work to prepare for exercising powers under an envisaged legislative regime (not yet finalised) which will put this body on a statutory footing and confer additional (and potentially far-reaching) regulatory powers upon it. The precise scope of these powers is still under consideration at the governmental level. Alongside these institutional developments, there have been a number of further announcements by the CMA, the Digital Markets Taskforce and the central government, which are of significance to the regulation of the digital economy. These are considered below.

10.1.1 The Furman Report (2019)

In the UK, the Furman Report has been seminal in the development of policy thinking on the regulation of the digital economy. The report found that many digital markets are global and dominated by one or two of the five largest digital companies: Amazon, Apple, Facebook, Google and Microsoft. Whilst digital markets have brought enormous benefits, there are now heightened concerns about the advantages enjoyed by the largest companies, including economies of scale and scope, access to data, network effects, limited switching possibilities, access to finance and so on. The resulting barriers to entry mean that many digital platform markets are not freely contestable, and the incumbents have significant market power over users, resulting in user dependency, privacy and data security issues, lower levels of innovation and potentially unfair treatment of customers.

The Furman Report recommended establishing a Code of Conduct for digital companies enjoying a "strategic market status" (**SMS**), which would require a much greater degree of fairness, transparency and consistent behaviour by platforms towards consumers, business users and potential rivals, particularly as regards access and prominence of display, together with greater personal data mobility and the creation of open standards. This Code would be developed by a new Digital Markets Unit based within the CMA. Although outside the scope of this national report, the Furman Report also recommended closer scrutiny of digital mergers and the development of a new "balance of harms" test.

---

[2] *Digital Markets Taskforce: Terms of Reference,* available at: https://www.gov.uk/government/publications/digital-markets-taskforce-terms-of-reference/digital-markets-taskforce-terms-of-reference%2D%2D3. Accessed 31 August 2021.

10.1.2 The CMA's Digital Report (2020)

The CMA's Digital Report, made under the market investigation powers of the Enterprise Act 2002 (EA02), has developed the Furman Report's proposals in the light of more detailed information about the activities of Google and Facebook, which, according to the CMA, has reinforced the need for reform along the Furman Report's lines.

The CMA found that Google had over 90% of the UK general search market and that Facebook generated over 50% of UK display advertising revenues. Both Google and Facebook benefited from substantial network effects—in Google's case, particularly the scale of its click and query data, and in Facebook's case, its large network of connected users, raising significant barriers to potential rivals, subsequently reinforced in Google's case by default agreements, for example where mobile phone suppliers agree that Google Search occupies the default search position on the device. In terms of personal data, the CMA found that many platforms do not give users the ability to control their data, or they offer privacy policies that are so complicated that most users most simply accept the default choice they are presented with. In addition, their unique and vast sources of data enable Google and Facebook to offer advertisers and media agencies options for targeting consumers that other platforms cannot match—Google's mobile data even enables it to track users offline. Google and Facebook also interpret the UK General Data Protection Regulation (**GDPR**) favourably to themselves, according to the CMA.

The CMA also found that there was little transparency in the operation of online automated bidding for digital advertising inventory or in the opaque algorithms determining search results, as well as little ability for advertisers or publishers to determine whether they were receiving value for money. Further concerns related to the potential for leverage into other areas afforded by "ecosystems"—in Google's case, Chrome, Android, Double Click and YouTube, and Facebook with What's App, Instagram and various other activities. Finally, the CMA identified potential concerns with vertical integration and conflicts of interests, where, for example, Google has a very strong position in advertising intermediation generally, in supply-side platforms and in demand-side platforms, allowing Google to self-preference its own activities and reinforce its market power.

Against this background, the CMA came to the conclusion that Google and Facebook have such entrenched market power that a new pro-competition ex ante regulatory regime was needed for online platforms funded by digital advertising, thereby endorsing the findings of the Furman Report. As suggested by the Furman Report, this new regulatory regime would apply to platforms having SMS and would comprise of an enforceable Code of Conduct governing behaviour as well as pro-competitive interventions to promote competition. As to the Code of Conduct, each SMS would have its own code based on fair trading (to counter exploitative behaviour), open choices (to counter exclusionary behaviour) and trust and transparency (to ensure users are provided with sufficient information to make informed decisions). As to the pro-competitive interventions, these should include powers to increase consumer control over data, to mandate interoperability and third-party access to data and to mandate data separation where necessary. The CMA sees these powers, or some combination of them, as a toolkit to address such matters as the default positions enjoyed by Google on devices, allowing third-party search engines better access to data and thus improving their algorithms; increasing consumers' control over data and choice over what advertising they wish to receive; resolving conflicts of interests, if necessary through separating various vertically integrated functions; and mandating greater transparency and access to data in digital advertising markets, subject to GDPR requirements.

The CMA, like Furman, recommends legislation to implement these reforms and the establishment of a DMU to carry them out. The precise scope of these reforms has been further considered by the Digital Market Taskforce in a report published in December 2020 (discussed below).

Finally, it should be noted that the CMA's Digital Report has highlighted certain aspects of the interface between antitrust law and data protection law (including the GDPR and its post-Brexit equivalent), including the use of over-complex terms and conditions (**T & Cs**) to which users are effectively forced to adhere and the (mis)use of the GDPR to prevent the sharing of certain data with potential rivals. Although hitherto antitrust issues have not figured prominently in relation to data protection, which is the separate responsibility of the ICO, the data protection/antitrust interface is expected to come under closer scrutiny in the work of the DMU, in which, as stated in Sect. 10.1.1, the CMA, the ICO and Ofcom are all represented. Looking to the future, the CMA may well challenge the use of data protection laws in ways that weaken competition.

10.2 UK's Approach Towards the Market Power of Social Media Networks and Online Platforms

Unlike the Bundeskartellamt, there are no specific guidelines in the UK to assess the market power of social media networks and digital platforms. However, there is a significant discussion on how to determine market power based on the way in which a network or platform operates.

The Furman Report looked at the market shares held by various digital platforms and noted that in December 2018, over 92% of UK online page views that originated from a search engine were estimated to come from Google,[3] and in terms of page views originating from social media sites, over 73% were from Facebook or Instagram. That said, the CMA noted that a number of other metrics, such as level of concentration, user numbers, traffic generated, or time spent on each service, could be assessed as possible metrics. For example, 33.5% of all UK e-commerce

[3] Statcounter, GlobalStats data for UK search engine market share, available at: https://gs.statcounter.com/search-engine-market-share/all/united-kingdom. Accessed 20 April 2020.

transactions went through Amazon in 2018;[4] it is a source of continuing debate as to what the relevant market definition is for online retail or marketplaces.

The Furman Report also looked in detail at barriers to entry that exist in digital platform markets and the high degree of control and influence that some firms have over both (i) the relationship between buyers and sellers and (ii) the access that advertisers have to potential buyers. It noted that this gives platforms three distinct types of power: "the ability to control access and charge high fees; the ability to manipulate rankings or prominence; and the ability to control reputations".[5] As a result, a small number of companies have strategically important gateway positions, can control others' market access and can wield significant bargaining power over their users. It is this that the Furman Report considered as market power.[6]

The CMA's Digital Report took these considerations one step further and also used profitability analysis to assess the market power of digital platforms. It found that the global return on capital employed, for both Google and Facebook, has been well above any reasonable benchmarks for many years. It estimated that the cost of capital for both Google and Facebook in 2018 was around 9%, whereas their actual returns have been substantially higher, at least 40% for Google's search. It concludes that this level of profit is indicative of these platforms' ability to exploit market power.[7]

As already identified under Sect. 10.1.2, the CMA considered that the following factors may inhibit entry or expansion by an online platform's rivals and undermine competition: (i) network effects and economies of scale, (ii) consumer behaviour and the power of defaults (i.e. default behaviour by consumers and the way in which choices are presented to consumers), (iii) unequal access to user data, (iv) problems relating to a lack of transparency and (v) vertical integration and conflicts of interest.[8] Based on the Furman Report and the Digital Report, these are the main factors for reviewing the market power of networks and platforms in the UK.

[4] Plum Consulting report commissioned by DCMS, Online advertising in the UK (January 2019), available at: https://assets.publishing.service.gov.uk/government/uploads/system/uploads/attachment_data/file/777996/Plum_DCMS_Online_Advertising_in_the_UK.pdf. Accessed 31 August 2021.

[5] K. Sabeel Rahman, Regulating informational infrastructure: Internet platforms as the new public utilities (July 2018), available at: https://georgetownlawtechreview.org/regulating-informational-infrastructure-internet-platforms-as-the-new-public-utilities/GLTR-07-2018/. Accessed 31 August 2021.

[6] Furman Report, para 1.161.

[7] CMA, Online platforms and digital advertising Market study final report ('Digital Report') (July 2020), paras 2.77–2.81, available at https://assets.publishing.service.gov.uk/media/5efc57ed3a6f4023d242ed56/Final_report_1_July_2020_.pdf. Accessed 31 August 2021.

[8] Digital Report, para 21.

10.2.1 UK's Approach Towards Defining the Term "Platform" and Other Linked Concepts

The Terms of Reference[9] for the new Digital Markets Taskforce define digital platforms as online services that intermediate between different groups to buy, sell, share and exchange different goods and services, typically collecting and using vast amounts of data to deliver their services. The terms further cite the Organisation for Economic Co-operation and Development (OECD) definition[10] of online services, which is "a digital service that facilitates interactions between two or more distinct but interdependent sets of users (whether firms or individuals) who interact through the service via the Internet". According to the terms, "[t]his includes, but is not limited to, e-commerce platforms, peer-to-peer platforms, social media networks and search engines, which have become an everyday part of modern life".

The Furman Report and the Digital Report both note that online platforms facilitate interactions and transactions between users for no monetary cost. However, these platforms instead collect data from their users in return.

As outlined in the CMA's Digital Report, the integration of different services is also a key concept in the business model of online platforms because once such platforms have attracted customers by offering their core services for free, they then make money from another "side" of that same platform. In the case of Amazon, this money is made from the commission generated from retailers on its Marketplace, whilst the Apple App Store, for example, derives revenue from the charges to app developers. Equally, there is no clear separation on search engines and social media sites between the services offered by the platform and the advertisements that the end user views, for which the platform generates revenue.[11]

The Digital Report also found that online marketplaces, such as Amazon and eBay, or price comparison sites help consumers search for certain types of information by way of "specialist search", as well as offering their products for sale. From a consumer perspective, there are differences between general search and specialised search. General search engines help consumers with a wider range of queries, including many that are not served by specialised providers, whilst specialised search provides functionality that allows consumers to compare listed products across multiple dimensions.[12]

Within the UK, there is unlikely to be a formal approach to distinguish between different types of platforms or different functions within platforms: each situation will be judged on its own facts.

[9] CMA, Digital markets taskforce: terms of reference.

[10] Footnote 4, CMA, Digital markets taskforce: terms of reference.

[11] Digital Report, paras 2.3 and 2.4.

[12] Digital Report, paras 3.44–3.51.

10.2.2 Can There Be an Antitrust "Market" for Free Digital Services in the UK?

Antitrust agencies generally attempt to analyse markets by assessing their effects on end prices, but free services make it difficult to define separate markets, determine market power or assess the effects of a platform or transaction on competition.

It is well recognised, for example by the EC Digital Content and Digital Services Directive (2019),[13] that digital content and services are usually provided in exchange for personal data rather than for money. However, the first question to consider is whether such services are, in fact, offered for free.[14]

Various international jurisdictions have accepted that a consumer's personal data, preferences and user-generated content can have an economic value and is a valid form of consideration in return for a service.[15] This is also the view of the CMA, which does not consider the use of online platforms as free services. The Digital Report notes that "[a]lthough consumers do not pay money for these services, there is still an exchange that takes place between them and the platform. In exchange for searching the internet, watching videos, or communicating with friends, consumers provide their attention and data about themselves." The CMA notes that "[t]he advertising-funded business model is not novel, nor is it inherently problematic from a competition perspective. Newspapers have been generating revenue in the UK through advertising for several hundred years. On television, ITV provided the first alternative to the BBC in 1955, when it began its ad-funded broadcasting. Similarly, commercial radio stations have been generating revenue through advertising in the UK ever since the market was liberalised in 1973."[16] The CMA notes that the apparently "free" services of search and social media are paid for by digital advertising, the cost of which was around BPD14 billion in the UK in 2019.

Indeed, some economists have in fact argued that whilst online services appear to be free, the zero price paid is in fact too high and consumers could be extracting a greater value in return for their data.[17]

The CMA noted in its response to the House of Lords' consideration of the proposed EU regulation of online platforms[18] that whilst novel online business

---

[13]Directive (EU) 2019/770 of the European Parliament and of the Council of 20 May 2019 on certain aspects concerning contracts for the supply of digital content and digital services (The Digital Contents Directive 2019/770), available at: https://eur-lex.europa.eu/legal-content/EN/TXT/PDF/?uri=CELEX:32019L0770&from=EN. Accessed 31 August 2021.

[14]See recital 24 of the Digital Contents Directive 2019/770.

[15]CMA, Appendix A: The legal framework, Paragraph 68, available at: https://assets.publishing.service.gov.uk/media/5df9ebf0e5274a0910cb6d7c/Appendix_A_The_legal_framework.pdf. Accessed 31 August 2021.

[16]Digital Report, paras 2.5 and 2.6.

[17]See, for example, the discussion by Tim Harford, 'Treat social media like email and search engines', Financial Times, 27 April 2018.

[18]CMA, Select Committee on the European Union: Internal Market Sub Committee, CMA response to the call for evidence: Online Platforms and the EU Digital Single Market, 23 October

models may require existing legal and economic concepts to be applied in new contexts, these platforms do not create fundamentally different issues than those that have been addressed before (even where services are provided at no monetary cost).

10.3 Dominant Companies and Special Responsibility Status

As discussed under Sect. 10.1, the CMA's Digital Report essentially confirmed the Furman conclusions that some major companies in digital markets have excessive market power and that a special pro-competitive regime for companies distinguished as having SMS is urgently needed.

Essentially, the conclusion to be drawn from these reports is that existing tools such as the "special responsibility of dominant companies" are inadequate to deal with the situation and further legislation is required. Many other countries have expressed similar concerns in relation to the power of online platforms. For example, in April 2020, the Australian government directed the Australian Competition and Consumer Commission (ACCC) to develop a mandatory Code of Conduct to address bargaining power imbalances between digital platforms and media companies in response to the ACCC's Digital Platforms Inquiry final report.[19]

10.3.1 Policy Proposals in Response to the CMA's Digital Report

There have been a number of significant policy responses to the CMA's Digital Report. In November 2020, the government published a formal response to the CMA's recommendations (**"the Government Response"**).[20] A short time later, in December 2020, the Digital Markets Taskforce published its advice, at the request of the government, on the design and implementation of a code of practice and the shape of a new pro-competition regime for digital markets in the UK.[21] This latter report is likely to be very influential as regards the shape of future legislative reform in the United Kingdom in this area.

2015, available at: https://assets.publishing.service.gov.uk/government/uploads/system/uploads/attachment_data/file/502607/Response_to_House_of_Lords_inquiry_on_online_platforms.pdf. Accessed 31 August 2021.

[19] The Hon Paul Fletcher MP, ACCC mandatory Code of Conduct to govern the commercial relationship between digital platforms and media companies (Media Release, 20 April 2020), available at: https://minister.infrastructure.gov.au/fletcher/media-release/accc-mandatory-code-conduct-govern-commercial-relationship-between-digital-platforms-and-media-companies. Accessed 31 August 2021.

[20] See Response to the CMA's Market Study into Online Platforms and Digital Advertising, available at: https://assets.publishing.service.gov.uk/government/uploads/system/uploads/attachment_data/file/939008/government-response-to-cma-study.pdf. Accessed 31 August 2021.

[21] The Taskforce Report, and appendices, which contain further substantive analysis, are available here: https://www.gov.uk/cma-cases/digital-markets-taskforce#taskforce-advice. Accessed 31 August 2021.

10.3.1.1 Government Response to CMA's Digital Report

The Government Response essentially accepted the key recommendations in the CMA's Digital Report and laid down a number of proposed next steps. The response accepts that the inherent features of digital markets described in the CMA's Digital Report (such as network effects, economies of scale, barriers to entry and unequal access to data and financing) "means that they tend to tip in favour of a single incumbent" and that this is "leading to higher prices for goods and services, less innovation, and less choice and control for consumers".[22] As a result, the government also accepts that the UK's "existing competition tools are not equipped to address the systematic harms associated with dominant platforms" in the digital advertising, general search and social media markets.[23] Based on this analysis, the government agrees that substantial legislative and regulatory reform is now needed.

As to what this reform entails, the response endorses the establishment of an enforceable Code of Conduct to govern the behaviour of platforms funded by digital advertising that are designated as having strategic market status.[24] The government has asked the Digital Markets Taskforce to provide further advice on the design and implementation of a Code of Conduct and, more broadly, on the approach that should be adopted in respect of firms having substantial and enduring market power and therefore designated with "Strategic Market Status". This advice was provided by the Taskforce in December 2020 and is discussed further below.

Overall, the response accepts that the Code of Conduct must be mandatory and enforceable since, in the government's view, an unenforceable regime, as exists in Australia, would be insufficient to deter anti-competitive behaviour.[25] The government also agrees as to the need for a Digital Markets Unit with the task of introducing, maintaining and enforcing the Code of Conduct and exercising a range of powers to undertake "pro-competitive" interventions in the market for digital services. The DMU was established provisionally in April 2021 (albeit merely to undertake preparatory work for the time being pending the conferral of legislative powers).

The establishment of a Code of Conduct and the powers of the DMU to enforce this regime and the regime for pro-competitive interventions will require new legislation, which the government has undertaken to provide "as soon as parliamentary time allows".[26] Draft legislation has not yet been published but is expected in the course of 2021. The government sees these legislative reforms as part of a broader agenda of regulating the digital economy, explaining that "a new pro-competition" regime will form a key part of this broader agenda, with other

[22] Government Response, para 14.

[23] Government Response, para 15.

[24] Although beyond the scope of this Chapter, the response observes that the code should also address the relationships between online platforms and news publishers, "in order to ensure the sustainability of high-quality journalism and news publishing" Government Response, para 17.

[25] Government Response, para 25.

[26] Government Response, para 22.

elements, including measures, to "boost innovation, build public trust through greater safety and security for users of digital tech, and promote a democratic, open society". As part of this agenda, draft legislation in the form of the *Online Safety* bill has been introduced in Parliament, with the aim of "establish[ing] a new regulatory regime to address illegal and harmful content online, with the aim of preventing harm to individuals in the United Kingdom", imposing duties of care on Internet service providers.[27]

10.3.1.2 Further Advice from Digital Markets Taskforce

The Digital Markets Taskforce December 2020 Advice ("**the Taskforce Report**") provides influential advice to the government on a number of key aspects of the newly envisaged regulatory regime, including procedural and substantive aspects of the strategic market status test; the purpose, form, content and powers associated with the code of practice for designated firms; arrangements for pro-competitive interventions in respect of designated firms; and recommendations on other important matters, including merger control.

The strategic market status designation is at the core of the proposed regime.[28] As regards this designation, the Taskforce recommends that this status should require a finding that a firm has "substantial, entrenched market power in at least one digital activity, providing the firm with a strategic position in respect of that activity".[29] The report defines a "digital activity" as "a group of products or services that can be considered together" because they "can reasonably be described as having a similar function or can reasonably be described as fulfilling, in combination, a specific function".[30] The report therefore proposes a flexible approach and deliberately eschews reliance on markets and detailed analysis for the purpose of market definition. Indeed, the report observes that "[a]n activity could consist of products and services that would be identified as belonging to separate relevant markets under a traditional competition law approach".[31] It is proposed that the strategic market status designation would be assessed by reference to these specific activities (rather

---

[27] Online Safety Bill, Explanatory Notes, para 1, available at: https://assets.publishing.service.gov.uk/government/uploads/system/uploads/attachment_data/file/985031/Explanatory_Notes_Accessible.pdf. Accessed 31 August 2021.

[28] The Taskforce does not recommend that a statutory duty be placed on the Digital Markets Unit to designate a firm some of whose activities meet the test. Instead, the report recommends that designation be a discretionary power, which would be challengeable by way of judicial review rather than a merit-based appeal. *See* Taskforce Report, paras 11 and 12 and § 4.114.

[29] Appendix B § 56, Taskforce Report. Appendix B describes the proposed Strategic Market Status test in detail. It is available here: https://assets.publishing.service.gov.uk/media/5fce72c58fa8f54d564aefda/Appendix_B_-_The_SMS_regime_-_designating_SMS_firms.pdf. Accessed 31 August 2021.

[30] Appendix B, para 15.

[31] Appendix B, para 18.

than all activities of the company as a whole), and the code of practice would then apply to these specific activities.[32]

The Taskforce recommends that the test focus on whether the power of the business in respect of specific digital activities is "entrenched", in other words that it is expected to "persist over time and is unlikely to be competed away in the short term".[33] The report explains that whilst the temporary or transitory attainment of substantial market power is of less concern, once a firm's position becomes entrenched, the likelihood of a rival emerging and taking a substantial share of the market is low. In such circumstances, it is likely that prices will be persistently higher, whilst quality, investment and innovation will be persistently lower than would otherwise be the case, to the long-term detriment of consumers. The report recommends that the assessment of market power should focus on direct evidence of market power, specifically evidence of substitutability, competitive rivalry and barriers to entry and expansion.

A further important distinguishing feature of the designation is that a firm has a "strategic position" as a result of its "entrenched and substantial" market power. It is explained that a "strategic position" in respect of an activity is one where the firm's market power is particularly "widespread or significant".[34] The report identifies various factors or indicia that may inform this assessment, including the size and scale of the company, whether it represents an important "gateway" or access point to consumers, whether the firm has the ability to extend its market power from one activity to others, whether it has an "ecosystem" of products, whether the firm can use its market power to "determine the rules of the game" and whether the firm has impact on activities with broader social or cultural importance.[35]

Once the strategic market status test is met, consistent with the Furman Report, the Taskforce recommends that the Digital Markets Unit should establish a legally enforceable Code of Conduct for the firm in relation to its "designated activities". It is proposed that a tailored code will be established for each firm by the Digital Markets Unit, specifically in respect of its designated digital activities.[36] As can be seen, the concept of "designated activities" is, therefore, of particular importance to the operation of the scheme. The report explains that the Code will provide a set of *ex ante* principles for designated firms to follow, with the aim of preventing consumers and businesses from being exploited and preventing practices by firms that could undermine fair competition.[37] It is envisaged that each code will provide objectives, principles and guidance. The objectives would be laid down by Parliament in legislation. The principles lay down a set of standards with which the firm must comply, whilst the guidance sets out information as to how the principles will be

[32] Appendix B, paras 13–20.

[33] Taskforce Report, paras 28–30.

[34] Taskforce Report, paras 4.18.

[35] Taskforce Report, para 4.19.

[36] Taskforce Report, paras 4.47–4.51.

[37] Taskforce Report, para 4.33.

interpreted. This approach is adopted in favour of either rule-based or outcome-based regulation, each of which is seen to have drawbacks (inflexibility and over-prescription in the case of rules and undue ambiguity in the case of outcome regulation).[38] This proposed model is, in part, informed by regulation by Ofgem and the FCA in the UK's energy and financial service sector, where principles and guidance are used to regulate enterprises.[39] The report recommends that the over-arching objectives of the code should be fair trading, open choices for consumers, trust and transparency.[40] It is also recommended that the Digital Markets Unit be given the power to open formal investigations into a breach of the code and, in the event of a breach, to impose substantial penalties with a significant deterrent effect. It is proposed that a breach should be either deliberate or negligent[41] and could result in a financial penalty of up to 10% of global turnover.[42]

As foreshadowed above, a new regime of "pro-competitive interventions" is also recommended. The Taskforce argues that pro-competitive interventions are an important tool to enable the Digital Markets Unit to intervene in markets to promote dynamic competition and innovation. It is envisaged that whilst the *Code of Conduct* will address concerns created by the strategic market power of certain forms, *the intervention regime* will address matters that have given rise to ineffective competition in the first place.[43] This regime is analogous to the Significant Market Power regime operated by Ofcom and other telecom regulators in the EU which allows for proactive interventions to ensure greater competition in the field of communications.[44] The report gives the following examples of powers which it suggests should be available to the Digital Markets Unit by way of pro-competitive interventions: data-related interventions, interoperability and common standards, consumer choice and default interventions, obligations to provide access on fair and reasonable terms and separation remedies.[45] Such interventions would also be subject to further principles, including effectiveness and proportionality, and should avoid wider effects on the designated firm's business beyond designated activities.

Although beyond the scope of this chapter, it is worth noting that the Taskforce Report also recommended change in the field of merger control as regards companies with strategic market status. In particular, a designated firm would be required to report all mergers (rather than simply report those where the transaction value or share of supply crosses a particular threshold). Affected transactions would be subject to a prohibition on the completion of the transaction prior to regulatory approval. Finally (and importantly), the report recommends that the current regime

[38] Appendix C, paras 17–27.

[39] Appendix C, para C9.

[40] Taskforce Report, para 4.38.

[41] Taskforce Report, para 4.95.

[42] Taskforce Report, para 4.96.

[43] Taskforce Report, para 4.61.

[44] Taskforce Report, para 4.62.

[45] Appendix D, para 17.

for merger intervention be lowered so that the CMA is permitted to intervene where there is a "realistic prospect" that a transaction will result in a substantial lessening of competition.[46]

Overall, the Taskforce Report is hugely detailed and contains a number of quite far-reaching recommendations. The government is presently considering the Digital Report with a view to drawing up a legislative framework for the work of the Digital Markets Unit. Draft legislation is still under consideration and, as yet, no draft bill has been published, but the government has committed to consult on a draft bill in 2021. In the meantime, in February 2021, the CMA published its *Digital Markets Strategy: 2021 Refresh*.[47] The strategy was updated in light of developments, including the Taskforce Report. The strategy explains that a key priority for the CMA in the coming year is to work closely with the government to consider and operationalise the organisational design for the Digital Markets Unit, including its funding, governance and decision-making structures and how it will interact with existing CMA functions (as well as deal with practical matters such as staffing, skills, information technology support and so forth). The CMA has also established a Data and Technology Analytics Unit. According to the CMA this unit has available to it sophisticated data engineering technology to facilitate its monitoring and enforcement functions over businesses which deliver services in digital markets.

10.4 Digital Networks and Online Platforms: The Interaction Between Competition Law and Data Protection Regulation

The behaviour of digital social network operators continues to be subject to scrutiny. Recently, the ICO has reviewed how advertisements across the Internet are bought and sold.[48] The ICO's findings were that, contrary to data protection law, information and sensitive personal data relating to consumers are regularly being used to serve adverts without the explicit consent of those consumers. The ICO did not single out any specific network operators but instead offered a grace period of 6 months during which the entire adtech industry must reform its practice and make improvements to comply with the GDPR. After the grace period, the ICO will undertake a further industry review. The ICO's probe appears to be having an impact in the UK because, since the review, companies such as Google and the industry body for digital advertising called Internet Advertising Bureau UK have committed to make changes to the way they use and share consumer data. However,

[46] Taskforce Report, para 4.153.

[47] Digital Markets Strategy: 2021 Refresh, available at: https://www.gov.uk/government/publications/competition-and-markets-authoritys-digital-markets-strategy/the-cmas-digital-markets-strategy-february-2021-refresh. Accessed 31 August 2021.

[48] Information Commissioner's Office, Update Report into adtech and real time bidding, 20 June 2019, available at: https://ico.org.uk/media/about-the-ico/documents/2615156/adtech-real-time-bidding-report-201906-dl191220.pdf. Accessed 31 August 2021.

the ICO had admitted there are still shortcomings and is continuing to investigate the industry.[49]

10.4.1 Developing Tensions Between Data Protection and Competition Law

The CMA's Digital Report is concerned that, by virtue of their market power, large platforms such as Google and Facebook increasingly appear to be acting in a quasi-regulatory capacity in relation to data protection considerations, setting the rules around data sharing not just within their own ecosystems but also for other market participants. It cites Google's recent announcement that, with the support of the ICO, Google was phasing out support for third-party cookies on the Chrome browser, for data protection reasons. According to the CMA, this potentially restricts publishers' ability to offer personalised advertising. Whilst Google itself would still be able to use the insights it obtains from users' activities on Google Search and YouTube to select personalised ads on Google's properties, publishers, advertisers and intermediaries that are currently reliant on third-party data obtained via data management platforms and data brokers are likely to have less granular data with which to target personalised advertising.[50] The CMA's concerns in this area are not yet resolved.[51] On 8 January 2021, the CMA announced that it was opening an investigation into Google's "Privacy Sandbox" browser changes, by which it proposed to remove third-party cookies and other functionalities from its Chrome browser. The investigation will assess whether the proposals could cause advertising spend to become even more concentrated on Google's ecosystem at the expense of its competitors and whether this may constitute abuse by Google of its market power.[52]

More generally, regulators in the UK are increasingly alive to the potential for tension between competition law and data protection principles in the context of digital markets. With this in mind, the CMA, ICO and Ofcom (which is the UK's communications regulator) jointly established the Digital Regulation Cooperation Forum (**DRCF**) in July 2020. The DRCF's role is broad, namely "to support cooperation and coordination between the CMA, the ICO and Ofcom on online regulatory matters". In practical terms, the regulators explain that through the forum,

[49] Simon McDougall, Information and Commissioner's Office, Blog: Adtech - the reform of real time bidding has started and will continue, ICO News and Events, 17 January 2020, available at: https://ico.org.uk/about-the-ico/news-and-events/news-and-blogs/2020/01/blog-adtech-the-reform-of-real-time-bidding-has-started/. Accessed 31 August 2021.

[50] Digital Report, paras 5.313–5.328.

[51] See Digital Report, para 5.328, where the CMA points out that measures that enhance an aspect of consumer privacy in the near term may have a negative impact on consumer welfare, e.g. by concentrating personal data amongst fewer providers.

[52] *See* CMA Press Release, CMA to investigate Google's 'Privacy Sandbox' browser changes, 8 January 2021 available at: https://www.gov.uk/government/news/cma-to-investigate-google-s-privacy-sandbox-browser-changes. Accessed 31 August 2021.

they hope, *inter alia*, to "[c]ollaborate to advance a coherent regulatory approach", "inform regulatory policy making" and "anticipate future developments".[53] Whilst the underling tensions (for example between competition law and data protection) will not easily be resolved, it is to be hoped that the existence of the forum will better enable each regulator to identify concerns falling within the specialist expertise of the other promptly and to provide a forum in which views can be exchanged and coordinated solutions discussed.

10.4.2 Personalised Digital Advertisements and Data Protection

As indicated in Sect. 10.4.1, the CMA Digital Report is deeply concerned by large digital platforms' collection of data from multiple applications and the use of this to generate (highly lucrative) personalised advertisements. The CMA finds that it is particularly important that platforms give consumers the opportunity, if they want it, to make informed choices about whether to allow their personal data to be used for advertising. Survey evidence is clear that consumers want to be able to control the use of their data and that they feel that they lack this control.[54] Some platforms do not allow consumers to turn off personalised advertising at all. For these consumers, there is no choice but to provide their personal data for advertising if they wish to use the service. Even where they are given a choice about the use of their data, consumers are only able to judge whether they are content to share their data if they are provided with clear and fairly presented information and options about the implications for them. Moreover, they can only truly make an informed choice if they can act on their preferences and withhold or share data as they wish.[55] The CMA concludes that the low levels of actual consumer engagement reflect, to a great extent, the limitations and choice architecture put in place by platforms. As a result, the CMA thinks that many consumers may use services and share more data than they might otherwise have decided to.[56]

The CMA (and the European Commission) have raised specific concerns regarding the collection and use of advertising data by Facebook. In June 2021, the CMA announced that it was opening an investigation under Chapter II of the Competition Act 1998 (**CA98**) into whether Facebook has unfairly used the data gained from its advertising and single sign-on to benefit its own services.[57] The European

[53] See DCRF Launch Document, July 2020, available at: https://assets.publishing.service.gov.uk/government/uploads/system/uploads/attachment_data/file/896827/Digital_Regulation_Cooperation_Forum.pdf. Accessed 31 August 2021.

[54] Digital Report, para 4.228.

[55] Digital Report, para 4.230.

[56] Digital Report para 4.234.

[57] CMA Press Release, 4 June 2021, available at: https://www.gov.uk/cma-cases/investigation-into-facebooks-use-of-data. Accessed 31 August 2021.

Commission simultaneously launched an investigation into Facebook's use of data on the same date.[58]

It should also be noted that in applying data protection law, the ICO has forced WhatsApp not to share its users' data with its parent company, Facebook. In 2018, the ICO found that WhatsApp had no legal basis for the sharing of such data and had it shared such, it would have acted unlawfully and in contravention of the first and second principles of the Data Protection Act 2018. WhatsApp was therefore required to sign an undertaking to prevent it from sharing data with Facebook prior to the implementation of the GDPR in May 2018.[59]

10.4.3 The Increasing Overlap Between Competition Law and Data Protection

Within the UK, the ICO is responsible for scrutinising compliance with both the UK Data Protection Act 2018 and the GDPR. Nevertheless, the CMA is increasingly carrying out work in this area from an antitrust perspective, as described in Sect. 10.4.2.

The CMA's Digital Report found that some digital platforms do not give consumers any choice about the use of their data for personalised advertising. Where digital platforms do offer a choice to consumers, the way it is done is not transparent to consumers and has the effect of promoting the interests of the platform.[60]

The CMA is particularly concerned with how some digital platforms require consumers to agree to the use of their data as part of their initial use of the platform—in particular through the use of "clickwrap" agreements and consent which does not provide sufficiently granular choices to consumers (i.e. inappropriately aggregating consumer consent).[61] It found that the use of clickwrap agreements and aggregated consent (i.e. the collection of data for use on multiple platforms) without the option of more granular controls is likely to mean that consumers do not engage in consideration of the terms they are signing up to when they register for a specific digital platform.[62]

The CMA's Digital Report has also proposed a "Fairness by Design" obligation on digital platforms to maximise users' ability to make informed choices about the

[58] Commission Press Release, 4 June 2021, Antitrust: Commission opens investigation into possible anticompetitive conduct of Facebook available at: https://ec.europa.eu/commission/presscorner/detail/en/ip_21_2848. Accessed 31 August 2021.

[59] See Information and Commissioner's Office, Blog: "A win for the data protection of UK consumers", ICO News and Events, 14 March 2018, available at: https://ico.org.uk/about-the-ico/news-and-events/news-and-blogs/2018/03/blog-a-win-for-the-data-protection-of-uk-consumers/. Accessed 24 August 2020.

[60] Digital Report, paras 4.145 and 4.146.

[61] Digital Report, para 4.155.

[62] Digital Report, para 4.167.

use of personal data, and the CMA is planning to work with the ICO to ensure this complements the data protection by design obligation under the GDPR.[63]

These and other recommendations will be taken forward by the Digital Markets Taskforce, expected to report in December 2020.[64]

10.4.4 CMA's Treatment of General Terms and Conditions in Relation to Data Collection

As discussed in Sect. 10.4.3, the CMA is concerned about digital platforms not allowing consumers to opt out of personalised advertising, forcing them to share considerable amounts of personal data as a condition for using the service.

The CMA Digital Report found that digital platform default settings have a profound impact on the shape of competition in both search and social media. First, default settings play a very important role in influencing consumers' use of search engines, and second, default settings and the way in which choices are presented to consumers have a strong influence on the ability of platforms—particularly social media platforms—to collect data about their users and the ability of users in turn to control the use of their data. In particular, Google's extensive default positions across devices and browsers, and in particular on almost all mobile devices in the UK, act as a barrier to expansion for other search engines. Second, the CMA notes that some digital platforms operate a take it or leave it model, where they do not give their users the ability to control their data. This is particularly prevalent across most social media platforms, including Facebook and Instagram, whose users are unable to turn off personalised advertising whilst continuing to use the service.[65]

Specifically, the CMA found that digital platforms' privacy terms and conditions were long and complicated, typically stretching to many thousands of words. It did not consider that it is reasonable for platforms to expect consumers to have read and understood all of these, often complex, terms before signing up to use a service. Research has shown that very few consumers read privacy policies when signing up for an online service, and the evidence gathered confirms this: for example, in a recent 28-day period, the average visit to the Google privacy page was just 47 s, with 85% of visits lasting less than 10 s. The upshot of this is that users understandably simply agree to the default choices they are presented with. These are set by the platforms, and it is hard to be confident that they will adequately balance users' preferences about the use of their personal data against the substantial benefits to the platform.[66]

The Daily Mail Group told the CMA that "while the GDPR inter alia aimed at placing some limits on the way digital platforms collect and process personal data,

---

[63] Digital Report, paras 94–97.

[64] CMA, Digital markets taskforce: terms of reference.

[65] Digital Report, paras 31 to 36.

[66] Digital Report, paras 39 to 40.

these platforms turned this regulation to their advantage". The News Media Association submitted that publishers have struggled to resist attempts by global tech companies to force unfair terms on them. A further example of the imbalance of bargaining power, cited by several publishers, was the approach taken by Google to updating its terms and conditions shortly before the introduction of the GDPR. Publishers consider that these changes were made in a non-negotiable way and that they had no choice but to accept this update to the T & Cs.[67]

10.5 CMA's Wide Competition and Consumer Protection Powers and Their Application to the Digital Economy

The CMA's market study into online platforms and digital advertising, resulting in the Digital Report, gives the CMA very broad powers to investigate sectors of the economy that it considers may not be working competitively (CMA's markets regime). This regime is the model for the new Digital Tool, which the European Commission is proposing to adopt to tackle structural competition issue, principally in digital markets.[68]

The CMA's market regime is divided into two phases:[69]

- *Phase 1 "market study"*: during this phase, the CMA decides whether to refer a market for a Phase 2 "market investigation".
- *Phase 2 "market investigation"*: this is a more detailed examination into whether there is an "adverse effect on competition" (AEC) in the referred market(s).

In the case of the Digital Report, the CMA did not proceed to a full market investigation and considered that these issues would need to be dealt with under primary legislation.

In relation to price comparison websites, the Private Motor Investigation referred to in Sect. 10.6 was a full market investigation and led to an order prohibiting wide MFNs in motor insurance.

Despite the shortcomings of market investigations, it is still the preferred mechanism for addressing digital markets in the UK, at the present stage. It is thought unlikely that limited competition law tools will be adequate for the task.

[67] Digital Report, paras 5.313 to 5.320.

[68] See European Commission, Press Release: "Antitrust: Commission consults stakeholders on a possible new competition tool", European Commission Press Corner, 2 June 2020, available at: https://ec.europa.eu/commission/presscorner/detail/en/ip_20_977. Accessed 31 August 2021.

[69] Section 5 EA02.

10.5.1 CMA's Consumer Protection Powers Under Market Investigations

The CMA also has wide powers: the sphere of consumer protection.[70] Specifically in relation to price comparison portals, apart from the specific investigations mentioned in the paragraph below, the CMA's Digital Comparison Tools market study (2017) reported on a number of aspects of digital comparison tools (DCTs) from the consumer protection perspective and made a number of recommendations to ensure that consumers were treated fairly. The CMA requires that price comparison websites be clear, accurate, responsible and easy to use. Using its consumer powers, the CMA has taken a number of actions to enforce consumer standards, particularly banning misleading sales practices, for example in the hotel bookings sector,[71] and ensuring the accurate quotation of charges and related matters in the car hire sector.[72]

The CMA has carried out market studies and investigations under the CA98 into the following digital sectors:

- *Online platforms and digital advertising market study* (the Digital Report)[73]
- *Retail banking market investigation*, which resulted in the introduction of "Open Banking", the first major digital interoperability remedy in the world[74]
- *Digital comparison tools market study*, which led in turn to both consumer and competition enforcement action[75]
- *The "posters" cartel between sellers on Amazon Marketplace* enacted through a pricing algorithm[76]

[70]The Unfair Terms in Consumer Contracts Regulations 1999, The Consumer Protection from Unfair Trading Regulations 2008.

[71]CMA, Major overhaul of hotel booking sector after CMA action, Press Release, 13 September 2019, available at: https://www.gov.uk/government/news/major-overhaul-of-hotel-booking-sector-after-cma-action. Accessed 31 August 2021.

[72]CMA, Car hire sites to provide full costs upfront after CMA action, Press Release, 29 March 2018, available at: https://www.gov.uk/government/news/car-hire-sites-to-provide-full-costs-upfront-after-cma-action. Accessed 31 August 2021.

[73]CMA, Online platforms and digital advertising market study, Online Platforms and Digital Advertising Study, September 2020, last updated 01 July 2020, available at: https://www.gov.uk/cma-cases/online-platforms-and-digital-advertising-market-study. Accessed 31 August 2021.

[74]CMA, CMA paves the way for Open Banking revolution, Press Release, 9 August 2016, available at: https://www.gov.uk/government/news/cma-paves-the-way-for-open-banking-revolution. Accessed 31 August 2021.

[75]Online Platforms and Digital Advertising Study.

[76]Online sales of posters and frames, CMA decision of 12 August 2016, Case 50233, available at: https://assets.publishing.service.gov.uk/media/57ee7c2740f0b606dc000018/case-50223-final-non-confidential-infringement-decision.pdf. Accessed 31 August 2021.

- *Commitments from an online auction service provider* to change practices that may have prevented rivals from being able to compete effectively[77]
- *Investigation into the use of "MFN" clauses* by a price comparison website in relation to home insurance products.[78]

10.6 The UK's Approach to MFNs

The legal framework for analysing MFNs in the UK has been the EA02 and CA98. The CMA also has wider consumer protection powers, under the Unfair Terms in Consumer Contracts Regulations 1999 and The Consumer Protection from Unfair Trading Regulations 2008.

The CMA's Private Motor Insurance Market investigation (2014) concluded that "wide MFNs"[79] in contracts between motor insurers and price comparison websites limited competition between comparison websites.[80] Consequently, in relation to private motor insurance, the CMA prohibited providers of price comparison websites from entering "wide MFNs", or equivalents, where comparison sites make more than 300,000 sales per year. However, it considered that "narrow MFNs" were unlikely to have the same impact on competition as they allowed premiums to vary between different comparison sites.[81]

The CMA's DCT market study (2017) also found that "wide MFNs" soften competition between digital comparison tools and competing channels by reducing incentives to compete on commission and to innovate.[82] The CMA also considered that "narrow MFNs" could restrict competition in some circumstances, particularly clauses that are broader than necessary for achieving the claimed efficiency benefits,

[77] CMA case page, Auction services: anti-competitive practices, last updated 13 July 2017, available at: https://www.gov.uk/cma-cases/auction-services-anti-competitive-practices. Accessed 31 August 2021.

[78] CMA case page, Price comparison website: use of most favoured nation clauses, available at: https://www.gov.uk/cma-cases/price-comparison-website-use-of-most-favoured-nation-clauses. Accessed 30 August 2021. The CMA published a non-confidential infringement decision on 9 February 2021.

[79] Wide MFNs specify that a supplier sets a price that is no higher than the price offered through its own website or through any other sales channel. Narrow MFNs specify that a supplier sets a price that is no higher than the price offered through its own website but does not stipulate conditions for sales via other channels.

[80] CMA, Private motor insurance market investigation Final Report, 24 September 2014, paras 59–60, available at: https://assets.publishing.service.gov.uk/media/5421c2ade5274a1314000001/Final_report.pdf. Accessed 31 August 2021.

[81] CMA, Private motor insurance market investigation Final Report, 24 September 2014, paras 8.27–8.123, 12.3, 12.128, available at: https://assets.publishing.service.gov.uk/media/5421c2ade5274a1314000001/Final_report.pdf. Accessed 31 August 2021.

[82] CMA Digital Comparison Tools Market Study, Final Report, 26 September 2017, para 4.91, available at: https://assets.publishing.service.gov.uk/media/59c93546e5274a77468120d6/digital-comparison-tools-market-study-final-report.pdf. Accessed 31 August 2021.

such as preventing free riding, but the CMA has not to date taken any action in relation to narrow MFNs.[83]

On 19 November 2020, the UK CMA imposed a fine of £17.9 million on a price comparison website operator for imposing MFNs in contracts with home insurers selling through the well-known website comparethemarket.com.[84] This decision is presently subject to an appeal by BGL (Holdings) Ltd and others before the Competition Appeal Tribunal.

10.7 Conclusion

The UK has been at the forefront of proposals to regulate the digital economy. However, their proposed new regulatory regime for platforms having "strategic market status" borrows extensively from the EU's existing, *ex ante*, regulatory regime for electronic communications operators with significant market power.[85] Whilst, thus far, the government has proven receptive to (and perhaps even welcoming of) the idea of regulatory reform of the digital economy, there are many competing demands for Parliamentary time and government attention. The government has committed formally to consulting on legislation this year and bringing into force legislation when "Parliamentary time allows". However, it remains to be seen whether regulatory reform in this area will continue to be prioritised, given the complexity of the task at hand and the trade-offs involved in such reform. More broadly, there is also a real question as to whether (and, if so, how) the UK will seek to coordinate a new regulatory regime with that in the EU. With the UK having left the EU, there is likely to be increased legal divergence and regulatory competition between the UK and EU as the UK seeks to establish a distinct competition and regulatory identity, separate from that of its larger European neighbour. From a policy perspective, there is certainly potential for tension between robust regulation of businesses with substantial market power in the digital economy and providing an attractive environment for such businesses to invest and develop. Whether, in this context, there will be a political appetite for the kind of robust regulatory reform that the CMA and Digital Markets Taskforce have proposed remains to be seen.

[83] Ibid 12, paras 4.99–4.10.

[84] The CMA's infringement decision, and information on its investigation, is available here: https://www.gov.uk/cma-cases/price-comparison-website-use-of-most-favoured-nation-clauses. Accessed 31 August 2021.

[85] Regulatory framework for electronic communications, available at: https://eur-lex.europa.eu/legal-content/EN/TXT/HTML/?uri=LEGISSUM:l24216a&from=EN. Accessed 31 August 2021.

Part II

Legal Framework for Influencers, Native Advertising and Control over the Use of AI in Marketing

11 International Report

Marta Delgado Echevarría

11.1 Introduction

This report deals with the legal challenges and opportunities that come with new trends in commercial practices and the increased role played by artificial intelligence (AI) in product marketing. It is based on National Reports from Austria, Belgium, Brazil, Germany and Hungary.

There is a breadth of new commercial practices emerging, thanks to the infinite new possibilities for marketing and product and service provision brought about by artificial intelligence and more sophisticated software systems and also a breadth of new types of consumer practices and uses of technology. The delimitation of concepts such as consumer, seller and promoter becomes less clear as new hybrid types of commercial practices and participants enter into the product/service marketing arena.

The author thanks the National Reporters for their excellent contributions—from Austria: Adrian Kubat and Arno Scharf; Belgium: Louis Hoffreumon and Benjamin Docquir; Brazil: Felipe Barros Oquendo; Germany: Thomas Hoeren and Hannah Hoffmann; and Hungary: Adam Liber, Lili Albert and Judit Firniksz. The International Report is in part a summary of the National Reports. Verbatim quotations from the National Reports are therefore not always identified as such. Here and there, the author has added her own critical comments. The National Reports are based on a questionnaire that was made available to the National Reporters. The International Report has based its systematisation and summary of the National Reports on the systematics of the questionnaire. Other special features that go beyond the questionnaire and were mentioned in the National Reports were taken into account if, in the opinion of the National Reporter, they were of general significance or of particular interest.

M. D. Echevarría (✉)
Jones Day, Madrid, Spain
e-mail: Mdelgado@jonesday.com

B. Kilpatrick et al. (eds.), *Antitrust in Data Driven Markets & Legal Framework for Influencers, Native Advertising and Control over the Use of AI in Marketing*, LIDC Contributions on Antitrust Law, Intellectual Property and Unfair Competition, https://doi.org/10.1007/978-3-031-07422-6_11

New uses of technology to improve customer service and enhance production and marketing processes are a result of much-welcomed progress, but their real or potential effects on both consumers and market players need to be considered carefully by legislators. Likewise, new marketing and advertising practices mostly arising on social networks may create confusion and as a result turn the market into an uncertain playing field, something that is far removed from the goal of national and EU governments.

In dealing with the question at stake, we consider two main blocks: on the one hand, the matter of new marketing trends through influencers and new types of advertising strategies, such as native advertising. On the other hand, we consider the impact that artificial intelligence and more robust and sophisticated software systems have on marketing. These have been considered in the National Reports under the perspective of consumer law, unfair competition law, antitrust law and data protection law.

As usually happens with innovation, these new trends proceed at a faster pace than regulation and existing laws, which become unable to respond adequately to the needs that arise, most notably, although not only, for the protection of consumers, of fair competition in the market or of the free market. Furthermore, as has been noted[1] for practitioners, there is a growing interaction between the laws protecting private interests (such as contracts, fair competition, e-commerce) and the law that is of public order (such as antitrust, data protection and eventually artificial intelligence): a General Data Protection Regulation (GDPR) consent form that must comply with the Unfair Terms Directive[2] when obtaining consent from a data subject[3] is a well-known example.

11.2 Influencers, Native Advertising and Other Practices of Disguised Advertising

11.2.1 Identification of Practices of Disguised Advertising

The exponential growth of social media in the past decade has dramatically changed the media landscape and, with it, marketing and advertising practices. Some argue that the partly negative perception of traditional advertising, which is increasingly making viewers feel uncomfortable, has made it necessary to rethink marketing.[4] Many new marketing practices have emerged, and some of them are in no man's land in terms of their regulation and supervision. Some examples of such newly developed advertising strategies are native advertising, influencer marketing and

[1] Belgian National Report, Section 1.

[2] Council Directive 93/13 of 5 April 1993 on unfair terms in consumer contracts, OJ 1993, L 95, p. 29.

[3] GDPR, Recital 42.

[4] Austrian National Report, Section 2.1.

advertorials (discussed below). These are all types of disguised advertising, that is, commercial communications that present themselves as non-commercial, in a way that they "blend in" with other content published by users on online social media: they tend to look like non-commercial content and appear in the same places on the platform where non-commercial content appears; also, they aim to appear as non-commercial as possible by ensuring that they look like content posted by non-commercial users.[5]

Native Advertising In simple terms, native advertising describes a type of marketing practice that aims to blend in with non-commercial content to the highest extent possible,[6] that is, content that bears a similarity to the news, feature articles, product reviews, entertainment and other materials that surround it online. The key distinguishing characteristic of native advertising content is that it aims to mimic user-generated content in order to increase the likelihood of capturing consumers' attention. This practice differs from display advertising, which is often displayed in a different format and is spatially separated from non-commercial content, making it easier for consumers to identify it as commercial in nature. This form of advertising not only offers the advertiser the advantage of being able to reach the desired target group but is also usually much cheaper than offline advertising.[7]

In a typical native advertising scenario, video advertisements are embedded in a text, leading to the impression that the video belongs to the editorial section of a report. Also, the graphic overlay of such an editorial article with associated advertising messages is becoming increasingly common.[8] Users can interact with native ads in the same way they interact with any other user-generated content: native ads on Twitter can be retweeted, liked or commented on; Facebook native ads can be liked, reacted to, commented on and shared; and Pinterest native ads can be pinned, shared and commented on.

Influencer Marketing This type of advertising technique refers to a form of social media marketing used by so-called influencers, which is increasingly being used by

[5] See Final Report on Behavioral Study in Advertising and Marketing Practices in Online Social Media, a study produced by GfK Belgium commissioned by the European Commission under the EU Consumer Programme (2014-220), 2018 ("Behavioral Study in Advertising and Marketing Practices in Online Social Media") p. 30. Available at https://ec.europa.eu/info/publications/behavioural-study-advertising-and-marketing-practices-social-media-0_en. Accessed 27 October 2021.

[6] B. W. Wojdynski, Native Advertising: Engagement, Deception, and Implications for Theory, In: The New Advertising: Branding, Content and Consumer Relationships in the Data-Driven Social Media Era, R. Brown, V. Jones, B. Ming Wang (eds), 2016.

[7] M. Specht and E. Theobald, Broadcast Your Ad, Nomos 2010, pp. 27f.

[8] See several examples in Behavioral Study in Advertising and Marketing Practices in Online Social Media, pp. 31, 32, available at https://ec.europa.eu/info/sites/default/files/osm-final-report_en.pdf. Accessed 27 October 2021. Other examples are Mini's advertising strategy on the news portal Buzzfeed at A. Wiebe and O. Kreutz, Native Advertising – Alter Wein in neuen Schläuchen? (Teil 1), WRP 2015(9), pp. 1053–1060 (1056).

advertisers.[9] Influencers are persons who have many followers on social media and have an extraordinary reach capability and impact on the market area in which they act as influencers (e.g. beauty products, sports, health, etc.). Influencer marketing relies on influencers' reputation and reach to promote products or services. The practice involves the creation and promotion of content that features specific brands or products, focusing on the influencer's positive experience with the product, and having them recommend the product to their connections or followers through social networks. In comparison with native advertising, the advantage of influencer marketing is that it bears even fewer of the characteristics that make it possible for consumers to identify an advertisement: the content is published by an individual person—not a business—and it is typically presented as a personal endorsement rather than a direct and clearly identifiable promotion of a product.[10] Due to their sometimes enormous impact, but especially by building up an almost friendship-like relationship with their viewers, influencers also gain considerable value for the advertising industry as testimonials.

According to a recent study, about one in five people aged 16–64 has already made a purchase decision based on the recommendation of an influencer. In the age group of 16–24, every second person did so.[11] This type of marketing action has, however, recently received strong criticism and, in turn, has resulted in a backlash to celebrities' reputation.[12]

A research study published in Harvard Business Review suggests that the disclosure by social media influencers of commercial relationships with the brands they promote may actually increase rather than compromise their perceived expertise and authenticity.[13] However, influencer marketing continues to be the preferred advertising technique of brands.

Advertorials Advertorials represent a form of paid editorial content created to promote a product without making it clear to the consumer that the content is sponsored. They often appear to consumers as an objective, informative publication (e.g. a news article or report).

[9] G. Yodel, What Is Influencer Marketing? HUFFPOST of December 6, 2017. Available at https://www.huffpost.com/entry/what-is-influcner-marketing_b_10778128. Accessed 27 October 2021.

[10] See several examples in Behavioral Study in Advertising and Marketing Practices in Online Social Media, 2018, p. 33. Available at https://ec.europa.eu/info/sites/default/files/osm-final-report_en.pdf. Accessed 27 October 2021.

[11] BVDW, Umfrage zum Umgang mit Influencern 2020, available at https://www.bvdw.org/der-bvdw/news/detail/artikel/mehr-als-jeder-fuenfte-verkaeufe-durch-influencer-marketing-nehmen-laut-bvdw-studie-2020-nochmal-zu/. Accessed 08 June 2021.

[12] Celebrities lose their gloss, Financial Times, 28 May 2020; Fake Famous" and the Tedium of Influencer Culture, The New Yorker, 20 February 2021; The rise and fall of influencers: Is the age of giveaways and #ad posts coming to an end?, The Tab 4 May 2021.

[13] A. Audrezet and K. Charry, Do Influencers Need to Tell Audiences They're Getting Paid?, 29 August 2019, Harvard Business Review.

Hidden Advertising in Search Results Higher ranking or any prominent placement of commercial offers within online search results by the providers of online search functionality is another type of disguised advertising and has an important impact on consumers. Ranking refers to the relative prominence of the offers of traders or the relevance given to search results as presented, organised or communicated by providers of online search functionality, including one resulting from the use of algorithmic sequencing, rating or review mechanisms, visual highlights, other saliency tools or combinations thereof.[14]

A key issue common to all these types of advertising is that often customers are unaware of the commercial nature of the input they receive, and in fact it is often the purpose not to make it obvious that the content is an advertisement. Another common feature is that labels used typically by social media to identify the content as advertising (disclosure tags) are not perceived by consumers as such and get lost among other cues that are more salient and more likely to capture consumers' attention.[15]

This issue is particularly relevant for vulnerable groups of consumers, such as children, senior consumers or disabled people.[16] Studies have shown that children are receptive to ads by digital influencers (for example, through social media, "vlogging" or gaming platforms), who create their own content and with whom children can identify more easily than with "traditional celebrities," resulting in a lack of consciousness that their "interaction" with the influencer consists in receiving commercial communication around a product or service.[17]

11.2.2 Legal Framework Applicable to Disguised Advertising

Advertorials are banned by the Unfair Commercial Practices Directive 2005/29, which requires traders to make a clear differentiation between editorial and advertorial content. In addition to point No. 11 of Annex I, which explicitly bans advertorials if it is not made clear to the consumer that they include paid promotions of products, the prohibitions on hidden marketing in Article 7(2) and point No. 22 of Annex I to the Directive are relevant.

[14]Directive 2019/2161 of the European Parliament and of the Council of 27 November 2019 amending Council Directive 93/13/EEC and Directives 98/6/EC, 2005/29/EC and 2011/83/EU of the European Parliament and of the Council as regards the better enforcement and modernisation of Union consumer protection rules ("Omnibus Directive"), Recital 19.

[15]Behavioral Study in Advertising and Marketing Practices in Online Social Media, 2018, p. 44.

[16]Belgium National Report, Section 3.2.

[17]V. Verdoodt and N. Feci, Digital Influencers and vlogging advertising: Calling for awareness, guidance and enforcement, A & M, 2018-2019/1, pp. 11–22. See, from the same author: V. Verdoodt, Kinderrechten en reclamewijsheid in het digitale tijdperk - Richting een empowerend regelgevend kader voor commerciële communicatie, D.C.C.R., 2019/3, n° 124, pp. 3–13.

Conversely, *influencer marketing* and *native marketing* are not expressly regulated at an EU level as such. Although it is prohibited for traders to falsely present themselves as consumers,[18] influencers do not necessarily fall within this notion.

Finally, *hidden advertising in search results* was not regulated until recently.[19]

Allegedly, all these types of disguised advertising practices could fall under the concept of "misleading advertising" of Directive 2006/114 concerning misleading and comparative advertising,[20] which captures any advertising that in any way, including its presentation, deceives or is likely to deceive the persons to whom it is addressed or whom it reaches and, by reason of its deceptive nature, is likely to affect their economic behaviour or, for those reasons, harms or is likely to harm a competitor.

A high level of consumer protection in this area is also ensured by Article 6(a) of Directive 2000/31 and Article 9(1)(a) of Directive 2010/13/EU, which require commercial communications to be clearly identifiable as such in information society services and audiovisual media services, respectively.

However, this pre-existing regulation regarding advertising and disclosure obligations was clearly not sufficient: there is an abundance of disclosure practices fragmented across devices, jurisdictions and providers, while the legislative framework is open as to how and how much disclosure must be provided.[21] On the one hand, the disclosure of commercial content should be clear enough to comply with the existing legal requirements. On the other, the key added value of disguised advertising lies in blurring the line between what is commercial and non-commercial. To this end, the Omnibus Directive[22] has delivered additional rules to address (some of) these concerns.

First, the Omnibus Directive introduces a requirement for online marketplaces that invite to purchase to inform consumers, if not already apparent from the context, whether the third party offering the products is a trader or not (i.e. whether it is a "professional" trader and is therefore within the ambit of the consumer protection legislation or whether it is a private individual, who would not be) on the basis of the declaration of that third party to the provider of the online marketplace.[23] Additionally, online marketplaces must make clear what consumer protection provisions will apply to any given transaction and how responsibility for compliance with consumer law is shared between the seller and the online marketplace.

[18] Annex I No. 22 of the Unfair Commercial Practices Directive.

[19] Through the Omnibus Directive, as we discuss further below.

[20] Of 12 December 2006.

[21] Behavioral Study in Advertising and Marketing Practices in Online Social Media, p. 46.

[22] Directive 2019/2161 of the European Parliament and of the Council of 27 November 2019 amending Council Directive 93/13/EEC and Directives 98/6/EC, 2005/29/EC and 2011/83/EU of the European Parliament and of the Council as regards the better enforcement and modernisation of Union consumer protection rules.

[23] New Art. 7(4)(f) of the Unfair Commercial Practices Directive.

Second, the Directive bans[24] practices where a trader provides information to a consumer in the form of search results in response to the consumer's online search query without clearly disclosing any paid advertising or payment specifically for achieving a higher ranking of products. When a trader has directly or indirectly paid the provider of the online search functionality for a higher ranking of a product within the search results, the provider of the online search functionality should inform consumers of that fact in a concise, easily accessible and intelligible form.[25] Since consumers typically review the top-rated search results (the first or second page at most), it is expected that this gives more transparency to consumers and assists them in getting what they are searching for.

Traders should also inform consumers about the default main parameters determining the ranking of offers presented to the consumer (e.g. price, rating, purchase history or a combination of the same) as a result of the search query and their relative importance as opposed to other parameters.

The Member States have until 28 November 2021 to transpose the Omnibus Directive, and their national provisions should start to be applied from 28 May 2022.

Looking next at the national regulatory landscape, in Austria,[26] the current focus is on the area of labelling obligations. While the declaration of the commercial purpose of paid advertising is a well-known problem, other issues remain, especially where products are purchased and recommended free of charge. Time will tell whether the first judgments in Austria, which are probably unavoidable, will follow those in Germany and consider this to be a commercial practice. However, the fundamental question of the limits to the freedom of expression arises in this context, so particular caution is called for. A distinction should therefore be made in the approach adopted here. Editorial contributions in which the focus lies on the informational aspect fall within the private sphere of influencers, whereas purely company-related contributions, which in particular show a clear advertising overlap, must be labelled.

One issue that the jurisdiction has clarified, at least temporarily, concerns product liability. According to the Court of Justice of the European Union (CJEU), the strict product liability regulations do not apply to services such as advice, even if they are

[24] Annex I, new No. 11(a) Unfair Commercial Practices Directive.

[25] The Omnibus Directive explains that indirect payment could be in the form of acceptance by a trader of additional obligations towards the provider of the online search functionality of any kind, which has a higher ranking as its specific effect. The indirect payment could consist of increased commission per transaction as well as different compensation schemes that specifically lead to a higher ranking. Payments for general services, such as listing fees or membership subscriptions, which address a broad range of functionalities offered by the provider of the online search functionality to the trader, should not be considered payment for specifically achieving a higher ranking of products, provided that such payments are not dedicated to achieving a higher ranking. Online search functionality can be provided by different types of online traders, including intermediaries, such as online marketplaces, search engines and comparison websites. See Recital 20.

[26] Austrian National Report, Section 2.5.

printed in newspapers.[27] For influencer marketing, this initially brings relief. Of course, the last word has not been spoken yet as there is still recourse to general tort law. Therefore, it remains to be seen how the jurisdiction will resolve these two controversial issues in the interests of legal certainty.

In the United States, since 2017, the Federal Trade Commission (FTC) "has invited" companies and influencers, with "endorsement guides",[28] to disclose brand relationships, displaying the hashtag #ad or the mention "paid partnership" in a prominent position. Moreover, in 2019, the FTC published a Guide for Business in native advertising to help businesses understand the regulation and possible enforcement action against native advertising.[29]

In Brazil, in order to protect the consumer, the Brazilian Consumer Protection Code[30] has provided since 1990 that the offer of goods and services must obey some rules, such as sufficient precision and clarity of information—including characteristics, qualities, quantity, composition, origin, price, risks to health and safety, among others—and that any condition mentioned in an offer be included in the final agreement.

The National Counsel for Autoregulation of Advertisement (Portuguese acronym: CONAR), which is very influential, albeit private, has issued specific rules for advertising on the Internet through digital influencers.[31]

The Belgian regional decrees implementing the Audiovisual Media Services Directive[32] contain clear and precise rules regarding content (the exclusion of specific advertising techniques, products[33] or features) and also specific types of

[27] CJEU, case C-65/20, *VI v KRONE-Verlag Gesellschaft mbH & Co KG*, ECLI:EU:C:2021:471.

[28] Guides Concerning the Use of Endorsements and Testimonials in Advertising, FEDERAL TRADE COMMISSION, 16 CFR Part 255.

[29] Native Advertising: A Guide for Businesses. Available at https://www.ftc.gov/tips-advice/business-center/guidance/native-advertising-guide-businesses. Accessed 27 October 2021.

[30] Federal Law No. 8,078 of 11 September 1990, Articles 30 and 31.

[31] CONAR/CBAP, Digital Influencer Advertising Guidelines, 2021. Available at http://conar.org.br/pdf/CONAR_Digital-Influencer-Advertising-Guidelines_2021-03-11.pdf. Accessed 30 August 2021.

[32] Three different set of legislations implementing the Audiovisual Media Services Directive 2018/1808, pertaining to the Dutch-speaking community (*Decreet van 27 maart 2009 betreffende radio-omroep en televisie,* Belgian Official Gazette, 30 April 2009, p. 34470, and the amending decree of 19 March 2021 implementing the new AVMS Directive, Belgian Official Gazette, 29 April 2021, p. 40930), French-speaking community (*Décret coordonné du 26 mars 2009 sur les services de médias audiovisuels,* Belgian Official Gazette, 24 July 2009, p. 50609, now replaced by the *Décret du 4 février 2021 relatif aux services de médias audiovisuels et de services de partage de vidéos*, Belgian Official Gazette, p. 29036), and German-speaking community (*Dekret vom 27. Juni 2005 über den Rundfunk und die Kinovorstellungen*, Belgian Official Gazette, 6 September 2005, p. 38869, now replaced by the *Dekret vom 1. März 2021 über die Mediendienste und die Kinovorstellungen*, Belgian Official Gazette, 12 April 2021, p. 32156).

[33] Such as the advertising of cigarettes, alcohol or prescription medicines – see article 9 of the Audiovisual Media Services Directive.

communication (sponsorship,[34] product placement,[35] on-demand programs,[36] television advertising[37]).[38] For other types of content, self-regulation is also encouraged,[39] such as for accessibility, as well as direct marketing, profiling and behaviourally targeted advertising for minors,[40] alcoholic beverages and "junk food",[41] in all audiovisual commercial communications, including video-sharing platforms.[42] These rules will, however, be limited to media services falling under the scope of the Audiovisual Media Services Directive.

Moreover, even if applied to media covered by the Audiovisual Media Services Directive, rules on user-generated content (in which will lie posts by vloggers, influencers or influ-gamers, for instance) are only applicable to the extent that the communication is made with commercial intent and either the platform or the poster has editorial responsibility, which, in the latter case, is less evident, considering the number of intervening parties in creating the content.[43]

In Hungary, the practice of influencer marketing is developing more slowly than in other EU member countries, such as Germany.[44] No decision has been published by the courts on this matter so far. The possible reason for this is that the Hungarian Competition Authority (Gazdasági Versenyhivatal (GVH)) has only dealt with influencers in six cases up to this point. In the first three out of six cases, the authority has accepted commitments from the influencers and companies; i.e., the

---

[34] Article 10 Audiovisual Media Services Directive.

[35] Article 11 Audiovisual Media Services Directive.

[36] Articles 12 and 13 Audiovisual Media Services Directive.

[37] Articles 19 to 26 of the Audiovisual Media Services Directive and Article 27 of the Audiovisual Media Services Directive, which specifically deals with the protection of minors.

[38] See Belgian National Report, section 3.2.

[39] According to the EU Commission, experience has shown that both self- and co-regulatory instruments, implemented in accordance with the different legal traditions of the Member States, can play an important role in delivering a high level of consumer protection. See Recitals 13 and 14 of the AVMS Directive.

[40] Article 6a Audiovisual Media Services Directive.

[41] Articles 9 (3) and (4) Audiovisual Media Services Directive. Under "junk food", we are referring to the description in the AVMS Directive "foods and beverages containing nutrients and substances with a nutritional or physiological effect, in particular fat, trans-fatty acids, salt or sodium and sugars, of which excessive intakes in the overall diet are not recommended. Member States and the EU Commission can themselves adopt codes of conduct in these areas.

[42] Article 28b of the revised Audiovisual Media Services Directive. Regarding this latter category, the EU Parliament has recently issued guidelines trying to clarify which audio-visual media services were captured under the provisions applicable to video-sharing platform services when the latter constitutes their "essential functionality". See EU Commission Guidelines on the practical application of the essential functionality criterion of the definition of a 'video-sharing platform service' under the Audiovisual Media Services Directive (2020/C 223/02), 7 July 2020, available at: https://eur-lex.europa.eu/legal-content/EN/TXT/HTML/?uri=CELEX:52020XC0707(02)&from=EN. Accessed 25 September 2020.

[43] See on this topic V. Verdoodt and N. Feci, Digital Influencers and Vlogging Advertising: Calling for awareness, guidance and enforcement, A & M, 2018-2019/1.

[44] Hungary National Report, Section 3.5.

authority has not imposed a fine. In two cases, the authority found that the influencers and companies had not sufficiently implemented their commitments; therefore, the GVH imposed a fine. In the most recent case, in June 2020, the GVH punished the influencer and also the company for unmarked advertisements. In all three cases, the authority fined both the influencers and the companies. The highest fine was imposed in case VJ/49/2018: five million HUF (approx. 14,000 EUR).

At the same time as publishing its first decision in connection with influencers, the GVH also published a guideline that provides useful advice on how influencers should publish sponsored content in order to make it legally compliant.

11.2.3 Fake Reviews and Fake Endorsements

A different type of practice, but one that is also connected with new forms of advertising, social networks and unfair practices, is that of fake reviews and endorsements. Consumers rely increasingly on consumer reviews and endorsements when making transactional decisions, but these are not always truthful.

The phenomenon of fake reviews[45] is an increasingly prevalent problem that has attracted the attention of legal scholars and has raised the question of how such practices should be assessed under unfair competition law.[46] The issue is a complex one: consumers can be the victims and sometimes also the source of fake online reviews (e.g. by posting fake reviews in exchange for benefits in the form of discounts or gifts[47]). Fake reviews usually make false claims about the characteristics of the goods or service or create a false impression about a product or service or make unsupported, subjective claims regarding the quality or effectiveness of a product that might mislead the average consumer to make a purchase they

[45]Various terms exist to describe misleading online activity involving reviews. Some of them are opinion or review spamming, buzz marketing (encouraging consumers to voluntarily promote products in return for coupons or discounts), stealth campaigns (promotion of products by "consumers" whose identities are masked and may be faked), astroturfing (hiding or masking the sponsors of a particular message to give the appearance that it is supported by and originates from ordinary online users), briefing from the European Parliament, Online consumer reviews, The case of misleading or fake reviews, October 2015.

[46]In detail S. Holzweber, Kundenbewertungen im Internet, ÖBl 2021(3), pp. 100–105; H. Krieg and J.D. Roggenkamp, Astroturfing – rechtliche Probleme bei gefälschten Kundenbewertungen im Internet, K & R 2010(11), pp. 689–694; also A. Wilkat, Bewertungsportale im Internet, Nomos 2013; A. Anderl and A. Seling, Social Media – Rechtssicherheit im Unternehmen, ecolex 2018(6), pp. 535–538; A. Kulka, Das Fingieren von (größerer) Beliebtheit eines Unternehmens im Wettbewerbsrecht, ecolex 2012(2), pp. 148–151.

[47]See A. Asenjo and J. Ravindran, Reviewers revealed how they leave false reviews on Amazon, keep the products, and get their money back, Business Insider España, 21 February 2021. Available at https://www.businessinsider.com/facebook-telegram-paypal-false-guidelines-algorithm-detection-jeff-bezos-scam-2021-2. Accessed 27 October 2021.

would otherwise not have made.[48] Fake reviews and false ratings are more likely to occur on shopping platforms. Some noteworthy recent scandals have involved Amazon (a 2020 investigation from the Financial Times identified "suspicious activity" involving nine out of Amazon's top ten UK reviewer accounts)[49] and eBay (like an investigation by Which! found in 2020).[50] In October 2021, the Federal Trade Commission sent a notice to over 700 companies warning them that they could face steep fines for using misleading endorsements like fake reviews.[51]

Misleading or fake review factories have plagued the Internet for years to date and have undermined consumer confidence in the integrity of online reviews, to the consumers' detriment. This type of practice was not expressly regulated at the EU level[52] until, quite recently, the Omnibus Directive has included some amendments to the Unfair Commercial Practices Directive in order to tackle this problem. Under the provisions of the Directive, when traders provide access to consumer reviews of products, they should inform consumers whether processes or procedures are in place to ensure that the published reviews originate from consumers who have actually used or purchased the products.[53] If such processes or procedures are in place, traders should provide information on how the checks are made and provide clear information to consumers on how reviews are processed, for example, if all reviews, either positive or negative, are posted or whether those reviews have been sponsored or influenced by a contractual relationship with a trader.

Moreover, it will also be considered an unfair commercial practice to mislead consumers by stating that reviews of a product were submitted by consumers who actually used or purchased that product when no reasonable and proportionate steps were taken to ensure that they originate from such consumers.[54] Such steps could include technical means to verify the reliability of the person posting a review, for example by requesting information to verify that the consumer has actually used or

[48] European Consumer Centre Ireland, All You Need to Know on Fake Reviews and False Ratings. Available at https://www.eccireland.ie/fake-reviews-and-false-ratings-online-shopping/. Accessed 27 October 2021.

[49] Amazon deletes 20,000 reviews after evidence of profits for posts, Financial Times, 4 September 2020. Available at https://www.ft.com/content/bb03ba1c-add3-4440-9bf2-2a65566aef4a. Accessed 27 October 2021.

[50] By Hannah Walsh, How eBay's review system is promoting fake, counterfeit and even dangerous products, Which! 13 Mar 2020. Available at https://www.which.co.uk/news/2020/03/ebay-customer-reviews/ - Which?. Accessed 27 October 2021.

[51] Fines of up to USD 43,792 per violation. The FTC clarifies that a recipient's presence on the list of companies receiving the notice does not in any way suggest that it has engaged in deceptive or unfair conduct. See press release and notice at https://www.ftc.gov/news-events/press-releases/2021/10/ftc-puts-hundreds-businesses-notice-about-fake-reviews-other. Accessed 27 October 2021.

[52] See an interesting discussion in the Austria National Report on whether the pre-existing black list of the Unfair Commercial Practices Directive covered this already and to what extent, Section 2.4.

[53] Article 7.c) 6 of the Unfair Commercial Practices Directive.

[54] Annex I, b) 23(b) of the Unfair Commercial Practices Directive.

purchased the product.[55] Likewise, submitting or commissioning another legal or natural person to submit false consumer reviews or endorsements or misrepresenting consumer reviews or social endorsements in order to promote products is considered an unfair commercial practice.[56]

The fake review or endorsement situation described above should not be confused with the common and legitimate advertising practice of making exaggerated statements or statements that are not meant to be taken literally.[57] However, the Omnibus Directive does capture similar fake endorsement situations, such as "likes" on social media or commissioning others to do so in order to promote their products, as well the publication of only positive reviews and the deletion of negative ones or the extrapolation of social endorsements, where a user's positive interaction with certain online content is linked or transferred to different but related content, creating the appearance that the user also takes a positive stance towards the related content.[58]

11.3 The Use of Artificial Intelligence in Marketing

Algorithms are exact sequential sets of commands that are performed over a designed input to generate an output in a clearly defined format. Algorithms can be represented in plain language, diagrams, computer codes and other languages.[59] Examples of traditional well-known algorithms include food recipes, instruction manuals and music sheets, while more complex examples include matching algorithms of online dating platforms, ranking algorithms of search engines and linear programming solvers for process optimisation.[60] When an algorithm is programmed to amend its own decision-making rules to account for past experiences, it becomes a self-learning algorithm.[61] Artificial intelligence can be defined as a sort of modern algorithm.

Artificial intelligence is a challenge in every single sector of the economy due to the unavoidable growth of its use owing to the benefits it can provide, such as long-term savings to companies' budgets—despite the expensiveness of its implementation in the short term—and celerity and improvement to the companies' varied processes. What will define whether a company is acting in compliance with the law or not is not the simple use of this technology itself but the purpose for which it uses artificial intelligence and the means of obtaining data for artificial intelligence models. Society as a whole will be left behind by artificial intelligence if we do not

[55] Recital 47 of the Omnibus Directive.

[56] Annex I, b) 23(c) of the Unfair Commercial Practices Directive.

[57] Recital 48 of the Omnibus Directive.

[58] Recital 49 of the Omnibus Directive in connection with Annex I, b) 23(c) of the Unfair Commercial Practices Directive.

[59] OECD, DAF/COMP/M(2017)1/ANN3/FINAL.

[60] OECD, DAF/COMP/M(2017)1/ANN3/FINAL.

[61] Algorithms and collusion, Notes from the European Union, OECD, DAF/COMP/WD(2017)12.

find a way to adapt to this technology, and the same applies to companies that are capable of operating with artificial intelligence mechanisms and technologies, which will certainly have better competitive conditions and cannot be prevented from doing so as long as they comply with the law.[62]

Artificial intelligence is not regulated by European Member States' legislation, although scholars opine that the existing rules on contract law, next to those on consumer and privacy laws, should apply to the use of AI, subject to a few specific adaptations.[63] The European Commission's proposal for an Artificial Intelligence Act[64] as the world's first response to the issue of regulating artificial intelligence is a step in the right direction, but it is to be seen if and how it is going to be approved.

11.3.1 Price Discrimination and Artificial Intelligence Under Consumer Law

From an economic perspective, price discrimination is when two similar products, which have the same marginal cost to produce, are sold by a firm at different prices. A key characteristic is that the price that is charged is based partly on the value of the good to the customer rather than just on the cost of producing the good.[65] There are a number of different ways that firms can price discriminate; for example, it can be based on the characteristics of the buyer (referred to as group pricing or third-degree discrimination), the behaviour of the buyer (behaviour-based discrimination) or the characteristics of the product (versioning or second-degree discrimination).[66]

For consumers, price discrimination can be frustrating and seem unfair, and policy makers may address (and have addressed) this concern when it leads to unfair commercial practices through anti-discrimination or consumer protection laws. Certain companies have, however, managed to keep customers satisfied despite the fact that price is always changing. This is the case of Uber, whose dynamic pricing has conditioned users not to expect a stable or regular price.[67] While Uber claims that its algorithms set prices to reflect real-time supply and demand, initial research

---

[62] Brazilian National Report, Section xx.

[63] See for instance H. Jacquemin and J.-B. Hubin, L'intelligence artificielle et le droit, Collection du Crids, 2017, Brussels, Larcier ; Y. Poullet, Le RGPD face aux défis de l'intelligence artificielle, Collection du Crids, 2020, Brussels, Larcier; C. Vanleenhove and J. De Bruyne, Artificial Intelligence and the Law, Collection Centrum voor Verbintenissenrecht, 2021, Brussels, Intersentia.

[64] EU Commission, Proposal for a Regulation of the European Parliament and of the Council laying down harmonised rules on artificial intelligence ("Artificial Intelligence Act") and amending certain union legislative acts, COM (2021) 206 final of 21 April 2021.

[65] OECD, Roundtable on Price Discrimination held at the 126th meeting of the Competition Committee of the OECD DAF/COMP/M(2016)2/ANN5/FINAL.

[66] OECD, Roundtable on Price Discrimination held at the 126th meeting of the Competition Committee of the OECD DAF/COMP/M(2016)2/ANN5/FINAL.

[67] L. M. Khan, Amazon's Antitrust Paradox, The Yale Law Journal, Volume 126, 2016-2017, Number 3, January 2017564-907.

has found that the company manipulates the availability of both.[68] Moreover, it routinely gives away discount coupons to select users, effectively charging users different prices, even for the same service at the same time.

In this section, we analyse price discrimination under a consumer law approach. The analysis under competition law is carried out further below.

Personalised pricing is facilitated by the use of algorithms that allow undertakings to collect vast amounts of personal data. The creation of "digital user profiles" enables companies to better predict consumers' behaviour and preferences and to better identify the reserve prices of their customers, i.e. their expected willingness to pay. Personalised pricing is to be distinguished from mere dynamic pricing, where prices are flexibly set based on objective criteria (like stock, popularity and demand of the product, time of day, day of the week, etc.) and are uniformly applied to all customers.[69] It is a reality nowadays that online platforms often price discriminate among consumers. Platforms also apply dynamic pricing structures based on customer segmentation. In practice, this means that the consumers will see different prices depending on the browser type and browsing history, the type of end-user device, the number and timing of website visits, the geographic location, and similar user behaviour data. A relevant question thus arises with the use of algorithms for marketing: whether price personalisation is a practice to be forbidden, controlled or allowed freely.

Before getting into the subject, though, we need to distinguish between types of price personalisation through algorithms: price personalisation can take the shape of an algorithm showing to the consumer the products whose prices are more compatible with his/her profile, or it can be a straightforward change of price of a given product according to the profile of the person or even a denial to offer a product to a given consumer profile. When this discrimination is based on the consumer's location or nationality, this is known as "geo pricing" or "geo blocking".

Personalised prices—as is the case with other tools provided by the use of artificial intelligence—*a priori* represent a combination of benefits to both the companies and the consumers as long as they are used reasonably and in compliance with the law. Thus, to some, the benefits of personalised prices would make it unreasonable to prohibit its use in certain areas while ensuring that this technology is compatible with the principles that protect consumers, data subjects and advertising guiding principles, such as transparency, promotional identification and the right to choose whether or not to use a given technology.[70]

---

[68] See T. Hwang & M. Clare Elish, The Mirage of the Marketplace: The Disingenuous Ways Uber Hides Behind Its Algorithm, Slate (July 27, 2015, 6:00 AM). Available at http://www.slate.com/articles/technology/future_tense/2015/07/uber_s_algorithm_and_the_mirage_of_the_marketplace.html [http://perma.cc/B5UR-P9PN]. Accessed 27 October 2021.

[69] Austrian National Report, noting also that prominent examples include gas and travel prices (e.g. flight tickets); however, dynamic pricing may be combined with elements of personalised pricing.

[70] Brazil report, Section 3.5.

In the European Union, personalised pricing as such is not forbidden in a general way, although some aspects of personalised pricing have been subject to limitations. For instance, discrimination between EU customers to segment markets along national borders and increase profits to the detriment of foreign customers is considered unjustified geo-blocking and is forbidden by Regulation 2018/302 on addressing unjustified geo-blocking and other forms of discrimination based on customers' nationality, place of residence or place of establishment within the internal market. Likewise, pursuant to this Regulation, traders cannot apply different pricing to transactions across the Member States or grant clients different conditions of access to goods or services for reasons related to nationality, place of residence or place of establishment unless this is objectively justified. It must be noted, however, that actions by traders established in third countries fall outside the scope of the Directive.

But these provisions of the Regulation on geo-blocking were aimed at safeguarding the single market rather than protecting consumers' rights and were not enough to avoid potential discrimination against consumers. Recently, however, the Omnibus Directive[71] has set two new requirements for personalised prices that will need to be implemented under national laws.

First, Article 6 of the Unfair Terms Directive is updated to include as information to be provided prior to the conclusion of a "distance contract" the fact that the price was personalised based on automated decision-making.[72] When it does apply, this rule allows consumers to consider the potential risks in their purchasing decision. However, as noted by certain National Reporters,[73] this requirement can be satisfied with two issues. The first issue is the degree of detail of the information that must be provided to the consumer, while ensuring that such information is provided to them in a clear and understandable manner, or the extent to which the enterprise should explain the functioning of the algorithms funding the automated decision, remains unclear and should be subject to further guidance. The second issue is that this new requirement does not apply if personalisation was achieved through means other than automated processing, such as "real-time" or "dynamic" pricing. These techniques involve changing the price in a highly flexible and quick manner in response to market demands when those techniques do not involve personalisation based on automated decision-making.[74] Therefore, a company that does not resort to personalised automated decision-making but still adapts its prices using other real-time technologies such as cookies, trackers or other indicators can continue

[71] Directive 2019/2161 of the European Parliament and of the Council of 27 November 2019 amending Council Directive 93/13/EEC and Directives 98/6/EC, 2005/29/EC and 2011/83/EU of the European Parliament and of the Council as regards the better enforcement and modernisation of Union consumer protection rules.

[72] Article 4 (4) (a) (ii) of the Omnibus Directive, inserting a pt (ea) under article 6 (1) of the Unfair Terms Directive.

[73] Belgian Report, Section 4.1.1.

[74] Recital 45 of the Omnibus Directive.

practising different prices, and potential discrimination between consumers would not be covered by this obligation.

Second, Article 7 of the Unfair Commercial Practices Directive, dealing with misleading omissions, provides for a list of material information to be provided to the consumer. This list has been updated by the Omnibus Directive[75] to include a disclosure of the main parameters determining the ranking of products presented as a result of the consumer's search query when browsing an enterprise's website as well as the relative importance of those parameters. "Ranking" is defined as the relative prominence given to products, as presented, organised or communicated by the enterprise, irrespective of the technological means used for such presentation, organisation or communication.[76] This addition may be of limited use when it comes to artificial intelligence as the Omnibus Directive requires that such disclosure be without prejudice to the protection that enterprises may enjoy under the Trade Secrets Directive.[77] If artificial intelligence and the algorithms designed for its functioning are protected by trade secrets, the amount of information that an enterprise would have to provide without breaching any law would be limited. Moreover, the Omnibus Directive adds that companies should not be required to disclose the detailed functioning of their ranking mechanisms, including algorithms, but could rather limit the information to a general description of the main parameters determining the ranking and their relative importance as opposed to other parameters.[78]

It has also been noted[79] that the EU Regulation on platform-to-business relations[80]—which applies to all operators of the specified online services—prohibits, *inter alia*, the arbitrary and discriminatory treatment of business users. It follows that any acts against these rules can be deemed to be an unfair commercial practice that may trigger responsibility on the part of the undertaking concerned.

From an economic perspective, it has been said that personalised pricing is not harmful *per se*.[81]

[75] Article 3 (4) (b) of the Omnibus Directive, inserting a paragraph 4a under article 7 of the Unfair Commercial Practices Directive.

[76] Article 3 (4) (b) of the Omnibus Directive, inserting a point (m) under article 2 of the Unfair Commercial Practices Directive.

[77] Directive 2016/943 of the European Parliament and of the Council of 8 June 2016 on the protection of undisclosed know-how and business information (trade secrets) against their unlawful acquisition, use and disclosure, OJ, 2016, L 157, p. 1.

[78] Recital 23 of the Omnibus Directive.

[79] Austrian National Report, Section 3.3.3.

[80] Regulation 2019/1150 of the European Parliament and of the Council of 20 June 2019 on promoting fairness and transparency for business users of online intermediation service (P2B Regulation), OJ 2019 L 186, p. 57.

[81] Austrian National Report, Section 3.4., citing Cerre, Big Data and Competition Policy: Market Power, personalised pricing and advertising 2017, p. 8, available at https://www.cerre.eu/sites/cerre/files/170216_CERRE_CompData_FinalReport.pdf. Accessed 7 September 2021.

At the national level, in Germany, the prohibition of misleading advertising in section 5 of the German Unfair Competition Act (Gesetz gegen den unlauteren Wettbewerb (UWG)) does not constitute a real obstacle to personalised pricing because of the principle of price freedom.[82] The amount of the price, price fairness and the underlying calculation of prices are not subject to any legal control and do not have to be disclosed.[83] Therefore, prices set by an entrepreneur can be increased or decreased at any time.[84]

The personalised prices are unfair that can have relevance under the German Equal Treatment Act (Allgemeines Gleichbehandlungsgesetz (AGG)).[85] The provision in section 19 of the AGG constitutes a market conduct rule within the meaning of section 3a UWG, so that an infringement would be sanctionable under unfair competition law.[86] However, there is no general principle of equal treatment, so that prices do not have to be the same for everyone,[87] nor should an offer need to say that the goods will be offered to every customer for the same price.[88] Nevertheless, certain discriminations, such as those based on age, gender and religion, are prohibited (section 19 AGG). Although section 20 AGG provides for the possibility to justify different treatment, the higher willingness to pay does not justify different pricing.[89]

Generally, the General Equal Treatment Act does not cover personalised prices in principle, although discrimination may well take place in individual cases, in the event of which the AGG would apply. For this reason, at least the criterion on which pricing is based should be disclosed as it is essential for identifying and starting a complaint regarding any discrimination.[90] Such obligation does not currently exist and would, according to some,[91] need to be regulated by law. However, this would bring with it certain challenges related to the difficulties of disclosing algorithms.[92]

In Austria, the Local Supply Act (*Nahversorgungsgesetz*)[93] contains a general prohibition on discrimination in relationships between suppliers and authorised

[82] T. J. Tillmann and V. Vogt, Personalisierte Preise im Big-Data-Zeitalter, VuR 2018, pp. 447–455.

[83] F. Hofmann, Der maßgeschneiderte Preis, WRP 2016, pp. 1074–1081; D. Tietjen and F. Flöter, Dynamische und personalisierte Preise: Welche lauterkeitsrechtlichen Schranken gelten für Unternehmen?, GRUR-Prax 2017, pp. 546–548.

[84] BGH, Judgement of 13 March 2003 – I ZR 212/00, BGHZ 154, 306.

[85] T. J. Tillmann and V. Vogt, Personalisierte Preise im Big-Data-Zeitalter, VuR 2018, pp. 447–455.

[86] BT-Drs. 16/1780, 48/49.

[87] E. I. Obergfell, Personalisierte Preise im Lebensmittelhandel – Vertragsfreiheit oder Kundenbetrug?, ZLR 2017, pp. 290–301.

[88] F. Hofman, Der maßgeschneiderte Preis, WPR 2016, pp. 1074–1081.

[89] F. Hofman, Der maßgeschneiderte Preis, WPR 2016, pp. 1074–1081.

[90] H. *Zander-Hayat, L.A. Reisch and C. Steffen,* Personalisierte Preise – Eine verbraucherpolitische Einordnung, VuR 2016, pp. 403–410.

[91] T. J. Tillmann and V. Vogt, Personalisierte Preise im Big-Data-Zeitalter, VuR 2018, pp. 447–455.

[92] T. J. Tillmann and V. Vogt, Personalisierte Preise im Big-Data-Zeitalter, VuR 2018, pp. 447–455.

[93] Bundesgesetz vom 29. Juni 1977 zur Verbesserung der Nahversorgung und der Wettbewerbsbedingungen (Nahversorgungsgesetz), BGBl 392/1977, Sec 2 para 1.

resellers. A supplier that—without objective justification—grants or offers different conditions to authorised resellers may be required by court order to cease and desist.[94] This may also apply to algorithmic-based price discrimination. Preferential treatment of individual resellers is only feasible where an economically equivalent consideration exists in return. Finally, the preferential treatment must be appropriate and necessary to achieve the objective pursued (proportionality test).[95] The Local Supply Act applies to all undertakings, irrespective of their (dominant) market position.[96] However, the Supreme Court has left open the question of whether the undertaking concerned must at least have a certain amount of market power.[97]

In Belgium, the Discrimination Act forbids forms of discrimination, in the public and private sector, among others in relation to the access to goods and services put at the public's disposal or the access, participation, and the exercise of any economic, social, cultural or political accessible to the public, which is based on age, sexual orientation, civil state, birth, fortune, religious of philosophical belief, handicap, physical or genetic features and social origin, unless justified for legitimate purposes and where appropriate and necessary means for their realisation have been put in place.[98]

According to the National Report, this is not a mere theoretical statement as Belgian courts already ruled that offering specific promotions or reductions to a category of consumers according to their age,[99] job category or location[100] was not objectively justified and was therefore in breach of the Belgian Discrimination Act. Such breach, in turn, was deemed an unfair commercial practice by the business from which the promotions originated.

In Hungary, the use of search engines, digital comparison tools, personalised advertisements and personalised offers (personalised targeting) raises the question of

---

[94] For details see N. Gugerbauer, In: Gugerbauer (ed), KartG und WettbG[3], Verlag Österreich 2017, Sec 2 Local Supply Act paras 1ff; on the Local Supply Act in general, see e.g. W. Barfuß, Das Bundesgesetz zur Verbesserung der Nahversorgung und der Wettbewerbsbedingungen "NVG", ÖZW 1978(1), pp. 10–14.

[95] N. Gugerbauer, In: Gugerbauer (ed), KartG und WettbG, Verlag Österreich 2017, Sec 2 Local Supply Act para 7.

[96] See also OGH 9 October 2000, 16 Ok 8/00.

[97] OGH 16 July 2008, 16 Ok 3/08, *Sägerundholz*.

[98] Act of 10 May 2007 aiming at combating certain forms of discrimination, Belgian Official Gazette, 30 May 2007, p. 29016.

[99] For instance, this has been the case for reductions offered by a glass retailer, which went higher the older the customer was. See Court of appeal of Antwerp, 26 April 2007, and J. Stuyck, Geen goedkope brillen voor ouderen - de onverwachte invloed van de anti-discriminatiewet, Annuaire Pratiques du commerce & Concurrence, 2007, pp. 478–482. Other courts have upheld the same ruling, see President of the Commercial Court of Turnhout, 7 November 2008, Annuaire Pratique du commerce & Concurrence, 2008, p. 289; President of the Commercial Court of Brussels, 17 September 2008, Annuaire Pratiques du commerce & Concurrence, 2008, p. 57; President of the Commercial Court of Brussels, 15 October 2008, Annuaire Pratiques du commerce & Concurrence, 2008, p. 272.

[100] See Court of Appeal of Brussels, 4 May 2010, D.C.C.R., 2011, n°90, p. 103.

whether the consumer needs to understand the logic used by search engines and direct-to-consumer (DTC) business models and, if the answer is positive, to what extent the consumer must have an insight into the operation of search engines. Under the Hungarian regulatory framework, search engines and personalised targeting based on consumer profiles would be assessed as a factor influencing consumers' decision-making process by limiting the range of available choices; therefore, this issue would be treated in the framework of the Unfair Commercial Practices Act.[101]

The current regulatory framework includes the fundamental legal requirements to enable enforcement as regards personalised targeting. Competent authorities, however, meet a challenge when it comes to investigating the technicalities of personalisation and providing proper evidence for legal assessment.

Lastly, in Brazil, personalised price discrimination is forbidden in certain circumstances.[102] In the same line as the above arguments, there is a need for a specific regulation to guarantee that the programming process of algorithms complies with all objectives and principles of law, especially the most recent ones, such as the Brazilian laws No. 12.965/2014 and No. 13.709/2018 and the recent alterations to consumer law, which arose exactly from the need to keep up with technological evolution in order to prevent discrimination.

11.3.2 The Use of Artificial Intelligence in Marketing and Possible Implications Under Competition Law: Risks of Collusion and Abuse of Dominance

When undertakings delegate their pricing (or other relevant commercial) decisions to algorithms, the potential risks to the competitive process need to be assessed. Some say that it is necessary to assess whether the current legal framework can effectively deal with this phenomenon.[103] There is no doubt that the use of algorithms generates certain efficiencies on both the supply and the demand sides.[104] Algorithms allow suppliers to better respond to consumer demand, better allocate their resources and save on human capital.[105] Algorithms reduce transaction costs by providing

[101] The Hungarian implementation of the Directive 2005/29 of the European Parliament and of the Council of 11 May 2005 concerning unfair business-to-consumer commercial practices in the internal market.

[102] Brazilian Consumer Protection Code, articles 4°, caput, items I and III; 6°, items II, III and IV, 37, §2°, and 39, items II, IX and X.

[103] Austrian Report, Section 3.1.

[104] See e.g. CMA, Pricing Algorithms 2018, paras 4.1.ff., and OECD Executive Summary of the 127th roundtable, DAF/COMP/M(2017)1/ANN3/FINAL.

[105] M. Gal, Algorithms as Illegal Agreement, Berkeley Technology Law Journal 2019(67:34), pp. 67–118 (70); OECD, Algorithms and Collusion – Competition Policy in the Digital Age 2017, pp. 14f.

consumers with a wealth of information[106] that in turn allows them to efficiently compare offers and enables them to find products that fit them better, for example.[107] Increased market transparency achieved through reduced information asymmetries may intensify competition to the benefit of consumers.

However, the adoption of algorithms can have far-reaching impacts on the forms of and conditions for competitive business conduct and might lead to undesirable, anti-competitive results.[108] The consequences of such behaviour may be ultimately borne by the consumers in the form of excessive prices or a loss in total consumer welfare. Data-driven markets—notably those in the digital economy—with few competitors, homogenous products and high entry barriers are particularly vulnerable. The use of algorithms in the commercial process (most notably in marketing, including pricing) may under certain circumstances lead to a breach of competition law. First, algorithms may facilitate and accelerate the coordination of prices between undertakings, thereby potentially violating the ban on restrictive agreements. Second, algorithms may enable undertakings to offer different prices to different consumers according to the information they hold about them, and under certain circumstances, such personalised pricing may qualify as price discrimination amounting to abuse of dominance.

11.3.2.1 Algorithmic Collusion

There is a growing number of companies using computer algorithms to improve their pricing models, customise services and predict market trends. Algorithmic pricing allows companies to achieve many relevant efficiencies such as reduction in costs, improvement of their offering through optimal price discovery and reduction of barriers to entry.[109] However, the widespread usage of algorithms could also result in anti-competitive effects by making it easier for firms to achieve and sustain collusion without any formal agreement or human interaction. Collusion can be facilitated by the implementation of algorithms: price agreements can be enforced more easily, their compliance can be monitored better and deviations can be detected and punished more quickly.[110] In general, due to the characteristics of the digital world, there is less need for *ex ante* communication. However, as it has been

[106] However, it should not be overlooked that users may be confronted with "too much" information and that the costs related to the sorting and selection of this new variety of information may constitute a form transaction costs itself, see in particular B. Schwartz, The Paradox of Choice - Why More Is Less, 1st ed, Harper Perennial 2004.

[107] M. Gal and N. Elkin-Koren, Algorithmic Consumers, Harvard Journal of Law & Technology 2017(30:2), pp. 309–353 (318).

[108] P. Picht and G. Loderer, Framing Algorithms – Competition law and (Other) Regulatory Tools, MPI Research Paper 2018(24), pp. 1–35 (2).

[109] G. Shier, The risks of using algorithms in business: artificial price collusion, Oxera Agenda, November 2020.

[110] M. Gal, Algorithms & Competition Law: Interview of Michal Gal by Thibault Schrepel, e-Competitions Algorithms, Concurrences 2020, Art. N° 93929, pp. 1–12.

noted,[111] algorithms are also helpful to detect cartels and provide a useful tool for competition authorities worldwide to monitor markets.

Some have argued[112] that algorithmic collusion is a fundamentally unimportant subject for antitrust and competition law, for two reasons. First, empirical studies documenting the frequency of the phenomenon in the real world remain to be produced (one cannot find any quantification of algorithmic collusion in official publications coming from antitrust and competition agencies). Second, even if algorithmic collusion were already, or could become, a frequent practice, it would remain old wine in new bottles. Indeed, such collusion is simply a more elegant way of implementing the same practices known for centuries. Whether they are algorithmic or not, the nature of anti-competitive collusion remains identical.

However, the reality is that competition authorities worldwide are studying the implications of the widespread use of algorithms by companies as a tool to set their offerings. In questioning whether we can rely on traditional antitrust notions and principles, the European Commission has noted[113] that while we may indeed rely on traditional tools, we may need to adjust them to new market realities. This is because it is necessary to consider new trends in markets: online prices are highly transparent, and monitoring can be done at short time intervals and at a low cost. Firms can use software to automatically monitor competitors' prices almost in real time and adopt prices (or "reprice") accordingly. In the e-commerce sector inquiry, half of the retailers reported tracking prices of online competitors, and most of them (about 70%) acknowledged doing so through software. Moreover, a large majority of respondents are using software to re-price accordingly.

The European Commission's report titled Competition Policy for the Digital Era[114] holds that the basic competition law framework under Articles 101 and 102 of the Treaty on the Functioning of the European Union (TFEU) provides a sound basis for protecting competition in the digital economy. However, it notes that the features of platforms, digital ecosystems and the data economy may need to be adopted when looking at market power and defining relevant markets.[115]

---

[111] M. Huber and D. Imhof, Machine learning with screens for detecting bid-rigging cartels, International Journal of Industrial Organization, Volume 65, 2019, Pages 277–301, ISSN 0167-7187. Available at https://doi.org/10.1016/j.ijindorg.2019.04.002. Accessed 27 October 2021.

[112] T. Schrepel, The Fundamental Unimportance of Algorithmic Collusion for Antitrust Law, February 07, 2020- in Collusion by Blockchain and Smart Contracts, 33 Harv. J.L. & Tech. 117 (2019), available at https://papers.ssrn.com/sol3/papers.cfm?abstract_id=3315182. Accessed 27 October 2021.

[113] OECD, Summary of Discussion of the Roundtable on Algorithms and Collusion, Annex to the Summary Record of the 127th meeting of the Competition Committee, DAF/COMP/M(2017)1/ANN2/FINAL.

[114] European Commission, Competition Policy for the digital era, Final report (April 2019). Available at: https://ec.europa.eu/competition/publications/reports/kd0419345enn.pdf.

[115] See, further, Competition Policy for the digital era (fn 32), ch 3.

The United Kingdom's Competition and Markets Authority[116] (CMA) has questioned whether the legal tools currently available to the CMA are capable of tackling all the challenges presented by the rise of the algorithmic economy, such as self-learning algorithms.

The Brazilian competition authority, Conselho Administrativo de Defesa Econômica (CADE), has also recently stated that the issue of collusion achieved by algorithms should be more thoroughly examined.[117]

Various scenarios can be envisaged in which algorithms may promote restrictive agreements. Without being exhaustive,[118] we can see, first, a scenario where competitors use algorithms as a simple tool to implement an existing price agreement, which relates to the alignment of prices as well as the use of algorithms as a means of coordination.[119] Such collusion constitutes a conventional "agreement" within the meaning of Article 101 TFEU.

A second scenario is one in which competitors autonomously decide to use (the same or different) pricing algorithms, leading to price alignments. In this scenario, the cartel price is merely based on the software independently used by the competitors, i.e. without explicit communication between the undertakings. In such cases, it may be hard to differentiate between mere (legitimate) parallel behaviour and anti-competitive conduct contravening the ban on cartels. The existence of collusive behaviour is particularly difficult to prove where (self-learning) algorithms interact with each other.[120]

---

[116] Through its former Chairman, David Currie: Pricing algorithms, Economic working paper on the use of algorithms to facilitate collusion and personalised pricing (8 October 2018) (CMA94) ('Pricing algorithms'), available at: https://assets.publishing.service.gov.uk/government/uploads/system/uploads/attachment_data/file/746353/Algorithms_econ_report.pdf. Accessed 27 October 2021.

[117] Mercados de Plataformas Digitais, p. 133. Available at https://cdn.cade.gov.br/Portal/centrais-de-conteudo/publicacoes/estudos-economicos/cadernos-do-cade/plataformas-digitais.pdf. Accessed 31 August 2021.

[118] For a good overview of the possible scenarios, see the Austrian National Report and read Algorithmic Pricing Collusion: When Does Liability Arise? In: A. Yaholnyk and A. Zeleniuk (eds), A Handbook for Foreign Clients 2020, 14 September 2020.

[119] Zimmer, in: Immenga and Mestmäcker (eds), Wettbewerbsrecht, 6th ed, C. H. Beck 2019, Art. 101 para 1 TFEU para 77. These considerations were illustrated in the Competition and Markets Authority's poster decision (12 August 2016, Case 50223, *Trod Ltd*) where the undertakings concerned agreed not to undercut each other on prices for certain posters and frames sold only by the two of them on the Amazon UK marketplace. This agreement was implemented by using (different third-party) pricing software. In a similar case investigated by the US Department of Justice (US District Court of Northern California, 6 April 2015, CR 15-00201 WHO, *US/Topkins*), a seller in the Amazon marketplace was charged for coordinating the prices of posters sold online. The parties concerned designed and shared dynamic pricing algorithms that were programmed to act in conformity with their pre-existing explicit agreement. Both cases constitute traditional horizontal price-fixing cartels where the agreement was implemented by the use of algorithms.

[120] Algorithms may be designed to detect the market behaviour of competitors and rationally follow price leadership, thereby reducing strategic uncertainty and facilitating a tacitly coordinated outcome. This type of algorithm price-setting category does not involve any prior contact between the

Third, in a vertical context, for example, a producer can use algorithms to monitor the enforcement of a resale price maintenance (RPM) arrangement by retailers.[121]

A fourth possible scenario would be outsourcing pricing decisions to a common agent (e.g. to a service provider) who would act as a *hub and spoke*.[122]

In truth, very few precedents exist so far finding that algorithms have either facilitated or implemented collusion.

A debate concerning the competition implications of algorithms has focused in large part on tacit collusion. Is algorithmic pricing making tacit collusion more pervasive and more effective, and if so, what would be the appropriate response from a competition enforcement perspective?[123]

Regarding possible defences as in any traditional analysis of restrictive agreements, an agreement or a concerted practice can be justified if the conduct is associated with specific efficiencies mentioned in Article 101 para 3 TFEU. This individual exemption also applies to algorithmic collusion. Potential (countervailing) efficiencies associated with the algorithm need to be balanced with its allegedly negative effects, which might reinforce the ascertained

---

undertakings concerned or via a hub as a third party. Rather, the alignment might result from a mere interaction of computers based on the parallel use of individual algorithms. In the absence of an agreement, this qualifies as tacit coordination, which may not be caught by the ban on cartels. Rather, such unilateral conduct is covered by the right of any undertaking to adapt its market behaviour intelligently to the existing or anticipated conduct of its competitors.

[121] For example, in 2018, the European Commission held Asus, Denon & Marantz, Philips and Pioneer liable for resale price maintenance, which was facilitated by the use of price comparison websites and a special pricing program, which helped producers trace the prices of resellers, detect deviation and maintain the level of retail prices – Asus (Case No. 40465) Commission Decision C (2018) 4773 as of 24 August 2018.

[122] In this category, competitors delegate certain strategic (pricing) decisions to a third party, which then takes the decisions using an algorithm (Bundeskartelamt and Autorité de la Concurrence, Working Paper – Algorithms and Competition 2019, p. 32 ff). In the *Eturas* case (CJEU, case C-74/14, ECLI:EU:C:2016:42), the CJEU considered a possible hub-and-spoke conspiracy facilitated by an online system. Several Lithuanian travel agencies (spokes) used a common travel booking system (hub). The system administrator sent a message to the participating travel agencies proposing a cap on discount rates (max 3%). According to the CJEU, this would constitute a concerted practice if it were proven that the travel agencies were aware of that message (para 44). However, the focus of this case lay on the awareness of the message and not on the subsequent implementation of a software rule that limits the possibility of granting higher discounts. Another example of an algorithm-driven hub-and-spoke conspiracy concerns the car service platform *UBER* (US District Court of Southern NY, 31 March 2016, 15 Civ. 9796 (JSR), Meyer/Kalanick). The latter connects drivers and passengers via a smartphone application; the prices are determined by Uber's automated price-setting algorithm. The industry-wide use of a single pricing algorithm (hub-Uber) used by competing drivers (spokes-cars) may lead to a form of horizontal coordination that eliminates price competition between allegedly independent undertakings.

[123] Algorithms and Collusion - Note from the European Union to the OECD Roundtable on Algorithms and Collusion.

anti-competitive practice.[124] In this context, it is key to analyse the algorithm and its role in the collusive scheme.

Companies should also be aware of how the algorithm works. In terms of the attribution of liability for antitrust infringements, the European Commission treats an artificial intelligence decision-maker in the same way that a human and business cannot escape liability by simply attributing conduct to a machine. According to some views, this policy and enforcement resemble the approach taken in relation to competition law compliance generally where a business is expected to take appropriate steps to train staff with a view to preventing anti-competitive conduct and cannot generally absolve itself from liability when those measures fail.[125] Although the detection of algorithmic facilitated collusion remains difficult in practice, undertakings that create and build algorithms that directly affect the market process are well advised to keep competition law compliance in mind.[126]

11.3.2.2 Abuse of Dominance

The use of algorithms or artificial intelligence in the marketing of products can also pose problems of unilateral conduct or abuse. Several possible situations of abuse can arise, such as the refusal to supply access to or information relating to algorithms; discriminatory pricing, given the ability of dominant firms to commit discriminatory behavior; and self-preferencing of a dominant company *vis á vis* its competitors and customers.

Refusal to Supply Access/Information Relating to Algorithms A dominant company's refusal to supply a competitor with information relating to its algorithms could potentially constitute an exclusionary abuse.[127] According to established case law, such a finding would require a determination that information relating to the algorithm is indispensable to someone wishing to compete and that a refusal to grant access would lead to the elimination of effective competition.[128] The 2004 Microsoft case (in which the Commission concluded that Microsoft had a dominant position in the work group server operating system market and had abused said position by refusing to provide a competing operating system vendor with information enabling it to design its system for seamless integration into Microsoft's group server system) is cited as a similar early precedent.[129] In that case, the Commission held that the

[124] Bundeskartelamt and Autorité de la Concurrence, Working Paper - Algorithms and Competition 2019, p. 29.

[125] S. Rab, Artificial intelligence, algorithms and antitrust, Competition Law Journal, 2019, Vol. 18, No. 4.

[126] S. Vezzoso, In: Lundqvist and Gal (eds), Competition Law for the Digital Economy, Edward Elgar 2019, p. 100.

[127] Bundeskartelamt and Autorité de la Concurrence, Algorithms and competition -Working Paper, November 2019.

[128] See Whish/Bailey, Competition Law, 9th edn., 2018, pp. 716, 818 et seq.

[129] Bundeskartelamt and Autorité de la Concurrence, Algorithms and competition - Working Paper, November 2019.

interoperability disclosures in question were indispensable even though certain limited open industry standards, limited options for reverse engineering and limited protocol licensing programs existed.[130]

Discriminatory Pricing Pricing to customers on a discriminatory basis, when carried out by a dominant company, may constitute an abuse of dominance falling under the typology of excessive pricing or unfair terms and conditions. Price discrimination is the practice of charging different prices to customers in the same position without justification or charging uniform prices to customers whose circumstances are different. Such customers may be competitors or consumers of goods/services.

In their joint analysis of algorithms and competition, the German and French competition authorities hold that the overall effect of price discrimination between consumers is unclear, i.e. depending on the characteristics of the case at hand. In particular, countervailing effects on consumer welfare can occur; e.g., some consumers might be better off under price discrimination, while others might be worse off.[131]

It is evident that price discrimination can benefit consumers by reducing prices for some and by increasing the number of consumers that are served.[132] However, it can also have an adverse impact on consumers. For clarity on the enforcement of competition law, the Organisation for Economic Co-operation and Development (OECD) distinguishes three different potential anti-competitive effects: (a) the exclusion of rivals, leading to the exploitation of consumers; (b) the exploitation of consumers; and (c) in upstream markets, the exploitation of intermediate customers and the creation of distortionary effects that harm consumers in downstream markets. In each case, under competition law, it is the effect on consumers, and not the fairness of the discrimination, that determines the acceptability of the discrimination.[133]

When assessing potential antitrust concerns arising from personalised prices, it is important to distinguish between, firstly, personalised pricing for the same product and, secondly, differential pricing for customised products and services. Where pricing reflects the costs and specific features of customised products and services, including the efficiencies arising from selling to that customer, it is difficult to see

[130] Commission, Decision of 24.03.04 (Microsoft), Case COMP/C-3/37.792, paras. 666 et seq.

[131] Bundeskartelamt and Autorité de la Concurrence, Algorithms and competition - Working Paper, November 2019, p. 24.

[132] OECD, Roundtable on Price Discrimination held at the 126th meeting of the Competition Committee of the OECD DAF/COMP/M(2016)2/ANN5/FINAL, Executive summary of the roundtable on discrimination.

[133] OECD, Roundtable on Price Discrimination held at the 126th meeting of the Competition Committee of the OECD DAF/COMP/M(2016)2/ANN5/FINAL, Executive summary of the roundtable on discrimination.

how this should be an antitrust concern at all.[134] A more interesting situation has been held to be where there is personalised pricing for the same product or service and whether this may lead to customers being nudged into exploitative transactions.[135] When those customers are competitors in that market of the dominant company, the effect may be the exclusion of such customers from the market.

Certain businesses using algorithms to categorise individuals to determine prices could abuse their dominant position. This is particularly the case of major players (such as the so-called digital gatekeepers: Google, Facebook, Amazon and Apple) pooling their data and enriching each of their own artificial intelligence devices, by creating unfairly discriminative prices under Article 102 TFEU. However, such concerns should not be exaggerated as, according to some, a business enjoying a dominant position would be in a better position to set prohibitive prices without personalisation than it would be with personalisation since by following personalisation, everyone would be exactly paying the price that is adapted to them.[136] However, EU competition law remains an interesting tool for dealing with the issue of price discrimination as it is flexible and evolves depending on the specific situation at stake and allows the legislator to determine objectively which behaviours can be deemed anti-competitive in an economy based on algorithms.[137] In this connection, national competition authorities as well as the European Commission have tended to give preference to cases of exclusionary abuses. In *MEO*, AG Wahl pointed out that exploitative cases of price discrimination are "extremely rare".[138] An exclusionary type of abuse through discriminatory pricing may be the case where the market-dominant supplier offers more favourable conditions to its own customers in an attempt to prevent them from switching to rivals, i.e. to foreclose rivals on the upstream market.[139]

It has been said that we still lack a full understanding of the nature of harm arising from dynamic or personalised pricing—more generally and where artificial intelligence is involved. Some are of the view that, to the extent that there is an enforcement issue with tackling personalised pricing, this probably relates to deficiencies in the antitrust framework for dealing with price discrimination and not any special feature of artificial intelligence.[140] As with any other case of abuse, in order to

[134] S. Rab, Artificial intelligence, algorithms and antitrust, Competition Law Journal, 2019, Vol. 18, No. 4.

[135] S. Rab, Artificial intelligence, algorithms and antitrust, Competition Law Journal, 2019, Vol. 18, No. 4.

[136] See, on this particular topic N. Petit, Algorithmes tarifaires et droit européen de la concurrence, in l'Europe au présent!, Bruxelles, Bruylant, 2018, pp. 167–176.

[137] N. Petit, Algorithmes tarifaires et droit européen de la concurrence, in l'Europe au présent!, Bruxelles, Bruylant, 2018, pp. 167–176.

[138] AG Wahl, case C-295/17, *MEO*, ECLI:EU:C:2018:413, pt 80.

[139] I. Graef, Algorithms and fairness – What role for competition law in targeting price discrimination towards end consumers, Columbia Journal of European Law 2018(24:3), pp. 541–559 (542f).

[140] S. Rab, Artificial intelligence, algorithms and antitrust, Competition Law Journal, 2019, Vol. 18, No. 4.

appropriately assess the potential of algorithms to lead to an abuse of dominance, it is important to distinguish the nature of the harm arising and whether this is a harm to competition or whether concerns relate more to consumer protection.[141] In this regard, the OECD has analysed new forms of abuse in digital markets and specifically whether any new theories of harm arise from these new forms of competition.[142]

In this context, the OECD has proposed a (relatively) new theory of harm, which brings us to the third type of abuse connected with the use of algorithms by dominant undertakings: the potential of a dominant firm active in multiple related markets to engage in abusive leveraging or discriminatory leveraging.

Discriminatory Leveraging or Self-preferencing Ranking algorithms could feature a ranking bias by preferring a company's own services, to the competitors' detriment. This could also be a hypothetical situation for a dominant company holding a ranking algorithm to exert pressure on other companies, e.g., by threatening to de-rank suppliers/customers in order to induce them to engage in certain anti-competitive conduct.

Instead of appropriating a competitor's innovations, abusive leveraging theories of harm focus on ways in which a firm can use (or leverage) its dominant position in one market to favour its products in a related market.[143] This type of conduct, which can take the form of self-preferencing (for example providing platform access advantages to its own product), has been identified as a potential exclusionary abuse of dominance by some competition authorities.[144]

To date, there have been relatively few cases, albeit some very high-profile cases, in digital markets focusing on this theory of harm. The European Commission's Google Shopping case is one example. The European Commission fined Google, finding that the more favourable positioning and display by Google, on its general search result pages, of its own comparison-shopping service compared to competing comparison-shopping services infringed Article 102 TFEU.[145] The Commission explained that Google's service was not subject to these dedicated algorithms, which reduced the rankings in Google's general search pages. By contrast, the US Federal Trade Commission and the Turkish Competition Authority did not find a

[141] S. Rab, Artificial intelligence, algorithms and antitrust, Competition Law Journal, 2019, Vol. 18, No. 4.

[142] OECD (2020), Abuse of dominance in digital markets. Available at www.oecd.org/daf/competition/abuse-of-dominance-in-digital-markets-2020.pdf, p. 53. Accessed 27 October 2021.

[143] OECD (2020), Abuse of dominance in digital markets. Available at www.oecd.org/daf/competition/abuse-of-dominance-in-digital-markets-2020.pdf, p. 54. Accessed 27 October 2021.

[144] OECD (2020), Abuse of dominance in digital markets. Available at www.oecd.org/daf/competition/abuse-of-dominance-in-digital-markets-2020.pdf, p. 54. Accessed 27 October 2021.

[145] Commission decision of 27 June 2017 relating to a proceeding under Article 102 of the Treaty on the Functioning of the European Union and Article 54 of the EEA Agreement (Case AT.39740 — Google Search (Shopping).

competition law infringement in relation to this conduct.[146] In addition, the Polish Office of Competition and Consumer Protection (UOKiK) opened in 2019 an investigation (which is ongoing) into an online shopping platform called Allegro with the concern that it may have obtained advantages through better information about the platform's algorithm and enjoyed preferential treatment in terms of sales, promotions and advertisement campaigns. Finally, the French competition authority considered also this theory of harm in a 2018 decision in which a website creation intermediation company, Interactive Lab, complained of discriminatory and exclusionary practices, constitutive of an abuse of a dominant position, which were allegedly implemented by Google in its AdWords service. Google was accused of manipulating the results of AdWords' auction system in order to maximise the number of customers and revenues of its service and consequently to exclude Interactive Lab from the intermediation market for the creation of websites on which it operates. This case was rejected by the competition authority due to a lack of evidence.[147]

11.3.2.3 Can Businesses Be Discriminated Due to the Use of Artificial Intelligence?

From the point of view of businesses offering their products, the increased use of artificial intelligence by businesses in marketing may lead us to question whether certain businesses can be discriminated against because, on the one hand, unequal competitive conditions may result for those who cannot afford to use artificial intelligence (and whether there would be a need for intervention) and, on the other hand, the algorithms used by an advertisement campaign that considers past purchase decisions may make a selection of businesses, leaving some businesses out.

On the first topic, whether the use of artificial intelligence in marketing can create unequal competitive conditions, the general consensus among National Reporters is that the use of artificial intelligence as such, even if it creates unequal competitive positions, does not constitute an infringement of competition law. The right to free competition allows each individual company to decide for itself whether to use artificial intelligence. Even before artificial intelligence appeared on stage, companies could decide to implement different marketing strategies, how much to spend on them, what tools to use, etc. While it is clear that companies with large databases can often use artificial intelligence more effectively, this situation is not different from the past, when—before algorithms were being used to better target sales—a retailer had a list of costumers and potential customers and targeted its marketing campaigns to them. Instead of challenging this situation under competi-

[146] OECD, 2018, p. 38[105].

[147] Decision 18-D-13 of 20 July 2018, regarding practices implemented by Google in the sector of online advertising https://www.autoritedelaconcurrence.fr/fr/decision/relative-des-pratiques-mises-en-oeuvrepar-google-dans-le-secteur-de-la-publicite-en-ligne. Accessed 27 October 2021.

tion law, some consider that it should be decided whether access to data needs to be (re-)regulated.[148]

On the second topic—potential discrimination in how algorithms can analyse previous purchase decisions and suggest advertising on that basis—the factual hypothesis is that one of the main purposes of algorithms is filtering and ranking information based on factors included in their code due to the numerous data online and the limited capacity of an individual to absorb it all. Personalised marketing thus represents a way to facilitate purchase research mechanisms and help consumers to have access to products that match his or her profile. However, algorithms that evaluate previous purchase decisions and then place new advertisements would not aim—except when there is proof to the contrary, of course—to delete or hide information or benefit any company for unreasonable motivations that would constitute a discriminatory practice. Basing new advertisements to be received by consumers on previous purchase decisions seems a rational type of ranking, especially considering they are intended to be personalised according to one's interests.

It could be held that this type of algorithm could make it more difficult for new businesses or struggling companies to stand out, but this argument would not go a long way. The evaluation of previous purchase decisions is reflected in the profile created of each individual purchaser. Also, new companies can get access to this profile and thus benefit from it. The fact that a new company has not yet been considered (its goods have not been sold and cannot have been thus considered in an analysis of past purchases) does not conflict with the principle of fair competition, as the company can pay to be considered showing that it is a similar product/service as those in past purchase decisions of the profile. As such, for the new company, it would not be different from a barrier to entry in the form of an investment in publicity. Evidently, not all companies operating in the market have the same means and opportunities, and that is precisely one feature of a free market economy.

Also, it has been held by some National Reporters[149] that these companies might follow other kinds of marketing strategies, increasing their purchase numbers, which will enable them to start being considered by the described algorithms.

The conclusion is that while the possibility is there for discrimination, the current situation does not allow setting forth that an intervention in the market is necessary (unless of course the company exercising such discrimination is a dominant company, in which case the practice could have the potential to fall under the prohibition of abuse of dominance, Article 102 TFEU).

[148] German National Report, section 4.2.5.

[149] Brazilian report, Section 11.

11.3.3 The Use of Artificial Intelligence in Marketing and Possible Implications Under Data Protection Laws

Marketing activities are usually addressed to consumers, i.e. individuals or data subjects as defined in the GDPR.[150] When dealing with individuals or data subjects, the observance of GDPR provisions is of great importance for companies, moreover taking into account the increase in the enforcement activities of the different EU data protection regulators and the potential amount of sanctions imposed in infringement procedures in the Member States.

Among the obligations established in the GDPR relating to the processing of personal data, two issues appear as key: the lawfulness of processing personal data and transparency obligations. In this section, we will examine how personal data should be lawfully processed, i.e. using an adequate legal basis established in the GDPR for the processing of personal data, and how the transparency principle must also be observed.

11.3.3.1 Lawful Processing of Personal Data

One of the first questions that a data controller must face when processing persona data is the legal basis to be used in order to process personal data. However, it is not only necessary to choose one legal basis established in Article 6 (1) of the GDPR: it is also necessary to determine the appropriate legal basis for each processing in order to lawfully process personal data. If the regulators find that the legal basis used for the processing of personal data is not the correct one, the company could face an infringement procedure and eventually be fined for such infringement.

This was the case, for example, for PricewaterhouseCoopers (PwC) in Greece, where the Hellenic Data Protection Authority fined the company with 150,000 euros for alleged violations of the GDPR. The Hellenic Data Protection Authority found that PwC did not process the data of its employees in a compliant manner. While PwC had obtained the consent to process its employees' information, the Hellenic Data Protection Authority held that consent had not been freely given by the staff members and that the appropriate legal basis would have been the performance of a contract to which the data subject is party, as provided in Article 6 (1) (b) GDPR.[151]

The National Reports consider both consent[152] and legitimate interest[153] of the company as potentially adequate legal bases for the processing of personal data.

[150] Article 4 (1) of the GDPR defines data subject as "an identifiable natural person is one who can be identified, directly or indirectly, in particular by reference to an identifier such as a name, an identification number, location data, an online identifier or to one or more factors specific to the physical, physiological, genetic, mental, economic, cultural or social identity of that natural person".

[151] A summary of the Decision (in English) is available on https://www.dpa.gr/en/enimerwtiko/prakseisArxis/parabiasi-arhon-epexergasias-dedomenon-prosopikoy-haraktira-ergazomenon. Accessed 27 October 2021.

[152] Article 6 (1) (a) GDPR.

[153] Article 6 (1) (f) GDPR.

Another legal basis that could have been considered is the performance of a contract.[154] However, this legal basis would probably not be adequate in all cases, given that in most marketing activities, there is no performance of a contract to which the data subject is party, and the processing of personal data does not take place at the request of the data subject prior to entering into a contract.

Consent seems to be the most adequate legal basis for the processing of personal data. However, consent could be a burdensome legal basis for companies: in order for consent to be considered valid under GDPR provisions, this should be freely given, specific, informed and unambiguous according to Article 4 (11) GDPR. Moreover, valid consent requires a positive (opt-in) action; the use of pre-ticked boxes or any other method of default consent has not been considered valid anymore since the entry into force of the GDPR.

However, some voices question the adequacy of consent to ensure an appropriate balance between the data sovereignty of individuals and the economic interests of businesses. Even if the user is aware of all the circumstances of the data processing, he or she may be forced to accept the extensive processing of his or her personal data only because there are no comparable alternatives. The negligence of users represents another major risk: some users may not take the time to read through data protection declarations but may carelessly consent to the use of their data by clicking on them.[155] This risk could be reduced by the use of warning symbols or the implementation of a uniform European certification system, which classifies a website or app according to its truthfulness in terms of data processing.[156] Furthermore, these proposals would not constitute a one-sided consumer protection regulation to the disadvantage of the digital economy: if data protection and data security were correctly understood as a competitive strength, the European digital economy could gain an advantage over international competition precisely through transparent and user-oriented business models.[157]

Another issue is the expiry date of consent. Especially in the case of non-written requests, it might be difficult for the company processing the data to know when consent is invalid or has been revoked,[158] and it might be even more difficult for the user to know whether the company is complying.

In any event, when we are dealing with the use of cookies for advertising purposes, it is necessary to recall that the processing of personal data through the use of cookies can only be lawful based on the consent of the individual. Cookies are small blocks of data created by a web server while a user is browsing a website and

---

[154] Article 6 (1) (b) GDPR.

[155] P.-L. Krüger, Datensouveränität und Digitalisierung, ZRP 2016, pp. 190–192.

[156] S. Kraska, Datenschutz-Zertifizierungen in der EU-Datenschutzgrundverordnung, ZD 2016, pp. 153–154; G. Hornung and K. Hartl, Datenschutz durch Marktanreize – auch in Europa? - Stand der Diskussion zu Datenschutzzertifizierung und Datenschutzaudit, ZD 2014, pp. 219–225.

[157] P.-L. Krüger, Datensouveränität und Digitalisierung, ZRP 2016, pp. 190–192.

[158] E. M. Frenzel. In Paal and Pauly (eds), Datenschutzgrundverordnung, Bundesdatenschutzkontakt, 2nd ed, C.H. Beck 2018, Art. 6 DS-GVO, para. 11.

are placed by the user's web browser on the user's computer or other devices. Cookies are placed on the device used to access a website, and more than one cookie may be placed on a user's device during a session. Cookies consequently store personal data, which means that the obligations under the GDPR for the processing of such personal data must thus be observed by the company. Although the GDPR allows the processing of users' data as long as they receive consent or there is a legitimate interest, Article 5 (3) of the ePrivacy Directive[159] and the relevant transposition legislation in the Member States establish that personal data in the context of the use of cookies can only be done with an informed, clear, affirmative and unambiguous indication of their intentions, which constitutes a valid consent from data subjects to have their personal data processed by cookies. Thus, companies are not allowed to activate non-necessary cookies on their websites that process personal data until users have given their explicit consent. For instance, two different types of consent would be needed for analytic cookies and for functional cookies.[160] This requirement also prevents the data controller from relying on consent for other purposes than those for which it was originally obtained.

The alternative legal basis for the processing of personal data in these cases would be the *legitimate interest* of the company. However, it is difficult to use this legal basis for the processing of personal data for marketing purposes. Although Recital 47 of the GDPR states that direct marketing may be a legitimate interest of the data controller, such interest cannot override the interests or the fundamental rights and freedoms of the data subjects. This condition implies the need to analyse on a case-by-case basis the balance between the legitimate interest of the data controller and the interests or the fundamental rights and freedoms of the data subjects. This could lead to uncertainty for the data controller regarding the possibility to use the legitimate interest for certain processing of personal data, also taking into account the relevant country where the legitimate interest was to be used as legal basis, and the different approaches of the various regulators regarding the use of legitimate interest as legal basis, e.g. the Spanish regulator has not typically favoured the legitimate interest as a valid legal basis.

It is also necessary to take into account that the useful regulation regarding the need for consent for the use of cookies in the users' computers could become outdated in the near future. Companies like Google are developing new technologies, such as Federated Learning of Cohorts (FLoC), which is a type of web tracking through federated learning. It groups people into "cohorts" based on their browsing history for the purpose of interest-based advertising.[161] Apparently, FLoC will not be processing personal data in the same manner as cookies, and this

[159] Directive 2002/58 of the European Parliament and of the Council of 12 July 2002 concerning the processing of personal data and the protection of privacy in the electronic communications sector (Directive on privacy and electronic communications).

[160] BDPA Recommendation on Direct Marketing, p. 63.

[161] D. Bohn, Privacy and ads in Chrome are about to become FLoCing complicated, The Verge, March 30, 2021.

could lead to a more flexible scenario for advertising activities with customers. However, this technology is not being used in the EU, and compliance with the GDPR provisions is under suspicion.

11.3.3.2 Transparency Principle for the Processing of Personal Data

The principle of transparency requires that all information and communication regarding data processing must be accessible and easy to understand. Thus, simple and clear language must be used, without resorting to long texts with legal terminology or small print. On the contrary, the data controller must highlight the problems and risks involved in the processing of user data. Only in this way would the data subjects be giving informed consent and only in this way would the principle of transparency be fulfilled.

In addition to detailed information on the risks, data subjects will also need to have the contact of the data controller. The principle of transparency also mentions that the data subjects must also know the rules, safeguards and rights related to the processing of personal data, as well as how to exercise them.

On the other hand, the principle of transparency should not be limited solely to the legal basis of consent since it is not the only one under which personal data can be processed. The GDPR provides that the principle of transparency, as well as the duty to inform, will be taken into account whenever there is a processing of personal data.

To comply with the GDPR, the information provided to the data subjects must be presented in a "granular" way. That is, it must be fragmented by layers or levels. At first, the data subject will be able to see at a glance the basic and summarised information, with an understandable language and in which they can understand all the processing activities of their personal data. And, on a second level, the same information must be offered in more detail, in case the data subject wants to go deeper and learn more about what the person in charge is doing with his or her personal data.

The requirement that consent must be specific to be valid reinforces this information obligation. The data subject needs to be able to distinguish between different purposes for the processing of his or her personal data and choose to which ones they want to agree (the "granularity of consent").[162]

The result is that in consumers face a substantial amount of information and steps to be taken when surfing the Internet, on each merchant website they visit, which is necessary to obtain their consent for the enterprise's processing purposes. The Belgian Data Protection Authority (BDPA),[163] the European Data Protection

[162] EDPB Guidelines 05/2020 of 4 May 2020 of the EDPS, on consent under Regulation 2016/679, n°56, p. 13. Available at https://edpb.europa.eu/sites/edpb/files/files/file1/edpb_guidelines_202005_consent_en.pdf. Accessed 10 May 2020.

[163] *Ibid.*, p. 67.

Board (EDPB)[164] and Belgian scholars share the view that this could lead to a loss of control over the personal data communicated by the consumers as they will tend to be less vigilant and experience a real "privacy fatigue", where they will mechanically consent or refuse processing.

A more effective and reasoned approach to consent, however, is possible but would rest on the enterprise's willingness to communicate that prior information to the consumer in an effective and friendly manner: the works of R. Ducato[165] have shown that "informed consent" is not the issue as such but rather the means that the controller uses to convey information to the consumer. Privacy fatigue could be avoided by giving consumers a sense of control and agentivity over their personal data if techniques other than the provision of lengthy privacy policies are used: relying on new technology (videos, multi-layered menus, chatbots or other types of electronic assistants) or techniques that require a more active interplay with the consumer (questionnaires or "privacy" mini-games).[166] The BDPA and EPDB also suggest alternative techniques where the data subject makes a specific movement with the mouse or finger on the controller's webpage or repeats a phrase orally using a microphone. However, in the new version of its guidelines on consent, the EDPS specified that based on Recital 32, actions such as scrolling or swiping through a webpage or similar user activity would not be sufficient since such actions are difficult to distinguish from other activities or interactions by a user.[167]

We should bear in mind that the transparency principle is also very relevant in cases where automatic decision-making is involved. Article 22 GDPR[168] establishes the right not to be subject to automated decisions. It should be noted that this provision has confusing wording. Article 22 is part of the set of powers that are recognised for holders, which are derived from the fundamental right to data protection, since such provision is part of Chapter III, titled "Rights of the Interested Party",[169] and also, specifically within this chapter, this article must be related to the

[164] Referring to "Click-fatigue", EDPB Guidelines 05/2020 of 4 May 2020 on consent under Regulation 2016/679, available at https://edpb.europa.eu/sites/edpb/files/files/file1/edpb_guidelines_202005_consent_en.pdf, consulted on 10 May 2020., n°87-89, p. 18.

[165] See R. Ducato and A. Strowel, Information duties between consumer and data protection in the "Internet of Platforms": promoting awareness by design, D.C.C.R., 2019, n°122-123, pp. 123–149; A. Rossi, R. Ducato, H. Haapion and S. Passera, When Design Met Law: Design Patterns for Information Technology, D.C.C.R., 2019/1-2, n°122-123, p. 79-12.

[166] See also, sharing the same opinion: F. Jacques, Personnalisation des prix - Regard européen sur la pratique, R.D.T.I., n°78-79, 2020; H. Jacquemin, Le big data à l'épreuve des pratiques du marché et de la protection du consommateur, R.D.T.I, 2018/1, n°70, pp. 75–91.

[167] It would also not be possible to allow the data subject to withdraw his or her consent as easy as he or she has given it, see Guidelines 05/2020 of 4 May 2020 of the EDPS, on consent under Regulation 2016/679, n° 86, p. 18. Available at https://edpb.europa.eu/sites/edpb/files/files/file1/edpb_guidelines_202005_consent_en.pdf. Accessed 10 May 2020.

[168] Article 22 (1) GDPR: "The data subject shall have the right not to be subject to a decision based solely on automated processing, including profiling, which produces legal effects concerning him or her or similarly significantly affects him or her."

[169] Articles 15 to 22 GDPR.

right of opposition since both are part of Section 4, called "Right of Opposition and Automated Individual decisions". Such a formal fit has led part of the doctrine and control authorities to consider that Article 22 only or centrally recognises a variant of the right of opposition, specifically the right of opposition to automated decisions or the right not to be subjected to this type of decision. However, if together with this formal analysis of the provision we add a complete study of all the content covered by Article 22, it can be concluded that the article we are analysing hides much more than a right to object to automated decisions. What is really being recognised in Article 22 is the specific regulation of a specific personal data processing, that is, automated decision-making, including the creation of profiles that produce legal effects or significantly affect the interested party in a similar way.

In the field of fully automated individual decisions, the GDPR provides a series of clarifications related to the information that, where appropriate, must be provided by those responsible. Such clarifications on the principle of transparency are materialised in Articles 13 (2) (f) and 14 (2) (g) GDPR (Information Rights), Article 15 (1) (h) GDPR (Right of Access) and Recital 71 GDPR (Right of Explanation).

Articles 13 (2) (f) and 14 (2) (g) GDPR oblige the data controller to inform the interested party about "the existence of automated decisions, including profiling, referred to in Article 22, sections 1 and 4, and, at least in such cases, significant information on the logic applied, as well as the importance and expected consequences of said data processing for the data subject". From this provision arises a series of transparency duties: (i) to inform about the existence of automated decisions, (ii) to provide significant information about the applied logic of the treatment and (iii) to indicate the importance and expected consequences of such data processing.

With regard to the information on the existence of automated decisions, the European standard provides that any data subject who is subjected to the processing defined in Article 22 should be aware of such automated decisions. In this sense, the information is a reflection of the principle of purpose limitation embodied in Article 5 (1) (b) GDPR; that is, the purpose of the data processing must be clearly determined and made explicit, as well as, consequently, the existence of this processing. Now, the GDPR not only requires data controllers to communicate whether they are carrying out the data processing defined in Article 22, but it also obliges them to report on certain specific characteristics of the said data processing, that is, that they provide meaningful information on the applied logic of such processing, as well as the importance and expected consequences of this processing.[170]

In relation to the information related to the applied logic of the treatment, it must be sufficient so that the individual who is subjected to said decisions at least can get

[170] B. Goodman and S. Flaxman, EU Regulations on Algorithmic Decision-Making and a 'right to Explanation', AI Magazine, Vol 38, No. 3, 2017. Available at: http://arxiv.org/abs/1606.08813. S. Watcher, B. Mittelstadt, L. Floridi, Why a Right to Explanation of Automated Decision Making Does Not Exist in the General Data Protection Regulation, International Data Privacy Law, Volume 7, Issue 2, 1 May 2017, Pages 76–99, Available at: https://doi.org/10.1093/idpl/ipx005.

an idea of how the computer program can work based on the individual's personal data. Thus, the first thing that should be included in this information is the data or categories of data that, where appropriate, have been used to train/feed the algorithm.[171] The data can be both personal and non-personal. That is why it will be essential to know what data are considered to be most relevant and, where appropriate, to indicate why.[172] Along with this, in certain models, it may also become possible to report on estimated logic, that is, how the inputs (personal data of the individual) are transformed into a certain result by the algorithm (automated decision). In these cases, it is convenient to provide patterns that are derived from the daily practice of the computer program itself. For example, the input data could be a customer's purchase history, and the output could be products that the customer is likely to purchase in the future. As a result of this, the individual will not be subsequently surprised by the decision since he or she will be able to foresee the possible results as a guide, whether or not they are harmful.

Specifically, with the aim of favouring a better understanding of this applied logic and the foreseen consequences, the GDPR imposes in general terms a series of formal requirements in relation to how the said information should be provided; thus, Article 12 (1) requires that such information be provided in a concise, transparent, intelligible and easily accessible form, with clear and simple language. Once these requirements are transferred to the data processing that we are analysing, it will be understood that such information should be flexible, not excessively technical and, where appropriate, functional; that is, it must be information that is useful for the data subject who is immersed or is going to be immersed in automated decision-making[173] so that the individual has certain tools to later challenge the decision.

Moreover, the EU Commission's proposal for an Artificial Intelligence Act[174] also contains transparency obligations for high-risk artificial intelligence systems (Article 13) and mitigated transparency obligations for certain artificial intelligence systems (Article 52). According to Article 52 of the proposal for an Artificial Intelligence Act, artificial intelligence systems that interact with natural persons must inform the natural person that he or she is interacting with such a system unless it is obvious. This exception offers room for interpretation, which—if this is how the final regulation is approved—will be a task for the responsible supervisory authorities and courts. Furthermore, it remains unclear how the duty to inform should look in practice. On the one hand, supplying very basic information to the

[171] Article 29 Data Protection Working Party Guidelines on Automated individual decision-making and Profiling for the purposes of Regulation 2016/679, page 35.

[172] M. Brkan, AI-supported decision-making under the general data protection regulation, Protection Regulation. In Proceedings of ICAIL '17, London, United Kingdom, June 12–16, 2017, pp. 3–8, p. 5.

[173] A. D. Selbst and J. Powles, Meaningful information and the right to explanation, International Data Privacy Law, 2017, Vol. 7, No. 4, page 236.

[174] EU Commission, Proposal for a Regulation of the European Parliament and of the Council laying down harmonised rules on artificial intelligence and amending certain union legislative acts, COM (2021) 206 final of 21.4.2021.

affected person about the artificial intelligence system does not help, since many people will ignore the information. On the other hand, it is generally assumed that only an informed person is able to take responsibility for himself or herself and take necessary protective measures. In the context of artificial intelligence systems, often the persons concerned are not in a position to recognise the indirect consequences, which may lie in the future. Instead of referring to the use of an artificial intelligence system, it would be preferable to provide information about the criteria used by an artificial intelligence system and their possible consequences.

A comprehensive right to explanation, which the data subject can invoke, is currently not enshrined in the proposal. The existing provisions must be further specified by the legislator—or at least by the courts. However, Articles 13–15 and 22 of the GDPR and Article 13 of the proposal for an Artificial Intelligence Act indicate that there is a duty to explain algorithmic decisions. Nevertheless, transparency obligations and the right to explanation conflict with other rights, especially the protection of trade secrets, and face technical barriers, which will need to be adequately addressed.

Acknowledgments The author wishes to thank Irene Robledo for her valuable contribution to this report.

Austria

12

Adrian Kubat and Arno Scharf

12.1 Introduction

The digital economy has not only led to the emergence of new data-driven business models but also changed advertising and pricing strategies. Although the increasing digitisation undoubtedly creates efficiencies for both consumers and suppliers, it raises new legal challenges. In this paper, some of the most topical issues will be discussed from a competition law perspective. The first part deals with new advertising strategies under Austrian unfair competition law, in particular native advertising, influencer marketing and related labelling obligations (see Sect. 12.2). The second part assesses innovative pricing strategies under Austrian antitrust law, namely the use of artificial intelligence (AI) and algorithms as a means for anticompetitive collusion as well as personalised pricing (see Sect. 12.3).[1]

12.2 Legal Framework in Influencer Marketing and Native Advertising

12.2.1 Terminology

In the course of digitisation, the media landscape has changed over the past years and decades. This has also been followed by a change in the way products are advertised. In particular, the partly negative perception of traditional advertising, which is

[1] Adrian Kubat has written Sect. 12.2, whereas Sect. 12.3 has been written by Arno Scharf.

A. Kubat (✉) · A. Scharf
University of Vienna, Vienna, Austria
e-mail: adrian.kubat@univie.ac.at

B. Kilpatrick et al. (eds.), *Antitrust in Data Driven Markets & Legal Framework for Influencers, Native Advertising and Control over the Use of AI in Marketing*, LIDC Contributions on Antitrust Law, Intellectual Property and Unfair Competition,
https://doi.org/10.1007/978-3-031-07422-6_12

increasingly making viewers feel uncomfortable, has made it necessary to rethink marketing. A prominent example of modern marketing strategies includes the so-called *native advertising*. In simple terms, native advertising describes a situation where the advertisement is embedded in an environment familiar to the viewer.[2] The advertised party shall be addressed individually and should not get the feeling of being affected by the advertisement in his reading flow or other activities. This form of advertising not only offers the advertiser the advantage of being able to reach the desired target group but is also usually much cheaper than offline advertising.[3]

In a typical native advertising scenario, video advertisements are embedded in a text, leading to the impression that the video belongs to the editorial section of a report. Also, the graphic overlay of such an editorial article with associated advertising messages is becoming increasingly common.[4] If, for example, an online daily newspaper reports on the advantages of electric mobility and this editorial article is embraced by partly animated advertising messages from a manufacturer of renowned electric cars, then this advertising message does not disturb the viewer's reading flow but rather shows the reader possible sources of supply. In a separate view, this form of advertising may seem just as intrusive as "classical" advertising. Yet it is the interaction of advertisement with information and the embedding in a familiar environment that relieve native advertising of its disruptive effect and make it more attractive for the viewer.

Naturally, however, native advertising is dynamic and constantly adapts to new circumstances. Meanwhile, one can therefore subsume a multitude of other advertising strategies and problem areas under it, especially *influencer marketing*. The latter refers to a form of social media marketing used by so-called influencers. Influencers characterise the use of a social media channel that primarily focuses on oneself but also advertises products and campaigns.[5] Due to their sometimes enormous impact, but especially by building up an almost friendship-like relationship with their viewers, influencers also gain considerable value for the advertising industry as testimonials. As elaborate and convincing a company's commercial may be, it still has the nimbus of a beautified advertisement hovering over it. Influencer marketing can fade this stigma into the background as the focus instead lies on the recommendation of the influencer as an alleged friend. According to a recent study, about one in five people aged between 16 and 64 has already made a

[2] In detail A. Wiebe and O. Kreutz, Native Advertising – Alter Wein in neuen Schläuchen? (Teil 1), WRP 2015(9), pp. 1053–1060.

[3] M. Specht and E. Theobald, Broadcast Your Ad, Nomos 2010, pp. 27f.

[4] See for example the very fitting example of Mini's advertising strategy on the news portal Buzzfeed at A. Wiebe and O. Kreutz, Native Advertising – Alter Wein in neuen Schläuchen? (Teil 1), WRP 2015(9), pp. 1053–1060 (1056).

[5] On their job profile S. Lettmann, Schleichwerbung durch Influencer Marketing - Das Erscheinungsbild der Influencer, GRUR 2018(12), pp. 1206–1211.

purchase decision based on the recommendation of an influencer. In the age group of 16–24, every second person did so.[6]

The legal framework for this young phenomenon is certainly not new but sometimes leads to differentiations based on the target group and the digital environment. In Austria, the respective legal provisions can, in particular, be found in the E-Commerce Act (ECG),[7] the Audiovisual Media Services Act (AMD-G),[8] the Media Act (MedienG)[9] as well as the Act Against Unfair Competition (UWG).[10] They regulate the behaviour of influencers by means of specific advertisement restrictions and also genuine competition law standards. Three problem areas that currently dominate the discussion in Austria will be presented in the following: (1) the *obligation to label advertising* contributions, (2) the *applicability of product liability standards to influencers* who provide false information about products and (3) the problem that occurs with *fake product ratings*.

12.2.2 Labelling Obligations in Influencer Marketing

The current body of law contains numerous special labelling obligations, the violation of which is primarily sanctioned by administrative penalties. In general, with the exception of the MedienG (up to EUR 20,000), these are not particularly high but play an important role in the context of unfair competition law.[11] Common to these provisions is that they are linked to the payment of remuneration for the determination of the contributions to be labelled. Even if the definitions in the respective legal provisions overlap to some extent, there are some (significant) differences that will be elaborated on in more detail below.

12.2.2.1 Labelling Obligations According to Sec 26 MedienG

The first of these obligations is set out in Sec 26 of the Media Act (MedienG), which requires that paid publications be labelled, unless the advertising character of the contribution itself is evident. The Austrian Media Act, which was originally designed for print media, now also applies to *periodic electronic media*, provided that these have a corresponding distribution, namely broadcasting at least four times a year. Thus, the vast majority of channels used by influencers fall within the scope

[6] BVDW, Umfrage zum Umgang mit Influencern 2020, available at https://www.bvdw.org/der-bvdw/news/detail/artikel/mehr-als-jeder-fuenfte-verkaeufe-durch-influencer-marketing-nehmen-laut-bvdw-studie-2020-nochmal-zu/. Accessed 08. June 2021.

[7] Bundesgesetz, mit dem bestimmte rechtliche Aspekte des elektronischen Geschäfts- und Rechtsverkehrs geregelt werden (E-Commerce-Gesetz – ECG) BGBl I 152/2001.

[8] Bundesgesetz über audiovisuelle Mediendienste (Audiovisuelle Mediendienste-Gesetz – AMD-G) BGBl I 85/2011.

[9] Bundesgesetz vom 12. Juni 1981 über die Presse und andere publizistische Medien (Mediengesetz – MedienG) BGBl 314/1981.

[10] Bundesgesetz über den unlauteren Wettbewerb 1984 – UWG BGBl 448/1984.

[11] See Sect. 12.2.2.1.

of application of the MedienG. This includes those channels that are not publicly accessible but remain hidden behind a paywall or whose access depends on the consent of the channel owner.[12]

In addition, a labelling obligation under Sec 26 requires a remuneration to be paid to the influencer. The definition of the consideration is still subject to vivid discussions as, meanwhile, mere monetary fees have been replaced by a variety of different methods of payment. Especially, products or invitations to events are often made available in return for the insertion of advertising. This kind of consideration is certainly covered by Sec 26, as has become now an old-fashioned monetary payment. In general, an asset of any kind is to be subsumed under the concept of remuneration in the Media Act. The provision of goods also leads to an increase in the recipient's financial resources, which is why a contribution inserted for this purpose must be labelled accordingly.

Nevertheless, cases of so-called *favouritism reporting* (*Gefälligkeitsberichterstattung*) have always been controversial and problematic. Favouritism reporting is characterised by a labelled advertisement and a (corresponding) editorial contribution that almost merge into each other. Typical for favouritism reporting is media coverage, where half of a page consists of an advertisement provided and paid for by the advertising company, while the other half provides an article that emphasises the advantages of the same company to a great extent. Since the accompanying advertisement was actually paid for, one could now be tempted to assume that also the corresponding editorial contribution must be labelled. However, the Supreme Court of Justice (OGH)[13] assumes, in accordance with the explanatory text of the law,[14] that Sec 26 requires a synallagmatic relationship between the payment and the article. This means that the payment of remuneration must have been the very cause of the publication of the contribution. Yet this is not the case if the advertising company and the publisher only agree to pay for an advertisement but the additionally disclosed contribution is not covered by this agreement but is published merely out of courtesy or gratitude.

According to the jurisdiction of the OGH, in such cases, a label must therefore be applied only to the advertisement but not to the editorial contribution, no matter how enthusiastic it may be about the company or product. In doing so, the OGH uses the term *editorial contributions with advertising surplus*[15] and focuses exclusively on the underlying agreement, thus leaving other aspects out of consideration. As a result, which has even been criticised by media associations themselves,[16] the OGH

---

[12] A. Koukal, In: Berka, Heindl, Höhne and Koukal (eds), MedienG, 4th ed, LexisNexis 2019, Sec 1 MedienG para 7.

[13] OGH 4 Ob 60/16a, *Gefälligkeitsberichterstattung*, ÖBl 2017/25 (L. Wiltschek).

[14] AB 743 BlgNR 15.GP, p. 11.

[15] Authentic version "*redaktionelle Beiträge mit werblichem Überschuss*".

[16] See for example the press release of the PR Ethics Council, available at www.prethikrat.at/wp-content/uploads/2016/11/Presseinfo_PR-Ethikrat_OGH-Urteil_2016-11-15.pdf. Accessed 8 September 2021.

leaves it up to the parties to decide on the labelling obligation. Although such a result may well be congruent with the history of the origins of Sec 26 MedienG, it remains to be seen whether it is also desirable from a legal policy point of view. To prevent the protective concept of the labelling obligation from being undermined, it would in any case be more consistent to strive for a de lege ferenda regulation that also includes the problem of favouritism reporting in the scope of Sec 26 Media Act. The concept of remuneration could, for example, be amended by abandoning the synallagmatic relationship between the report and the payment and instead subject the report to a separate review. If this review reveals that payment would typically have had to be made for such a report, then this should also be sufficient to require a labelling obligation. The fact that this is not only possible from a legal point of view but can also be implemented in practice is shown in particular by the standards of the AMD-G and ECG, which will be discussed in a moment.

12.2.2.2 Labelling Obligations According to Sec 31 AMD-G

The Audiovisual Media Services Act (AMD-G) can be traced back to the EU Audiovisual Media Services Directive (AVMS Directive)[17] and, similar to Sec 26 MedienG outlined above, aims at increasing market transparency by providing information to the consumer. While the MedienG is indisputably applicable to most channels of influencer marketing, this was anything but clear for the AMD-G in the past. In order to fall within the scope of application, and hence possible labelling obligations, of the AMD-G, the existence of an audiovisual media service is necessary in the first place. Audiovisual media services are characterised by six elements, as defined in the AVMS Directive. Accordingly, it must be a service within the meaning of Articles 56 and 57 Treaty on the Functioning of the European Union (TFEU), which is under the editorial responsibility of a media service provider. Furthermore, the service must be operated for the principal purpose of providing programmes intended to inform, entertain or educate the general public through electronic communications networks.[18] Recently, the element of the main purpose has been subject to controversial discussions. In an eagerly awaited recent judgement, the Court of Justice of the European Union (CJEU)[19] ruled that the YouTube channel of a company on which new cars are advertised does not constitute an audiovisual media service as the advertising purpose of the channel would clearly be in the foreground. However, as the ruling deals with a channel that exclusively provides advertisement content, while typical influencer marketing is characterised by a mixture of editorial content and advertising contributions, it would not apply to

[17]Directive 2010/13 of the European Parliament and of the Council of 10 March 2010 on the coordination of certain provisions laid down by law, regulation or administrative action in Member States concerning the provision of audiovisual media services (Audiovisual Media Services Directive), OJ 2010 L 95, p. 1.

[18]In detail C. Handig, Wenn "Influencer" zum Beruf wird, stellt sich ua die berufsrechtliche Frage, ecolex 2019(6), pp. 547–550.

[19]CJEU 21 February 2018, case C-132/17, *Peugeot Deutschland*, ECLI:EU:C:2018:85.

influencer marketing. Apart from this, the amended AVMS Directive,[20] which has not yet been implemented into national law, now explicitly includes *video-sharing platforms*. By doing so, the European legislator takes account of changes in consumer behaviour and subjects video blogs to regulation.[21]

Although the labelling obligation under Sec 31 AMD-G is, in general, similar to the one laid down in Sec 26 MedienG, it differs in important aspects. What they have in common is that commercial communication as such must be easily recognisable.[22] However, the notion of remuneration in the AMD-G differs significantly from the one used in the MedienG described above. What is relevant to the concept of remuneration is not merely the donation of an asset but an objective standard that focuses on the question of whether or not a remuneration would typically have to be made for such a contribution. The assessment of a contribution is based on the perception of the targeted consumers. The decisive question is whether a fee would typically have to be paid for that very special contribution.[23] If this is the case, the labelling obligation of Sec 31 AMD-G applies. However, this means that the labelling obligation also applies to those contributions that the OGH describes as *editorial contributions with advertising surplus*.[24] If the author of the editorial contribution exceeds the limit of objectivity and slides into almost euphoric calls to buy, this will, in the perception of the consumers, generally represent a paid advertisement, which must be labelled.

12.2.2.3 Labelling Obligations According to Sec 6 ECG

Thirdly and lastly, a labelling provision can also be found in the E-Commerce Act (ECG), which in part, however, goes beyond the requirements of its related provisions. Section 6 ECG requires not only a clear and unambiguous identification of commercial communication but also the disclosure of the client.[25] This obligation applies to the provider of a so-called *information society service*, defined as a *service, usually against payment, provided electronically at a distance on individual demand*.[26] Hence, this definition also covers platforms typically used by influencers in the vast majority of cases. Indeed, the concept of remuneration is understood

[20] Directive 2018/1808 of the European Parliament and of the Council of 14 November 2018 amending Directive 2010/13 on the coordination of certain provisions laid down by law, regulation or administrative action in Member States concerning the provision of audiovisual media services (Audio Media Services Directive) in view of changing market realities, OJ 2018 L 303, p. 69.

[21] In detail P. Zurth and V. Pless, #transparenz: Die Kennzeichnung nutzergenerierter Werbung in sozialen Netzwerken unter der neuen AVMD-Richtlinie Teil 1, ZUM 2019(5), pp. 414–424 (416f).

[22] See e.g. P. Zurth and V. Pless, #transparenz: Die Kennzeichnung nutzergenerierter Werbung in sozialen Netzwerken unter der neuen AVMD-Richtlinie Teil 2, ZUM 2019(6), pp. 457–466 (464f).

[23] Inter alia VwGH 26.2.2016, Ra 2016/03/0021; 1.9.2017, Ra 2017/03/0007; see also M. Kogler, M. Traimer and M. Truppe, Österreichische Rundfunkgesetze, 4th ed, LexisNexis 2018 Sec 2 AMD-G p. 434.

[24] OGH 4 Ob 60/16a, *Gefälligkeitsberichterstattung*, ÖBl 2017/25 (L. Wiltschek).

[25] In detail M. Ciresa, In: Schwimann and Kodek (eds), ABGB, 4th ed, Manz 2015, Sec 6 ECG.

[26] Sec 3 ECG; e.g. M. Ciresa, In: Schwimann and Kodek (eds), ABGB, 4th, Manz 2015, Sec 3 ECG.

broadly and does not require payment for the service by the end customer but rather includes self-promotion and third-party financing.[27] In addition to the elements of electronic retrieval and distance selling, which are inherent to influencer marketing, individual retrieval is also generally given. This requires users to be able to choose when and what they would like to consume from a programme catalogue, something that is undoubtedly possible on usual video platforms, such as YouTube or Instagram, and which leads to the overall applicability of the ECG.[28]

A broad understanding of the concept of remuneration is also appropriate with regard to the definition of commercial communications. If the influencer's communication *directly or indirectly* promotes the sale of goods or services or the image of a company, then the commercial character of a contribution is affirmed.[29] This means that all common forms of advertising by influencers fall within the scope of application of Sec 6 ECG, particularly since it is not a question of the transfer of an asset, but an objective standard, such as discussed in regard to the AMD-G,[30] must be applied.[31] Independent statements about companies, in particular those made without any consideration, are explicitly excluded from the scope of application.[32] However, the relevance of this exception is doubtful for influencer marketing. After all, the narrowly defined requirements can only be met by factual tests or descriptive statements about the company since, otherwise, there is always at least an indirect promotion of sales or appearance.[33] Hence, in business relations between advertising companies and influencers, it is strongly recommended to keep in mind all relevant regulations. Special attention should be given to the disclosure obligation, whereby a link to the company's web content is normally sufficient.[34]

12.2.2.4 Relevance to Unfair Competition Law and Transparency Provisions Within the UWG

While the special legal provisions described above are primarily sanctioned by administrative penalties, the Act Against Unfair Competition (UWG) undoubtedly plays the most important role in unfair competition law. This is for two reasons. First, the UWG functions as a gateway for special legal norms through the category of breach of law in its Sec 1, which can also be enforced by law-abiding competitors

---

[27] In detail M. Ciresa, In: Schwimann and Kodek (eds), ABGB (ed), 4th ed, Manz 2015, Sec 3 ECG paras 6f; W. Zankl, E-Commerce-Gesetz, 2nd ed, Verlag Österreich 2016, Sec 3 paras 48f.

[28] M. Ciresa, In: Schwimann and Kodek (eds), ABGB, 4th ed, Manz 2015, Sec 3 ECG para 15; A. Kubat, Kennzeichnungspflicht im Influencer-Marketing, ÖBl 2020(1), pp. 8–16 (12f).

[29] M. Ciresa, In: Schwimann and Kodek (eds), ABGB, 4th ed, Manz 2015, Sec 3 ECG paras 6f; W. Zankl, E-Commerce-Gesetz, 2nd ed, Verlag Österreich 2016, Sec 3 paras 48f.

[30] See Sect. 12.2.2.2.

[31] See P. Raffling and H. Wittmann, Rechtsfragen am Beispiel des "Influencer Marketings", MR 2017(4), pp. 163–168 (166f).

[32] Sec 3 Nr 6 lit b ECG.

[33] W. Zankl, E-Commerce-Gesetz, 2nd ed, Verlag Österreich 2016, Sec 3 para 74.

[34] Again M. Ciresa, In: Schwimann and Kodek, ABGB (eds), 4th ed, Manz 2015, Sec 6 ECG para 11.

through actions for injunction. Second, the UWG itself also contains substantive transparency provisions in Sec 2 para 4 and in Nr 11 of its Annex. The telos of these provisions, in general, coincides with the rules discussed above but in part also opens up room for interpretation.

Acting in the Course of Trade

Before potential labelling obligations can be discussed, it must be clarified whether and under what conditions the UWG applies to influencer marketing. The UWG requires to "act in the course of trade".[35] According to the jurisdiction of the OGH, this applies to any independent, acquisition-oriented activity that goes beyond purely private or official activities.[36] It is obvious that entrepreneurs who use influencer marketing generally fall within the scope of the UWG. However, it must be noted that also those who intervene in somebody else's commercial activity and try to promote or hinder the competitive process are addressees of the UWG.[37] Thus, also influencers themselves might fall within the scope of application and must therefore comply with the respective provisions of the UWG.

However, it is anything but clear when influencers are bound by the UWG and under which circumstances their behaviour is covered by the *constitutionally guaranteed freedom of expression*.[38] In this regard, also the German courts do not follow a uniform approach but rather diverge considerably in certain aspects. The German jurisdiction identifies at least three categories of contributions. *Firstly*, the influencer receives a monetary advantage from the applicant for the contribution. In this case, it is undoubtedly assumed that the commercial activity of the applicant is encouraged. This is not a new phenomenon but merely a further manifestation. The commercial activity is equally given to the advertising influencer and to the advertised company, as would be the case with "traditional" advertising methods. However, *secondly*, the matter is more difficult if the influencer purchases the advertised product himself and receives no consideration for the advertisement. In this case, the allocation depends to a great extent on the content of the contribution. If the content is limited to pure commercial purposes, which typically leads to euphoric calls to buy, it can be assumed that the influencer is acting in the course of trade in the form of promoting the commercial activity of somebody else. However, *thirdly*, the situation is different if the content contains information related to the advertised product. Particular problems have arisen in this respect with the use of hyperlinks. By setting a link, the viewer can be forwarded to the social media profile of the advertised company with a single click.

This can be illustrated with the following example: *Influencer A* runs a fashion blog on Instagram, where she regularly reports on her daily routine and the latest trends. She is invited by *Company B* to an event in New York to report on their new

[35] Handeln im geschäftlichen Verkehr.

[36] Inter alia OGH 4 Ob 2007/96t, *Clinclowns*, ÖBL 1996, 191.

[37] H.G. Koppensteiner, Wettbewerbsrecht, 3rd ed, Orac 1997, Sec 23 para 7.

[38] In detail A. Kubat, Kennzeichnungspflicht im Influencer-Marketing ÖBl 2020(1), pp. 8–16 (12f).

collection. In return for reporting, *Company B* finances her journey and stay in New York. On her flight home, *Influencer A* publishes several photos on her blog. First, a posting from the event she visited appears, with *Company B's* application. This posting must be clearly marked as *Influencer A* receiving an asset for this contribution from *Company B*. Furthermore, she publishes a photo of a headphone she bought herself, which cannot be attributed to *Company B*. However, in the text accompanying the photo, she links to *Company B* and reports once again on the pleasant journey. According to a convincing view, this must be seen as promoting the commercial activity of *Company B* since the link is not related to the photo and does not cover any informational needs of the viewer.

However, the fact that the influencer has previously received payment for another entry only plays a secondary role. If one follows the view of the German *KG*[39] in this respect, promotion of commercial activity can be assumed if the link does not represent a connection to the contribution or the need for information. Consequently, the third photo published by *Influencer A* on her journey, namely a self-portrait with a link to the brands belonging to the company of the clothing worn, does not constitute an act in the course of trade. The fact that viewers of a fashion blog as well as readers of a fashion magazine expect not only photographs of new items of clothing but also information about them is almost inherent to the idea of a fashion blog and cannot in itself constitute a promotion of commercial activity. On the contrary, this is covered by freedom of expression since the link to the companies here undoubtedly covers the informational needs of the viewers.

This is convincing, especially since the Austrian VfGH also differentiates in its jurisdiction between commercial and editorial contributions. Although the first of these are also protected under the principle of freedom of expression, the scope for legislative action goes further, particularly with regard to the proportionality test.[40] The draft regulation of the German Ministry of Justice,[41] which—in consultation with the European Commission (EC)—is intended to clarify the UWG accordingly, also moves in this direction. In particular, the draft regulation subjects only those contributions to the UWG in which the company-related aspects of the contribution clearly outweigh the need for information. If there is therefore a surplus of advertising in individual cases, then such a contribution should also be marked for transparency reasons. In the meantime, the Austrian federal government does not intend to

[39] KG 5 U 83/18 MMR 2019, 175.

[40] E.g. VfGH B 658/85 MR 1986, 32 (B. Binder); see also W. Berka, Die Kommunikationsfreiheit in Österreich, EuGRZ 1982(19), pp. 413–427.

[41] Regelungsvorschlag des Bundesministeriums der Justiz und für Verbraucherschutz zur Abgrenzung nichtkommerzieller Kommunikation zur Information und Meinungsbildung von geschäftlichen Handlungen 2020 available at https://www.bmjv.de/SharedDocs/Gesetzgebungsverfahren/Dokumente/Regelungsvorschlag_Influenzer.pdf?__blob=publicationFile&v=1. Accessed 19 November 2020.

make any modifications but rather considers the existing legal framework to be sufficient.[42]

Transparency Provisions Within the UWG

If the above-mentioned conditions are fulfilled, the UWG is applicable, including several obligations on transparency. First of all, *Nr 11* Annex UWG renders business practices unfair if advertising is disguised as information by using editorial content to promote a product, where the company has paid for the promotion and this fact is not clearly evident for the consumer (e.g. from the content, images or sounds used). In view of the per se nature of the prohibition, these conditions must be interpreted narrowly. In general, Nr 11 Annex UWG requires both editorial content and payment. Editorial content can, in particular, be found on YouTube channels as these are typically characterised by recurring and thematically related contributions. As regards payment, the term "fee" again covers every asset-related benefit, including product mailings and invitations to events. Hence, the favouritism reporting discussed above cannot be subsumed under Nr 11 Annex UWG.

The situation is different for *misleading conduct* in the meaning of Sec 2 para 4 UWG. This provision is a lex generalis and also includes business practices that are not unfair under Nr 11 Annex UWG, e.g., because they lack editorial content or proof of payment. Under Sec 2 para 4 UWG, such business practices are unfair if they hide their commercial purpose. However, it is not defined by Sec 2 para 4 UWG in more detail when this is the case or how the commercial purpose is to be understood. It seems appropriate to apply an objective standard for this purpose as the legislator could not have intended to leave the determination of what is a commercial character to the hands of those against which the consumer should actually be protected under the UWG. Therefore, also favouritism reporting may be subject to a labelling obligation under Sec 2 para 4 UWG. Although mainly dealing with the Media Act, the OGH[43] ruled against such an interpretation. Following the approach of the BGH,[44] a review seems desirable in order to ensure the transparency necessary to protect the average informed consumer.

Finally, the category of infringement of law in the general clause of Sec 1 also provides a gateway for the special labelling provisions of the law.[45] A market participant should not be able to achieve competitive advantages by infringing the law. Consequently, such breaches of law are equally unfair business practices to the extent that they could potentially cause a shift in demand. In this context, violations of labelling regulations serve as prime examples since the targeted consumer group is more likely to deal with content that has not been embellished through the eyes of the advertising industry. The advantages of the UWG for law-abiding competitors

[42] See response 1358/AB (XXVII. GP) of the Federal Ministry of Digitisation and Economic Affairs to enquiry 1390/J (XXVII. GP).

[43] OGH 4 Ob 60/16a, *Gefälligkeitsberichterstattung*, ÖBl 2017/25 (L. Wiltschek).

[44] BGH I ZR 205/11, *Preisrätselgewinnauslobung* V, GRUR 2013, 644.

[45] For the special labelling provisions see Sects. 12.2.2.1, 12.2.2.2 and 12.2.2.3.

compared to the special legal standards should not be underestimated. While in the case of the latter it is up to the competent administrative authorities to initiate proceedings and to impose administrative penalties, the UWG grants a subjective right to affected competitors to claim injunctive relief and, if necessary, damages.

12.2.2.5 Exemptions from the Labelling Requirement on the Basis of Perceptibility

Irrespective of the relevant standard, the provisions dispense with the need for labelling if the *advertising character* is *apparent at first glance* to the average attentive and knowledgeable observer.[46] This may be the case if the contribution stands out by its presentation, for example by using colours different to those used in the editorial part or by its placement in a separate, exclusively commercial channel.[47]

Recently, the OGH has given the average viewer a particularly critical appraisal and considers an exuberant application within an editorial contribution to be recognisable, as the average viewer would not assume that editorial contributions are always objective.[48] Of course, there is still no supreme court decision in Austria regarding influencer marketing, but as far as the recognisability of commercial contributions is concerned, first conclusions can be drawn from recent decisions in Germany. The *LG München I*[49] has recently assessed a free (!) contribution by a well-known fashion blogger as an advertisement subject to mandatory labelling, only to row back to the extent that this contribution is clearly recognisable as an advertisement for the average viewer anyway due to the enormous reach of the influencer.

Nevertheless, it must be stated that a connection between the reach of an influencer and labelling obligations is not very helpful. In the case mentioned, the defendant influencer had 485,000 followers. Based on this wide reach, the LG München I concluded that the commercial purpose of the entire channel would be clear to *any viewer, however uniformed.* A person with such an impact, whose channel is public and not private, would undoubtedly always pursue commercial interests. However, this vehemently fails to recognise the supporting pillars of social interaction and ultimately leads to the idea of transparency ad absurdum. Rather, it is inherent to influencer marketing that private and public life merge more and more in order to generate appropriate attention. As a result, the protection against

[46] Inter alia OGH 4 Ob 62/09k, *Promotion II*, MR 2009, 258 (G. Korn); OGH 4 Ob 113/08h, *Medium T*, MR 2008, 261 (P. Burgstaller).

[47] See e.g. A. Koukal, In: Berka, Heindl, Höhne and Koukal (eds), Mediengesetz, 4th ed, LexisNexis 2019, Sec 26 para 3; A. Kubat, Kennzeichnungspflicht im Influencer-Marketing, ÖBl 2020(1), pp. 8–16 (11).

[48] OGH 4 Ob 60/16a, *Gefälligkeitsberichterstattung*, ÖBl 2017/25 (L. Wiltschek).

[49] LG München I 4 HK O 14312/18 GRUR-RR 2019, 332 (P. Zurth and V. Pless).

surreptitious advertising cannot be less for such influencers with a large reach than for those with a small reach.[50] It remains to be seen whether this view will prevail.[51]

12.2.2.6 Form of Labelling

If one ultimately comes to the conclusion that a contribution is subject to compulsory labelling, the latter must be done in a certain form. According to the standards laid down in legislation, the labelling must be more or less identical to "clear and unambiguous". What is meant by this has always been the subject of discussion and is now also the object of dispute in various court decisions in the field of influencer marketing.

If one initially sticks to the wording of the law, the terms *Anzeige*, *entgeltliche Einschaltung* and *Werbung* can be found in Sec 26 MedienG as suitable forms of identification. The expressions chosen by the legislator itself must therefore undoubtedly be suitable to ensure adequate transparency in influencer marketing. However, also German-speaking influencers are increasingly using English expressions or abbreviations, and in some cases, there is a general attempt to push the name into the background. The problem that the latter can quickly lead to conflicts can be seen from the very fact that the law requires clear and unambiguous disclosure. If, to give just a few examples, the marking is therefore written in a minimum font size, marked in colours that coincide with the background in such a way that the font becomes almost illegible or the text is even hidden behind other characters,[52] this is generally not sufficient marking. In one case concerning print media, the OGH[53] considered a marking to be inadequate because it was placed next to the advertisement in such a small font size that it was not easily recognisable even on closer inspection. Therefore, as a general rule, the basic considerations that apply to the recognisability of advertising, which from the outset excludes an obligation to label, also apply to the type of labelling and must therefore be based on the specific presentation and design of the individual case.

There are various opinions on *English language labelling*, although a tendency seems to be emerging in jurisdiction. This applies in particular to Germany, where

[50] See A. Kubat, Kennzeichnungspflicht im Influencer-Marketing, ÖBl 2020(1), pp. 8–16 (16); P. Zurth and V. Pless, Schleichwerbung in Posts von Influencern durch Verlinkung – Cathy Hummels, GRUR-RR 2019(7), pp. 332–336.

[51] The OLG München appealed to in the second instance confirmed the decision of the LG München I at least in this respect (although without necessity, since it had previously already denied the applicability of the UWG for lack of commercial activity). The case is now pending at the BGH. At the time of submitting this report, no ruling of the BGH has yet been proclaimed.

[52] Popular in this context is the *hiding* behind another text or the profile picture of the influencer. The Instagram platform allows, for example, the posting of so-called *stories*, which, when called up by the viewer, show the name and profile picture of the influencer in the upper left corner. If the influencer places the label on the photo or video published in the story in this corner, it is almost impossible for the viewer to see it at first glance. Only when the viewer actively pauses the story would the name and the picture of the influencer disappear in the upper left corner and reveal what is hidden behind it.

[53] OGH 4 Ob 172/90 MR 1991, 75.

the legal situation with regard to labelling requirements is essentially comparable to Austria, and it is already possible to rely on relevant judgements regarding influencer marketing. For example, the terms *#ad* and *#sponsoredby* have now been judged inadmissible. While the Higher Regional Court of Celle (OLG Celle)[54] with regard to "#ad" referred to the placement of the hashtag in the middle of a "hashtag cloud"[55] and thus denied a clear and unambiguous recognisability, it left open whether the term in itself would suffice as identification. With regard to the abbreviation and the associated mental intermediate step for the viewer, namely the correct assignment of the abbreviation, this will tend to be negated at present. However, it must be admitted that the consistently young circle of consumers seems to become more and more accustomed to this abbreviation and that this therefore becomes increasingly valid. A change from the previous judicature thus seems to be at least within the realm of possibility, provided that the trend towards influencer marketing remains constant.

The same applies to the terms *#sponsoredby* and *#promotion*. The OLG Celle,[56] and also the OGH,[57] explained that an English language term would, in general, not reveal the commercial purpose of the contribution to a German-speaking audience. This seems to be doubtful, especially since the average addressee of influencer marketing is familiar with the digital environment, where users have to deal with English language expressions on a daily basis. If this circle of addressees is thus acknowledged to have at least a rudimentary knowledge of English, it will have to be permissible to label them by the terms mentioned, especially since the OGH was dealing with a much older circle of addressees in its decision cited above.[58] It should also be pointed out that the recurrent marking of contributions by "unpaid advertising" *de lege lata* has no legal basis. Insofar as the reference to "unpaid" is intended to indicate that no payment was made for the insertion of the contribution but merely that a product was provided, this is in any case contrary to the legal situation. The aim of the legal situation is not the payment of a fee but rather the inclusion of all valuable assets in the obligation to label. If, however, no consideration of any kind has actually been received and the contribution contains at least a minimum of information that goes beyond commercial content, there is no need to label it as *unpaid advertising* because there is no contribution to be labelled at all.

[54] OLG Celle 13 U 53/17 WRP 2017, 1236.

[55] What is meant by this is the sequence of a large number of so-called hashtags, which are represented by the #-symbol and are intended to facilitate the search within the network.

[56] OLG Celle 13 U 53/17 WRP 2017, 1236.

[57] OGH 4 Ob 62/09k, *Promotion II*, MR 2009, 258 (G. Korn).

[58] See A. Anderl and A. Seling, Social Media – Rechtssicherheit im Unternehmen, ecolex 2018(6), pp. 535–538.

12.2.3 Influencer Marketing and Product Liability

In times of allegations of *fake news*, the question of liability for false information and advice is becoming increasingly important. Especially when products are advertised, it is important not to overstretch the boundaries of what is permissible. However, if false product characteristics are promised, this is not only a problem of unfair competition law (so-called misleading business practices) but may also give rise to liability for damages incurred. In order to deal with this problem, the legislator has enacted specific liability rules, which can be found above all in the Austrian *Product Liability Act* (PHG).

The PHG standardises a *strict liability of the manufacturer* for the products manufactured by him *without setting any requirements regarding fault*. The purpose of this liability is not to have to refer injured parties exclusively to delictual claims as, in this case, the fault of the manufacturer would generally be ruled out and an attribution would only be possible under the strict conditions of the liability of assistants.[59] The claimants are not only the direct purchaser of the product but also third parties (*innocent bystanders*) who suffer personal injury or damage to property as a result of the product. However, a particularly relevant question arises in the area of influencer marketing, namely how the product term of the PHG is to be understood.

In this respect, the text of the law itself refers in Sec 4 to the *product as a movable physical object*, including energy and such objects that are connected with other movable or immovable objects.[60] Strictly according to the wording and to stay with the textbook example that is often used, it is therefore at first glance obvious (and probably inconsistent) that the architect's construction plan is not subject to strict liability under the PHG but the executed construction project based on this very plan is.[61] In terms of influencer marketing, at first sight, this would mean that influencers are liable not for *wrong advice or indications* but for the dangerousness of self-made products, such as cosmetics. Nevertheless, the discussion in Austria was recently spurred on by a reader's lawsuit against a daily newspaper to which she subscribed. The defendant newspaper printed a health tip to use freshly torn horseradish as a medical salve against rheumatic pain. This paste—and this is the decisive point of the lawsuit—could, according to the author's opinion, be left on the skin for *quite two to five hours*. However, this does not correspond to the facts; rather, the medically recommended duration of applying such a paste on the skin is well below 2–5 h, namely 2–5 min (!).

---

[59] E.g. C. Rabl, PHG, LexisNexis 2017, Vorbem paras 11f; the standard of Sec 1315 ABGB, which is relevant to the area of delictual claims, attributes fault to assistants if the manufacturer uses "*an unfit or knowingly dangerous person*". This is, of course, verifiable in the very least cases.

[60] In detail W. Posch and U. Terlitza, In: Schwimann and Kodek (eds), ABGB, 4th ed, Manz 2015, Sec 4; C. Rabl, PHG, LexisNexis 2017, Sec 4.

[61] R. Welser and B. Zöchling-Jud, Grundriss des bürgerlichen Rechts II, 14th ed, Manz 2015, para 1628.

Once the reader noticed a painful, toxic reaction on the ankle after about 3 h, she removed the paste and eventually filed a lawsuit against the daily newspaper under the PHG. She explained that the author's advice took shape in the form of a printed copy of the newspaper and thus represented a movable, physical object. While the first and second courts rejected the action, the OGH now referred the question to the CJEU for a *preliminary ruling*. In doing so, the OGH[62] pointed out that a newspaper is bought by the reader not as a stack of paper but obviously because of the printed content.[63] Of course, the textbook example of the faulty plan mentioned above could not be omitted in the reasoning in a modified form. The OGH stated that it is inconsistent if the author of a cookbook who gives a harmful dose of an ingredient is not subject to product liability, whereas the manufacturer of a finished product who mistakenly adds the same dose to his product is. However, the protective purpose of the PHG, which explicitly refers to the danger of the object itself and not to the advice or information, contradicts the view of the OGH.[64]

Undoubtedly, it is also questionable how far this desired consequence should go and where it can find a limit. After all, would it not also be inconsistent if product liability depended on the *format*? A factual differentiation between the faulty recipe printed in book form and the one published online on a blog seems to be hardly possible. For influencer marketing, which by definition takes place in a digital context, this very question is decisive. If the CJEU extends the scope of the product concept to include intellectual performances, then the influencer who gives faulty advice in his blog is ultimately also subject to strict liability. However, if one sticks to the wording of the law as the starting point for interpretation, Sec 4 PHG explicitly refers to physicality and mobility. A subsumption of incorporeal, intellectual achievements, which never take shape in physical form, seems constructed under the current law, although the result is admittedly inconsistent.[65] Taking these arguments into account, the CJEU[66] ultimately also answered the question referred by the OGH by stating that a service that is incorporated into a physical, movable object does not belong to the inherent characteristics of this object. Accordingly, the incorrect health tip is not a product within the meaning of the PHG, even if it was printed in a daily newspaper. This must therefore apply even more to the subject of influencer marketing, according to which a service is not subject to the strict product liability standards even if it is offered in a digital context.

[62]OGH 1 Ob 163/19f AnwBl 2020, 278 = ecolex 2020/210.

[63]The OGH had already had the opportunity to address this issue in 2007 but did not do so, citing the lack of compensatory damages. The matter in question was a legal calendar that contained erroneous time sheets and on the basis of which a lawyer suffered damage; OGH 6 Ob 256/06z MR 2007, 144 (G. Streit) = ecolex 2007/176 = Zak 2007/275 (D. Beig).

[64]See e.g. W. Posch and U. Terlitza, In: Schwimann and Kodek (eds), ABGB, 4th ed, Manz 2015, Sec 1 PHG para 1.

[65]C. Rabl, PHG, LexisNexis 2017, Sec 4 para 52.

[66]CJEU, case C-65/20, *VI v KRONE-Verlag Gesellschaft mbH & Co KG*, ECLI:EU:C:2021:471.

12.2.4 The Issue with "Astroturfing"

Global networking and the corresponding impact of contributions on the Internet give consumers the opportunity to review products or companies and to exchange experiences with like-minded users. Various technical solutions are available for this purpose, but the possibility of awarding stars on a scale of 1–5 for a product and reviewing it is particularly popular in this context. These reviews are usually publicly available and sometimes offer customers an important reason for their purchase decision. After all, the evaluation of a private person seems to be much more reliable than the promotion of the product by the manufacturer. Especially when there are multiple reviews from different people and most of them highlight the product positively, this can be the decisive factor for the purchase after comparing several similar products. The same applies in the other direction, of course. A product that has been given a negative evaluation by many people will also at least cause scepticism among the viewers and ultimately discourage the purchase.

However, there is no legal provision for any kind of verification of the reviewer or a requirement to use his real name. For example, one of the largest trading platforms also offers the possibility of publishing reviews under a pseudonym. Obviously, it is therefore also an opportunity for the manufacturer himself to give his products or company a positive evaluation on this platform or to invite employees to do so. Yet this leads the idea of objective reviews ad absurdum as the viewer is deceived about the commercial purpose of the review and the author's business character. This increasingly prevalent problem,[67] which is referred to as *astroturfing*, has also attracted the attention of legal scholars and has raised the question of how such practices should be assessed under unfair competition law.[68]

Obviously, this issue is once again a manifestation of the problem discussed above with regard to the labelling requirements in influencer marketing. If the commercial nature of a contribution is not adequately clarified to the reader, the consumer risks buying on a false basis and is therefore misled. Hence, the legal framework of astroturfing is broadly identical to that governing labelling obligations, and reference may therefore be made to our comments on the latter.[69] In particular, the UWG is generally applicable to fake ratings since the author of such ratings typically either promotes his own product (and thus also promotes his own commercial activity) or rates competing products correspondingly poorly (and therefore

[67]Der Standard, Amazon: Riesiger Betrug mit Fake-Bewertungen aufgeflogen (2020), available at https://www.derstandard.at/story/2000119826429/amazon-riesiger-betrug-mit-fake-bewertungen-aufgeflogen. Accessed 18 November 2020.

[68]In detail S. Holzweber, Kundenbewertungen im Internet, ÖBl 2021(3), pp. 100–105; H. Krieg and J.D. Roggenkamp, Astroturfing – rechtliche Probleme bei gefälschten Kundenbewertungen im Internet, K&R 2010(11), pp. 689–694; also A. Wilkat, Bewertungsportale im Internet, Nomos 2013; A. Anderl and A. Seling, Social Media – Rechtssicherheit im Unternehmen, ecolex 2018(6), pp. 535–538; A. Kulka, Das Fingieren von (größerer) Beliebtheit eines Unternehmens im Wettbewerbsrecht, ecolex 2012(2), pp. 148–151.

[69]See Sect. 12.2.2.

engages in conduct that is objectively capable of hindering somebody else's commercial activity). Advertising the products of a third-party company, with which, for example, a cooperation or even a group of companies is established, also constitutes a part of trade since it promotes competition among other companies.

However, the picture is more differentiated in terms of *applicable standards*. If one initially examines the per se bans of the black list, one will find what appears to be the case with *Nr 11* (disguising commercial content as editorial contributions), although this will generally not be applied. Neither does the occasional evaluation of products constitute editorial content within the meaning of Nr 11, nor does the author necessarily receive any consideration for this contribution, especially since own products are often advertised.[70] There is also a disagreement about the application of *Nr 22*,[71] according to which a false claim or the creation of an incorrect impression that the trader is not acting for purposes related to his trade, business, craft or profession or a false appearance as a consumer is in any case unfair. At least in Germany (where the relevant provision can be found in Nr 23 of the Annex), this is sometimes denied since fake evaluations do not constitute a direct offer to buy and therefore do not serve the purpose of protection.[72] The situation is different in Austria, where the presumably prevailing doctrine assumes that at least the false appearance as a consumer is realised in the submission of fake evaluations.[73] In our view, this has to be endorsed since the manipulation of the basis of information prior to the conclusion of the contract affects the consumer's business decision in the same way as it does directly when the contract is concluded.

Nonetheless, there is agreement that such assessments are to be subsumed under the provision of *Sec 2 para 4 of the UWG* covering misleading conduct.[74] Admittedly, this requires that the rating is suitable to affect the protective interests of

---

[70] The situation is different, however, in the case of those contributions which are similar to editorial contributions in specialist journals due to their content conception and for which the author is paid by the applicant. See H. Krieg and J.D. Roggenkamp, Astroturfing – rechtliche Probleme bei gefälschten Kundenbewertungen im Internet, K&R 2010(11), pp. 689–694 (690); also A. Anderl and A. Seling, Social Media – Rechtssicherheit im Unternehmen, ecolex 2018(6), pp. 535–538 (537).

[71] In detail A. Anderl and A. Appl, In: Wiebe and Kodek (eds), UWG, 2nd ed, Manz 2016, Annex to Sec 2 paras 217f.

[72] Inter alia O. Sosnitza in Ohly and Sosnitza (eds), Gesetz gegen den unlauteren Wettbewerb, 7th ed, C. H. Beck 2016, Annex to Sec 3 para 3 at para 64; also H. Krieg and J.D. Roggenkamp, Astroturfing – rechtliche Probleme bei gefälschten Kundenbewertungen im Internet, K&R 2010(11), pp. 689–694 (691); for an applicability however H. Köhler, In: Köhler, Bornkamm and Feddersen (eds), UWG, 38th ed, C. H. Beck 2020, Annex to Sec 3 para 3 at para 23.1.

[73] A. Anderl and A. Appl, In: Wiebe and Kodek (eds), UWG, 2nd ed, Manz 2016, Annex to Sec 2 para 217; also A. Anderl and A. Seling, Social Media – Rechtssicherheit im Unternehmen, ecolex 2018(6), pp. 535–538 (537); contrary A. Kulka, Das Fingieren von (größerer) Beliebtheit eines Unternehmens im Wettbewerbsrecht, ecolex 2012(2), pp. 148–151 (150).

[74] A. Anderl and A. Appl, In: Wiebe and Kodek (eds), UWG, 2nd ed, Manz 2016, Sec 2 para 517; H. Köhler, In: Köhler, Bornkamm and Feddersen (eds), UWG, 38th ed, C. H. Beck 2020, Sec 5a para 7.80; in detail H. Krieg and J.D. Roggenkamp, Astroturfing – rechtliche Probleme bei gefälschten Kundenbewertungen im Internet, K&R 2010(11), pp. 689–694 (691).

market participants to a noticeable extent.[75] While suitability will generally not pose a problem, as customers will at least make decisions in reliance on a secure information basis, which does not exist due to falsified rating, this must be considered in a differentiated manner with regard to *noticeability*. If the individually falsified rating falls significantly behind the bulk of the real ratings, it is difficult to argue that precisely this one falsified rating was the decisive factor for the purchase decision and was therefore noticeable in the sense of the UWG.[76]

Ultimately, however, these can only be exceptional examples and not those cases that proceed systematically and order a large number of fake ratings via agencies, for example.[77] This also applies to a variation of astroturfing, which does not directly deal with fake ratings but falsifies the information provided in such a way that negative ratings are hidden in order to prevent the viewer from seeing them.[78] For example, if a company advertises its products with a broad rating system and publishes only those—genuine—positive ratings immediately while negative ratings are first subjected to an internal review, then the viewer is misled by the unfiltered rating basis.[79] Ultimately, it is not only the legal problems that have far-reaching consequences but also the public perception that typically causes lasting damage to the company's reputation.

12.2.5 Conclusion

Native advertising is becoming more and more important in the marketing world, especially when using influencers. It is thus hardly surprising that the associated legal issues are increasingly coming into focus, In Austria, they are currently focussing primarily on the area of labelling obligations. While the declaration of the commercial purpose of paid advertising is a well-known problem, other issues remain, especially where products are purchased and recommended free of charge. Time will tell whether the first judgements in Austria, which are probably unavoidable, will follow those in Germany and consider this to be a commercial practice. However, the fundamental question of the limits to freedom of expression arises in this context, so particular caution is called for. A distinction should therefore be made in the approach adopted here.[80] Editorial contributions in which the focus lies on the informational aspect fall within the private sphere of influencers, whereas

[75] A. Anderl and A. Appl, In: Wiebe and Kodek (eds), UWG, 2nd ed, Manz 2016, Sec 2 paras 45f.

[76] H. Krieg and J.D. Roggenkamp, Astroturfing – rechtliche Probleme bei gefälschten Kundenbewertungen im Internet, K&R 2010(11), pp. 689 – 694 (691).

[77] Such an approach must, of course, also take into account the criminal law component, since a major fraudulent activity must, at least in all probability, be considered to be likely to cause damage to the consumer.

[78] A. Anderl and A. Seling, Social Media – Rechtssicherheit im Unternehmen, ecolex 2018(6), pp. 535–538 (536f).

[79] BGH I ZR 252/14, *Kundenbewertungen im Internet*, GRUR 2016, 828.

[80] See Sect. 12.2.2.4.1.

purely company-related contributions, which in particular show a clear advertising overlap, must be labelled according to Sec 2 para 4 UWG. One issue that the jurisdiction has clarified, at least temporarily, concerns product liability. According to the CJEU, the strict product liability regulations do not apply to services such as advice, even if they are printed in newspapers.[81] For influencer marketing, this initially brings relief, especially since it already takes place in the digital sphere and would therefore have been difficult to align with the product concept of the PHG, which is based on physicality and mobility. Of course, the last word has not been spoken yet as there is still recourse to general tort law. Therefore, it remains to be seen how the jurisdiction will resolve these two controversial issues in the interests of legal certainty.

12.3 Algorithmic Collusion and Personalised Pricing Under Competition Law

12.3.1 Introduction

The advent of the digital economy allowed transactions to be carried out to an increasing extent on the Internet, where pricing is more and more based on algorithms. In case undertakings delegate their pricing decisions to algorithms, the question arises whether such practices pose a risk to the competitive process and whether the current legal framework can effectively deal with this phenomenon. There is no doubt that the use of algorithms generates efficiencies, both on the supply and on the demand sides.[82] Algorithms allow suppliers to better respond to consumer demand, better allocate their resources and save on human capital.[83] Algorithms reduce transaction costs by providing consumers with a wealth of information[84] that in turn, e.g., allows them to efficiently compare offers and enables them to find products that fit them better.[85] Increased market transparency achieved through reduced information asymmetries may intensify competition to the benefit of consumers. However, the adoption of algorithms can have far-reaching impacts on the forms of and conditions for competitive business conduct and might lead to

[81] CJEU, case C-65/20, *VI v KRONE-Verlag Gesellschaft mbH & Co KG*, ECLI:EU:C:2021:471.

[82] See e.g. CMA, Pricing Algorithms 2018, paras 4.1.ff.

[83] M. Gal, Algorithms as Illegal Agreement, Berkeley Technology Law Journal 2019(67:34), pp. 67–118 (70); OECD, Algorithms and Collusion – Competition Policy in the Digital Age 2017, pp. 14f.

[84] However, it should not be overlooked that users may be confronted with "too much" information and that the costs related to the sorting and selection of this new variety of information may constitute a form of transaction costs itself; see in particular B. Schwartz, The Paradox of Choice - Why More Is Less, 1st ed, Harper Perennial 2004.

[85] M. Gal and N. Elkin-Koren, Algorithmic Consumers, Harvard Journal of Law & Technology 2017(30:2), pp. 309–353 (318).

undesirable, anti-competitive results.[86] The consequences of such behaviour may be ultimately borne by the consumers in the form of excessive prices or a loss in total consumer welfare. Data-driven markets—notably those in the digital economy—with few competitors, homogenous products and high entry barriers are particularly vulnerable. First, algorithms may facilitate and accelerate the coordination of prices between undertakings, thereby potentially violating the ban on cartels (Art 101 TFEU; Sec 1 Cartel Act) (see Sect. 12.3.2). Second, algorithms may enable undertakings to offer different prices to different consumers according to the information they hold about them. Such personalised pricing may qualify as price discrimination contravening the ban on abusive practices (Art 102 TFEU; Sec 5 Cartel Act) (see Sect. 12.3.3).

12.3.2 Algorithmic Collusion

Big data[87] and big data analytics[88] enable undertakings to implement software-based automated price adjustments. While algorithms can be an efficient means to achieve price optimisation, commentators also consider anti-competitive effects in connection with collusion made possible or facilitated by the implementation of algorithms: price agreements can be enforced more easily, their compliance can be monitored better and deviations can be detected and punished more quickly. In general, due to the characteristics of the digital world, there is less need for *ex ante* communication.[89] Rather, algorithms can coordinate actions and may enable new forms of coordination that were not observed nor even possible before. This is referred to as "algorithmic collusion".[90]

In general, two different scenarios in which algorithms may promote collusion can be distinguished. First, competitors use algorithms as a simple tool to implement an existing price agreement, which relates to the alignment of prices as well as to the

[86] P. Picht and G. Loderer, Framing Algorithms – Competition law and (Other) Regulatory Tools, MPI Research Paper 2018(24), pp. 1–35 (2).

[87] Big Data is characterised by the famous four V's made possible by technological progress: the *volume* of data processed, the *variety* of data aggregated, the *velocity* at which data is collected and used and the *value* of the information found in the data, see A. De Mauro/M. Greco/M. Grimaldi, A formal definition of Big data based on its essential features, Library Review 2016(65:3), pp. 122–135; the former "3 V"-definition was originally introduced by D. Laney, 3D Data Management: Controlling Data Volume, Velocity, and Variety, available at https://blogs.gartner.com/doug-laney/files/2012/01/ad949-3D-Data-Management-Controlling-Data-Volume-Velocity-and-Variety.pdf. Accessed 7 September 2021.

[88] Big Data Analytics is the process of collecting, organising and analysing Big Data to discover patterns and other useful information.

[89] M. Gal, Algorithms & Competition Law: Interview of Michal Gal by Thibault Schrepel, e-Competitions Algorithms, Concurrences 2020, Art. N° 93929, pp. 1–12.

[90] OECD, Algorithms and Collusion - Competition Policy in the Digital Age 2017, p. 19.

use of algorithms as a means of coordination.[91] Such collusion constitutes a conventional "agreement" within the meaning of Art 101 TFEU (Sec 1 Cartel Act). Second, competitors autonomously decide to use (the same or different) pricing algorithms, leading to price alignments. In this scenario, the cartel price is merely based on the software independently used by the competitors, i.e. without explicit communication between the undertakings. In such cases, it may be hard to differentiate between legitimate mere parallel behaviour and anticompetitive conduct contravening the ban on cartels. The existence of collusive behaviour is particularly difficult to prove where (self-learning) algorithms interact with each other.

12.3.2.1 Ban on Cartels

Article 101 para 1 TFEU (Sec 1 para 1 Cartel Act) prohibits agreements and concerted practices between undertakings that have as their object or effect the prevention, restriction or distortion of competition. The ban on cartels thus requires some element of "collusion" between independent undertakings. The different types of collusion are distinguishable from each other only by their intensity.[92] An agreement requires "the existence of a concurrence of wills"[93] of the participating undertakings "to conduct themselves on the market in a specific way".[94] The concept of concerted practices is designed to provide a safety net, catching looser forms of collusion. The latter is defined as "a mental consensus whereby practical cooperation is knowingly substituted for competition".[95] This applies in particular to "any direct or indirect contact between such operators, the object or effect whereof is either to influence the conduct on the market of an actual or potential competitor or to disclose to such a competitor the course of conduct which they themselves have decided to adopt or contemplate adopting on the market".[96] As opposed to an agreement, the concept of a concerted practice requires both concertation between undertakings and subsequent conduct on the market, and a relationship of cause and effect between the two.[97] However, there is a presumption that concertation has been followed by conduct if the undertakings involved in the concerted action remained active on the market.[98]

---

[91] Zimmer, In: Immenga and Mestmäcker (eds), Wettbewerbsrecht, 6th ed, C. H. Beck 2019, Art. 101 para 1 TFEU para 77.

[92] A. Jones and B. Sufrin, EU Competition Law, 5th ed, Oxford University Press 2014, p. 149.

[93] EGC, case T-41/96, *Bayer*, ECLI:EU:T:2000:242, pt 96.

[94] CJEU, case C-49/92, *Anic Partecipazioni*, ECLI:EU:C:1999:356, pt 130.

[95] R. Wish and D. Bailey, Competition Law, 9th ed, Oxford University Press 2018, p. 116; see also CJEU, case C-8/08, *T-Mobile Netherlands*, ECLI:EU:C:2009:343, pt 26.

[96] CJEU, case C-40/73 ao, *Suiker Unie*, ECLI:EU:C:1975:174, pt 174.

[97] CJEU, case C-199/92, *Hüls*, ECLI:EU:C:1999:358, pt 161.

[98] CJEU, case C-199/92, *Hüls*, ECLI:EU:C:1999:358, pt 162; see also CJEU, case C-8/08, *T-Mobile Netherlands*, ECLI:EU:C:2009:343, pt 52, stating "that the presumption of a causal connection stems from Article 81(1) EC, [...] and [...] consequently forms an integral part of applicable Community law."

Whether an agreement or concerted practice is the result of human communication or algorithmic interaction does not change the legal assessment of the conduct. Most importantly, Art 101 TFEU (Sec 1 Cartel Act) requires each undertaking to determine independently the policy that it intends to adopt.[99] However, this does not deprive the undertakings of their right to adapt themselves intelligently to the existing or anticipated conduct of their competitors.[100] Such mere parallel behaviour or tacit collusion does constitute neither an agreement nor a concerted practice and is permitted under Art 101 TFEU (Sec 1 Cartel Act).[101] It follows that the implementation of pricing algorithms is legitimate as long as they unilaterally observe, analyse and react to publicly observable behaviour of competitors (e.g. to pricing decisions). This is considered as an intelligent adaption to the market rather than an illicit coordination.[102]

Despite the existence of a restriction of competition, an agreement or a concerted practice can be justified if the conduct is associated with specific efficiencies mentioned in Art 101 para 3 TFEU (Sec 2 para 1 Cartel Act). This individual exemption also applies to algorithmic collusion. In this context, it might be useful to analyse the algorithm and its role in the collusive scheme. Potential (counteracting) efficiencies associated with the algorithm need to be balanced with its allegedly negative effects, which might reinforce the ascertained anti-competitive practice.[103]

12.3.2.2 Application to Different Scenarios

Algorithms as Tools to Implement Pre-existing Agreements

Messenger Scenario

The first category considers the so-called messenger scenario. In this case, humans agree to collude and decide to use computers as simple tools to implement, monitor and police existing collusion.[104] As a result, this kind of collusion constitutes an anti-competitive agreement caught by Art 101 TFEU (Sec 1 Cartel Act).

These considerations were illustrated in the Competition and Markets Authority's (CMA's) poster decision.[105] The undertakings concerned agreed not to undercut each other on prices for certain posters and frames sold only by the two of them on the Amazon UK marketplace. This agreement was implemented by using (different

[99] CJEU, case C-40/73 ao, *Suiker Unie*, ECLI:EU:C:1975:174, pt 173.

[100] CJEU, case C-40/73 ao, *Suiker Unie*, ECLI:EU:C:1975:174, pt 174; CJEU, case C-8/08, *T-Mobile Netherlands*, ECLI:EU:C:2009:343, pt 33.

[101] CJEU, case 89/85, *Woodpulp*, ECLI:EU:C:1993:120.

[102] BKartA and Autorité de la Concurrence, Working Paper - Algorithms and Competition 2019, p. 56.

[103] BKartA and Autorité de la Concurrence, Working Paper - Algorithms and Competition 2019, p. 29.

[104] A. Ezrachi and M. Stucke, Virtual Competition, Harvard University Press 2016, p. 36.

[105] CMA, 12 August 2016, Case 50223, *Trod Ltd*; a similar case was undertaken in the US: US District Court of Northern California, 11 August 2016, CR 15-0419 WHO, *US/Trod Ltd.*

third-party) pricing software. In a similar case undertaken by the US Department of Justice (DoJ),[106] a seller in the Amazon marketplace was charged for coordinating the prices of posters sold online. The parties concerned designed and shared dynamic pricing algorithms, which were programmed to act in conformity with their pre-existing explicit agreement. Both cases constitute traditional horizontal price-fixing cartels, where the agreement was implemented by the use of algorithms.

Hub-and-Spoke Scenario

In this category, competitors delegate certain strategic (pricing) decisions to a third party, which then takes the decisions using an algorithm.[107] This concept is often related to the established "hub and spoke" doctrine. Platforms or computer algorithms of one supplier execute the function of the "hub" to facilitate collusion between competitors, referred to as "spokes". Although the competitors do not interact directly with each other, their strategic (pricing) principles are coordinated or aligned via the hub. In the extreme, the effects of a horizontal hard-core cartel can be achieved.

In its *Eturas*[108] case, the CJEU was considering a possible hub-and-spoke conspiracy facilitated by an online system.[109] Several Lithuanian travel agencies (spokes) used a common travel booking system (hub). The system administrator sent a message to the participating travel agencies proposing a cap on discount rates (max 3%). According to the CJEU, this would constitute a concerted practice if it were proven that the travel agencies were aware of that message.[110] However, the focus of this case laid on the awareness of the message and not on the subsequent implementation of a software rule that limits the possibility of granting higher discounts.[111] Another example of an algorithm-fuelled hub-and-spoke conspiracy concerns the car

---

[106] US District Court of Northern California, 6 April 2015, CR 15-00201 WHO, *US/Topkins*.

[107] BKartA and Autorité de la Concurrence, Working Paper - Algorithms and Competition 2019, p. 32ff further distinguish between situations, in which the respective companies (i) do or (ii) do not know that they use the same or somehow coordinated algorithmic solutions.

[108] CJEU, case C-74/14, *Eturas*, ECLI:EU:C:2016:42.

[109] See, however, AG Spunar, case C-74/14, *Eturas*, pt 65, who refuses to qualify the *Eturas* case as a hub-and-spoke conspiracy since the latter only applies to constellations in which the exchange of information may also be considered a legitimate commercial practice, whereas the limitation of discount rates in the *Eturas* case would be *a priori* inadmissible; such a narrow interpretation of the hub-and-spoke notion is not necessary – rather, it is decisive that the coordination does not take place directly between competitors but via a third party; see A. Heinemann, Algorithmenbasierte Kartelle, p. 19, footnote 73, available at https://www.ius.uzh.ch/dam/jcr:6baf23a8-6b0d-4b11-95a9128bc5c686bd/Heinemann_Algorithmenbasierte_Kartelle.pdf. Accessed 7 September 2021; and J. Safron, The Application of EU Competition Law to the Sharing Economy, Stanford-Vienna European Union Law Working Paper 2018(27), p. 32, https://www-cdn.law.stanford.edu/wp-content/uploads/2018/02/safron_eulawwp27.pdf. Accessed 7 September 2021.

[110] CJEU, case C-74/14, *Eturas*, ECLI:EU:C:2016:42, pt 44.

[111] European Union, Note to the OECD on Algorithmic Collusion, DAF/COMP/WD(2017)12, para 25.

service platform *UBER*.[112] The latter connects drivers and passengers via a smartphone application; the prices are determined by Uber's automated price-setting algorithm. The industry-wide use of a single pricing algorithm (hub) used by competing drivers (spokes) may lead to horizontal coordination, which eliminates price competition between allegedly independent undertakings.[113]

Parallel Use of Individual Algorithms with No Pre-existing Agreement

Predictable Agent Scenario

In the third collusion scenario, undertakings unilaterally design and implement an algorithm, which reacts to external factors in a predictable way.[114] Algorithms may be designed to detect the market behaviour of competitors and rationally follow price leadership, thereby reducing strategic uncertainty and facilitating a tacitly coordinated outcome. As opposed to the messenger and hub-and-spoke scenario, this category does not involve any prior contact between the undertakings concerned or via a hub as a third party. Rather, the alignment might result from a mere interaction of computers based on the parallel use of individual algorithms. Such mere parallel "algorithmic" behaviour, i.e. in the absence of an agreement, qualifies as tacit coordination, which is not caught by the ban on cartels. Rather, such unilateral conduct is covered by the right of any undertaking to adapt its market behaviour intelligently to the existing or anticipated conduct of their competitors.[115]

Digital Eye

The most complex and subtle way in which algorithms may facilitate collusive outcomes is the use of machine-learning and deep-learning technologies (this scenario is also referred to as "digital eye"). These kinds of algorithms are not explicitly designed to tacitly collude, but do so itself through self-learning.[116] However, to date, it is "not clear how machine learning algorithms may actually reach a collusive outcome".[117] In particular, it remains unclear whether algorithms would actually manage to align strategic decisions tacitly or whether the coordination would still

[112] US District Court of Southern NY, 31 March 2016, 15 Civ. 9796 (JSR), *Meyer/Kalanick*.

[113] A. Ezrachi and M. Stucke, Note to the OECD on Algorithmic Collusion, DAF/COMP/WD (2017)25, para 32; A. Ezrachi and M. Stucke, Virtual Competition, Harvard University Press 2016, p. 52; critical on this issue with regard to potential pro-competitive effects of Uber's business model U. Salaschek and M. Serafimova, Preissetzungsalgorithmen im Lichte von Art. 101 AEUV, WuW 2018(1), pp. 8–17 (12f).

[114] CMA, Pricing Algorithms 2018, para 5.22.

[115] CJEU, case C-40/73 ao, *Suiker Unie*, ECLI:EU:C:1975:174, pt 174; CJEU, case C-8/08, *T-Mobile Netherlands*, ECLI:EU:C:2009:343, pt 33; see also German FCA, Fuel Sector Inquiry 2011, p. 21: "Mutual price monitoring without communication does not raise competition concerns", available at https://www.bundeskartellamt.de/SharedDocs/Publikation/DE/Sektoruntersuchungen/Sektoruntersuchung%20Kraftstoffe%20-%20Zusammenfassung.pdf?__blob=publicationFile&v=5. Accessed 7 September 2021.

[116] CMA, Pricing Algorithms 2018, para 5.24.

[117] OECD, Algorithms and Collusion - Competition Policy in the Digital Age 2017, p. 30,

require some kind of communication between the algorithms.[118] As already explained above, any direct or indirect prior exchange of information would constitute an agreement or concerted practice within the meaning of Art 101 TFEU (Sec 1 Cartel Act). In this context, a specific form of such a communication may be signalling practices.

Signalling Scenario

Signalling describes situations in which undertakings publicly announce their intent to change a relevant parameter of competition, such as the price.[119] In contrast to the predictable agent and digital eye scenario, the alignment of prices occurs not "automatically" but rather by means of sending price signals to other undertakings with the prior intent to collude. According to the EC's Horizontal Co-operation Guidelines, "genuinely public" announcements do not amount to a concerted practice within the meaning of Art 101 TFEU.[120] However, in a situation where such an announcement was followed by public announcements by other competitors, the possibility of finding a concerted practice cannot be excluded.[121] Signalling is inevitably associated with certain costs and risks for the signalling firm. If the competitors do not receive or intentionally not react to the signal, the signalling firm might lose sales and profit. To reduce or eliminate these costs and risks, the undertakings concerned might set up algorithms that establish and negotiate the terms of collusion.[122] Algorithms might, e.g., enable the companies to automatically set very fast actions that cannot be exploited by the consumers but are still visible to competitors.[123]

Alternatively, the respective signals might be conveyed in a masked way; i.e., the publicly disclosed data are used as a code to propose and negotiate price increases.[124] This was observed in the US *Airline Tariff Publishing* case,[125] where airline companies used first ticket dates to signal intended future prices using a

[118] BKartA and Autorité de la Concurrence, Working Paper - Algorithms and Competition 2019, p. 43.

[119] BKartA and Autorité de la Concurrence, Working Paper - Algorithms and Competition 2019, p. 53; for an overview of the (scarce) EU case law on signaling practices see B. Meyring, Signalling – When Should We Worry? JECLAP 2020(7), pp. 335–345.

[120] EC, Guidelines on the applicability of Art 101 TFEU to horizontal co-operation agreements, OJ 2011 C 11, p. 1, para 63.

[121] EC, Guidelines on the applicability of Art 101 TFEU to horizontal co-operation agreements, OJ 2011 C 11, p. 1, para 63.

[122] German Monopolies Commission, XXII. Biennial Report – Competition 2018, para 188.

[123] OECD, Algorithms and Collusion – Competition Policy in the Digital Age 2017, p. 30, e.g. "snapshot price changes during the middle of the night".

[124] BKartA and Autorité de la Concurrence, Working Paper - Algorithms and Competition 2019, p. 55.

[125] US District Court of Columbia, 21 December 1992, 92-2854, *US/Airline Tariff Publishing Company*.

third-party centre. Algorithms enabled them to negotiate higher fares and to monitor and punish deviations.

12.3.2.3 Attribution and Responsibility

Undertakings are liable for algorithmic price behaviour if the conduct in question constitutes an agreement or a concerted practice within the meaning of Art 101 TFEU (Sec 1 Cartel Act), which has as their object or effect a restriction of competition. This is the case if algorithms are merely used as tools to implement pre-existing explicit agreements, namely in the messenger and hub-and-spoke scenario. The same might apply for signalling practices where price signals are sent to competitors with the intent to collude.[126] On the contrary, the independent use of algorithms by a number of undertakings to monitor and react intelligently to the market behaviour of competitors qualifies as unilateral conduct not caught by the ban on cartels. Therefore, given the absence of other means of communication, the predictable agent scenario does, in general, not raise competition concerns.

In the digital eye scenario, the undertaking's accountability for the behaviour of its (self- or deep-learning) algorithms is much less clear.[127] In general, undertakings are obliged to know the competition rules and act accordingly.[128] Such an obligation to ensure antitrust compliance also requires undertakings to design algorithms in a compliant way[129] to "prevent collusion in the first place".[130] As the algorithm remains under the firm's control, the firm cannot avoid liability on the ground that they outsourced their pricing decision to algorithms.[131] This must also be true for self-learning algorithms.[132]

Although the detection of algorithmic facilitated collusion remains difficult in practice, undertakings that create and build algorithms that directly affect market processes are well advised to keep competition law compliance in mind.[133] As the

---

[126] R. Hoffer, In: Binder Grösswang (ed), Digital Law, LexisNexis 2018, p. 199.

[127] In academia, different standards have been outlined, for an overview see BKartA and Autorité de la Concurrence, Working Paper - Algorithms and Competition 2019, p. 57ff with further references.

[128] CJEU, case 19/77, *Miller/Commission*, ECLI:EU:C:1978:19, pt 17ff; CJEU, case 27/76, *United Brands*, ECLI:EU:C:1978:22, pt 299, 301.

[129] M. Vestager, Algorithms and competition, Speech at the Bundeskartellamt's 18th Conference on Competition, Berlin, 16 March 2017; see also H. Schweitzer, In: Kühling and Zimmer (eds), Neue Gemeinwohlherausforderungen – Konsequenzen für Wettbewerbsrecht und Regulierung, Nomos 2020, p. 61.

[130] J. Laitenberger, Competition at the digital frontier, Speech at the Consumer and Competition Day, Malta, 24 April 2017, p. 8, available at https://ec.europa.eu/competition/speeches/text/sp2017_06_en.pdf. Accessed 7 September 2021.

[131] European Union, Note to the OECD on Algorithmic Collusion, DAF/COMP/WD(2017)12, para 25; BKartA 20 May 2018, Case Summary B9-175/17, *Lufthansa/Air Berlin*, p. 4, although ultimately no proceedings were initiated against Lufthansa based on excessive pricing.

[132] R. Hoffer, In: Binder Grösswang (ed), Digital Law, LexisNexis 2018, p. 199ff; M. Becka, Preisabsprachen zwischen selbstlernenden Preisalgorithmen, ÖZK 2019(1), pp. 9–16 (16).

[133] S. Vezzoso, In: Lundqvist and Gal (eds), Competition Law for the Digital Economy, Edward Elgar 2019, p. 100.

distinction between collusion and mere parallel behaviour permitted under Art 101 TFEU concerns the very essence of the ban on cartels, the Member States should exercise a degree of caution. According to Art 3 para 2 Reg 1/2003, national rules on the ban on cartels may not be stricter than those laid down at the European Union level. As a prerequisite, EU competition law must be applicable; cases of the digital economy will regularly fulfil this condition. Therefore, we recommend awaiting for more clarity or guidelines at the EU level related to algorithmic facilitated collusion before implementing regulatory or legislative measures in Austria.[134]

12.3.3 Personalised Pricing

The practice of personalised pricing describes a situation where an undertaking charges different prices to different consumers for a given product or service based on the consumers' willingness to pay. The latter is determined by the consumers' personal characteristics and subjective criteria (e.g. age, sex, occupation, income, geo-location, etc.) and conduct (e.g. off-/online behaviour). Such person-specific pricing can, in particular, be observed in online commerce, where companies, e.g., charge different prices depending on the customers' search queries or past online purchases. Personalised pricing is facilitated by the use of algorithms that allow undertakings to collect vast amounts of personal data. The creation of "digital user profiles" enables companies to better predict consumers' behaviour and preferences and to better identify the reserve prices of their customers, i.e. their expected willingness to pay.[135] Personalised pricing is to be distinguished from mere dynamic pricing, where prices are flexibly set based on objective criteria (like stock, the popularity and demand of a product, time of day, day of the week, etc.) and are uniformly applied to all customers.[136] Although personalised pricing does not allow a margin of savings, it is, in general, not designed to rip off those customers in a state of need; it rather matches the highest value a customer is willing and able to pay for a particular product or service.[137]

12.3.3.1 Abuse of a Dominant Position

Such person-specific pricing may qualify as abusive price discrimination in the meaning of Art 102 TFEU (Sec 5 Cartel Act). This may apply to constellations

[134] See also Opinion of the Studienvereinigung Kartellrecht on the Austrian FCA's Position Paper on digitisation and competition law, para 14, available at https://www.bwb.gv.at/fileadmin/user_upload/Digitalisierung_und_Wettbewerbsrecht_Thesenpapier.pdf. Accessed 7 September 2021.

[135] M. Gal, Algorithms as Illegal Agreements, Berkeley Technology Law Journal 2019(67:34), pp. 67–118 (91); for an overview of different forms of price discrimination in digital markets, see A. Ezrachi and M. Stucke, Virtual Competition, Harvard University Press 2016, p. 106ff.

[136] Prominent examples include gas and travel prices (e.g. flight tickets); however, dynamic pricing may be combined with elements of personalised pricing.

[137] M. Maggiolino, Personalized Prices in European Competition Law, Bocconi Legal Studies Research Paper No. 2984840, pp. 1–24 (2).

where two transactions of the same good or service occur at different prices despite having the same cost.[138] Unlike algorithmic collusion, where computer programs are used to increase market transparency in order to coordinate prices (see Sect. 12.3.2), this part focuses on unilateral strategies to limit price transparency and to use algorithms as a vehicle for price discrimination. Constellations, where two or more undertakings agree to apply the same personalised prices to their customers, do not fall within the remit of Art 102 TFEU (Sec 5 Cartel Act) but rather constitute a collusive cartel behaviour prohibited by Art 101 para 1 lit d TFEU (Sec 1 para 2 Nr 4 Cartel Act); the ban on abusive practices only covers personalised prices that have been unilaterally imposed.[139] In this context, commentators suggest that algorithmic collusion is more difficult where personalised pricing was implemented;[140] this is due to the increased number of factors that must be taken into account.[141]

Article 102 TFEU (Sec 5 Cartel Act) prohibits dominant companies from abusing their position on a particular market. According to case law, market dominance is defined as a position of economic strength that confers to an undertaking "the power to behave to an appreciable extent independently of its competitors, customers and ultimately of its consumers".[142] Such independence is mainly reflected by the undertaking's power to profitably maintain prices above the competitive level for a significant period of time.[143] The assessment of dominance will take various factors into account, including the competitive structure of the market (in particular market shares),[144] the structure of the company and the market behaviour of the undertaking concerned. Unlike in the United States,[145] to be market dominant is not in itself considered illegal.[146] However, a market dominant company has the special

[138] I. Lianos, V. Korah and P. Siciliani, Competition Law, Oxford University Press 2019, p. 1231.

[139] M. Maggiolino, Personalized Prices in European Competition Law, Bocconi Legal Studies Research Paper No. 2984840, pp. 1–24 (17ff).

[140] H. Schweitzer, In: Kühling and Zimmer (eds), Neue Gemeinwohlherausforderungen – Konsequenzen für Wettbewerbsrecht und Regulierung, Nomos 2020, p. 60; see also CMA, Pricing Algorithms 2018, paras 7.27ff.

[141] M. Gal, Algorithms & Competition Law: Interview of Michal Gal by Thibault Schrepel, e-Competitions Algorithms, Concurrences 2020, Art. N° 93929, pp. 1–12.

[142] CJEU, case 27/76, *United Brands*, ECLI:EU:C:1978:22, pt 65.

[143] EC, Guidance on the Commission's enforcement priorities in applying Article 82 of the EC Treaty to abusive exclusionary conduct by dominant undertakings, OJ 2009 C 45, p. 7, para 11; EGC, case T-321/05, *Astra Zeneca*, ECLI:EU:T:2010:266, pt 267; O. Blanco, Market Power in EU Antitrust Law, Hart 2012, p. 47.

[144] Different presumptions of dominance might apply under national and EU Law: While the Austrian Cartel Act provides a rebuttable presumption of dominance for undertakings with a market share of at least 30% (Sec 4 para 2 Nr 1 Cartel Act), under EU law such a presumption applies at a 50% market share, see CJEU, case C-62/86, *AKZO*, ECLI:EU:C:1991:286, pt 60.

[145] Sec 2 Sherman Act: "Every person who shall monopolize, or attempt to monopolize, or combine or conspire with any other person or persons, to monopolize any part of the trade or commerce among the several States, or with foreign nations, shall be deemed guilty of a felony [. . .]."

[146] CJEU, case C-209/10, *Post Danmark/Konkurrencerådet*, ECLI:EU:C:2012:172, pt 21 with further references from case law.

responsibility not to distort competition[147] to the detriment of final consumers (exploitative abuse) or competitors (exclusionary abuse). Examples of behaviour that may amount to abuse include personalised pricing, which may be facilitated by the use of algorithms.

12.3.3.2 Application to Different Scenarios

Algorithmic assisted personalised pricing may take the form of either exploitative or exclusionary abuse, depending on whether final consumers or business partners are affected.

Exploitative Abuse

In the first (exploitative) scenario, the supplier gives preferential treatment to some customers and not to others. If the undertaking concerned has market power, the firm would be able to set different prices for different customer groups it has identified based on the data collected.[148] So far, the courts have never fully clarified the question of whether and to what extent Art 102 TFEU (Sec 5 Cartel Act) could sanction cases of price discrimination harming final customers.[149] In *MEO*, *AG Wahl* pointed out that exploitative cases of price discrimination are "extremely rare".[150] However, it is conceivable that this scenario may be treated under the general abuse of dominance clause of Art 102 TFEU (Sec 5 para 1 Cartel Act) or as a form of excessive pricing or unfair trading conditions within the meaning of Art 102 lit a) TFEU (Sec 5 para 1 Nr 1 Cartel Act). This may be the case when personalised pricing has led to higher prices which would have prevailed in the absence of the algorithmic manipulation.[151] Some commentators argue that personalised pricing may as well fall within the remit of Art 102 lit c) TFEU (Sec 5 para 1 Nr 3 Cartel Act),[152] which prohibits the application of dissimilar conditions to equivalent transactions. Although the latter provision explicitly targets only trading partners (for details, see below), it is argued that the ban on abusive practices may—based on its functional interpretation—be applied beyond the strict letter of the law.[153]

---

[147] CJEU, case C-322/81, *Michelin*, ECLI:EU:C:1983:313, pt 57.

[148] BKartA and Autorité de la concurrence, Competition Law and Data 2016, p. 22.

[149] Affirmative C. Townley, E. Morrison and K. Yeung, Big Data and Personalized Price Discrimination in EU Competition Law, Yearbook of European Law 2017(36), pp. 683–748.

[150] AG Wahl, case C-295/17, *MEO*, ECLI:EU:C:2018:413, pt 80.

[151] I. Lianos, V. Korah and P. Siciliani, Competition Law, Oxford University Press 2019, p. 1235.

[152] I. Graef, Algorithms and fairness – What role for competition law in targeting price discrimination towards end consumers, Columbia Journal of European Law 2018(24:3), pp. 541–559 (549); P. Akman, To Abuse or not to Abuse: Discrimination between Consumers, ELR 2007(32), pp. 492–512 (498f).

[153] See also EC, COMP/C-1/36.915, *Deutsche Post AG*, pt 133 noting that "consumers are affected negatively by having to pay prices for these services which are higher than those charged to other senders and by having their mailings delayed significantly".

Following this approach, Art 102 lit c) TFEU (Sec 5 para 1 Nr 3 Cartel Act) could sanction cases of price discrimination harming final customers too.[154]

Exclusionary Abuse

Under certain circumstances, algorithmic-based price discrimination may also amount to exclusionary abuse. Article 102 lit c TFEU (Sec 5 para 1 Nr 3 Cartel Act) explicitly forbids the discrimination of business trading partners who compete with one another.[155] This may be the case where the market dominant supplier offers more favourable conditions to its own customers in an attempt to prevent them from switching to rivals, i.e. to foreclose rivals on the upstream market.[156] The most important element of Art 102 lit c TFEU (Sec 5 para 1 Nr 3 Cartel Act) is that the application of dissimilar conditions must place trading partners at a "competitive disadvantage". According to case law, this term does not require proof of actual quantifiable deterioration in the competitive situation but must lead to the conclusion that—based on all relevant circumstances of the case[157]—the behaviour affects "the costs, profits or any other relevant interest"[158] of one or more trading partners. However, the EC and the National Competition Authorities (NCAs) face a high burden of proof as the competition enforcers are required to prove the existence of a competitive disadvantage suffered by one of the discriminated customers.[159]

12.3.3.3 Digression: Austrian Local Supply Act (Nahversorgungsgesetz)

Apart from the provisions on price discrimination stipulated in the TFEU and the Cartel Act, Sec 2 para 1 of the Austrian Local Supply Act (*Nahversorgungsgesetz*)[160] contains a general prohibition on discrimination in relationships between suppliers and authorised resellers. A supplier that—without objective justification—grants or offers different conditions to authorised resellers may be sued for a cease-and-desist order.[161] Of course, this may also apply to

[154] P. Akman, To Abuse or not to Abuse: Discrimination between Consumers, ELR 2007(32), pp. 492–512 (498) argues that consumers could also be considered as trading partners in the meaning of Art 102 lit c) TFEU.

[155] CJEU, case C-525/16, *MEO*, ECLI:EU:C:2018:270, pt 27.

[156] I. Graef, Algorithms and fairness – What role for competition law in targeting price discrimination towards end consumers, Columbia Journal of European Law 2018(24:3), pp. 541–559 (542f).

[157] For an overview of elements to be considered see CJEU, case C-525/16, *MEO*, ECLI:EU:C:2018:270, pt 31; see also CJEU, case C-413/14 P, *Intel*, EU:C:2017:632, pt 139.

[158] CJEU, case C-525/16, *MEO*, ECLI:EU:C:2018:270, pt 37.

[159] M. Botta and K. Wiedemann, To discriminate or not to discriminate? Personalised pricing in online markets as exploitative abuse of dominance, EJLE 2020(50:3), pp. 381–404 (392); C. Ritter, Price discrimination as an abuse of dominant position under Article 102 TFEU, CMLR 2019(56:1), pp. 259–274 (267).

[160] Bundesgesetz vom 29. Juni 1977 zur Verbesserung der Nahversorgung und der Wettbewerbsbedingungen (Nahversorgungsgesetz), BGBl 392/1977.

[161] For details see N. Gugerbauer, In: Gugerbauer (ed), KartG und WettbG^3, Verlag Österreich 2017, Sec 2 Local Supply Act paras 1ff; on the Local Supply Act in general, see e.g. W. Barfuß, Das

algorithmic-based price discrimination. Preferential treatment of individual resellers is only feasible where an economically equivalent consideration exists in return. Finally, the preferential treatment must be appropriate and necessary to achieve the objective pursued (proportionality test).[162] The Local Supply Act is to be viewed as a supplementary law to the Cartel Act, which is also executed by the Cartel Court.[163] However, it must be noted that the Local Supply Act, including the provision cited, applies to all undertakings, irrespective of their (dominant) market position.[164] However, the Supreme Court left open the question of whether the undertaking concerned must at least have a certain amount of market power.[165]

Similarly, the EU Regulation on platform-to-business relations[166] applies to all operators of the specified online services, i.e. not only to dominant undertakings. The P2B Regulation sets out guidance on the equitable delivery of online services and prohibits, *inter alia*, the arbitrary and discriminatory treatment of business users. Any acts against these rules are deemed to be an unfair commercial practice that may trigger responsibility from the undertaking concerned. Besides the Austrian Economic Chamber (*Wirtschaftskammer Österreich*) and the Association Against Unfair Competition (*Schutzverband gegen unlauteren Wettbewerb*), since 10 September 2020, also the Austrian FCA has had a standing to bring proceedings before the competent court.[167]

12.3.4 Conclusion

The relevance of collusive behaviour and person-specific pricing is reinforced by the increasing use of algorithms and the greater availability of market data. The first part analysed the effects of algorithms on the ability of competitors to coordinate their conduct. Algorithmic collusion may take place either explicitly, i.e. simple collusion where computers limit competition through an agreement or concerted practice within the meaning of Art 101 TFEU (Sec 1 Cartel Act), or tacitly, where collusion is achieved through subtler means not amounting to a hard-core cartel. Whether algorithmic coordination constitutes an agreement or concerted practice largely depends on the specific circumstances of the respective case (market structure,

---

Bundesgesetz zur Verbesserung der Nahversorgung und der Wettbewerbsbedingungen "NVG", ÖZW 1978(1), pp. 10–14.

[162]N. Gugerbauer, In: Gugerbauer (ed), KartG und WettbG^3, Verlag Österreich 2017, Sec 2 Local Supply Act para 7.

[163]Sec 6 Local Supply Act.

[164]See also OGH 9 October 2000, 16 Ok 8/00.

[165]OGH 16 July 2008, 16 Ok 3/08, *Sägerundholz*.

[166]Regulation 2019/1150 of the European Parliament and of the Council of 20 June 2019 on promoting fairness and transparency for business users of online intermediation service (P2B Regulation), OJ 2019 L 186, p. 57.

[167]List of designated organisations, associations and public entities drawn up and published pursuant to Article 14 para 6 P2B Regulation, OJ 2020 C 300, p. 2.

supply-and-demand-related factors, etc.).[168] While unilaterally observing, analysing and reacting to publicly observable behaviours of competitors through algorithms does not contravene the ban on cartels, constellations involving (algorithmic) the exchange of information can be deemed unlawful. In this context, one of the main risks of algorithms is that they may enable firms to replace explicit collusion with new forms of tacit coordination.[169] Therefore, competition enforcers are well advised to expand their range of detection tools for cartel activity, possibly complemented by setting up new (regulatory) agencies specialised in information technology (IT) and AI.[170]

The second part argued that price discrimination facilitated by algorithms may constitute an abuse of dominance within the meaning of Art 102 TFEU (Sec 5 Cartel Act). On the one hand, this may apply to the unequal treatment between commercial customers (exclusionary abuse). On the other hand, the question of whether and, if so, to what extent Art 102 TFEU (Sec 5 Cartel Act) can also sanction personalised pricing that harms final consumers is less clear. However, as suggested by some commentators, broadening the scope of Art 102 TFEU (Sec 5 Cartel Act) is certainly worth to be discussed. In this regard, it should be considered that the digital economy introduces new forms of exploitative business strategies that directly harm final consumers.[171] From an economic viewpoint, personalised pricing is not harmful *per se*.[172] While some consumers, i.e. those with a high willingness to pay, end up paying a higher price, personalised pricing can also favour others, i.e. those with a lower willingness to pay. These consumers who value less the products would receive better prices than in the absence of discrimination.[173] Although personalised pricing reallocates wealth in favour of undertakings, it makes goods or services accessible which would otherwise be inaccessible by the poorest.[174] It follows that the overall effect of price discrimination on welfare is ambiguous and requires a case-by-case analysis under competition law.[175] Besides this, the detrimental effects

[168] German Monopolies Commission, XXII. Biennial Report – Competition 2018, para 197.

[169] OECD, Algorithms and Collusion – Competition Policy in the Digital Age 2017, p. 25.

[170] As suggested by the Brics Competition Law and Policy Center, Digital Era Competition 2019 (Brics Report), pp. 671ff.

[171] See e.g. the proceedings against Facebook in Germany where the theory of harm is based on a lack of users' choice, for an overview see R. Podszun, Der Verbraucher als Marktakteur – Kartellrecht und Datenschutz in der "Facebook"-Entscheidung des BGH, GRUR 2020(12), pp. 1268–1276; *Scharf*, Exploitative business terms in the era of big data – the Bundeskartellamt's Facebook decision, ECLR 2019(7), pp. 331–339.

[172] CERRE, Big Data and Competition Policy: Market Power, personalised pricing and advertising 2017, p. 8, available at https://www.cerre.eu/sites/cerre/files/170216_CERRE_CompData_FinalReport.pdf. Accessed 7 September 2021.

[173] BKartA and Autorité de la concurrence, Competition Law and Data 2016, p. 21.

[174] M. Maggiolino, Personalized Prices in European Competition Law, Bocconi Legal Studies Research Paper No. 2984840, pp. 1–24 (3, 13ff).

[175] M. Botta and K. Wiedemann, To discriminate or not to discriminate? Personalised pricing in online markets as exploitative abuse of dominance, EJLE 2020(50:3), pp. 381–404 (388, 399).

of personalised pricing on competition would have to be demonstrated, including the absence of a proportionate justification.[176] Undertakings may, e.g., argue that price discrimination enables them to increase output to the benefit of (many) consumers, which in turn may enhance the aggregate consumer welfare.[177]

[176] BKartA and Autorité de la concurrence, Competition Law and Data 2016, p. 21; U. Salaschek and M. Serafimova, Preissetzungsalgorithmen im Lichte von Art. 101 AEUV, WuW 2018(1), pp. 8–17 (12); for an overview of challenges for competition enforcers in assessing personalised pricing strategies in digital markets see M. Botta and K. Wiedemann, To discriminate or not to discriminate? Personalised pricing in online markets as exploitative abuse of dominance, EJLE 2020(50:3), pp. 381–404 (392).

[177] C. Townley, E. Morrison and K. Yeung, Big Data and Personalized Price Discrimination in EU Competition Law, Yearbook of European Law 2017(36), pp. 683–748 (730).

Belgium

13

Louis Hoffreumon and Benjamin Docquir

13.1 Introduction

The recent years have seen many new forms of marketing practices from businesses that target consumers, including the increasing use of artificial intelligence (AI) and other computational or profiling technologies. These practices do not only concern direct marketing, as such, but also advertising, promotions, pricing, and also entering into an agreement with the consumer, with the final objective to sell goods or procure a service.

This shift towards an increasingly virtualised environment raises questions regarding the appropriateness of the current legal protections and remedies offered to the consumer, under Belgian and EU laws, as the target of such practices, a contractual counterparty and an individual whose personal data are processed by a data controller.

In this new paradigm, practitioners are facing a growing interaction between the laws protecting consumers (trade practices, consumer contracts, e-commerce and unfair competition), mostly issued through EU directives (as further described hereunder), and the rules on the protection of privacy, including the General Data Protection Regulation (GDPR), adopted in 2016. A GDPR consent form, which must comply with the Unfair Terms Directive[1] when obtaining consent from a data subject,[2] appears as the most obvious and well-known example.

[1] Council Directive 93/13 of 5 April 1993 on unfair terms in consumer contracts, OJ 1993, L 95, p. 29.

[2] GDPR, Recital 42.

L. Hoffreumon (✉) · B. Docquir (✉)
Osborne Clarke, Brussels, Belgium
e-mail: Louis.Hoffreumon@osborneclarke.com; Benjamin.Docquir@osborneclarke.com

B. Kilpatrick et al. (eds.), *Antitrust in Data Driven Markets & Legal Framework for Influencers, Native Advertising and Control over the Use of AI in Marketing*, LIDC Contributions on Antitrust Law, Intellectual Property and Unfair Competition, https://doi.org/10.1007/978-3-031-07422-6_13

Moreover, when commenting on the New Deal for Consumers initiative, the European Data Protection Supervisor (EDPS) appreciated the fact that both consumer and data protection laws share a common goal of redressing the imbalances of informational and market power, and suggested that, together with competition law, they needed to work in a consistent manner to ensure that people are treated fairly.[3] We will use this common thread throughout this contribution.

Under Belgium's domestic law, there currently is no unique legislation dealing specifically with this new paradigm. Consequently, at the time of this contribution, issues concerning consumer rights continue to be approached through unfair competition and economic law rules, while issues on the processing of consumers' personal data are dealt with according to the GDPR and its domestic implementations, each being supervised by somewhat separate regulators.[4]

In this contribution, we provide a description of the current Belgian legal framework on advertising and marketing (Sect. 13.2) and discuss two currently debated topics. First, this contribution questions whether the current scope of legislation on "commercial communication" and "identifiability" could be improved to address new forms of marketing and advertising (Sect. 13.3). Second, we assess the interactions between consumer and privacy laws illustrated by the concept of consumer personalisation in an online environment, whether or not automated means such as AI are used (Sect. 13.4).

We will use the terminology applied in Belgian law and speak of "enterprise" when referring to a trader (a "professional" targeting consumers, as defined under the EU directives) or data controller (the same trader acting as an advertiser, advertising network or data source or collecting consumers' personal data for further processing in order to send them personalised advertising and, subsequently, goods and/or services). On the other hand, we will refer to "consumer" when talking about the other party to that relationship, the one receiving the ads and purchasing the goods and/or services at the end of the chain, also the data subject under the GDPR whose data are collected and processed.

[3]Opinion 8/2018 of the EDPS, on the legislative package "a New Deal for Consumers", available at: https://edps.europa.eu/sites/edp/files/publication/18-10-05_opinion_consumer_law_en.pdf, pts 6–7, p. 8. The same idea can be found in the Communication from the Commission to the European Parliament, the Council, the European Economic and Social Committee and the Committee of the Regions, "A European Strategy for data", COM(2020) 66 final, 19 February, available at https://eur-lex.europa.eu/legal-content/EN/TXT/HTML/?uri=CELEX:52020DC0066&from=EN. Accessed 5 March 2021.

[4]Namely, the Belgian ministry of economy, for the most part, regarding compliance by enterprises with consumer rights set out under Book VI of the Code of Economic Law, in Belgium, and the Belgian Data Protection Authority (BDPA), concerning issues related to privacy and consumers' personal data.

13.2 The Belgian Legislative Landscape

13.2.1 Unfair Commercial Practices

13.2.1.1 General Legislation

The relevant legislation implementing (1) the Advertising Directive,[5] (2) the Unfair Commercial Practices Directive, (3) the Unfair Terms Directive, (4) the Directive on Electronic Commerce, (5) some provisions of the e-Privacy Directive[6] and (6) the Consumer Rights Directive[7] is contained in Books VI (Consumer Law and Trade Practices) and XII (Digital Economy) of the Belgian Code of Economic Law (CEL).[8]

First, Book VI CEL deals with consumer information on the market,[9] unfair commercial practices[10] (defined as "any act, omission, course of conduct or representation, commercial communication including advertising and marketing, by a business, directly connected with the promotion, sale or supply of a product to consumers")[11] and business-to-consumer (B2C) contracts.[12,13]

A commercial practice can be deemed unfair and may be subject to civil or criminal sanctions, following the widely accepted "three-time reasoning" by Belgian scholars and practitioners under Article 5 of the Unfair Commercial Practices Directive and their Belgian counterpart: the fairness of a practice is first considered under a general norm; a commercial practice is unfair if it is contrary to the requirements of professional diligence and influences the economic behaviour of a consumer. Then other practices will be deemed unfair under a "semi-general" norm if they are misleading or aggressive. Finally, specific practices are deemed unfair in all circumstances if they are contained in a "black-list".[14] As practitioners ourselves,

[5] Directive 2006/114 of the European Parliament and of the Council of 12 December 2006 concerning misleading and comparative advertising, OJ, 2006, L 376, p. 21.

[6] Directive 2002/58 of the European Parliament and of the Council of 12 July 2002 concerning the processing of personal data and the protection of privacy in the electronic communications sector (Directive on privacy and electronic communications), OJ 2002, L 201, p. 37.

[7] Directive 2011/83 of the European Parliament and of the Council of 25 October 2011 on consumer rights, amending Council Directive 93/13/EEC and Directive 1999/44 of the European Parliament and of the Council and repealing Council Directive 85/577 and Directive 97/7 of the European Parliament and of the Council.

[8] Act of 28 February 2013 creating the Code of Economic Law, Belgian Official Gazette, 28 February 2013, p. 19975.

[9] Articles VI.1/1 to VI. 36 CEL.

[10] Articles VI.92 to VI.117 CEL.

[11] Article I.8, 23° CEL.

[12] Articles VI.37 to VI.91 CEL.

[13] These provisions will evolve in the future following the upcoming implementation under Belgian law of the contents of the Omnibus Directive and the Digital Content Directive, both adopted in the framework of the New Deal for Consumers, and which will be considered later in this contribution where relevant.

[14] Annex I of the Unfair Commercial Practices Directive.

we prefer a dynamic and risk-oriented approach, which is more appropriate in terms of legal design, where the order of such reasoning is reversed, as described below.

First, a practice will be tested against the "blacklist" of specific commercial practices deemed unfair in all circumstances for consumers (and therefore forbidden) because they are considered either misleading or aggressive.[15]

Second, if the practice is not captured by this blacklist, one needs to assess if it falls under the "semi-general norm". Under this second set of rules, a practice is deemed unfair if it is misleading[16] or aggressive[17] and affects the consumer in such a way as to cause him or her to take a transactional decision[18] that he/she would not have taken otherwise.

A practice will be considered misleading when it contains false information or, where no false information is conveyed, if it is deceitful or likely to deceive the average consumer with regard to some of its elements, including the main characteristics of the product, endorsements, prices and methods of price calculation or the identity of the business. Parasitic competition and the omission of material information[19] that the consumer needs to be aware of are other examples of misleading commercial practices.

A commercial practice will be considered aggressive if, in its factual context and taking into account all its features and circumstances, by means of harassment, coercion (including the use of physical force) or undue influence, it significantly impairs or is likely to significantly impair the average consumer's freedom of choice or conduct with regard to the product, causing the consumer to take a transactional decision that he or she would not have taken otherwise.[20]

---

[15] Respectively Articles VI.100 and VI.103 CEL, implementing Annex I of the Unfair Commercial Trade Practices Directive.

[16] Article VI.97 to 99 CEL (misleading commercial practices towards consumers); Article VI.105 and 105/1 CEL (misleading commercial practices towards enterprises).

[17] Article VI.101–102 CEL (aggressive commercial practices towards consumers); Article VI.109/1 to 109/3 CEL (aggressive commercial practices towards enterprises).

[18] Article I.8, 28° CEL. A "transactional decision" means any decision concerning whether, how and on what terms to, on the part of the consumer, purchase, make, retain or dispose of a product or to proceed to payment in whole or in part for such product or to exercise a right in relation thereto.

[19] Such material information is listed under Article VI.99 §4 CEL and includes (1) the main characteristics of the product, in the appropriate measure and taking into account the means of communication and the product concerned; (2) the geographic address and the trader's identity and, if applicable, the geographical address and identity of the enterprise on which account the trader is acting; (3) the price (all taxes included) or, when it can be calculated in advance, the way that the price is calculated, together with, if applicable, all additional transport, shipping and delivery costs or, when these costs cannot be calculated in advance, a mention that these costs can be charged to the consumer; (4) the process for payment, delivery and complaints, if they are different from professional standards; and (5) the existence or not of a withdrawal right. Under Article VI.99 §5 CEL, it is also considered material all mandatory information in relation to commercial communication under the directives listed in Annex II of the Unfair Commercial Practices Directive, as well as other mentioned European legislation.

[20] For completeness, Article VI.102 CEL set outs the elements to take into account when assessing the presence of harassment; coercion, including the use of physical force; or undue influence: (1) its timing, location, nature or persistence; (2) the use of threatening or abusive language or behaviour;

If the envisaged practice can be deemed neither misleading nor aggressive, a third stage will consist in assessing whether the enterprise complies with the "general norm" for unfair commercial practices. Under this norm, a commercial practice is unfair if it is contrary to the requirements of professional diligence and materially distorts or is likely to materially distort the economic behaviour of the average consumer whom it reaches or to whom it is addressed regarding the product or of the average member of the group when a trade practice is directed to a particular group of consumers.[21] In the latter case, the enterprise must assess the practice from the perspective of the average member of a clearly identifiable group of consumers that is particularly vulnerable because of mental or physical infirmity, age or credulity.

Aside from the commercial practices directed to consumers, Book VI CEL contains rules applicable to commercial practices detrimental to other businesses. While not envisaged in this contribution, a central provision in that regard is Article VI.104, which forbids commercial practices by a business that are detrimental to the professional interests of one or several businesses. Under this rule, unanimous case law considers that a breach of any legislation or adopting a behaviour contrary to the requirements of professional diligence was sufficient to establish the existence of such forbidden practice.[22] As a result, any practice that infringes upon consumers' rights to privacy or that otherwise does not comply with statutory rules on the processing of personal data and the recording of, or access to, electronic communication data can justify an application for injunction or a cease-and-desist measure on the part of a competing enterprise.

Read together, it appears that the general norm has the potential for a broad application and can be of use to qualify any practice in breach of the law as an unfair commercial practice.

Two other sets of rules under this Book are worth mentioning. First, Article VI.17 CEL describes the rules applicable to comparative advertising, which are similar to the ones contained under Article 4 of the Advertising Directive. Second, the recent so-called B2B Act[23] introduces a similar set of rules in relation to commercial practices in a business-to-business (B2B) context.

---

(3) the exploitation by the trader of any specific misfortune or circumstance of such gravity as to impair the consumer's judgment, of which the trader is aware, to influence the consumer's decision with regard to the product; (4) any onerous or disproportionate non-contractual barriers imposed by the trader where a consumer wishes to exercise rights under the contract, including rights to terminate a contract or to switch to another product or another trader; (5) any threat to take any action that cannot legally be taken.

[21] Article VI.93 CEL.

[22] For a complete overview of the principles based on this article, see J. Stuyck and B. Keirsbilck, Artikel VI.104 WER: oneerlijke mededinging in J. Stuyck, B. Keirsbilck, Handels- en economisch recht, Serie "Beginselen van het Belgisch Privaatrecht", n° 14, Kluwer, Malines, 2019, pp. 204–302.

[23] Law of 4 April 2019 amending the Code of Economic Law regarding abuses of economic dependence, abusive clauses and unfair market practices between undertakings, Belgian Official Gazette, 24 May 2019, p. 50066. This law is intended to protect SMEs against certain unfair commercial practices used by bigger and international businesses. The commercial practices covered by

Title I of Book XII[24] deals with advertising and marketing when they are provided by means of information society services,[25] and implements, *inter alia*, the transparency and information obligations contained herein (Article XII.6 and 7), together with technical provisions on agreements,[26] the identifiability requirements for advertising and promotions (Article XII.12), the (opt-in) rules on electronic marketing by e-mail (Article XII.13 and 14) and the rules on editorial responsibility applicable to intermediaries (Article XII.17–20).

13.2.1.2 Specific Legislation

In relation to advertising and marketing, in addition to general legislation, some sectors or products are governed by specific legislation at the Belgian or EU level (among others, gambling,[27] medicine,[28] alcohol and tobacco,[29] food[30] and cosmetics[31]).

---

the Act are similar to those covered by consumer laws: according to the drafters of the Act, SMEs should enjoy the same protection as consumers. See Draft Act amending the Code of Economic Law regarding abuse of a significant dominant position, 13 November 2015, Doc. Parl. Ch., 2015–2016, n°54-1451/1, p. 6.

[24] For completeness, Book XII CEL contains a Book II in relation to the rules applicable to trust services, as provided under the eIDAS Regulation (Regulation 910/2014 of the European Parliament and of the Council of 23 July 2014 on electronic identification and trust services for electronic transactions in the internal market and repealing Directive 1999/93, OJ, 2014, p. 73).

[25] The Directive on electronic commerce refers to the definition of "information society services" given under Directive (EU) 2015/1535 of the European Parliament and of the Council of 9 September 2015 laying down a procedure for the provision of information in the field of technical regulations and of rules on Information Society Services, OJ, 2015, L 241, p. 1.

[26] Articles XII.8 to 11, 15 and 16 CEL.

[27] Act of 7 May 1999 on Gambling, Betting, Gambling Establishments and Gambler Protection, Belgian Official Gazette, 30 December 1999, p. 50040 (and subsequent amendments) and the Royal Decree of 25 October 2018 on Modalities for the Operation of Games of Chance and Betting Operated by Means of Information Society Instruments, Belgian Official Gazette, 31 October 2018, p. 82744.

[28] Article 9 of the Act of 25 March 1964 on Medicines, Belgian Official Gazette, 17 April 1964, p. 4206, and its subsequent executing decrees. This Law also implemented the rules set out under Directive 2001/83.

[29] Act of 24 January 1977 on Consumer Health Protection in relation with Foodstuffs and other Products, Belgian Official State Gazette, 8 April 1977, Article 7.

[30] See, among others, Article 16 of Regulation No 178/2002 of the European Parliament and of the Council of 28 January 2002 laying down the general principles and requirements of food law, establishing the European Food Safety Authority and laying down procedures in matters of food safety, OJ 2002, L 31, p. 1; Regulation 1169/2011 of the European Parliament and of the Council of 25 October 2011 on the provision of food information to consumers, OJ 2011, L 304, p. 18; and Regulation 1924/2006 of the European Parliament and of the Council of 20 December 2006 on nutrition and health claims made on foods, OJ 2006, L 404, p. 9. For the aspects not covered by European legislation, the abovementioned Act of 27 January 1977 and its executing decrees remain applicable.

[31] Article 20 of Regulation No 1223/2009 of the European Parliament and of the Council of 30 November 2009 on cosmetic products, OJ 2009, L 342, p. 59.

The rules applicable to advertising in audiovisual media services, as defined under the Audiovisual Media Services (AVMS) Directive,[32] have been implemented at the regional level and are subject to three different regional decrees.[33] Stricter rules apply to advertising and its identifiability, but only insofar as it is conveyed on specific "audiovisual" media and, since 2018, on "video-sharing platform services" falling under the scope of the AVMS Directive. This may create issues for the provider of online services who needs to determine if its service falls under the scope of these rules. If so, this provider would then need to consider whether it is necessary to inform users, and in some cases monitor actively their compliance with these rules.

13.2.1.3 Self-Regulation

Next to mandatory legislation, the legal framework of advertising is characterised by self-regulation, supervised in Belgium by an independent non-governmental body, BEJA (*Jury d'éthique publicitaire*; *Ethische Jury voor Ethische Praktijken inzake Reclame*).[34]

BEJA is a non-profit association vested with the competence to render decisions on the conformity of advertisements with legal and ethical rules on the protection of consumers based on either mandatory legislation or self-regulatory codes (primarily the International Code Council Marketing and Advertising Communications Code

[32] Directive 2010/13/EU of the European Parliament and of the Council of 10 March 2010 on the coordination of certain provisions laid down by law, regulation or administrative action in Member States concerning the provision of audiovisual media services, OJ 2010, L 95, p. 1. Directive 2018/1808 of the European Parliament and of the Council of 14 November 2018, amending the AVMS Directive, has recently been implemented under Belgian law. See footnote n°36.

[33] In Belgium, audiovisual media services are regulated at the "communities" (Belgian territories divided depending on language and culture) level, resulting in three different sets of legislation implementing the AVMS Directive, that is, pertaining to the Dutch-speaking community (*Decreet van 27 maart 2009 betreffende radio-omroep en televisie*, Belgian Official Gazette, 30 April 2009, p. 34470, and the amending Decree of 19 March 2021 implementing the new AVMS Directive, Belgian Official Gazette, 29 April 2021, p. 40930), French-speaking community (*Décret coordonné du 26 mars 2009 sur les services de médias audiovisuels*, Belgian Official Gazette, 24 July 2009, p. 50609, now replaced by the *Décret du 4 février 2021 relatif aux services de médias audiovisuels et de services de partage de vidéos*, Belgian Official Gazette, p. 29036) and German-speaking community (*Dekret vom 27. Juni 2005 über den Rundfunk und die Kinovorstellungen*, Belgian Official Gazette, 6 September 2005, p. 38869, now replaced by the *Dekret vom 1. März 2021 über die Mediendienste und die Kinovorstellungen*, Belgian Official Gazette, 12 April 2021, p. 32156). Moreover, media falling under the scope of this Directive are governed by three different regulators also operating at the community level. Audiovisual media services not attached to any of these regions remain governed by a supplementary federal law, the Law of 5 May 2017 relating to the audiovisual media services in the bilingual region of Brussels-Capital, Belgian Official Gazette, 23 May 2017, p. 58970.

[34] As a self-regulatory advertising body, the BEJA is also a member of the European Advertising Standards Alliance (EASA).

(ICC Code) and the guidelines and rules adopted based on it).[35] Its decisions are not legally enforceable, and it cannot impose fines or specific sanctions.

Here again, the scope of this self-regulation is limited to specific media (television, cinema, radio and also "digital" media, such as an advertiser's own website, pop-ups, search results, in-app advertising or "advergaming"). Questions relating to other commercial practices, Internet Protocol (IP) or aspects of direct marketing that do not relate to advertising *stricto sensu* (including databases, the processing of personal data, behavioural advertising) fall outside of its scope.[36]

13.2.2 Privacy and Personal Data

The GDPR is part of the Belgian legislative framework and is completed by specific rules under the Data Protection Act[37] and the Act creating the (Belgian) Data Protection Authority (BDPA).[38] For instance, Article 7 of the Data Protection Act implements Article 8 GDPR and sets the age (13 or more) at which an underage data subject is able to consent to the processing of its personal data without intervention from its legal guardians.

The BDPA, when issuing decisions, is required to take into account not only the rights and freedoms of individuals in relation to the processing and circulation of their personal data but also the country's consumer protection policy.[39] Building on this obligation, the BDPA made direct marketing, a practice having an impact on consumers directly, one of its priority sectors for the next 5 years, and further scrutiny is to be expected of enterprises processing personal data for these purposes.[40]

[35] Article 1 of the BEJA's international regulations as at 1 January 2020, hereafter the "BEJA Regulations" (consulted on 28 May 2020). Available (in French) at: https://www.jep.be/sites/default/files/inline-media/reglement_jep_fr_-_2020_cc.pdf. Examples of self-regulatory codes are the Advertising Code for food products, issued with the Food Industry Federation, available at: https://www.jep.be/sites/default/files/rule_reccommendation/fevia_fr.pdf (the "FEVIA Code"); the Convention on advertising and marketing of alcoholic beverages of 25 January 2013, amended on 2 September 2019, available at: https://www.jep.be/sites/default/files/rule_reccommendation/convention_alcool_-_fr_-_2019.pdf; the Advertising Code for advertising and marketing of cosmetic products, revised on May 2020, available at: https://www.jep.be/sites/default/files/rule_reccommendation/code_publicite_produits_cosmetiques_detic_2020.pdf; and the Code of ethical and responsible advertising by the national lottery, available at: https://www.jep.be/sites/default/files/rule_reccommendation/code_voor_ethische_reclame_fr.pdf. Accessed 10 September 2021.

[36] A complete list of the scope of action of the BEJA is set under Article 2 of the BEJA Regulations.

[37] Act of 30 July 2018 on the protection of natural persons with regard to the processing of personal data, Belgian Official Gazette, 5 September 2018, p. 68616.

[38] Act of 3 December 2017 creating the BDPA, Belgian Official Gazette, 10 January 2018, p. 989. The BDPA has a dedicated website: https://www.dataprotectionauthority.be (for the English version).

[39] Article 52, §1 of the Act creating the BDPA.

[40] See Strategy Plan of the BDPA, p. 23, available (in French or Dutch) at https://www.autoriteprotectiondonnees.be/publications/plan-strategique-2020-2025.pdf. Accessed 12 January 2021.

The rules applicable to cookies, implementing Articles 5 and 6 of the e-Privacy Directive, are contained in a separate legislation under Article 129 of the Belgian Electronic Communications Act.[41] However, this article clearly links cookies to compliance with privacy legislation.[42]

While not binding, the (endorsed) opinions and guidelines issued by the Article 29 Working Party, now the European Data Protection Board (EDPB), together with the recommendations issued by the BDPA, remain a useful tool for providing a comprehensive interpretation of some key provisions of the GDPR and related legislation. Recently, the BDPA has issued a Recommendation on Direct Marketing,[43] inspired by various EDPB guidelines.

13.2.3 AI

AI is not regulated by a specific Belgian law, although scholars opine that the existing rules on contract law, next to those on consumer and privacy laws, should apply to the use of AI, subject to a few specific adaptations.[44]

The need to amend existing legislation or to adopt a specific text relating to AI has also been discussed at length at the EU level, taking into account the fact that some specific features of AI (such as the opacity of its functioning) can make the application and enforcement of existing legislation more difficult.[45] At the date of this contribution, the EU Parliament has issued a first proposal regulation relating to AI liability,[46] providing a first tangible legal initiative on the matter.

[41] Act of 13 June 2005 on Electronic Communications, Belgian Official Gazette, 20 June 2005, p. 28070.

[42] Article 129 of the Electronic Communications Act refers to both the information of the consumer and his or her consent, having to comply with the previous Belgian Privacy Act, which must be read as referring to the GDPR.

[43] BDPA, Recommendation 01/2020 of 17 January 2020 in relation to the processing of personal data for direct marketing purposes. Available (in French) at https://autoriteprotectiondonnees.be/publications/recommandation-n-01-2020.pdf. Accessed on 5 May 2020.

[44] See for instance H. Jacquemin and J.-B. Hubin, L'intelligence artificielle et le droit, Collection du Crids, 2017, Brussels, Larcier Y. Poullet, le RGPD face aux défis de l'intelligence artificielle, Collection du Crids, 2020, Brussels, Larcier; C. Vanleenhove and J. De Bruyne, Artificial Intelligence and the Law, Collection Centrum voor Verbintenissenrecht, 2021, Brussels, Intersentia.

[45] European Commission White Paper on Artificial Intelligence - A European approach to excellence and trust, 19 February 2020, p. 10. Available at: https://ec.europa.eu/info/sites/info/files/commission-white-paper-artificial-intelligence-feb2020_en.pdf. Accessed 10 May 2020.

[46] Proposal for a Regulation on the liability for the operation of Artificial Intelligence-systems, contained in the European Parliament resolution of 20 October 2020 with recommendations to the Commission on a civil liability regime for artificial intelligence (2020/2014(INL)). Available at: https://www.europarl.europa.eu/doceo/document/TA-9-2020-0276_EN.html. Consulted 25 November 2020.

13.3 Identifiability and the Precise Scope of Advertising Laws

Advertising is no longer the reserved playing field for "for profit" actors selling goods or services, and this raises the question of whether "non-commercial" advertising should also be approached by the law. Also, the development of new forms of marketing challenges the identifiability requirement in an environment where new technologies evolve faster than the law.

13.3.1 Non-commercial Advertising

Book VI and Book XII CEL each have their own definition of "advertising", which covers communications with the direct or indirect purpose of promoting goods or services. Moreover, under Book XII CEL, the communication can also be made to promote the image "of a company, organisation or person having a commercial, industrial or regulated activity".[47] The latter results from the implementation of the definition contained in the Directive on electronic commerce.

Information giving access to the activities of an undertaking or natural person (such as domain name or e-mail address) and independently developed communications (among others, unpaid communications) are excluded from this definition. Meanwhile, the Court of Justice of the European Union (CJEU) ruled that advertising could also include metatags and domain names when used together and when referring to products, services or the commercial name of a company.[48]

The practice has shown that both definitions tend to be used indistinctively by the courts, and their practical differences, due to the way they are interpreted, seem to be of little relevance.[49]

These definitions used to determine the scope of application of Book VI and Book XII CEL, contain a requirement that the communication must originate from an organisation or company having a lucrative purpose, and promoting the goods, services, or their own image. Consequently, "non-profit" advertising, such as communications from political candidates or parties or religious or philosophical associations, remains out of the scope of these rules, and the consumers do not enjoy protection as to their misleading or aggressive character. However, specific sanctions

[47] In this interpretation, Book XII CEL contains rules that will come on top of those set out under Book VI when an ad or commercial communication is provided with the use of information society services.

[48] CJEU, 11 July 2013, case C-657/11, *Belgian Electronic Sorting Technology*, ECLI:EU:C:2013:516. However, the court confirms that the registration of a domain name, as such, is not encompassed by this definition. See also A. Bochon, To be a communication or not to be: the influence of IT innovation on the material scope of EU advertising law, R.E.D.C., 2014/25, p. 289–308.

[49] As reported in J. Vandendriessche, Elektronische handel – reclame online in Praktijkboek Recht en ICT, 2015, Bruges, Vanden Broele, p. 19.

under criminal law could apply as this practice could constitute offences such as harassment, proselytising or breach of electoral laws.[50]

The same idea is found in the AVMS Directive,[51] which defines audiovisual commercial communication as images (with or without sound) designed to promote, directly or indirectly, the goods, services or image of a natural or legal person pursuing an economic activity qualifying as a service under the Treaty on the Functioning of the European Union (TFEU).[52]

The exclusion from these rules of non-commercial communication can be put into question due to the general use of social media by all stakeholders, notably the development of promotional communication by political parties or non-profit associations, which could also convey misleading or aggressive messages. Among many examples, we mention here the increasing appearance of fake news or unchecked facts used to support a political opinion or comment regarding the efficiency of vaccines or the actual effects of a pandemic, as well as the use of specific technologies, such as deepfakes or other sorts of sophisticated montages that a recipient would be unable to identify as being "promotional" communication.

This exclusion can raise concerns since these non-commercial forms of advertising could constitute a threat to individuals far beyond their consumer rights: one can wonder how someone can safeguard his or her freedom of expression, right to vote or freedom of religion if he/she is unknowingly influenced to waive some of his/her personal rights and adopt specific voting decisions or opinions that he/she would not make in the absence of the "non-commercial" practice at stake.

A reflection should be led at the legislative level to consider an extension of the rules applicable to fair commercial practices to other forms of promotional practices that do not have as a main goal the sale of goods or services, taking into account their specific nature. Building on the definition of the "material distortion of a consumer's economic behaviour",[53] we can think, for instance, of alternative criteria to determine how a non-commercial practice will have an influence on a consumer's decision, which for non-commercial practices will not be animated by economic, but rather "societal", purposes. One can imagine a definition such as "any decision taken by a recipient concerning whether, how and on what terms to make a decision

---

[50] See on this subject F. Coppens, Le spamming politique : affaire de harcèlement, de prospection et de traitement de données à caractère personnel?, Rev. dr. pén. Entr., 2010/4, p. 321–330.

[51] For a detailed analysis of the new rules introduced by the new AVMS Directive, see also P. Vlacke, N. Feci and V. Verdoodt, Herziening van de Richtlijn Audiovisuele Mediadiensten: Over Porsches en ezelwagens in tijden van convergentie, A&M, 2018-201, n°2,p. 218–235.

[52] Article 1 (1) (h) of the AVMS Directive. See also Recital 21 AVMS Directive where reference is made, when referring to an economic activity, to the definition of "services" as defined under the TFEU, Article 57: *Services shall be considered to be "services" within the meaning of the Treaties where they are normally provided for remuneration, in so far as they are no governed by the provisions relating to freedom of movement for goods, capital and persons. "Services" shall in particular include: (a) activities of an industrial character; (b) activities of a commercial character; (c) activities of craftsmen; (d) activities f the professions (...).*

[53] Article 2 (e) of the Unfair Commercial Practices Directive.

impacting their individual rights and society, to waive or restrict their own freedoms or those of others".

Under privacy laws, this issue is less likely to arise when a similar communication is realised following the processing of personal data. In that case, the communication will qualify as "direct marketing", subject to the requirements of the GDPR, and enjoy a broader understanding that the BDPA defined as "[a]ny communication, whether solicited or unsolicited, aimed at promoting an organisation or an individual, services, products, whether paid or free, as well as brands or ideas, communicated by an organisation or an individual acting for commercial purposes or not, directly to one of more natural persons in a private or professional framework, by any means, and involving processing of personal data".[54]

In cases where a direct marketing communication is directed towards a particular individual, the above-mentioned lack of protection under consumer law could be mitigated by the rules set out under the GDPR as their sender will be subject to far-reaching information obligations (including the disclosure of the data controller's identity under Article 13 (1) (a) GDPR, the purposes for which their personal data are processed (c), the recipients or the categories of recipients of the personal data (e), the categories of personal data and their further processing purposes under Article 14 (1) (c) and (d)), especially if they intend to rely on consent as a legal base for processing or additionally, be informed of the right to object to processing for direct marketing purposes under Article 21 (1) GDPR if the sender relies on legitimate interests as a ground for processing. For example, the BDPA recently sanctioned a non-for-profit organisation for sending e-mails to subscribers without informing them about their right to reject the processing of their personal data, nor allowing for the effective exercise of such right, outside of any marketing of a product or service.[55]

If this communication includes the placement of cookies on a consumer's remote computer, he or she will have to be provided with the same clear and comprehensive information in accordance with the GDPR, and the sender will necessarily have to rely on the user's prior consent to proceed.[56]

The communication of political opinions and religious or philosophical beliefs may involve the processing of related personal data, which will fall under the definition of "sensitive data" (Article 9 GDPR). These will require explicit consent

[54]Freely translated from the French version of the BDPA Recommendation on Direct Marketing, *op cit*, p. 8. The BDPA rightfully points out that the term is used, e.g., in recital 47 or Article 21.2 of GDPR but never receives a proper definition.

[55]See Decision n°32/2020 of 16 June 2020 of the BDPA, available at: https://autoriteprotectiondonnees.be/publications/decision-quant-au-fond-n-32-2020.pdf. Accessed 18 August 2020. The BDPA doesn't specify the activities in which this NPO was involved. In this case, the sanction from the BDPA was also justified by the complete lack of cooperation from the data controller throughout the BDPA's prior investigation.

[56]Article 129 of the Electronic Communications Act.

from the data subject, which will give him or her additional protection as they will have to actively consent to such processing.[57]

However, for those non-commercial ads that are not directed to individuals using their personal data, the issue remains that the "consumers" will not enjoy protection from either the GDPR or the CEL, and this point should equally be further addressed. A first step is to be found in the proposed e-Privacy Regulation, which specifies in its recital 32 that advertising covers not only the offering of products and services for commercial purposes but also "messages sent by political parties that contact natural persons via electronic communications services in order to promote their parties" and "messages sent by other non-profit organisations to support the purposes of the organisation".[58]

13.3.2 Identifiability of the "Commercial" Communication

Consumers should be able to identify a communication, including the media on which it is conveyed, as constituting "advertising". Under recital 29 of the Directive on electronic commerce, it was already mentioned that such transparency requirements met the interests of consumer protection and fair trading, taking into account that such communication is essential for the financing of information society services and for developing a wide variety of new, charge-free services.

More recent legislation takes into account the fact that consumers rely increasingly on consumer reviews and endorsements when making transactional decisions and introduces requirements to provide that consumer with information on other consumers' reviews, such as their actual use of the product—or that the reviewed constitutes a paid endorsement. Moreover, enterprises are obliged to put appropriate checks in place to ensure the reality and veracity of such reviews.[59] Specific forms of audiovisual advertising, such as product placements, are regulated as well, if not forbidden, when used in programmes targeting children.[60]

This issue is particularly relevant to weaker groups of consumers, such as children, senior consumers or disabled people. Studies have shown that children

---

[57] However, for existing or former members of a political or religious organisation, or people having regular contact with it, that organisation could rely on the exception set out under Article 9 (2) (d) GDPR to process personal data in the course of its "legitimate" activities with appropriate safeguards, and provided that such collected personal data are not disclosed outside the organisation without the data subject's consent.

[58] Proposal for a Regulation of the European Parliament and the Council concerning the respect for private life and the protection of personal data in electronic communications and repealing Directive 2002/58 (Regulation on Privacy and Electronic Communications), COM/2017/010 final - 2017/03 (COD). The Article 29 Working Party noted however that this consideration was only mentioned in a recital and should be reproduced in proper articles dealing with direct marketing – see Article 29 Data Protection Working Party, Opinion 01/2017 on the Proposed Regulation for the ePrivacy Regulation (2002/58), adopted on 4 April 2017. Accessed on 25 May 2020. Available at: http://ec.europa.eu/newsroom/document.cfm?doc_id=44103.

[59] See Recital 47 of the Omnibus Directive; Recital 25 and Article 5 of the P2B Regulation.

[60] See recitals 32 and 34 of the AVMS Directive.

are receptive to ads from digital influencers (e.g. through social media, "vlogging" or gaming platforms), who create their own content and with whom children can identify more easily than with "traditional celebrities", resulting in a lack of consciousness that their "interaction" with the influencer consists in receiving commercial communication around a product or service.[61] Moreover, due to the increasing use of AI, concern has been raised that the same population groups create an emotional bond with a virtual machine, for instance embodying a friendly or reassuring figure, and lack judgment as to the communication and "instructions" that they will receive from it.[62]

However, the absence of a general and technology neutral norm on identifiability dealing with these new forms of advertising could create interpretation issues.

The Belgian implementing AVMS decrees, at regional level, [63] contains rules that appear more precise and stricter than those of the CEL, not only in terms of content (exclusion of specific advertising techniques, products[64] or features) but also regarding specific types of communication (sponsorship,[65] product placement,[66] on-demand programmes,[67] television advertising[68]). For other types of content, self-regulation is also encouraged,[69] such as for accessibility, as well as direct marketing, profiling and behaviourally targeted advertising for minors,[70] alcoholic

[61] V. Verdoodt and N. Feci, Digital Influencers and vlogging advertising: Calling for awareness, guidance and enforcement, A&M, 2018–2019/1, p. 11–22. See, from the same author: V. Verdoodt, Kinderrechten en reclamewijsheid in het digitale tijdperk - Richting een empowerend regelgevend kader voor commerciële communicatie, D.C.C.R., 2019/3, n° 124, p. 3–13.

[62] Pt 3 of European Parliament resolution of 16 February 2017 with recommendations to the Commission on Civil Law Rules on Robotics (2015/2103(INL)) (2018/C252/25), available at: https://eur-lex.europa.eu/legal-content/EN/TXT/HTML/?uri=CELEX:52017IP0051&from=EN. Accessed on 25 May 2020. On the need to protect children against personalised or behavioural advertising, Articles 6 (a) (2) and 28 (b) (3), in light of the revised AVMS Directive, provide that the processing of personal data of minors collected or otherwise generated by media service providers cannot be processed for commercial purposes, such as direct marketing, profiling and behaviourally targeted advertising. Such collection of children's personal data inevitably happens due to the obligation of media service providers to put child protection mechanism systems. As personal data will be processed for the sole purposes of the functioning of such systems, these providers should not reuse the children's personal data for commercial purposes (Recital 21 of the AVMS Directive).

[63] See footnote n°36.

[64] Such as the advertising of cigarettes, alcohol or prescription medicines – see Article 9 of the AVMS Directive.

[65] Article 10 AVMS Directive.

[66] Article 11 AVMS Directive.

[67] Articles 12 and 13 AVMS Directive.

[68] Articles 19 to 26 of the AVMS Directive and Article 27 of the AVMS Directive, which specifically deals with protection of minors.

[69] According to the EU Commission, experience has shown that both self- and co-regulatory instruments, implemented in accordance with the different legal traditions of the Member States, can play an important role in delivering a high level of consumer protection. See Recitals 13 and 14 of the AVMS Directive.

[70] Article 6a AVMS Directive.

beverages and "junk food"[71] in all audiovisual commercial communications, including now, video-sharing platforms.[72]

These rules will, however, be limited to media services falling under the scope of the AVMS Directive. The concern was raised that the EU policy continues taking an approach that specifies the audiovisual media services to which such rules apply, with a relative lack of clarity as to its exact scope, whereas a more general approach would achieve a better protection of the recipients of such ads. In the area of digital services, EU policy tends to work in silos, distinguishing between all types of services, whereas a rather horizontal, layered approach would be necessary with a regulation applying to all types of media indiscriminately.[73]

Moreover, rules on user-generated content (in which will lie posts by vloggers, influencers or influ-gamers, for instance) applied to a media within the scope of the AVMS Directive are only applicable to the extent that the communication is made with a commercial intent, and that either the platform or the poster has editorial responsibility, which, in the latter case, is less evident considering the number of intervening parties in creating the content.[74]

The same issue arise in the (self-regulatory) guidelines adopted by BEJA on influencers[75] and on native advertising,[76] for which an identification obligation only exists if (1) the "influencer" receives a counterpart from the business willing to have

[71] Articles 9 (3) and (4) AVMS Directive. Under "junk food", we are referring to the description in the AVMS Directive of "foods and beverages containing nutrients and substances with a nutritional or physiological effect, in particular fat, trans-fatty acids, salt or sodium and sugars, of which excessive intakes in the overall diet are not recommended". The Member States and the EU Commission can themselves adopt codes of conduct in these areas.

[72] Article 28b of the revised AVMS Directive. Regarding this latter category, the EU Parliament has recently issued guidelines trying to clarify which audiovisual media services were captured under the provisions applicable to video-sharing platform services when the latter constitutes their "essential functionality". See EU Commission Guidelines on the practical application of the essential functionality criterion of the definition of a 'video-sharing platform service' under the Audiovisual Media Services Directive (2020/C 223/02), 7 July 2020, available at: https://eur-lex.europa.eu/legal-content/EN/TXT/HTML/?uri=CELEX:52020XC0707(02)&from=EN. Accessed 25 September 2020.

[73] J.-F. Furnémont, Directive SMA : une gouvernance renforcée au service de règles dépassées?, A&M, 2017, n°3, p. 304–311. This author explains that such an approach is currently quite difficult due to the rigidity of the decisional processes of EU institutions, since various subjects are dealt with at different directorates within the EU Commission, different ministers at the European council and different commissions within the EU Parliament.

[74] See on this topic V. Verdoodt and N. Feci, Digital Influencers and vlogging advertising: Calling for awareness, guidance and enforcement, A&M, 2018–2019/1.

[75] Recommandations du Conseil de la Publicité en matière d'influenceurs en ligne (in Dutch: *Aanbevelingen inzake online influencers*), October 2018 (date not further specified), available at: https://www.jep.be/sites/default/files/rule_reccommendation/code_en_matiere_didentification_des_publicites_natives_et_communications_commerciales_connexes.pdf. Accessed 13 April 2020).

[76] Code en matière d'identification des publicités natives et communications commerciales connexes (in Dutch: *Code over herkenbaargeid van native advertising en aanverwante commerciële communicatie*), 12 December 2018 (published 4 January 2019), available at: https://

its goods or services promoted (either cash or any other advantage) and (2) the business has decisional power over the communication's content.

This means that in cases where a business is paying a user of a social network or audiovisual service to promote a product or service without making any decision on how this communication will take place, thus not exercising any editorial control, such promotion would not be subject to the identifiability requirements. Naïve influencers would bear full editorial responsibility for the commercial communication of products or services, although they do not specifically act as professionals and may not be even aware that their post constitutes an ad or has the purpose of realising a commercial communication.[77]

The fact that the general norm for unfair commercial practices contained under the CEL is broad enough could serve to mitigate such issues as any breach of legislation could constitute an unfair commercial practice, be it local laws beyond the rules of the CEL, international laws of self-regulatory codes of conduct. The issue with the latter category lies in the fact that the business should be a signatory of that code for there to be a breach and subsequent unfair trade practice.[78]

The "Black list" of commercial practices under the CEL includes the use of editorial content in the media—paid for by a business—to promote a product or service without making this clear in the content or through images or sounds easily identifiable by the consumer,[79] as well as the inclusion in an advertisement of a direct exhortation to children to buy advertised products or persuade their parents or other adults to buy advertised products for them[80] and also unauthorised online direct marketing.[81]

Finally, the rules on unfair competition could be useful to denounce practices that are, in appearance, legitimate but remain unfair when considering their specific target audience, mostly when children are involved, even though, in the latter case, some scholars plead for the adoption of a specific set of rules protecting this sensitive audience.[82]

www.jep.be/sites/default/files/rule_reccommendation/code_en_matiere_didentification_des_publicites_natives_et_communications_commerciales_connexes.pdf. Accessed 12 April 2020.

[77] See also V. Verdoodt, Aanbevelingen voor online influencer-marketing van de Raad voor Reclame, D.C.C.R., 2018/3–4, n°120–121, p. 241–243.

[78] G. Sorreaux, Publicité et étiquetage des denrées alimentaires, Bruxelles, Editions Larcier, 2016, p. 49–50.

[79] Article VI.100, 11° CEL.

[80] Article VI.103, 5° CEL.

[81] Article VI.103, 3° CEL. Under this bullet, the CEL is referring to unauthorised direct marketing as marketing not authorised by law, be it under Article VI.110 (unsolicited communications), XII.13 (direct marketing by e-mail), or a specific law or regulation authorising direct marketing in order to perform a contractual obligation.

[82] See H. Jacquemin, Chapitre 8 – Protection du Consommateur et numérique en droits européen et belge, in: Vulnérabilités et droits dans l'environnement numérique, Bruxelles, Editions Larcier, 2018, p. 244. See also, on this specific question, V. Verdoodt, Kinderrechten en reclamewijsheid in het digitale tijdperk - Richting een empowerend regelgevend kader voor commerciële communicatie, D.C.C.R., 2019/3, n° 124, p. 3–13.

For the identifiability requirement, the Omnibus Directive provides for additional rules that could address these concerns, without considering the type of media involved or the existing editorial liability of either the business or the intermediary.

On the one hand, the following practices are added into the "black list" of commercial unfair practices: (1) the fact that a business provides search results in response to a consumer's online search query without clearly disclosing any paid advertisement or payment specifically for achieving higher ranking of product within the search results, (2) stating that reviews of a product are submitted by consumers who have actually used or purchased the product without taking reasonable and proportionate steps to check that they originate from such consumers and (3) submitting or commissioning another legal or natural person to submit false consumer reviews or endorsements, or misrepresenting consumer reviews or social endorsement, in order to promote products.[83]

On the other hand, when a business gives access to consumer reviews of products, it should provide information about whether and how the business ensures that the published reviews originate from consumers who have actually used or purchased the product. Failure to do so will be deemed an omission of information and will qualify as a misleading commercial practice.[84] A similar obligation is provided for a business exploiting an online marketplace, which is required to specify whether the third party offering the product is a trader or not based on the declaration that the third party made to that business.[85]

13.4 Personalisation: Providing the Consumer with Tailor-Made Communication or Product, with or Without Automated Means

The increasing use of automated means, such as AI, to create personalised commercial communications, services and products shows a shift of the market towards a consumer-centric (or human-centric) approach, defined by the EU Commission as "an international approach that promotes the respect of fundamental rights, including human dignity, pluralism, inclusion, non-discrimination and protection of privacy and personal data".[86]

[83] See Article 3 (7) (a) and (b) of the Omnibus Directive adding, respectively, pts 11a, 23b and 23c to Annex I of the Unfair Commercial Practices Directive.

[84] Article 3 (4) (c) of the Omnibus Directive.

[85] Article 3 (4) (a) (ii) of the Omnibus Directive. On this latter obligation, one can wonder why the provider of online platform service does not have an active duty to verify the identity of such third party, as is the case for consumer reviews, or an obligation to put the appropriate checks in place.

[86] See European Commission White Paper on Artificial Intelligence - A European approach to excellence and trust, 19 February 2020. Accessed on 10 May 2020. Available at: https://ec.europa.eu/info/sites/info/files/commission-white-paper-artificial-intelligence-feb2020_en.pdf, p. 9. Under p. 13, the EU Commission considers that legislation could be improved especially regarding product safety, liability and transparency.

"Personalisation" in itself, with or without automated means, should be considered a commercial practice that can potentially have an influence on consumers' behaviour and their decision over a transaction.[87]

The legal literature recognizes that the performance of an agreement involving personal data using AI could lead to automated processing of such data under Article 22 GDPR if there is no human intervention at all. In its Proposal for a Regulation on the liability for the operation of AI systems, the EU Parliament confirms that AI is a form of "automated decision-making", defined as a "situation in which a user initially delegates a decision, partly or completely, to an entity, by means of software or a service. That entity, in turn, uses automatically executed decision-making models to perform an action on behalf of a user, or to inform the user's decision in performing an action."[88]

Both elements combined challenge again some of the existing rules on consumer and unfair competition laws (including consumer information) and data protection (including the rules applicable to the automated processing of personal data and the principles of accountability and privacy by design under the GDPR) and show that a convergence of both sets of legislation is necessary to guarantee consumers' empowerment (or "agentivity")[89] in the exercise of their rights, allowing them to maintain control over their decision-making and the communication of their personal data. This, in turn, ensures that the enterprise can achieve a proper personalisation of consumers' needs and process relevant quality data to achieve these purposes using automated means.

13.4.1 Consumer Information

13.4.1.1 Consumer Laws

In terms of consumer contracts, an enterprise is under obligation to inform the consumer of the total price of goods or services or, if such a price cannot be calculated in advance, the manner in which the price is to be calculated, as well as

[87] F. Jacques, Personnalisation des prix - Regard européen sur la pratique, R.D.T.I., n°78–79, 2020, p. 59; H. Jacquemin, Le big data à l'épreuve des pratiques du marché et de la protection du consommateur, R.D.T.I, 2018/1, n°70, p. 75–91.

[88] European Parliament report of 5 October 2020 with recommendations to the Commission on a civil liability regime for artificial intelligence (2020/2014(INL)), available at: https://www.europarl.europa.eu/doceo/document/A-9-2020-0178_EN.html, p. 16. Accessed 28 November 2020. The idea that a human facing the use of AI for a commercial transaction, whether or not involving the processing of its personal data, was already debated before. See Independent high-level Expert Group on Artificial Intelligence set up by the European Commission, Ethic Guidelines for Trustworthy AI, 8 April 2019, available at https://ec.europa.eu/newsroom/dae/document.cfm?doc_id=60419. Accessed 10 May 2020. p. 18.

[89] C. Zolynski, Protection + Agentivité, la nouvelle équation pour penser les relations entre consommateurs et intelligences artificielles, Ent. & L., 2019, n°6, pp. 331–341.

all additional charges and whether those are included in the price or not.[90] This information obligation could even be more general to the extent that the sold goods or services are personalised, with or without the use of automated means, taking into account the obligation of an enterprise to inform the consumer of the main characteristics of the goods or services.[91]

This set of rules reinforces the civil law rules on consent, which requires that the contractual counterparty, taking into account the possible imbalance of power[92] (because he or she is a consumer), is provided with the necessary information allowing him or her to engage "wilfully and knowingly".

If an enterprise fails to provide information to the consumer in a clear and understandable manner (because it is either missing, false or presented in a way that is confusing to the consumer), this will be considered a misleading commercial practice. More specifically, it seems that the business would be under a broad obligation to explain the rationale behind and the methods used in the personalisation of a good or service since not providing information in relation to its "(. . .) benefits, risks, execution, composition, (. . .) method and date of manufacture or provision (. . .), specification"[93] would constitute a misleading commercial practice.

Regarding the price, the lack of information or misleading information concerning the price or the manner in which it is calculated or the existence of a specific price advantage is also already considered a misleading commercial practice.[94]

It has been debated in the legal literature whether an enterprise's information obligation towards the consumer should cover the use of AI and include an obligation to disclose prior to a transaction that the entity with whom the consumer interacts is not human.[95] Such is the case when a consumer interacts with a virtual assistant or chatbot and may have the impression that he or she is communicating with a human of flesh and blood who is there to—and wants to—help him/her complete the transaction. This information should also contain a disclosure of the logic behind the algorithms used underlying this interaction between the consumer and non-human entity.

[90] Article VI.2, 3° CEL (general information obligation) and Article VI.45,3° CEL for distance contracts.

[91] Article VI.2, 1° CEL (general information obligation) and Article VI.45,1° for distance contracts.

[92] This seems to be systematically the case for a consumer navigating in an online environment, who does not have any leverage to negotiate or waive some of the terms of a business prior to entering the transaction.

[93] Article VI.97,2° CEL (misleading information) and VI.99, §4, 1° CEL (omission of information).

[94] Article VI.97, 4° CEL. (misleading information) and VI.99, §4, 3° CEL (omission of information).

[95] J.-B. Hubin and H. Jacquemin, Titre 3 – L'intelligence artificielle: vraie ou fausse amie du justiciable? - Enjeux du recours à l'IA par les avocats, assureurs et legaltechs, In: Le juge et l'algorithme: juges augmentés ou justice diminuée?, Editions Larcier 2019, p. 75–104.

Where Belgian law in its current state does not provide for such specific obligations, the Omnibus Directive provides two new requirements that will need to be implemented under the CEL.

First, Article 6 of the Unfair Terms Directive, implemented under Article VI.45 CEL, has been updated to include an obligation to provide information prior to the conclusion of a "distance contract" that the price was personalised based on automated decision-making.[96] This rule allows consumers to consider the potential risks of their purchasing decision. However, this requirement is met with two issues. First, the level of information detail that must be provided to the consumer while ensuring that such information is provided to them in a clear and understandable manner or the extent to which the enterprise should explain the functioning of the algorithms leading to the automated decision remains unclear and should be subject to further guidance.

Also, this new requirement does not apply if personalisation was achieved through means other than automated processing, such as "real-time" or "dynamic" pricing. For instance, these techniques can consist in changing the price in a highly flexible and quick manner in response to market demands and do not involve personalisation based on automated decision-making.[97] Therefore, an enterprise that does not resort to personalised automated decision-making but still adapts its prices using other real-time technologies, such as cookies, tracers or other indicators, can continue to use different prices, and potential discrimination between consumers would not be covered by this obligation.

Second, Article 7 of the Unfair Commercial Practices Directive, dealing with misleading omissions, provides for a list of material information to be provided to the consumers and is implemented under Article VI.99 §4 CEL. This list has been updated by the Omnibus Directive[98] to include a disclosure of the main parameters that determine the ranking of products presented as a result of a consumer's search query when browsing an enterprise's website and the relative importance of those parameters. "Ranking" is defined as the relative prominence given to products, as presented, organised or communicated by the enterprise, irrespective of the technological means used for such presentation, organisation or communication.[99] This addition may be of limited use to AI as the Omnibus Directive provides that such disclosure be without prejudice to the protection that enterprises may enjoy under the Trade Secrets Directive.[100] If the algorithms designed for the functioning of an AI

[96] Article 4 (4) (a) (ii) of the Omnibus Directive, inserting a point (ea) under Article 6 (1) of the Unfair Terms Directive.

[97] Recital 45 of the Omnibus Directive.

[98] Article 3 (4) (b) of the Omnibus Directive, inserting a paragraph 4a under Article 7 of the Unfair Commercial Practices Directive.

[99] Article 3 (4) (b) of the Omnibus Directive, inserting a point (m) under Article 2 of the Unfair Commercial Practices Directive.

[100] Directive 2016/943 of the European Parliament and of the Council of 8 June 2016 on the protection of undisclosed know-how and business information (trade secrets) against their unlawful acquisition, use and disclosure, OJ, 2016, L 157, p. 1.

solution are protected by trade secrets, the amount of information that an enterprise would have to provide without breaching any law would be limited. Moreover, the Omnibus Directive adds that enterprises should not be required to disclose a detailed functioning of their ranking mechanisms, including their algorithms, but rather could provide a general description of the main parameters determining their ranking and their relative importance as opposed to other parameters.[101]

Furthermore, F. Jacques[102] considers the fact that an enterprise can also undertake misleading commercial practices even if the information it provides to the consumer regarding the price or its means of calculation is factually correct. For example, when practicing personal rebates, a business could present a price as discounted specifically for the consumer whereas such price has been reduced beforehand based on personalisation using the consumer's shopping behaviour data. In doing so, the information provided on the price is prima facie correct but presented in a way that the consumer is misled as to its means of calculation.

His position is also that the concept of personalisation as a whole should be disclosed to the consumer and that not doing so would constitute an unfair commercial practice: consumers have shown that information regarding personalisation (the price and its means of calculation) could be considered as "material", the EU Commission further explained in its guidance document on the implementation of the Unfair Commercial Practices Directive that compliance with the rules on the protection of personal data needed to be considered when conducting the assessment, profiling or personalisation in breach of these rules amounting to an unfair commercial practice,[103] and the same guidance document treats the lack of consumer information regarding personal data processing going beyond the transaction with the enterprise, e.g. for marketing purposes, as the omission of a material information.[104] The lack of information on personalisation therefore undoubtedly results in the consumer making a commercial decision that he or she would not have taken had he/she been informed thereof.

---

[101] Recital 23 of the Omnibus Directive.

[102] F. Jacques, Personnalisation des prix - Regard européen sur la pratique, R.D.T.I., n°78–79, 2020, p. 60–62.

[103] European Commission Staff Working Document of 25 May 2016. Guidance on the Implementation/Application of Directive 2005/29/EC on Unfair Commercial Practices accompanying the document Communication from the Commission to the European Parliament, the Council, the European Economic and Social Committee and the Committee of the Regions – A comprehensive approach to stimulating cross-border e-Commerce for Europe's citizens and businesses; COM (2016) 320 final, SWD, 2016, 163 final, p. 24. Available at: https://eur-lex.europa.eu/legal-content/EN/TXT/?uri=celex%3A52016SC0163. Consulted on 18 August 2020.

[104] *Ibid.*, p. 63; N. Van Eijk, C. J. Van Hoofnagle and E. Kannekens, Unfair Commercial Practices: A Complementary Approach to Privacy Protection, E.D.P.L., n°3, 2017, p. 334: these authors see a strong overlap between the omission of material information under Article 7 of the Unfair Commercial Practices Directive and the information obligation from the data controller when collecting personal data under Articles 13 and 14 GDPR.

13.4.1.2 Under Data Protection Laws[105]

When using automated means, the fair and transparent processing of personal data under the GDPR requires the data controller to provide meaningful information about the logic involved in the decision made (Articles 13 (2)(f) and 14 (2) (g)).

The explanation should not necessarily be a complex overview of the algorithms used or a disclosure of the full algorithm. The information provided should, however, be sufficiently comprehensible for the data subject to understand the reasons for the decision made concerning him or her. The Article 29 Working Party adds that complexity is no excuse for failing to provide information to the data subject, which is particularly relevant in a technologically complex environment involving many actors.[106] This is undoubtedly the case when AI comes into play, and it can be noted that the business, the operator, the engineer and the software designer of the AI system used for a single consumer transaction are different people.

Unlike the specific requirements brought under by the Omnibus Directive, this appears to be a general rule that is not limited to specific categories of information such as price or ranking, and could reinforce a possible lack of information regarding the general rationale behind the algorithms used under consumer laws.

13.4.2 Privacy by Design and Accountability Principle

When using personalisation techniques, with or without automated means, an enterprise faces specific questions under privacy laws, based on the accountability principle (Article 5 (2) GDPR). For instance, the enterprise must ensure that its processing activity complies with its GDPR obligations, both at the time of determining the means of processing and during the processing itself, and it must implement appropriate technical and organisational measures designed to implement the data protection principles, such as data minimisation, in an effective manner and must integrate the necessary safeguards into the processing procedure in order to meet the requirements of the GDPR and to protect the rights of the data subjects.[107]

13.4.2.1 Finding the Appropriate Legal Basis for the Processing of Personal Data

An important question to consider on the part of the controller is which legitimate ground to use for the processing of personal data—the main legitimate grounds

[105] See also R. Steppe, Prijsdiscriminatie in het digitale tijdperk. Beschouwingen over de nieuwe Algemene Verordening Gegevensbescherming i), n R. Steppe, M. Storme, F. Helsen (eds.), Innovatie en disruptie in het economisch recht, 2017, pp. 105–149.

[106] Article 29 Data Protection Working Party, Guidelines on Automated individual decision-making and Profiling for the purposes of Regulation 2016/679, adopted on 3 October 2017 and last revised on 6 February 2018. Accessed on 15 May 2020. Available at: https://ec.europa.eu/newsroom/article29/document.cfm?action=display&doc_id=49826, p. 25.

[107] Articles 24 and 25 GDPR.

being consent (Article 6 (1) (a) GDPR) and legitimate interest (Article 6 (1) (f) GDPR).[108]

To rely on consent as a legitimate ground, the data controller will need to ensure that the data subject's consent is (1) freely given, (2) specific, (3) informed and (4) unambiguous.

Processing for the legitimate interests of the controller should be subject to the following conditions: (1) the interests of the controller must be legitimate,[109] (2) the processing of the personal data must be necessary for the realisation of the controller's objectives and (3) a necessary balance must take place between the interests, freedoms and fundamental rights of the data subject and the data controller's own interest.

Consent, under the strict conditions of the GDPR, has long been seen as the most appropriate way to ensure that the data subject is correctly informed and accepts the processing of his or her personal data with knowledge of things. Similar to consumer laws, the data subject must be provided with substantial pre-contractual information before consenting.

However, the need for a valid consent satisfying the conditions of the GDPR appears burdensome or no longer adapted to the current state of advertising technologies.

On the other hand, legitimate interest as a ground for processing personal data is not exempt from criticism either: it is seen as increasing an asymmetry of information between consumers, data subjects, and enterprises, data controllers and cannot be used as legal basis in some circumstances. It is therefore uncertain that it can provide an effective alternative to consent for personalisation.

Several issues arise that call into question the effectiveness of both grounds, and a reflection from the enterprise is needed to anticipate these issues.

[108] The performance of a contract (Article 6 (1) (b) GDPR) is another possible ground for processing but is not considered in detail in this contribution. We briefly touch upon it when discussing consent and the New Deal for Consumers below.

[109] Recital 47 GDPR states that direct marketing can be, in itself, a legitimate interest of the data controller. The WP29 specified that the interest is legitimate, as long as it is in compliance with data protection and other laws. Belgian scholars have deemed this approach too restrictive and the interests should also be assessed under the lights of fairness and transparency, which are required for any personal data. Moreover, these interests would necessarily include compliance with fundamental rights. Therefore, personalisation based on legitimate interests should take place in compliance with such fundamental rights and a data controller should avoid any breach of, inter alia, the anti-discrimination laws mentioned above. See Article 29 Working Party Opinion 06/2014 on the notion of legitimate interests of the data controller under Article 7 of the Directive 95/46, available at https://ec.europa.eu/justice/article-29/documentation/opinion-recommendation/files/2014/wp217_en.pdf. Accessed 9 May 2020; J.-F. Puyraimond, L'intérêt légitime du responsable du traitement dans le RGPD: in cauda venenum?, D.C.C.R., 2019/1–2, n°122–123, p. 52. See also D. Goens, Data protection bij financiële instellingen, 2018, Anvers - Cambridge, Intersentia, p. 246.

Privacy Fatigue

"Informed" consent requires that the data controller provide sufficiently transparent information, in a clear and intelligible language, on the processing of personal data, either at the time personal data is directly collected by the data controller (Article 13 GDPR) or within a reasonable term when that data is indirectly obtained (Article 14 GDPR).

Providing information to the data subject prior to obtaining his or her consent is essential so as to enable him/her to make an informed decision and exercise his/her rights. If the controller does not provide accessible information, user control becomes illusory and consent will be an invalid basis for processing.[110] On this aspect, the GDPR expressly refers to the requirements set out in the Unfair Terms Directive: information must be concise, easily accessible and in an easy-to-understand and plain language, and, additionally, where appropriate, visualisation should be used.[111] This is welcomed in situations where a lot of actors are involved and the technological complexity of a practice makes it difficult for the data subject to know and understand whether, by whom and for what purpose personal data relating to him/her are being collected, such as in the case of online advertising.[112] Moreover, the data subject should be able to give his/her consent by means of a written declaration (express or functionally similar, such as ticking a box), after being given the opportunity to read the information, usually contained in the terms and conditions, regarding the processing.[113]

Recently, the European Court of Justice (ECJ) explained that the same information obligation also applies to cookies and that, in order to guarantee transparent information, the controller needs to inform the data subject of the duration of such cookies and the possibility for third parties to have access to them.[114]

The BDPA also applied similar principles and imposed a fine on a website specialising in legal news, whose cookies policy was compliant with neither the GDPR nor Article 129 of the Act on Electronic Communications.[115]

[110] Guidelines 05/2020 of 4 May 2020 of the EDPS, on consent under Regulation 2016/679, available at https://edpb.europa.eu/sites/edpb/files/files/file1/edpb_guidelines_202005_consent_en.pdf. Accessed 10 May 2020, n° 62, p. 14.

[111] Recital 42 GDPR. Recital 38 and 58 GDPR specify that such information should be adapted to its target audience. Where children are targeted, such information should be in a clear and plain language that the child can easily understand, especially when direct marketing or the creation of personality or user profiles, is used to provide them services.

[112] Recital 58 GDPR.

[113] Getting the consumer to accept terms and conditions that they did not have the possibility to acknowledge would constitute an abusive contractual clause and, in turn, an unfair commercial practice under Article VI. 83, 26° CEL.

[114] CJEU, 1 October 2019, case C-673/17, *Planet49*, ECLI:EU:C:2019:801.

[115] Decision n° 12/2019 of 17 December 2019 of the BDPA, available at: https://www.autoriteprotectiondonnees.be/publications/decision-quant-au-fond-n-12-2019.pdf. Among other breaches, the website provided incomplete information to its users regarding cookies (Article 12 GDPR) and used pre-ticked boxes to obtain users' consent to the use of cookies, and the data controller used legitimate interest as a ground for the processing of data for the analysis of cookies.

The requirement that consent must be specific reinforces this information obligation. The data subject needs to be able to distinguish between different purposes for the processing of their personal data and choose to which ones they want to agree (the "granularity of consent").[116] In the case of cookies, for instance, two different types of consent would be needed: for analytic cookies and for functional cookies.[117] In this manner, the data controller is prevented from relying on consent for other purposes than those for which it was originally obtained. In terms of personalisation, a data controller would need to obtain a separate second consent from consumers to create specific profiles if it first obtained their personal data for direct marketing communication purposes.

As a result, the consumers are facing a substantial amount of information and steps to be taken when surfing the Internet, on each merchant website they visit, which is necessary to obtain their consent for the enterprise's processing purposes. The BDPA,[118] the EDPB[119] and Belgian scholars share the view that this could lead to a loss of control over the personal data communicated by the consumers as they will tend to be less vigilant and experience a real "privacy fatigue", wherein they will mechanically consent or refuse the processing of such data.

A more effective and reasoned approach to consent, however, is possible but would rest on the enterprise's willingness to communicate prior information to the consumer in an effective and friendly manner: the conclusions of the works of R. Ducato[120] suggest that "informed consent" is not the issue as such but rather the means that the controller uses to convey information to the consumer. Privacy fatigue could be avoided by giving consumers a sense of control and agentivity over their personal data by using other techniques than providing lengthy privacy policies: for example, using new technology (videos, multi-layered menus, chatbots or other types of electronic assistance) or techniques that require a more active interaction with the consumer (questionnaires or "privacy" mini-games).[121] The

On this last point, the BDPA points out that the current rule provides that only strictly necessary cookies are exempt from prior consent from the data subject but does not exclude that the rule could be extended to other types of cookies in the future, in the framework of the reform of the e-Privacy Directive.

[116] EDPB Guidelines 05/2020 of 4 May 2020 of the EDPS, on consent under Regulation 2016/679, available at https://edpb.europa.eu/sites/edpb/files/files/file1/edpb_guidelines_202005_consent_en.pdf. Accessed 10 May 2020, n°56, p. 13.

[117] BDPA Recommendation on Direct Marketing, p. 63.

[118] *Ibid.*, p. 67.

[119] Referring to "Click-fatigue", EDPB Guidelines 05/2020 of 4 May 2020 on consent under Regulation 2016/679, available at https://edpb.europa.eu/sites/edpb/files/files/file1/edpb_guidelines_202005_consent_en.pdf, consulted on 10 May 2020., n°87–89, p. 18.

[120] See R. Ducato and A. Strowel, Information duties between consumer and data protection in the "Internet of Platforms": promoting awareness by design, D.C.C.R., 2019, n°122–123, p. 123–149; A. Rossi, R. Ducato, H. Haapion and S. Passera, When Design Met Law: Design Patterns for Information Technology, D.C.C.R., 2019/1–2, n°122–123, p. 79–12.

[121] See also, sharing the same opinion: F. Jacques, Personnalisation des prix - Regard européen sur la pratique, R.D.T.I., n°78–79, 2020; H. Jacquemin, Le big data à l'épreuve des pratiques du marché et de la protection du consommateur, R.D.T.I, 2018/1, n°70, p. 75–91.

BDPA and EPDB also suggest alternative techniques, like where the data subject can make a specific movement with the mouse or with his/her finger on the controller's webpage or repeat a phrase orally using a microphone. However, in the new version of its guidelines on consent, the EDPS specified, based on recital 32, that actions such as scrolling or swiping through a webpage or any similar user activity would not be sufficient since such actions are difficult to distinguish from the other activities or interactions of a user.[122]

Meanwhile, relying on legitimate interest as a ground for the processing of personal data would also require the data subject to be informed under Articles 13 and 14 GDPR.

Freedom to Consent and Agreements Under the New Deal for Consumers

The data subject must be able to *freely accept* or refuse the processing of his or her personal data without being deprived of the intended service from or the performance of obligation by the data controller. The GDPR forbids that the performance of a service be conditioned upon a consumer's prior consent to have his/her personal data processed for purposes that are not necessary for the performance of an agreement (Article 7 (4) GDPR). Equally, the data subject should be able to withdraw consent without enduring negative consequences (such as exaggerated costs for withdrawal).[123] On the other hand, it seems possible for the data controller to continue providing specific advantages to those data subjects who did not withdraw their consent or for the latter to continue benefitting from additional services.[124]

Free consent also entails that the data subject is able to consent without any inappropriate pressure or influence from the data controller. The BDPA recently sanctioned a business that required its customers to share their electronic identity card and have its content read in the shops in order to create a fidelity card, without providing a valid alternative in case they refuse to do so. As a result, consumers could not validly consent to the processing of their personal data to obtain a fidelity card since if they refused to do so, they would not obtain the same advantages and promotions as the customers enjoying it.[125]

[122] It would also not be possible to allow the data subject to withdraw his or her consent as easy as he or she has given it, see Guidelines 05/2020 of 4 May 2020 of the EDPS, on consent under Regulation 2016/679, available at https://edpb.europa.eu/sites/edpb/files/files/file1/edpb_guidelines_202005_consent_en.pdf, consulted on 10 May 2020, n° 86, p. 18.

[123] Guidelines 05/2020 of 4 May 2020 of the EDPB on consent under Regulation 2016/679, available at https://edpb.europa.eu/sites/edpb/files/files/file1/edpb_guidelines_202005_consent_en.pdf, consulted on 10 May 2020, n°13., p. 6. The guidelines consist of a "slightly updated version" of the Article 29 Working Party Guidelines of 10 April 2018 on Consent under Regulation 2016/679 (WP259.01), which were endorsed by the EDPB.

[124] Y. Poullet, Consentement et RGPD: des zones d'ombre !, DCCR, 2019/1–2, n°122–123, p. 6.

[125] Decision n°06/2019 of 17 September 2019 of the BDPA, available (in Dutch) at: https://www.gegevensbeschermingsautoriteit.be/publications/beslissing-ten-gronde-nr.-06-2019.pdf (consulted on 15 March 2020). This decision was later annulled by the Belgian Market Courts as, *inter alia*, the BDPA had based its reasoning on the legislation on electronic ID, which was not yet in force at

The EDPS recently reviewed its guidelines on consent, which complements the ECJ's *Planet49* case,[126] and condemned the practice of "cookie walls" to obtain consent. Such situations occur when a data controller requires the user to agree to the use of cookies as a condition for the access to its services and functionalities.[127]

Under the upcoming New Deal for Consumers,[128] the European legislator introduces the notion of a consumer agreement the counterparty of which is not money. This cover the cases where the enterprise undertakes to deliver goods or services to the consumer "and the consumer provides or undertakes to provide personal data" to such business, except when such personal data are processed exclusively for the performance of that agreement.[129] The latter processing is subject to a different legal basis under Article 6 (1) (b) GDPR.

When reviewing earlier versions of the text, the EDPB had raised the concern that the grounds for processing under Article 6(1)(b) GDPR would be unduly relied upon, as a data controller would consider that personal data is simply the "counterpart" for receiving the goods or services, just as any payment in money, and rely on the consent to enter into an agreement rather than informed consent under Article 6 (1) (a) GDPR.[130] Also, this new kind of agreement would entice data controllers to take recourse to "tied in" or "bundled" consent, by which the performance of the agreement only occurs if personal data has been provided by the consumer.[131]

This clarification seems to inform enterprises that they cannot use personal data originally collected for the strict performance of an agreement, for further (marketing) purposes. However, it is doubtful that it will effectively remedy the issue related

---

the time of the breaching behaviour of the data controller; see Markets Court of Brussels, 19 February 2020, n°2019/AR/1600, available in Dutch at: https://www.autoriteprotectiondonnees.be/publications/arret-du-19-fevrier-2020-de-la-cour-des-marches-disponible-en-neerlandais.pdf.

[126] CJEU, 1 October 2019, case C-673/17, *Planet49*, ECLI:EU:C:2019:801. This case law did not envisage cookie walls as such but deals with the question of "unambiguous" consent, as defined under the GDPR, in relation to the placement of cookies and a tracking system on a user's device using pre-ticked consent boxes.

[127] Guidelines 05/2020 of 4 May 2020 of the EDPS, on consent under Regulation 2016/679, available at https://edpb.europa.eu/sites/edpb/files/files/file1/edpb_guidelines_202005_consent_en.pdf, consulted on 10 May 2020, n°39, p. 11.

[128] Namely, the Omnibus Directive, but also the Digital Contents Directive (Directive 2019/770 of the European Parliament and of the Council of 20 May 2019 on certain aspects concerning contracts for the supply of digital content and digital services, OJ, 2019, L 136, p. 1.) and the new Sales of Goods Directive (Directive 2019/771 of the European Parliament and of the Council of 20 May 2019 on certain aspects concerning contracts for the sale of goods, amending Regulation 2017/2394 and Directive 2009/22/EC, and repealing Directive 1999/44/EC, OJ, L 136, 2019, p. 28). The last directive is not envisaged in this contribution.

[129] Recital 31 Omnibus Directive, and Article 4 (2) inserting an Article 3 (1a) in the Consumer Rights Directive.

[130] Opinion 8/2018 of the EDPS, on the legislative package "a New Deal for Consumers", available at: https://edps.europa.eu/sites/edp/files/publication/18-10-05_opinion_consumer_law_en.pdf, pp. 14–16.

[131] *Ibid.*, p. 17; Recital 43 and Article 7 (4) GDPR.

to tied-in consent. Article 6 (1) (b) GDPR is limited in scope as it only covers processing that is "strictly" necessary for the performance of an agreement and therefore cannot be relied on if a data controller can reach the same purposes through other means.[132] In the cases where an enterprise cannot or does not need to rely on Article 6 (1) (b) GDPR to actually perform its agreement with a consumer, it can still be collecting personal data using consent, and be tempted to make it conditional on the provision of goods or services. In this understanding, the need for the consumer to provide consent and its reach can still be source of confusion if the consumer is not provided with proper information, since they still could be led to believe, wrongly, that the provision of their personal data is necessary to carry on a transaction with the enterprise.

To avoid any risk related to consumer confusion in relation to the basis for processing (consent or performance of agreement), a data controller could be tempted to rely on legitimate interest instead of consent. However, in the current state of the law, the issue linked to cookies remains since their placement on a consumer's/data subject's device still requires his or her consent, whether or not personal data are processed.[133] The same issue arises in the case of direct marketing communications sent by e-mail.[134] Moreover, "explicit" consent will also be mandatory when the data controller seeks to process sensitive personal data under Article 9 GDPR, such as race or ethnic origin, political opinions, religious or philosophical beliefs as well as health data. Processing sensitive personal data is forbidden unless the data controller can rely on one of the exceptions set out under Article 9 (2) GDPR, among which explicit consent, and this seems to be the only relevant exception in the framework of personalisation for promotional purposes, which are not covered by the other available exceptions available under that Article.

Imbalance of Power Between the Data Subject and the Data Controller

The use of legitimate interest as a ground for the processing of data remains doubtful as some authors raise concerns that it increases the information asymmetry between the controller and the data subject and, allegedly, does not offer the same protection as consent does.

[132] Guidelines 2/2019 of 8 October 2019 of the EDPS on the processing of personal data under Article 6(1)(b) GDPR in the context of the provision of online services to data subjects, available at https://edpb.europa.eu/sites/edpb/files/files/file1/edpb_guidelines-art_6-1-b-adopted_after_public_consultation_en.pdf., n°25, p. 8. Accessed on 22 August 2020. According to the BDPA, a possible case where a data controller could rely on such ground is where the data subject and the data controller have entered into an agreement for the sole purpose of receiving direct marketing communication, where personal data would then be processed solely for direct marketing purposes, i.e. the performance of the agreement. See BDPA Recommendation on Direct Marketing, p. 47.

[133] Article 129 of the Electronic Communications Act.

[134] The BDPA mentions as an example unsolicited communication by electronic means, which requires prior informed consent from the data subject and the "soft opt-in" exception related to such communications (see Article XII.13 CEL and its executing Decree of 4 April 2003 and Article 13(1) and (2) of the e-Privacy Directive.

While the data controller would be under the obligation to provide substantial pre-contractual information to the data subject when relying on consent, this alone, does not seem to eliminate such risk, just as is the case under consumer laws, since big business players enjoying a monopoly and *de facto* stronger power:[135] in most instances, not only will the text of the privacy policy (or the commercial terms and conditions) not be negotiable, but the consumer/data subject will feel compelled to consent either way without having control or leverage as to the information given. This puts into doubt the fact that the consumer has effectively "consented" to the business' commercial conditions and privacy policy.

Whereas the legal grounds of contract performance and of consent do not seem to solve this issue, the obligation for the data controller to provide specific safeguards when relying on legitimate interest could be seen as a sufficient trade-off allowing it to carry on with processing, without the "burden" of consent—which is not an efficient protection for the consumer anyway. The consumer, instead, in terms of safeguards is granted with an absolute right to oppose the processing of personal data for direct marketing purposes[136] and to oppose automated processing, or at least require a human intervention or review of the processing mechanics.[137] Moreover, an additional safeguard is found in the data minimisation and purpose limitation principles, under which a data controller will be prevented from reusing personal data obtained through another legal basis, and will have to assess whether another purpose for processing is compatible with the purpose for which data was originally collected.[138] As a matter of example, the BDPA issued several decisions sanctioning the processing of personal data for electoral purposes, whereas they had originally been collected for another purpose which was deemed incompatible, where one of the data controllers could even potentially obtain financial benefits which is an aggravating factor under Article 83 (2) (k) GDPR.[139]

If a data controller is compliant with GDPR, in terms of legal design, by providing the consumer/data subject with such remedies (more detailed under Sect. 13.4.3), this would appear to achieve a fairer balance of powers, combined

[135] Y. Poullet, Consentement et RGPD: des zones d'ombre !, DCCR, 2019/1–2, n°122–123, p. 6.

[136] Article 21 (2) GDPR.

[137] Article 22 (3) GDPR.

[138] Article 6 (4) GDPR. See also Guidelines 2/2019 of 8 October 2019 of the EDPS on the processing of personal data under Article 6(1)(b) GDPR in the context of the provision of online services to data subjects, available at https://edpb.europa.eu/sites/edpb/files/files/file1/edpb_guidelines-art_6-1-b-adopted_after_public_consultation_en.pdf. Accessed 22 August 2020, pp. 14–15: the EDPB considers rightfully that profiling or behavioural advertising is not necessary for the performance of an agreement, and their collection, or that further use cannot be justified by that fact that behavioural advertising funds the provision of the service.

[139] See Decision n° 04/2019 of 28 May 2019 of the BDPA, available at: https://www.autoriteprotectiondonnees.be/publications/decision-quant-au-fond-n-04-2019.pdf; Decision n°10/2019 of 25 November 2019 of the BDPA, available at: https://www.autoriteprotectiondonnees.be/publications/decision-quant-au-fond-n-10-2019.pdf; Decision n°11/2019 of 25 November 2019 of the BDPA, available at: https://www.autoriteprotectiondonnees.be/publications/decision-quant-au-fond-n-11-2019.pdf). All accessed 18 May 2020.

with a proper information of its rights as consumer/data subject, in a manner that he or she can easily understand.

13.4.2.2 Influence of Other Legislations

An enterprise should consider other applicable laws when designing a(n) automated personalisation solution, especially with regards to pricing.

First, Rules under EU competition law are relevant for price personalisation. Several businesses, using algorithms to categorise individuals to determine prices could, by doing so, proceed to collusion prohibited under Article 101 TFEU or abuse their dominant position. This is particularly the case of major players (such as the GAFAs) pooling their data and enriching each other AI devices, by creating unfairly discriminative prices under Article 102 TFEU. However, such concerns should not be exaggerated as there are doubts as to the actual capacities in the current state of technology of AI to enter into collusive strategies without human intervention. Moreover, there is doubt as to whether such practices would be reprehensible under EU competition law at all, because it is unlikely to happen, or even be beneficial to consumers. The specialists on this topic explain that a business enjoying a dominant position would be in a better position to set prohibitive prices without personalisation than with personalisation, since following personalisation everyone would be exactly paying the price that is adapted to them. However, EU competition law remains an interesting tool to deal with the issue of price discrimination as it is flexible and evolves depending on the specific situation at stake and, allow the legislator to determine objectively which behaviours can be deemed anti-competitive in an economy based on algorithms.[140]

Therefore, as automated personalisation could result in discrimination, businesses will have to contend with already far-reaching legislation. The Geoblocking Regulation[141] aims, in transnational situations, at preventing geo-blocking and other forms of discrimination based, directly or indirectly,[142] on the customers' nationality, place of residence or place of establishment; it also applies to cases when clients

[140] See, on this particular topic N. Petit, Algorithmes tarifaires et droit européen de la concurrence in l'Europe au présent!, Bruxelles, Bruylant, 2018, p. 167–176 and A. Ittoo and N. Petit, Titre 4 - Algorithmic Pricing Agents and Tacit Collusion: A Technological Perspective, in L'intelligence artificielle et le droit, Bruxelles, Editions Larcier, 2017, p. 241.

[141] Regulation 2018/302 of the European Parliament and of the Council of 28 February 2018 on addressing unjustified geo-blocking and other forms of discrimination based on customers' nationality, place of residence or place of establishment within the internal market and amending Regulations No 2006/2004 and 2017/2394 and Directive 2009/22, OJ, 2018, L 60l, p. 1., Article 1.

[142] Recital 6 Geo-Blocking Regulation: "Thus, it also seeks to cover unjustified differences of treatment on the basis of other distinguishing criteria which lead to the same result as the application of criteria directly based on customers' nationality or place of residence, regardless of whether the customer concerned is present, permanently or on a temporary basis, in another Member State, or place of establishment. Such other criteria can be applied, in particular, on the basis of information indicating the physical location of customers, such as the IP address used when accessing an online interface, the address submitted for the delivery of the goods, the choice of language made or the Member State where the customer's payment instrument has been issued."

are granted different conditions of access to goods or services, unless objectively justified (Article 4).[143]

Notwithstanding this obligation, Enterprises remain free to design country-specific websites, and choose specific products destined for a target audience, using e.g. personalised or dynamic pricing, provided that they treat the customers across the EU in a non-discriminatory manner.[144] Such will be the case for reasons "unrelated to nationality, place of residence or place of establishment",[145] or if the difference of treatment is provided under specific EU or law of a Member State compliant with EU law.[146]

The scope of the Geoblocking Regulation appears to be limited, however, as it does not apply to all services mentioned under Article 2 (2) of the Services Directive.[147] For instance, non-economic services of general interest, financial services, electronic communication services and networks, transports, audiovisual services or gambling are excluded from these rules.

At national level, the Belgian Discrimination Act forbids forms of discrimination, in the public and private sector, among others in relation to the access to goods and services put at the public's disposal or the access, participation, and the exercise of any economic, social, cultural or political accessible to the public, which is based on age, sexual orientation, civil state, birth, fortune, religious of philosophical belief, handicap, physical or genetic features and social origin, unless justified for legitimate purposes and where appropriate and necessary means for their realisation have been put in place.[148]

This is not a mere theoretical statement as Belgian courts already ruled that offering specific promotions or discounts to a category of consumers based on

[143] The Geo-Blocking Regulation distinguishes three types of access to goods or services: (a) the purchase of goods from a trader, where either those goods are delivered to a location in a Member State to which the trader offers delivery in the general conditions of access, or those goods are collected at a location agreed upon between the trader and the customer in a Member State in which the trader offers such an option in the general conditions of access; (b) the customer receives services supplied electronically by the trader, other than services the main feature of which the provision of access to and use of copyright-protected works or other protected subject matter, including the selling of copyright protected works or protected subject matter in an intangible form; (c) receives service from a trader, other than electronically supplied services, in a physical location within the territory of a Member State where the trader operates.

[144] Article 4 (2) of the Geo-Blocking Regulation.

[145] Recital 27 of the Geo-Blocking Regulation, listing e.g membership of a specific association, or contributions made to the trader.

[146] See E. Van Severen and E. Lievens, Online winkelen over grenzen heen: De nieuwe Verordening inzake de aanpak van ongerechtvaardigde geoblocking, D.C.C.R., 2018/2, n° 119, p. 3–16, in particular p. 13 referring to the examples of "regulated bookprices".

[147] Directive 2006/123 of the European Parliament and of the Council of 12 December 2006 on services in the internal market, OJ, 2006, L 376, p. 36.

[148] Act of 10 May 2007 aiming at combating certain forms of discrimination, Belgian Official Gazette, 30 May 2007, p. 29016.

their age,[149] job category or location[150] was not objectively justified and therefore in breach of the Belgian Discrimination Act. Such breach, in turn, was deemed an unfair commercial practice by the business from which the promotions originated.

13.4.2.3 Accountability and Transparency

To ensure that the enterprise, as a data controller, complies with its obligations regarding the processing of personal data, it would need to ensure that any processing is compliant with the above-mentioned laws dealing with consumer rights and anti-discrimination and that personal data are processed based on one of the grounds set forth in Article 6 (principle of lawfulness and fairness) and Article 5 (1) GDPR (transparency). The principles of data minimisation, purpose limitation and accuracy are also relevant in an AI environment and aim at ensuring that the automated processing solution is sufficiently "trained" to be compliant by the enterprise.

The data subject should be granted the opportunity to be informed of the limited and specific purposes of the processing and doing so, the data controller should provide distinct information on, on the one hand, the purposes of processing (offering a personalised price) and, on the other hand, the means of reaching such purposes (personalisation) (purpose limitation principle under Article 5 (1) (b) GDPR).[151] Adopting and informing the consumer clearly would prevent a possible blurring of purposes.[152]

In order to be transparent, a data controller would need to inform the consumer about the algorithms used in the framework of AI, which could be met with difficulties. The growing use and complexity of AI, using machine-learning or deep-learning, can make it challenging for a user to understand how an automated decision-making process or profiling works, and can lead to a "Blackbox phenomenon" where the AI solution would start making decisions independently, without its operator being able to understand or anticipate the rationales behind it anymore.

---

[149] For instance, this has been the case for discounts offered by a glass retailer, which went higher the older the customer was. See Court of appeal of Antwerp, 26 April 2007, and J. Stuyck, Geen goedkope brillen voor ouderen - de onverwachte invloed van de anti-discriminatiewet, Annuaire Pratiques du commerce & Concurrence, 2007, p. 478–482. Other courts have upheld the same ruling, see President of the Commercial Court of Turnhout, 7 November 2008, Annuaire Pratique du commerce & Concurrence, 2008, p. 289; President of the Commercial Court of Brussels, 17 September 2008, Annuaire Pratiques du commerce & Concurrence, 2008, p. 57; President of the Commercial Court of Brussels, 15 October 2008, Annuaire Pratiques du commerce & Concurrence, 2008, p. 272.

[150] See Court of Appeal of Brussels, 4 May 2010, D.C.C.R., 2011, n°90, p. 103.

[151] See also, on the concept of personalisation, Article 29 Data Protection Working Party, Guidelines on Transparency under Regulation 2016/679, adopted on 29 November 2017 and last revised on 11 April 2018. Accessed on 15 July 2020. Available at: https://ec.europa.eu/newsroom/article29/item-detail.cfm?item_id=622227.

[152] Guidelines 05/2020 of 4 May 2020 of the EDPS, on consent under Regulation 2016/679, available at https://edpb.europa.eu/sites/edpb/files/files/file1/edpb_guidelines_202005_consent_en.pdf, consulted on 10 May 2020, n°56, p. 13.

This issue should however not serve as an excuse to circumvent this obligation and the controller should find simple ways to tell the data subject about the rationale behind, or the criteria relied on in reaching the decision, especially as next to the information obligation under the GDPR, the consumer must be provided with clear and comprehensible pre-contractual information under Article VI.45 CEL. Achieving these requirements would, according to the EU Commission, improve consumer trust and welfare and empower the consumer to identify the processes and functioning of the AI, allowing them in turn to make informed decisions on their use, opt-out, or be able to check and correct any input given.[153]

Besides the transparency requirement, the quality of the automated decision is also a matter of good data governance: an AI operator should be able to "control" the behaviour of its equipment and to explain and use unbiased algorithms at all stages, which should be explainable not only to the consumer but also to all market surveillance authorities.[154]

For these purposes, the high-level EU expert group on AI suggests that an equipment involving AI should always be designed in a way allowing human oversight at all stages, with appropriate human involvement, before (pre-validation of data and algorithms used) and after production of the output, as well as the need for a human intervention in real time during functioning and, if needed, to deactivate the AI. This includes deciding whether or not to use an AI system in a particular situation, establishing levels of human discretion when using the system, ensuring the ability to override a decision made by a system and ensuring the ability of public enforcers to exercise oversight in line with their duty.[155]

Data controllers should also ensure that the AI system is used as it was designed to be and has undergone "training" in such a way that it will only process quality data, avoiding biased results that would undermine data subjects' fundamental rights and lead to, for instance, unfair discrimination between groups of individuals. Quality data are achieved when the data are as inclusive as possible and representing different population groups, taking into account potentially vulnerable persons and groups, such as workers, women, persons with disabilities, ethnic minorities, children, consumers or other groups at risk of exclusion.[156]

---

[153] European Parliament resolution of 12 February 2020 on automated decision-making processes: ensuring consumer protection and free movement of goods and services (2019/2915(RSP)), available at: https://www.europarl.europa.eu/doceo/document/TA-9-2020-0032_EN.pdf. Accessed on 28 March 2020; Independent High-Level Expert Group on Artificial Intelligence set up by the European Commission, Ethic Guidelines for Trustworthy AI, 8 April 2019, available at https://ec.europa.eu/newsroom/dae/document.cfm?doc_id=60419, consulted on 10 May 2020, p. 16.

[154] European Parliament resolution of 12 February 2020 on automated decision-making processes: ensuring consumer protection and free movement of goods and services (2019/2915(RSP)), *op cit.*, pts 13 and 14.

[155] Independent High-Level Expert Group on Artificial Intelligence set up by the European Commission, Ethic Guidelines for Trustworthy AI, available at https://ec.europa.eu/newsroom/dae/document.cfm?doc_id=60419, consulted on 10 May 2020, p. 16.

[156] *Ibid.*, p. 11. See also Article 29 Data Protection Working Party, Guidelines on Automated individual decision-making and Profiling for the purposes of Regulation 2016/679, adopted on

Facing these challenges, it appears that strict compliance with the data protection principles could achieve both goals. A data controller that needs to comply with the transparency requirements and provide consumers with easily understandable information, allowing them to make reasoned transactional decisions, must have entire control over its data processing solution using automated means and be able to intervene at all stages to bring necessary improvement and corrections. This way of working would allow the controller to assess on a permanent basis whether the processing mechanics behind the AI still satisfies data protection principles, thereby mitigating any risk of discriminating, or any harmful or unfair processing for the consumers/data subjects.

13.4.3 Consumer Rights and Remedies

As a third component, a better balance of power can be achieved if the consumers are granted specific rights they can easily enforce, allowing them to intervene in the decision-making leading to the creation of personalised elements offered to consumers, or contest such decision *ex post*. GDPR and consumer laws, read together, seem to bring comprehensive remedies to the consumer. A consumer will be able to successfully enforce these remedies provided that the enterprise complies with the elements above on its own accountability obligation, and the efficient provision of information to a consumer.

13.4.3.1 Remedies Under Consumer Law

Prior to entering into an agreement online, the consumer must receive specific information, including on the personalisation of the price that they will pay for goods or services. If the consumer can demonstrate that the provided information on personalisation constitutes misleading or aggressive commercial practices, influencing their transactional decision, they can claim a reimbursement of any amount paid to the enterprise while still keeping the purchased good or the benefit of the service, in some instances even without judicial intervention (Article VI.38 CEL).

Before placing an order, and therefore concluding an agreement with the enterprise, the consumer should be granted the technical means to analyse the errors in the data they seized into the system and correct them prior to submission (Article XII.8 CEL).

After concluding the agreement, the consumer benefits from a withdrawal right, which is to be exercised, for services, within a 14-day period from the date of the conclusion of the agreement or, for goods, from the date when the consumer

3 October 2017 and last revised on 6 February 2018. Accessed on 15 May 2020. Available at: https://ec.europa.eu/newsroom/article29/document.cfm?action=display&doc_id=49826, pp. 10–14. The data should also be up to date as "Decisions may be made on the basis of outdated data or the incorrect interpretation of external data. Inaccuracies may lead to inappropriate predictions or statements about, e.g., someone's health, credit, or insurance risk".

physically takes possession of the good. This right can in principle be exercised at no cost (Article VI.47 CEL).

However, when the agreement was concluded with the intervention of AI, there is uncertainty around the consumer being able to rely on his or her withdrawal right: this right is subject to several exceptions, and it will not apply to digital services where the consumer expressly agreed to start using it and acknowledged that by doing so, he or she would lose such right after the service is executed (Article VI.53, 13° CEL). Categories of online services are therefore excluded from this protection. Moreover, another exception denies withdrawal when the goods supplied are made according "to the consumer's specifications or are clearly personalised". If not only the price, but also the good purchased is personalised by means of AI, there would be room for the enterprise to argue that this exclusion applies (Article VI.53, 3° CEL).

Also, when the conclusion of the agreement required the communication of personal data by the consumer, the EDPB questioned how such right would articulate with the right to withdraw consent for the processing of its personal data under Article 7 (3) GDPR, and whether after exercising such right, the consumer would maintain the possibility to effectively enforce their data subject rights under the GDPR, and what type of reimbursement, or "fair compensation"[157] would be appropriate when the "counterpart" to the goods or services is consumer's personal data.

13.4.3.2 Remedies Under the GDPR

Article 22 GDPR contains a prohibition to rely on automated processing (subject to exceptions). When it is authorised, the data subject will always have the right to oppose such processing to the extent that this produces legal effects concerning the data subject or significantly affects him or her.[158]

"Legal effects" implies that the decision affects the data subject's legal rights or the legal status of his or her rights under a contract. As for the second criteria—that the processing would significantly affect the data subject—the decision must (1) have the potential to significantly affect the circumstances, behaviour or choices of the individual concerned; (2) have a prolonged or permanent impact on the data

[157] Opinion 8/2018 of the EDPS, on the legislative package "a New Deal for Consumers", available at: https://edps.europa.eu/sites/edp/files/publication/18-10-05_opinion_consumer_law_en.pdf, p. 17–18. See also R. Robert, Peut-on payer avec ses données personnelles? - La proposition de directive "contenu numérique" introduit le ver dans le fruit, JDE, 2017, pp. 356–358. This author explains the difficulties linked to giving a precise value to personal data by, inter alia, the fact that the consumer is himself unaware that he is using personal data as "money" in the framework of an equitable exchange, and professionals would not succeed either at determining a fair value. Moreover, the lack of clarity in some privacy policies cast doubt as to whether the collection of personal data is intended to create value in the first place.

[158] Article 22 (1) GDPR. Referring to Recital 71 GDPR, Article 29 Data Protection Working Party, Guidelines on Automated individual decision-making and Profiling for the purposes of Regulation 2016/679, adopted on 3 October 2017 and last revised on 6 February 2018. Accessed on 15 May 2020, p. 20: "Interpreting Article 22 as a prohibition rather than a right to be invoked means that individuals are automatically protected from the potential effects this type of processing may have."

subject; or, (3) at its most extreme, lead to the exclusion or discrimination of individuals. Under any of these circumstances, it appears that a data subject could successfully oppose the processing of his or her personal data. Under condition (1) examples could consist in cases where the use of AI will have a decisive effect on the decision to offer or not a contract to the data subject, for instance, and where the consumer accepts the offer made and created by the AI, which creates a legal obligation for the consumer to pay for the counterpart that he or she was seeking, and thus have legal effect.[159]

Under the condition (2), the AI could be used to influence the consumer's decision, even if not related immediately to an agreement, by means of personalised advertising or other marketing techniques, where the behaviour of the consumer would be impacted and he or she would make a decision that he or she otherwise would not have taken. Unfair commercial practices could reasonably belong to this category.[160]

Then, as a safeguard, the data subject is granted several rights, including at least the right to obtain human intervention on the part of the controller, to express his or her point of view and to contest the decision (Article 22 (3) GDPR). This allows the data subject to give the data controller input, which in turn can help it improving its processing processes.

The data subject also has the right to access their personal data at all times and to check see how it is processed and receive the same information about the "logic involved" (Article 15 (a), (b), (h) GDPR), and to require correction of the input if any errors or inaccuracies are found (Article 16 GDPR). While the extent to which a data controller must disclose the functioning of its algorithms remains unclear, these remedies reinforce the data subject's position as he or she would have the right to access such information, and assess whether it reveals that that data controller uses criteria in breach of anti-discrimination laws for instance.[161] Moreover, if sensitive

[159] F. Jacques, Personnalisation des prix - Regard européen sur la pratique, R.D.T.I., n°78–79, 2020, p. 78.

[160] Article 29 Data Protection Working Party, Guidelines on Automated individual decision-making and Profiling for the purposes of Regulation 2016/679, adopted on 3 October 2017 and last revised on 6 February 2018. Accessed on 15 May 2020, p. 22: "online advertising, which increasingly relies on automated tools and involves solely automated individual decision making. In many cases, targeted advertising based on profiling will not have a significant effect on an individual but, depending on particular characteristics of the case (intrusiveness, expectations and wishes, the way the advert is delivered, using the knowledge of the vulnerabilities of the data subjects targeted)." It will have little impact generally but can have a more significant effect on certain groups of society, such as minority groups or vulnerable adults. A significant effect would be found here if the automated decision-making results in differential pricing based on personal data or personal characteristics if, e.g., prohibitively high prices effectively barred some from certain goods or services.

[161] F. Jacques, Personnalisation des prix - Regard européen sur la pratique, R.D.T.I., n°78–79, 2020, p. 67 quoting L. Drechler and J.-C. Benito Sanchez, "The Price is (Not) Right: Data Protection and Discrimination in the Age of Pricing Algorithms", European Journal of Law and Technology, 2018/19 (3), p. 12.

personal data requiring explicit consent, under Article 9 (1) GDPR, is processed, the data controller should take suitable measures to safeguard the data subject's rights and freedoms and legitimate interests, and automated processing can only take place if these requirements are satisfied.[162] As this type of data is obtained through consent, the data subject can withdraw their consent at any time (Article 7 (3) GDPR).

The collection and processing of personal data should be limited to elements that are relevant and should not create any discrimination in view of personalisation, and allow the data subject to make a correct use of its rectification right to ensure that such data is accurate.[163] This would, in turn, ensure that the data minimisation principle and the quality of that data are complied with.

13.4.4 Synergies Between Consumer Law and Data Protection

Provided that, under the applicable rules, the enterprise provides a transparent and consumer-friendly environment, and the consumers are vested with effective rights, the legal arsenal in place could already serve as a base for protection and remedies facing the growing use of automated processing means such as AI. This is best achieved not through a legislative change, but based on the correct training, and willingness of the stakeholders to effectively comply with the rules in place, allowing the emergence of a renewed B2C model embracing new technologies. In this setting, the consumer would gain more trust in entering into transactions using automated systems if he or she could effectively enforce its rights giving him or her control over the personal data, provide data that he or she would be willing to disclose, and influence the processing of that data by using his or her right to corrections. This would, however, require that the enterprise, on its part, correctly provides information about the processing of data, knowing that incorrect or misleading information, in breach of data protection laws, could constitute an unfair commercial practice subject to sanctions.

13.5 Conclusion

In this contribution, we have seen that new challenges brought about by modern marketing techniques could be faced using technology-neutral legislation and advocating for an effective and combined reading of the rules under both consumer and privacy laws.

Further legislative efforts should be made to ensure that the rules covering commercial communication are subjecting all media to the same rules and that the identifiability requirement for a "promotion" is applicable in broader terms, that is

[162] Article 9 (2) (a) GDPR read with Article 22 (4) GDPR.

[163] Article 16 GDPR.

covering economic and non-economic purposes. This would allow consumers to understand, with a sufficiently critical mind and taking into account the average consumer criterion, the intent behind the numerous communications that they receive and prevent them from being subjected to undue influence during their decision-making process, be it for an economic transaction or social activity.

Personalisation, with or without the use of AI to facilitate the conclusion of an agreement, will be best captured by law, not through a legislative change but if the existing rules can be approached with a flexible application of consumer laws, on the one hand, and a reasoned approach towards privacy laws, on the other, taking into account consumers' fundamental rights as embodied in the anti-discrimination laws.

The existing consumer laws, and the upcoming changes brought about by the Omnibus Directive, can be applied and enforced in such a way that not only the price but also the personalisation of the entire transactional process can be captured under the rules applicable to unfair commercial practices.

Under *privacy* laws, the existing GDPR rules compels the enterprise to carefully assess the correct use of personalisation and the use of technology, materialised by a correct information to the consumer, in a manner allowing the consumer to understand their rights, whether personalisation is made using consent or legitimate interests as a legal basis. Moreover, strict compliance with the data protection principles and sufficient transparency on the means of processing used will help the enterprise achieve better results in terms of personalisation, with the collaboration of the consumer controlling its data, which in turn will allow for an improved delivery of goods or services.

These rules can be seen as a possible safety net, safeguarding the consumer's fundamental rights and anticipating further technological evolution. The way that the "general norm" under unfair commercial practices is interpreted under EU and Belgian laws is broad enough to be "future proof".[164] A judge or regulator applying this norm will have to acknowledge that any stakeholder involved in the personalisation and use of AI has to comply with the rules of the GDPR as well as the anti-discrimination laws. Even without breaching the law, any personalisation practice could be deemed contrary to professional diligence or could distort the economic behaviour of a consumer since personalisation, whether or not with the use of AI, is capable of influencing a consumer's transactional decisions.

This pleads for a reviewed approach at both the EU and Belgian levels, where the relevant unfair competition and privacy stakeholders should interact and adopt common solutions and guidelines to ensure an appropriate application of these rules. This will, in turn, help achieve a better balance of power on both sides. The consumer/data subject will be correctly informed of the promotional and personalised nature of the message conveyed to him or her and can enforce their rights to prevent any misuse of personal data threatening their fundamental rights or give them the choice not to be subject to personalisation or automated decision-

[164] N. Van Eijk, C. J. Van Hoofnagle and E. Kannekens, Unfair Commercial Practices: A Complementary Approach to Privacy Protection, E.D.P.L., n°3, 2017, p. 325–327.

making, allowing them to gain trust in the use of such new technologies. The enterprise/data controller willing to achieve a future-facing market position, taking a consumer-centric approach into account, will need to have the appropriate safeguards in place to ensure that they maintain control over their personalised communication and automated processes, which should remain identifiable and understandable at all time, knowing that breaches could expose them to regulatory or consumer enforcement under both unfair competition and privacy laws.

14 Brazil

Felipe Barros Oquendo

14.1 Personal Data Protection in Brazil and Current Challenges

Like the European Union and several other countries, Brazil has its own legal regulation on personal data, the General Data Protection Law (Portuguese acronym: LGPD). Enacted on 14 August 2018, the law entered into force in August 2020. Hence, since the mid-2020s, Brazil has had a Data Protection Law in place and a National Data Protection Authority (Portuguese acronym: ANPD) that regulates and supervises compliance with this law. The appointments for the board of the ANPD were approved by the Brazilian Senate on 20 October 2020, and the Authority is currently operational, including through its website, where anyone can file a complaint related to violations of personal data rights.[1]

As will be discussed throughout this report, numerous points of the LGPD are perfectly analogous to those of the General Data Protection Regulation 2016/679 (GDPR), by which Brazilian law was clearly inspired. The LGPD is therefore similar to the European model and different from the California Data Protection Act model.

When compared with European countries such as Italy, France, Germany and the Scandinavian countries, which benefited from legal acts such as the European Directive of 1995 and, in some cases, even with data protection authorities already existing in the 1990s, Brazil's history with regard to the legal protection of personal data is, however, less developed. This is relevant because while in the European Union the General Data Protection Regulation (GDPR) emerged as the culmination of coordinated efforts for harmonizing personal data protection by its Member

[1] See https://www.gov.br/anpd/pt-br. Accessed 3 September 2021.

F. B. Oquendo (✉)
Di Blasi, Parente & Associates, Rio de Janeiro, Brazil
e-mail: felipe.oquendo@diblasi.com.br

B. Kilpatrick et al. (eds.), *Antitrust in Data Driven Markets & Legal Framework for Influencers, Native Advertising and Control over the Use of AI in Marketing*, LIDC Contributions on Antitrust Law, Intellectual Property and Unfair Competition, https://doi.org/10.1007/978-3-031-07422-6_14

States, benefiting from the diversity and accumulated experience of national jurisdictions that developed spontaneously and independently, the introduction of LGPD in Brazil appears as something foreign, however important and welcome but strange to national issues and debates, as well as to the Brazilian legal culture. This dissonance can cause problems with the implementation of the law and the penalization or inspection of violations.

An honourable mention and exception to the general silence on data protection issues in Brazil must be made to Prof. Danilo Doneda's outstanding work in debating and writing about privacy and data protection in Brazil for many years before the discussions on the LGPD. Prof. Doneda's view was highly influenced by the works of Italian Prof. Stefano Rodotá, who came to Brazil several times as a speaker on this topic and greatly influenced the so-called civil-constitutional doctrine, especially in the Law School of the University of the State of Rio de Janeiro.

In Brazil, since at least the 1967 Constitution, the right to privacy and intimacy has been considered essential. The 1967 Constitution made its way to our current 1988 Constitution (Article 5, especially item X). The Civil Code of 2002 establishes, in Articles 11–20, several rights called "personality rights", among them the right to one's name, image, voice and privacy, which are considered non-transferable and cannot be renounced. The holder of an offended personality right may request the cessation of the offence and seek redress for material and moral damages caused by the offender.

Despite the progress represented by the Civil Code of 2002, there is no specific regulation on private and personal data, such as personal address, purchase history or travel history, this, even though in the 1990s there was already in Italy, whose laws greatly influenced the text of our Code, relatively well-developed legislation and authority to protect personal data, not to mention other European countries such as France and Spain.

On the other hand, for a long time, a right to secrecy has been established by sectorial laws on bank data, data related to taxes and income as well as telephone conversations, requiring a court order for such confidentiality to be exempted. Also worthy of note is the Brazilian Penal Code, which considers it a crime to violate correspondence; record and listen to telephone conversations; disclose secret information, including information covered by professional secrecy; and obtain, adulterate or destroy information on a computer without authorization. In fact, with regard to the protection of secrets, it is important to note that business secrets are expressly protected under the Industrial Property Law (LPI),[2] including test data delivered to regulatory bodies linked to the government.

It is clear from this brief and incomplete enumeration of rules that the obvious and clear violations of privacy and secrecy, as well as the use of data obtained in an improper and unauthorized manner concerning a person and his or her private life, are punishable by law. However, the use of data relating to shopping behavior or travel, telephone numbers, e-mail addresses, passwords, electronic documents and

[2]Federal Law N. 9,279 of 16 May 1996.

medical examination data, among others, is entirely unregulated, except for borderline situations where the bad faith is blatant.

In other words, there was a need not only to treat the protection of personal data in a punitive way but also to regulate it in more detail in order to avoid the violation of rights and to promote cooperation between those involved and not only repress it and compensate for the damages caused, a paradigm that generates a huge burden on the data subjects.

Technological advancements and the concentration of digital platforms in the hands of a few transnational players have greatly increased the possibilities of abusive use of personal data and violation of privacy on the Internet. This was one of the reasons for the promulgation of the Civil Rights Framework for Internet Act,[3] a Brazilian law that seeks to establish principles, guarantees, rights and duties for the use of the Internet in Brazil.

By the Civil Rights Framework for Internet Act, the protection of privacy and personal data is expressly provided as a principle that guides the use of the Internet in Brazil. This principle is the base of rights conferred by law, such as the inviolability of intimacy and private life; the inviolability and secrecy of the flow of communications over the Internet; the right to readily understandable and thorough information on the collection, use, storage, treatment and protection of users' personal data; or the deletion of personal data that the user has provided to a particular Internet application at the end of the relationship between the parties, except for cases of mandatory record keeping provided for by law.

The Civil Rights Framework for Internet Act established, therefore, rights very much in line with the current LGPD, but most of these principles were dependent on subsequent regulation, which was never issued. Nevertheless, proof of the affinity between the Civil Rights Framework for Internet Act and the protection of personal data is that the bill that subsequently became the LGPD was initially intended as a reform of the Framework Act. However, this initial wording of the bill was abandoned, mainly because the LGPD, like the GDPR, is extremely broad and goes beyond the scope of the Internet.

Furthermore, as will be seen, the LGPD confers a more specific and practical application of rights to the data subject, which goes beyond the mere judicial protection of rights that have already been violated. The intention of the LGPD is to regulate a stable relationship and prevent violations and abuses and not to leave to the judiciary the resolution of disputes between parties involved.

The LGPD has just entered into force, and the provisions on personal data and privacy of the Civil Rights Framework for Internet Act have not been implemented, which means that in Brazil, the debate is not as advanced as in the European Union. There is, however, an increasing number of judicial cases based directly on the LGPD, including a Civil Public Lawsuit filed by the Public Prosecutor of the Federal

[3] Federal Law N. 12.965, of 23 April 2014.

District.[4] The expectation is that the number of such cases will rise in the coming years, given the importance of issues related to the use of artificial intelligence (AI) in the collection and treatment of data, as well as in the offer of products and promotions over the Internet, with Brazil being the fifth largest nation worldwide in the number of Internet users.[5] It is also noteworthy that the ANPD has issued further guidance on the application of the LGPD[6] and that the National Council for the Autoregulation of Advertisement (Portuguese acronym: CONAR), a very influential albeit private body, has issued specific rules for advertising on the Internet through digital influencers.[7]

Finally, and without intending to exhaust the matter, the LGPD may represent an important advance in the Brazilian legal framework to regulate the processing of personal data along the lines of the GDPR. There is no doubt that Brazil has a path to follow with regard to the culture of data protection. However, we believe that the entry into force of the LGPD can assist in raising civil society's awareness of the importance of having clear rules that protect people and their data.

This report has been divided into two main parts: issues addressed in the light of already enacted legislation (positive law) (Sect. 14.2) and issues related to future developments or to points not expressly regulated by the law or the ANPD (Sect. 14.3).

14.2 Issues in the Light of Enacted Legislation

14.2.1 Search Engines, Artificial Intelligence and Freedom of Contract

A search engine is a piece of software designed to search for information on the Internet based on user input and to give the result at the end. The software may use numerous algorithms that make the search more efficient, such as the application of artificial intelligence, machine learning and other systems.

For the average user, understanding the intricacies of the mechanics employed in a search engine is often not feasible, especially if artificial intelligence is applied.

[4] A news piece of July 2021 reports over 600 decisions issued by Brazilian courts with a basis on the LGPD. See https://www1.folha.uol.com.br/mercado/2021/07/justica-ja-tem-600-decisoes-envolvendo-lei-de-protecao-de-dados.shtml. Accessed 22 July 2021.

[5] See https://www.statista.com/statistics/262966/number-of-internet-users-in-selected-countries/. Accessed 3 September 2021.

[6] ANPD, Guia Orientativo Para Definições dos Agentes de Tratamento de Dados Pessoais e do Encarregado, 2021. Available at https://www.gov.br/anpd/pt-br/assuntos/noticias/inclusao-de-arquivos-para-link-nas-noticias/2021-05-27-guia-agentes-de-tratamento_final.pdf. Accessed 31 August 2021.

[7] CONAR/CBAP, Digital Influencer Advertising Guidelines, 2021. Available at http://conar.org.br/pdf/CONAR_Digital-Influencer-Advertising-Guidelines_2021-03-11.pdf. Accessed 30 August 2021.

To further complicate the scenario, in addition to artificial intelligence technology, there is the application of "profiling" techniques, which aim to individualize the possible buyer based on the extraction of assumptions and predictions about its interests and behaviour. However, the profile may be defective and may suggest wrong information about the user, showing search results that do not correspond to his or her interests and expectations.

According to the Information Commissioner's Office (ICO),[8] "as regards erroneous algorithmic decisions, there are clear implications for the data protection principle of accuracy, such as inaccurate predictions based on biased profiling".

It is well known that a contract denotes the need for a defined and well-specified object, in addition to the autonomy, civil capacity and free will of the parties involved. The human will, not the law, is typically the nucleus, the source and what legitimates a contractual relationship. The force that compels the parties to perform the contract is based on will freely stipulated in the legal instrument, and the law is only responsible for ensuring the means that lead to the fulfilment of the obligation, thus having a suppletive position.

Should the search engine point the user to products that do not meet his/her expectations, it can be argued that there is a mathematical/theoretical flaw or simply the insertion of erroneous data on the profile of the user. Considering that the consumer uses a search mechanism (which may be vitiated) to find an object of desire, the consumer may be induced to buy another product that is not fully compatible with his/her goals and needs. The end result would be a defect in the will of the user, who would have been erroneously led into a deal he or she would not want if fully aware of the conditions of the search, the results and the offer.

By this logic, it appears that, in theory, the lack of knowledge of the underlying rationale of the search engine could suppress the consumer's contractual freedom by unduly inducing him/her to acquire a certain product.

However, we believe that the claim that there is a presumed restriction to contractual freedom due to erroneous profiling lacks basis. Depending on the specifics of the case, a balance must be stricken between the rights of the user and the rights of companies related to trade and business secrets, equally protected by the LGPD, which covers the way algorithms are programmed. In addition, due to the rapid advancement of technologies, there are now artificial intelligence structures that seek to give the choice back to the user.

In order to preserve the consumer, the Brazilian Consumer Protection Code[9] has provided, since 1990, that the offer of goods and services must obey some rules, such as sufficient precision and clarity of information—including characteristics, qualities, quantity, composition, origin, price and risks to health and safety, among

[8]ICO, Big data, artificial intelligence, machine learning and data protection, 2017. Available at https://ico.org.uk/media/for-organisations/documents/2013559/big-data-ai-ml-and-data-protection.pdf. Accessed 3 September 2021.

[9]Federal Law No. 8,078 of 11 September 1990.

others—and that any condition mentioned in an offer be included in the final agreement (Articles 30 and 31).

With regard to automated data processing, data subjects are entitled to request a review of the decisions taken solely on the basis of automated processing of personal data that affect their interests, including decisions designed to define their personal, professional, consumer and credit profile or aspects of their personality (Article 20 LGPD). The provision states that the controller must provide, whenever requested, clear and adequate information regarding the criteria and procedures used for the automated decision, observing commercial and industrial secrets. Failure to comply may lead to an audit for discriminatory aspects in the automated treatment of personal data.

Therefore, the mere lack of understanding by the user of the logic of search engines through artificial intelligence will not necessarily imply a restriction of contractual freedom, under penalty of rendering an important gain of society unfeasible in favor of a massive bureaucratization of business operations.

Notwithstanding the legal requirements and good practices related to the protection of personal data, which must be implemented and guaranteed to the data subject, it is essential to emphasize the importance of adopting an artificial intelligence governance model integrated with algorithmic ethics, which places the data subject as a protagonist in the management and development of automated analysis. In this way, companies will be able to guarantee to the data subject that the automatic decision-making process is as transparent, balanced and consistent with reality as possible.

In order to accomplish adopting an artificial intelligence governance model integrated with algorithmic ethics, according to the findings of the Personal Data Protection Commission—Singapore PDPC, here quoted as a thorough document that might be helpful to future regulations issued by the ANPD, organizations must consider: (1) the adaptation or creation of their organizational governance structure, to incorporate related values, risks and responsibilities in the algorithmic decision-making; (2) a methodology to assist them in defining their appetite risk for the use of artificial intelligence, that is, determining acceptable risks and identifying an appropriate level of human involvement in artificial intelligence decision making; (3) issues related to the development, selection and maintenance of artificial intelligence models, including correct data management; and, finally, (4) setting up communication and relationship strategies with stakeholders, mainly with the data subject.[10]

Regardless of this fine balance between free will in contracting and the need not to hinder technological implementations in the market, there is another side to this issue. Assessing whether an advertisement based on algorithm decision-making

[10] PDPC, Model Artificial Intelligence Governance Framework, 2020. Available at: https://www.pdpc.gov.sg/-/media/files/pdpc/pdf-files/resource-for-organisation/ai/sgmodelaigovframework2.pdf. Accessed 3 September 2021.

matches a consumer's interest is highly subjective—if not impossible to do in some situations.

First, it is because companies might be reluctant to publicly share their algorithms' decisions due to industrial or trade secrets. Second, some algorithmic decisions are unpredictable and are caused by factors that companies cannot fully justify as what was intended originally—the so-called black box effect.

Third, a conclusion in the sense that an advertisement exhibited due to algorithms' decision-making is not equivalent to someone's wishes depends on a deep knowledge of the said individual's interests: one should be able to prove that the information displayed to the individual did not correspond at all to his/her interests based on his/her previous behaviour expressed online. Depending on the case, an algorithm may be highly sensitive to any input it receives, to the point that one different search performed by a consumer may change completely the ads that are depicted to him/her—especially in cases involving a single account used by different people.

If a customer presents a specific behaviour online and suddenly changes it for any reason, it would be reasonable for algorithms to take the time (which could be longer or shorter depending on the case) to adjust their calculations in order to filter information and depict ads corresponding to the new interests of the consumer (which could also be temporary wishes, the following of trends, demanding temporary changes in the publicity material shown to the data subject).

On the other hand, it is not reasonable to state that the freedom of contract is unacceptably manipulated in this scenario, given that companies are interested in offering services—and depicting ads—that best suit their customers' interests to increase the chances of getting a deal closed. Businesses would gain nothing by violating their customers' contractual good faith. If a website invests in an algorithm that keeps showing users ads that do not correspond to their needs and expectations, it will not achieve profits and will consequently make the website lose space to other platforms. Internet application providers seem to be more interested in drawing the attention of people who share the same tastes and desires than to convince individuals who are not interested in their content or products.

For the above reason, it could be stated that there is less interest in manipulating a contract to display advertisements that do not correspond to someone's wishes—through the programming of an algorithm—than in filtering the correct costumers to show them individualized advertisements that would please each one of them.

14.2.2 Lawful Processing of Data by Chatbots (AI) upon the Closing of Contracts

Article 6 (1) b of the GDPR provides that the processing of personal data shall be lawful only if and to the extent that it is necessary for the performance of a contract to which the data subject is a party or in order to take steps at the request of the data subject prior to entering into a contract.

This legal disposition has a perfect equivalent in the LGPD, Article 7, item V.

In order for chatbots to be designed to lawfully process data under this legal provision, it is necessary to consider the data subject's proactive action of "opening the call", in the sense of requesting help from the chatbot.

This tool will also need to be developed and based, essentially, on the principles of purpose, adequacy, necessity and transparency, as established in Article 6, items I, II, III and VI, of the LGPD.

That is, the information handled will need to be adequate and stick to the purpose, which, in this case, is the performance of a contract. Likewise, there cannot be processing of data beyond what is necessary, in view of the principle of necessity, which corresponds also to the requirement of the principle "of minimization".

It is important to highlight that in order to comply with the principles of the LGPD, it is essential that the controller—responsible for the treatment of personal data and who determines the purposes for which the personal data will be treated—adopts appropriate technical and organizational measures. These measures must be designed to internalize the principles of protection of personal data, integrating the principles of Privacy by Design with new technologies, such as the chatbot, in order to safeguard the privacy and correct treatment of personal data from an organizational point of view, with the objective of extending privacy to the following, and comprehensive, triad: (1) information technology (IT) systems, (2) commercial and business practices and (3) IT infrastructure.

This is precisely what Article 46, §2, of the LGPD provides, that is, the duty to adopt security, technical and administrative measures capable of protecting personal data from unauthorized access and accidental or unlawful violation of personal data. Such requirements must be observed from the outset of the good or service design phase until commercialization.

Finally, with the exception of business secrets, the chatbot platform will need to be transparent regarding the treatment of data, especially if there is a possibility of transferring such data to third parties, in compliance with Article 9, item V, of the LGPD.

Likewise, the National Authority (ANPD) may issue complementary rules for the activities of communication and shared use of personal data (Art. 30 of the LGPD), which it has not done so far.[11]

Regarding the transparency principle, it is recommended that the platform uses a Data Protection and Privacy Policy that informs and guides the data subject on the manner and purpose(s) of data processing, the data retention periods and who the controller is, in addition to the responsibilities and rights, in compliance with Article 9 of the LGPD. As a good practice, a chatbot should use a clear and unambiguous language, aiming to interact with the user in a way that leaves no

[11] It has, however, published orientation regarding data protection and data leakage. See Cartilha de Segurança para Internet Fascículo Vazamento de Dados, 2021. Available at: https://cartilha.cert.br/fasciculos/protecao-de-dados/fasciculo-protecao-de-dados.pdf. Access 3 September 2021. And Cartilha de Segurança para Internet Fascículo Proteção de Dados, 2021. Available at https://cartilha.cert.br/fasciculos/vazamento-de-dados/fasciculo-vazamento-de-dados.pdf. Access 3 September 2021.

doubts about its functionalities and the impact that decisions or instructions made by the user while interacting with the chatbot will have on the personal data of the user.

As there is still no enacted regulation or jurisprudential guidance on the matter, Brazilian companies often resort to already consolidated understandings and guidelines from authorities in other countries.

Nevertheless, a draft regulation by the ANPD is under public consultation, and it proposes to relieve small companies, non-profit organizations, individuals and start-ups from having a data protection officer (DPO) and keeping registers of data treatment operations, as well as enabling these companies to have a simplified version of a data privacy policy. It must be noted that this proposed regulation is in line with the constitutional principle of special and proportionate treatment of small businesses (Brazilian Federal Constitution, Article 170, item IX). In any event, the draft makes an express exception to "high risk activities", namely the data processing of vulnerable groups, vigilance or control of areas accessible to the public, the use of novel technologies that may give rise to damages and, ultimately, the automated treatment of data affecting the interests of the data subjects, which is closely related (among others) to chatbot activity.

14.2.3 Use of Artificial Intelligence for Processing Personal Data and Data Protection Impact Assessment

Initially, it is important to clarify the concept of an impact assessment on the protection of personal data (AIPD), called by the LGPD the Impact Report on the Protection of Personal Data (RIPD).

Both the GDPR and the LGPD provide that the controller must produce the RIPD when the data processing activities that are carried out represent risks to civil liberties and the fundamental rights of data subjects.

The RIPD, according to Article 5, item XVII, of the LGPD, is documentation by the controller that contains a description of the procedure for the processing of personal data that may generate risks to civil liberties and fundamental rights, as well as measures, safeguards and risk mitigation mechanisms. It is, therefore, "a process that aims to establish risk mitigation mechanisms and demonstrate regulatory compliance, through a document".[12]

The main difference between the GDPR and the LGPD as regards this kind of report is that the GDPR establishes the occasions when the impact evaluation is mandatory, whereas the LGPD, at least in a literal interpretation, establishes that such a report is only needed when the National Authority (ANPD) demands it (Articles 10, §3, and 38). This aspect of the law remains to be regulated by the ANPD.

[12]LGPD Acadêmico. "Relatório de Impacto à Proteção de Dados Pessoais", p. 14. Available at http://wix.to/-UDYBak. Accessed 3 September 2021.

This fact, however, should in no way prevent or discourage companies, organizations and entities from maintaining a proactive stance and producing, based on international best practices, their respective impact reports in advance of any request by the National Authority, whenever the activity at hand may represent a risk to individual freedom and rights. As a matter of fact, the proactivity of companies in relation to risk mitigation is an important aspect that will be taken into account when calculating fines and other reprisals provided for by law in an alleged event of data leakage, as established in Article 52, §1, of the LGPD and its items, in special the good faith of the infringer (item III), the reiterated adoption of internal mechanisms and procedures capable of minimizing damages in the treatment of data (item VIII) and the general adoption of a policy of good practices and governance (item IX).

By carrying out an impact assessment, organizations will be complying with the principle of accountability, provided for by both European regulation and Brazilian legislation. Accountability consists in, as provided in Article 6, item X, of the LGPD, the demonstration by the controllers and processors that they have adopted effective measures capable of proving the observance of and compliance with the rules of data protection, including the effectiveness of these measures.

Under the LGPD, the use of artificial intelligence to process personal data to control interactions with the data subject will require the production of a personal data protection impact assessment whenever it may pose a high risk to the data subject, which may generate risks to civil liberties and fundamental rights, especially (1) when the processing is based on legitimate interest or (2) when sensitive data are processed.

Finally, Working Party 29—an independent European working group that, after the entry into force of the GDPR, was replaced by the European Data Protection Board (EDPB)—spoke on the issue of the impact assessment on data protection through guidelines published on 4 April 2017, which were endorsed by the EDPB. The legal opinion establishes nine criteria that must be observed for the elaboration of an RIPD, and among them, two apply to the subject at hand, which is already sufficient to justify the spontaneous adoption of the referred procedure, namely (1) systematic control by data subjects through (2) innovative solutions or the application of new technological solutions.

14.2.4 AI Software, Individualized Advertising and the Lawful Processing of Data Under the "Legitimate Interest" Clause

Article 7, item IX, LGPD, establishes that the processing of personal data can only be carried out when necessary to serve the legitimate interests of the controller or a third party, except in the event where the fundamental rights and freedoms requiring the protection of personal data of the data subject prevail. This article is very close to Article 6 (1) (f) GDPR on legitimate interest.

The legality of artificial intelligence software, along with the tracking of cookies, for the purpose of individualized advertising will be, as determined by Article 10, items I and II and §§ 1 and 2, directly conditioned on the following: (1) serving the legitimate interests of the controller or third parties, which may include supporting and promoting the activities of the controller; (2) the treatment of data that are strictly necessary (principles of necessity and minimization); (3) respecting the data subject's legitimate expectations and fundamental rights; (4) the adoption of measures to safeguard the transparency of the treatment.

Finally, as already mentioned, according to Article 10, §3, of the LGPD, the national authority may request from the controller an impact report on the protection of personal data, when the treatment is based on its legitimate interest, while safeguarding commercial and industrial secrets.

Therefore, there is the possibility that this type of data processing takes place in accordance with Brazilian law; however, the controller must carry out an assessment of the treatment to confirm that it is based on a legitimate interest and use administrative and technical measures of information security to protect the data subject.

The controller should also be transparent about the treatment of data and allow the data subject to opt out of receiving such advertisements as a measure of good market practice and in line with the Consumer Protection Code. Depending on how pricing, content and personalized advertisements will play out, the ANPD may mandate the labelling opt-out clauses for personalized content.

14.2.5 Profiling and Individualized Ads and Prices Under the LGPD

In tackling this issue, it is worth mentioning that the LGPD does not have a corresponding provision to Article 22 (1) of the GDPR on individual automated decisions nor defines the profile. Article 22 (1) GDPR provides that "the data subject has the right not to be subject to any decision taken exclusively based on automated treatment, including the definition of profiles, which has an effect on its legal sphere, or which significantly affects it in a similar way".

However, as already stated, the LGPD does establish in its Article 20 that the data subject has a right to request a review of decisions based on the automated treatment of personal data that affect their interests, including user profiling.

There is a subtle difference between the right not to be subject to any automated decision established under the GDPR and the right to request a review of decisions already taken on the basis of the LGPD. In the first case, the exercise of the right implies that the data subject will not be affected, having made an option in this sense, to any automated decisions, whereas in the text of the Brazilian law this will occur despite the data subject, but the decisions already taken may be reviewed at the request of the data subject.

Another aspect that deserves to be highlighted is that Article 20 LGPD in not clear on how an automated decision will be reviewed. This deserves attention because, when considering the legislative changes proposed by the Brazilian National Congress by means of Law N. 13,853/2019, which altered some articles of the LGPD, the President of the Republic decided to veto the provision requiring that this review must be carried out by a human being.

Therefore, and at least until further guidance is provided by courts or through ANPD regulation or decisions, the review of the automated decisions requested by the data subject may be redone by exclusively automated means, without any human intervention.

To the extent that there is a decision made solely on the basis of automated processing of data that affect the interests of the data subject, including decisions designed to define his or her personal, professional, consumer and credit profile, or even aspects of his/her personality, the data subject may (1) request a review of this decision, which may likewise be carried out using automated means, and (2) request from the controller information regarding the criteria and procedures used while safeguarding commercial and industrial secrets. If the controller does not provide such information, the ANPD may perform an audit to verify discriminatory aspects in the automated processing of personal data.

However, the ANPD has not yet issued regulations or guidelines on the LGPD. Thus, there are still no guidelines on the interpretation of that specific provision.

In view of the absence of specific guidelines in Brazil, in our view, one can refer to the definition of "profiling" in Article 4 (4) of the GDPR, given that the LGPD is inspired by the GDPR: the automated process used by the controller of personal data in its possession, aiming to extract assumptions and predictions about certain aspects of a person, such as his/her performance at work, economic situation, health status, personal preferences, interests, behaviour and other personal aspects.

It is important to understand that the practice of creating personalized advertisements and pricing is extremely dependent on the results of the profiling process, and the personal information extracted from the individual is the raw material for this practice.

According to Mendoza e Bygrave,[13] certain aspects must be present in the process for Article 22 of the GDPR to be applied, namely (1) the existence of a decision, based solely and exclusively on automated data processing, including the practice of profiling, and (2) that the decision causes legal or significantly relevant effects to the target person.

In the case of personalized prices, Article 22 is applied whenever (1) an algorithm sets automatically an individualized price for a customer and (2) the results extracted from the application of that algorithm are used to evaluate and measure the customer's personal aspects, such as his/her willingness to buy a particular item, as well as his/her economic situation.

These two conditions denote the existence of an individualized price that results directly from the automated processing of personal data.[14]

---

[13] I. Mendoza and L. A. Bygrave, L. A,, The Right Not to Be Subject to Automated Decisions Based on Profiling. In T. Synodinou, P. Jougleux, C. Markou, & T. Prastitou (eds.), EU Internet Law: Regulation and Online Price Discrimination and EU Data Privacy Law 365 Enforcement. Springer, 2017, Forthcoming; University of Oslo Faculty of Law Research Paper No. 2017– 20. Available at <https://papers.ssrn.com/sol3/papers.cfm?abstract_id=2964855>. Accessed 3 September 2121.

[14] Online Price Discrimination and EU Data Privacy Law, Frederik Zuiderveen Borgesius e Joost Poort. Access in <https://link.springer.com/content/pdf/10.1007/s10603-017-9354-z.pdf> Accessed 3 September 2021.

In the case of automated decisions and profiling, the EU regulation requires a kind of "extra transparency": Article 13 (2)(f) and Article 14 (2)(f) of the GDPR establish that the controller must provide useful information regarding the underlying logic of the profiling, as well as its importance and expected consequences for the data subject. The criteria and timing for providing this information vary depending on whether the personal data are provided directly by the data subject. In the negative case, the controller must provide the origin of the personal data obtained and, if necessary, whether such data come from publicly accessible sources.[15]

In view of the above, within the scope of the LGPD, the creation of personalized advertising or personalized prices generated based on decisions taken solely from the automated processing of personal data that affect the interests of the data subject should give the data subject the right to review such decisions, which review can also be automated, as well as the right to gain knowledge about the criteria and procedures used for the automated decision. In the event of the non-availability of information that does not constitute commercial and industrial secrets, the ANPD may perform an audit to verify discriminatory aspects (e.g. price discrimination) in the automated treatment of the personal data.

14.2.6 (In)sufficiency of Informed Consent to Data Processing

The fact that consent is merely "informed" is not enough to protect the data subject under Brazilian law. This is because consent must be composed of other essential attributes, which are inherent to its validity.

According to Recital 32 of the GDPR, consent should be given through a clear affirmative act establishing a freely given, specific, informed and unambiguous indication of the data subject's agreement to the processing of personal data relating to him or her.

Therefore, it is clear that in order to be considered valid, consent must be granted by the data subject within a proportional and balanced relationship, a fundamental requirement for compliance with the "freely" attribute of consent.

Furthermore, for the consent to be "specific", the purposes of the processing of personal data must be very clear to the data subject, so that there is no doubt about the purpose(s) of treatment to which the data subject is consenting.

Still, in relation to the fundamental characteristics and inherent to the validity of the consent, it must be "informed" to the data subject; therefore, the specifics of the personal data processing operations must be very clear and transparent to the data subject so that, when informed of the purposes, he or she can consent from an affirmative and unambiguous stance. Therefore, there is no such thing as tacit

[15] According to the GDPR, the controller must provide data subjects with concise, transparent, intelligible and easily accessible information on the processing of their personal data. For data collected directly from the data subject, this must be provided at the time of collection (Article 13) and for data obtained indirectly, the information must be provided within the time limits established in Article 14.

consent or consent by omission under Brazilian law. Underlying the thesis that consent must be "informed" is the obligation that the tools involved in the process of getting such consent must be explained in a clear and unambiguous language, aiming to give the data subject the correct and faithful portrait of the dynamics to which his or her personal data will be subject.

Besides Recital 32, Article 7(1) of the GDPR establishes that where processing is based on consent, the controller shall be able to demonstrate that the data subject has consented to the processing of his or her personal data.

Therefore, the controller must seek ways to collect the consent, which must be able to demonstrate the veracity of the concession made by the data subject so that, in this way, it fulfils the last and indispensable attribute of consent, which is to be "unambiguous". In addition, once consent is obtained for a (set of) purpose(s), it cannot be, at a later stage, used for a purpose that was not informed to the data subject.

All attributes, therefore, seek to re-balance the relationship between the agents of the personal data processing and the data subject. Conversely, the data subject would find himself or herself in an extremely unfavourable position to understand the purposes of the data processing activities.

Article 5, item XII, of the LGPD establishes that consent is a free, informed and unequivocal manifestation by which the data subject agrees to the treatment of his or her personal data for a specific purpose.

Therefore, similarly to the GDPR, Brazilian law aims to bring greater security to the activities of processing personal data carried out based on consent, not allowing the hypothesis of a too wide and too general consent, whose treatment is not assigned to a specific purpose and to which the data subject has not granted consent freely and unequivocally, under penalty of being considered void due to defect of consent (Article 8 §§ 3 and 4, of the LGPD).

In addition, the controller must use the means to operationalize the consent and the duties that emanate from it, considering the legal possibility of the consent being revoked at any time by the data subject or the expiry of a time-bound consent and the fact that the burden of proof that the consent has been lawfully obtained will be borne by the controller (Article 8, §§ 2 and 5, of the LGPD).

Such requirements were legally established to safeguard equity and harmony in data treatment, transparency and clarity in the provision of information and affirmative and unequivocal mechanisms of actions by the data subject, for specific and agreed purposes, in order to enable the controller to prove that it has lawfully collected consent.

Finally, both GDPR and LGPD aim at promoting the restructuring of the current model of concession and generic authorization for the activities of processing personal data, resuming the necessary and fundamental balance between the processing agents and the data subjects. This is undertaken especially bearing in mind the difficulty of the data subjects in understanding extensive and complex privacy policies and the technical intricacies of technology, programming and other technical aspects.

One of the best ways to operationalize consent, guaranteeing all the attributes inherent to its validity, is through *visual law*,[16] linked to granularized consent, which allows greater legal certainty to the two poles that are part of the same personal data processing relationship: the processing agent, who requires consent, and the data subject, who grants consent to the use of data for a certain purpose.

14.2.7 Consent, Cookie Policy and Access to Internet Services

Article 7 (4) of the GDPR establishes that conditioning the execution of an agreement or rendering of services to consent for treatment of personal data is only acceptable when this treatment is strictly needed for the rendering of services or execution of the agreement.

The LGPD does not contain such an obligation or, at least, one similar to the criterion set forth in Recital 43 of the GDPR. An integrated interpretation of the principles set out in the LGPD is needed to reach the same effects of Article 7 (4) of the GDPR.

According to the LGPD, consent is defined as "free, informed and unambiguous expression by which the data subject agrees with the treatment of his personal data for a given use" (Article 5, item XII).

However, when the processing of personal data is based on the performance of an agreement, the controllers will not be able to condition contractual performance on the consent of the data subject for the processing of additional personal data, except for those effectively needed to achieve the purpose of the agreement in question. This interpretation results from the application of the principles of necessity and purpose. The principle of necessity imposes the limitation of treatment to the minimum necessary for the realization of its purposes, with the scope of relevant data, proportional and not excessive in relation to the purposes of data processing (Article 6, item II, LGPD). In turn, the principle of purpose provides for "the treatment to be carried out for legitimate, specific, explicit and informed purposes to the data subject, without the possibility of further treatment in a manner incompatible with these purposes" (Article 6, item I, LGPD).

At the end of the day, we can say that it will be up to the scholars, court, or the National Data Protection Authority (ANPD) to construe the LGPD.

Despite the lack of court precedents on the subject, it must be said that it is still common for Brazilian companies to adopt privacy policies in a global and generic way, in which the data processing is "fully accepted", without the necessary request for granular consent for the processing of personal data. However, the LGPD prohibits the practice of generic and overly comprehensive policies[17] by providing

[16] P. Almeida. O que é Visual Law?. Available at https://www.jusbrasil.com.br/topicos/241689619/visual-law?ref=doc-topics. Accessed 3 September 2021.

[17] With the possible exception of small companies, individuals and NGOs established in the current draft regulation by ANPD, as already highlighted above.

that "consent must refer to specific purposes, and generic authorizations for the processing of personal data will be null" (Article 7, §4, LGPD). We emphasize that there is a gradual change of mindset and cultural change underway in Brazil.

As much as the Brazilian rule conditions the processing of public personal data upon goals, good faith and public interest, which justify making them publicly available (Article 7, § 3), this article still depends on further specific guidelines, which will eventually be formulated by the ANPD.

In this regard, the guidance of the Information Commissioner's Office (ICO) and Recital 43 of the GDPR is welcomed as a supplementary source of interpretation on the absence of ANPD regulation.

According to Brazilian law, if the use of any tool on the website is conditioned on the treatment of personal data, the data subject will have the right to receive this information in a prominent way. The controller must inform him or her about the consequences of the refusal to consent, in accordance with Article 9, §§ 3 and 18, item VIII, of the LGPD.

Some Brazilian companies have been adopting the guidelines of international organizations in their preparation processes for adaptation to the LGPD. Thus, one of the most interesting criteria is that contained in the already mentioned guide provided by the Information Commissioner's Office (ICO) in the United Kingdom. According to such guidance, the user must have full knowledge of cookie technologies or any other technology that has the function of collecting and storing personal information through websites.

In addition, the aforementioned technology and the reason for its use must be clearly explained, even though the information collected and processed is anonymous, it being necessary, in most cases, to obtain the user's consent in a clear, specific and unambiguous way.

Moreover, according to the European Data Protection Board, there are two exceptions to the use of cookies without the user's express consent being required. The first is when cookies are used to help, speed up or regulate the transmission of communication between services in the "electronic communications network" system, if communication is not possible without the use of cookies. The second is when cookies are strictly necessary for the activation of a service required by the user; that is, with cookies disabled, the service would not function properly.

Considering the time lapse between the publication of the LGPD and its entering into force, as well as the still absent regulation and guidance by the ANPD, there are currently no objective answers to issues that are somewhat only potential. However, our comments above reflect current trends and practices already adopted by some companies to (1) obtain granular consent, when necessary, and (2) not make the use of websites subject to the provision of consent, except when consent is strictly necessary for the provision of the service.

14.2.8 Black Box Phenomenon and the Transparency Principle

The "Black Box" phenomenon is typically observed in the realm of artificial intelligence, which occurs due to the controller's or even developer's lack of understanding of the principles and criteria underlying the decision-making of devices using artificial intelligence.

Artificial intelligence algorithms consist of a set of mathematical rules that will automate a process previously performed by a human. So technology, in fact, does not change reality but rather enhances the context of reality that already exists and automates the "status quo" since the initial data, which bring the machine's purpose and learning process to life, may have a bias in terms of neutrality, presenting themselves, albeit in a subtle way, as biased data.

Considering that the set of mathematical rules inserted in the machine essentially consists of trade secrets, it is possible to infer that the principle of transparency may have its efficiency suppressed in practice. However, there are already frameworks issued by personal data protection authorities that aim to guide a governance model in technologies such as artificial intelligence, aiming to safeguard the rights of data subjects and mitigate risks that involve the treatment of their information. This is the case, for example, of the Personal Data Protection Commission of Singapore (PDPC), which issued guidelines on the subject on 21 January 2020.[18]

According to the model suggested by the PDPC, organizations that use artificial intelligence should ensure that the decision-making process is explainable, transparent and fair and put individuals and data subjects as the main driver of technology; that is, centralize the individual in the development of the solution, aiming to protect individual rights and promoting their well-being and security.

There is, of course, a possibility that the system will be able to demonstrate to the user the process that resulted in the decision; however, this demand constitutes a notable challenge for the LGPD due to the provisions of the final part of Article 6, item VI, which conditions transparency on the safeguarding of commercial and industrial secrets.

That said, *prima facie*, the black box phenomenon is not compatible with the transparency principle of the Brazilian General Personal Data Protection Law (LGPD). However, if despite such phenomenon the controller is able to comply with the right to information of the data subject, then, in practice, the black box would not entail an infringement of rights.

The transparency principle means that processing agents shall provide data subjects with clear, accurate and easily accessible information about the personal data processing, including the processing chain of agents. Therefore, to be compatible with the transparency principle, data subjects need to have easy access and, in fact, understand information about the decision-making processes of the artificial intelligence technology. However, considering that algorithmic decisions are usually

[18] Personal Data Protection Comission – PDPC. Model AI Governance. Link: https://www.pdpc.gov.sg/model-ai-gov. Accessed 3 September 2021.

characterized by opacity, this is not an easy goal to achieve in practice, even in relation to supervised machine learning, in most circumstances.

Article 20 LGPD establishes that data subjects are entitled to request a review of decisions taken solely based on the automated processing of personal data that affect their interests, including decisions designed to define their personal, professional, consumer and credit profile or aspects of their personality. Article 20, §1, institutes the right to explanation, derived from the transparency principle, determining that the controller must provide data subjects, whenever requested, with clear and adequate information regarding the criteria and procedures used for the automated decision.

The LGPD does not stipulate the same exceptions provided in the GDPR regarding the non-applicability of the duty to provide information about the existence of automated decision-making and the meaningful information about the algorithms' logic and the envisaged consequences of such processing for the data subjects. Thus, the LGPD seems even stricter than the GDPR in this aspect. On the other hand, Article 20, §2, of the LGPD establishes that if the controller evokes commercial and industrial secrecy as a justification not to present such information, the ANPD may carry out an audit to verify discriminatory aspects in the automated processing of personal data. Such audit has not been regulated by the ANPD yet.

In view of the complexity of the artificial intelligence decision-making process, an interpretation in the sense that there is a prohibition on the use of machine learning algorithms whose decision-making activity is inscrutable could result from Article 20, §1, of the LGPD. However, such solution would be counterproductive and not feasible from an economic standpoint.

Alternatively, transparency can be interpreted under two aspects: accessibility and understandability.[19] Accessibility, interpreted as providing access to the program source code, may not be sufficient to provide an explanation about the automated decision-making processes since the source code only exposes the machine learning method used and not the decision criteria themselves. On the other hand, understandability about the criteria[20] used for the automated decision is quite desirable.

In general, algorithms can be understood when the human being is able to articulate the logic of a specific decision (without the need to know the details of how the system achieved the decision). Therefore, to eliminate or at least reduce the black box phenomenon in such a way as to make the decision-making processes of the artificial intelligence compatible with the transparency principle (and the rights

[19] B. D. Mittelstadt et al. The ethics of algorithms: Mapping the debate. Big Data & Society, 1-21, jul.-dez. 2016. https://journals.sagepub.com/doi/full/10.1177/2053951716679679. Accessed 22April 2020.

[20] The criteria would be: (i) what are the main factors that led to the decision; (ii) whether changing any of the determining factors would change the decision; and (iii) analyzing whether similar cases had different decisions and determining processes. see I. FERRARI. *Accountability de Algoritmos*: a falácia do acesso ao código e caminhos para uma explicabilidade efetiva. Available at https://itsrio.org/wp-content/uploads/2019/03/Isabela-Ferrari.pdf. Accessed 3 September 2021.

of access and to an explanation), it might be necessary to apply technical and entrepreneurial efforts supported by public policies in the conception of the algorithms, such as the following:[21]

(a) To train AI systems with human-interpretable terms and store data from each decision in order to probe the decision afterwards
(b) To establish different levels of controllability according to the different machine learning methods (*i.e.* supervised, unsupervised and reinforcement learning) and techniques (*e.g.* semantic networks and natural language processing, regression analysis, artificial neural networks, etc.)
(c) To prefer the adoption of machine learning methods and techniques that facilitate control and understandability, allowing for the artificial intelligence algorithms to use more complex logic only for a few cases that really need it (balancing transparency and business performance); and
(d) To design updatable AI systems to train proxies to verify whether predictions of terms that cannot be determined in advance through the learning inputs (such as in a litigation process) are correlated with the automated decisions generated by the AI system.

In Brazil, there are already initiatives for the establishing of large shared and open databases (*Serenata.ai* and *Colaboradados*) so that different algorithms can use them in their training, allowing biases to be identified and corrected by anyone.

The use of machines is expected to eliminate the prejudices or subconscious processes that affect human thought or, at least, to enable us to recognize them more easily whenever they appear.

14.2.9 On the Legality of Manufacturers' Prohibition on the Sale of Their Products on Certain Platforms in Order to Protect the Products' Image

According to Article 132, item III, of the Brazilian IP law, the trademark owner may not restrain the free circulation of products placed on the internal market by himself or herself or by another with his/her consent. Such order is related to the principle of exhaustion of trademark rights, which states that the trademark owner may not invoke exclusivity in order to restrain subsequent sales after the first sale of a product on the market.[22]

[21] These comments are made under the assumption that, in most cases, it is technically feasible to extract the kinds of explanations that are currently required of humans from artificial intelligence systems.

[22] Our analysis does not include comments with respect to the purchase of keywords, by distributors, containing trademarks for the sale of their products.

In *Adcos v. Mercado Livre*, the São Paulo Court of Appeals[23] rejected the plaintiff's appeal in order to restrain the defendant from disclosing Adco's products on its platform, which were being sold by third parties. The Sixth Chamber of Private Law ruled that the trademark owner may not justify its prohibition claim under the argument of protection of its product's image since it is grounded on future and uncertain damages that, if occurring, can be later repaired. Moreover, according to such a decision, "the supplier is liable for the ensuring of the safety of the product it places on the market, which should also consider the possibility of reselling the product after the exhaustion of trademark rights".

It is worth highlighting that between the supplier and distributor, nothing prevents the supplier, based on contracts that have exclusive or selective distribution clauses, from prohibiting its distributors from selling its products on certain platforms, such as in marketplaces. Thus, in a contractual sphere between the supplier and distributor, Brazilian law does not forbid such a prohibition.

However, once these official distributors launch the supplier's product on the national market, there is no way of preventing the free resale of this product by any third party, including on e-marketplaces. This understanding was adopted by the Superior Court of Justice in *Citizen Watch v. Mercado Livre*,[24] in which the Rapporteur Minister emphasized that drastic measures to control Internet content should only be used in extreme cases and should not be adopted as a rule, especially in the case of individual interests, without great serious risk of damages.

In specific cases in which effective damages are verified, although the trademark owner can file a lawsuit seeking compensatory damages against the liable party, the online platform itself cannot be liable for such practices. As duly mentioned in our Report of 2018,[25] Internet application providers will only be held liable if, after a court decision, they fail to take action to make unavailable the infringing content, as set forth in Articles 18 and 19 of the Civil Rights Framework for Internet.

This said, under Brazilian law, it is illegal for manufacturers to prohibit distributors from advertising the manufacturer's products on search engines inasmuch as the trademark owner does not have limitless rights regarding trademark use.

Brazil has an important precedent about this issue, applying this exception to cases of sponsored links, in the case of *L'Oréal vs.* (Interlocutory Appeal # 0089493-71.2012.8.26.0000; 1st Reserved Chamber of Business Law of the São Paulo Court of Justice, judgment date: 1 June 2012). The court has ruled that "it is

---

[23]Lawsuit # 0031284-65.2013.8.26.0068 – Adcos Indústria e Comércio LTDA vs. MercadoLivre.com Atividades de Internet Ltda. 06th Chamber of Private Law, Rapporteur Judge Eduardo Sá Pinto Sandeville. Court of Appeals' decision issued on 22 March 2018.

[24]Lawsuit # 1.383.354 – SP – Citizen Watch do Brasil S/A vs. MercadoLivre.com Atividades de Internet Ltda e Outro. Third Pannel, Rapporteur Minister Nancy Andrighi. Superior Court of Justice's decision issued on 27 August 2013.

[25]See also our past report solely on this subject in F. B. Oquendo, Chapter 17 (Brazil) of Këllezi, et. al. *Liability for Antitrust Law Infringements & Protection of IP Rights in Distribution*. Springer: Cham, 2018.

legal to use the Adwords tool to promote the sale of products made available to consumers on the websites of distribution companies".[26]

The Brazilian Superior Court of Justice also decided on 13 September 2016 (Special Appeal # 1.606.781/RJ) that, in cases involving sponsored links, the mere mention of the registered trademark on search engines is not considered unfair competition. Therefore:

(a) The owner of a trademark cannot prevent it from being used by distributors and being placed together with the distributors' trademark when the disputed use is made with legitimate promotional or commercial purposes.
(b) In such cases of promotion and commercialization of trademark owners' products in their physical establishments or over the Internet, it is not necessary for the distributors to obtain an authorization in view of the fair use exception referred above.

Notwithstanding, even in the context of promotion and commercialization by distributors, the use of the trademark is bound to some restrictions. It means that the trademark cannot be used in an unrestricted/indiscriminate way, without any limitation and based on the distributor's own criteria, or used to boost that distributor's business. The use of trademarks by distributors in advertising the manufacturer's products on search engines must be justified and loyal.

The distributors must make it clear to the public that they only sell the product and are not a trademark licensee or that there is no commercial relationship with the trademark owner other than distribution, so as not to constitute unfair competition.

In this sense, it is important to emphasize some relevant aspects so that it may be considered fair use according to the IP Law: the trademark owner may verify the real context of the trademarks used by the distributors when adverting its products on search engines; the trademark owner may determine if there is an intention by the distributor to cause confusion or undue association with the true owner to mislead consumers, for example when the distributor uses the trademark owner's logo or the same fonts and/or colours of the legitimate trademark owner, making an evident allusion to the owner's logo or its website layout; the distributor may not use the trademark is such a way as to try to misrepresent that it has a business relationship or association with the trademark owner, but its use is allowed as long as the standards outlined by the trademark owner are followed by the distributors, so as to avoid unfair competition or a violation of the trademark owner's rights set forth in Article 129 cited above; and the distributor of accessories that are used by or compatible with a third-party product and/trademark may not try to obtain undue advantage by misrepresenting that its accessories are manufactured or endorsed by the trademark owner of the main product. For example, the distributor cannot imply

[26] Available at https://tj-sp.jusbrasil.com.br/jurisprudencia/22004983/agravo-regimental-agr-894937120128260000-sp-0089493-7120128260000-tjsp/inteiro-teor-110494544?ref=juris-tabs. Accessed 3 September 2021.

that its accessory originated from or is associated with the original product's manufacturer.

14.2.10 Unequal Competitive Conditions and the Use of Artificial Intelligence by Businesses

Artificial intelligence is a challenge in every single sector of the economy due to the unavoidable growth of its use owing to the benefits it can provide, such as long-term savings to the companies' budget—despite the expensiveness of its implementation in the short term—as well as celerity and improvement to the companies' processes. It is true that the use of artificial intelligence is not forbidden in Brazilian legislation, including Brazilian competition law. What will define if the company acts in compliance with the law or not is not the simple use of this technology itself but the purpose for which it uses artificial intelligence and the means of obtaining data for the artificial intelligence models. It is undeniable that we all will be left behind by artificial intelligence if we do not find a way to adapt to this technology, and the same applies to companies that are capable of operating with artificial intelligence mechanisms and technologies, which will certainly have better competitive conditions and which cannot be prevented from doing so as long as they comply with the law.

So it is beyond doubt that employing AI in services will give a competitive edge to companies, and it is likely that doing so will not be deemed *per se* an infringement, at least from a current law point of view. How crucial AI will be in the future and the proportions of this competitive edge, however, may lead to added government control over companies' affairs to curb undue market control—from decisions that consider AI technologies an essential facility on some segments to regulations that determine the sharing of AI technologies through a network of regulated companies.

14.2.11 Algorithms and Discrimination Against Companies in the Placement of New Advertisements

One of the main purposes of algorithms is filtering and ranking information based on factors included in their code due to the numerous data online and the limited capacity of an individual to absorb them all. They are essential in a society that requires the organization of information received according to what would be more relevant to one person.

In this regard, algorithms that evaluate previous purchase decisions and then place new advertisements would not aim at deleting or hiding information or benefiting any company for unreasonable motivations—which would constitute a discriminatory practice. Basing new advertisements to be received by consumers on previous purchase decisions seems a rational type of ranking, especially considering they are intended to be personalized according to one's interests.

Even though this type of algorithm could make it more difficult for new businesses or struggling companies to stand out, it does not prevent them from having their advertisements exhibited to consumers in a proportional form according to their sales—even if in a less frequent way when compared to others. Also, these companies might address other kinds of marketing strategies, increasing their purchase numbers, which will enable them to start being considered by the described algorithms (again).

Of course, depending on the spread of the use of algorithms that place new ads based on previous purchase decisions by consumers/data subjects, eventual intervention in the market may be needed to avoid the tunnel effect, which could ultimately exclude new businesses. In other words, while the possibility is there, the current situation does not allow us to conclude that an intervention in the market to curb the activity of these types of algorithms is necessary or urgent.

14.3 Issues in the Light of Legal Principles and Propositions for Future Regulation

14.3.1 Transparency and Automated Data Collection

It is known that oftentimes the collection of data by artificial intelligence escapes the control of the companies that set the AI, resulting in the controller not knowing what data from third parties are ultimately being collected. This could affect the transparency principle, which, as already referred to in this report, is one of the guiding principles of the LGPD as well as of the GDPR.

It remains to be seen whether the companies will be held liable for acts undertaken by AI since there is no regulation, law or court precedent available in Brazil regarding this matter. We do believe, however, that liability will apply due to the fact that AI acts and their consequences should not be considered fortuitous or force majeure.

In these cases, it is recommended to adopt one or more of the best practices to ensure the safety of Internet users regarding personal data processing activities, namely (a) insertion of privacy notices on websites informing the public about the collection of data user's personal data, (b) insertion of privacy policy and cookie policy with clear and accessible language and (c) inserting the link to the privacy policy of any third party that has access to the data collected on the website.

In addition, it is important to mention that the Civil Rights Framework for the Internet establishes in its Article 15 that providers of Internet applications, constituted in the form of legal entity and which exercise this activity in an organized manner, professionally and for economic purposes, must keep records of access to their applications for a period of 6 (six) months, counted from the date of access to such applications.

Thus, even if the controller does not know exactly what third-party data are collected on his or her website, it is possible to infer that, due to a legal obligation, his/her website collects, at least, the records of access to his/her applications (the set of information referring to the date and time of use of a given Internet application from a given IP address). Therefore, in the privacy notices and/or privacy policy of the website, the need for such collection of personal data must be stated.

14.3.2 Need for Additional Legal Instruments to Restrict the Collection and Use of Data for the Personalization of Content, Advertising or Prices

With the entry into force of the LGPD, the processing of personal data in Brazil[27] can only occur in one of the ten cases provided for in Article 7, for "common" personal data,[28] or in the eight hypotheses described in Article 11,[29] for sensitive personal

[27] According to Article 5 of the LGPD, treatment of personal data is defined as "any operation performed with personal data, such as those referring to the collection, production, reception, classification, use, access, reproduction, transmission, distribution, processing, filing, storage, elimination, evaluation or control of information, modification, communication, transfer, diffusion or extraction".

[28] Art. 7 The processing of personal data can only be carried out in the following cases:

I - by providing consent by the data subject;

II - for the fulfillment of a legal or regulatory obligation by the controller;

III - by the public administration, for the treatment and shared use of data necessary for the execution of public policies provided for in laws and regulations or supported by contracts, agreements or similar instruments, subject to the provisions of Chapter IV of this Law;

IV - to carry out studies by a research body, guaranteeing, whenever possible, the anonymization of personal data;

V - when necessary for the execution of a contract or preliminary procedures related to a contract to which the data subject is a party, at the request of the data subject;

VI - for the regular exercise of rights in judicial, administrative or arbitral proceedings, the latter under the terms of Law No. 9,307, of September 23, 1996 (Arbitration Law);

VII - for the protection of the life or physical security of the data subject or of a third party;

VIII - for the protection of health, in a procedure carried out by health professionals or by health entities;

VIII - for the protection of health, exclusively, in a procedure performed by health professionals, health services or health authority;

IX - when necessary to serve the legitimate interests of the controller or third party, except in the event that the fundamental rights and freedoms of the owner prevail that require the protection of personal data; or

X - for credit protection, including the provisions of the relevant legislation.

[29] Art. 11. The processing of sensitive personal data can only occur in the following cases:

I - when the data subject or his legal guardian consents, in a specific and prominent way, for specific purposes;

II - without providing consent from the data subject, in the cases in which it is indispensable for:

a) compliance with legal or regulatory obligations by the controller;

b) shared treatment of data necessary for the execution, by the public administration, of public policies provided for in laws or regulations;

c) carrying out studies by a research body, guaranteeing, whenever possible, the anonymization of sensitive personal data;

d) regular exercise of rights, including in contract and in judicial, administrative and arbitration proceedings, the latter under the terms of Law No. 9,307, of 23 September 1996 (Arbitration Law);

data, i.e. personal data about racial or ethnic origin; religious belief; political opinion; union membership or organization of a religious, philosophical or political nature; data relating to health or sexual life; and genetic or biometric data, when linked to a natural person,[30] which are the main types of data collected and treated for the personalization of content, advertising and prices.

As a rule, the use of such hypotheses (legal bases) of treatment does not require that those responsible for the processing of personal data prepare other documents to guarantee the rights of the data subjects, except the notices and privacy policies that provide transparency to users about the activities involved in the processing of personal data.

However, the hypothesis of the legitimate interest of the company responsible for processing (controller) or a third party, provided for in item IX of Article 7, can be conditioned on the issuance of another legal document, namely the Impact Report on the Protection of Personal Data (RIPD),[31] documentation by the controller containing a description of the procedures for processing personal data that may generate risks to civil liberties and fundamental rights, as well as measures, safeguards and risk mitigation mechanisms,[32] which may be requested by the Brazilian Data Protection Authority when legitimate interest is used as a legal basis for the processing of personal data.

It should also be noted that, inspired by the recommendations of the European Union's data protection authorities and while the Brazilian Data Protection Authority is not, in fact, constituted, the Brazilian market has chosen to carry out a balance test between the interests of the controller and the rights of data subjects (legitimate interest test) to analyze and minimize the negative impacts on the privacy of the data subject, such as data leakage or unauthorized use of personal data.

In view of the above, combining the articles of the LGPD and the adoption of a proactive stance with the application of the balance of interests test seems to be sufficient to avoid the need for additional legal instruments to restrict the collection and use of data for the personalization of content/advertising/prices.

---

e) protection of the life or physical safety of the data subject or of a third party;

f) protection of health, in a procedure carried out by health professionals or by health entities; or

f) health supervision, exclusively, in a procedure performed by health professionals, health services or health authority; or

g) guarantee of fraud prevention and security of the data subject, in the processes of identification and registration authentication in electronic systems, safeguarding the rights mentioned in art. 9 of this Law and except in the event that the fundamental rights and freedoms of the data subject prevail that require the protection of personal data.

[30] Article 5, item II of the LGPD.

[31] Article 10, paragraph 3 of the LGPD.

[32] Article 5, item XVII of the LGPD.

It is worth mentioning that the Administrative Council for Economic Defense (CADE) has ruled on some interesting antitrust cases concerning Internet companies and has even quite recently launched specific guidelines concerning this segment.[33]

14.3.3 Creation of User Profiles Which Currently Also Use Data from International Third-Party Sources

Brazilian law places no restriction on the processing operation that involves personal data from outside the national territory and that are not the subject to communication, shared use of data with Brazilian processing agents or the object of international data transfer with another country other than that where the data was originated, provided that the country of origin provides a degree of protection of personal data in par with that provided for in the LGPD (Article 4, item IV).

However, if the processing operation is carried out in the national territory and aims to offer or supply goods or services or process data from individuals located in the national territory or the personal data object of the treatment has been collected in Brazil, this must be directly subject to the rules provided for in the LGPD (Article 3).

14.3.4 Price Collusion Through Artificial Intelligence Software

Algorithms are developed from an abstract idea to solve a problem. In other words, they are representations of various methods connected in stages and fulfil a function. In this sense, every time a solution is created, it becomes the basis for the creation and development of an algorithm. The complexity of algorithms is limited by creativity and can range from the simplest information and conditions, with a few variables, to the most complex structures and functions. An example of a complex algorithm is the automatic generation of response by natural language processing algorithms, employed to make chatbots operational. By classifying the meanings assigned to each set of words, sentences can have their meaning classified and their response estimated.

Algorithms that set the prices of products on the Internet work basically through three steps. First, the algorithm experiments with different pricing strategies over time. After the search, the algorithm determines which prices are profitable and learns the lessons. Ultimately, the algorithm's choice of pricing strategies changes gradually and relies on those that are most profitable. In other words, it works

[33] CADE, Mercados de Plataformas Digitais, 2021. Available at https://cdn.cade.gov.br/Portal/centrais-de-conteudo/publicacoes/estudos-economicos/cadernos-do-cade/plataformas-digitais.pdf. Accessed 30 August 2021.

through simple trial and error, learning as it goes what tends to turn into a profit and what does not.[34]

In the long run, there has been a high degree of investment in technology by companies in order to benefit from an "algorithmic competitive advantage" because if other companies in the sector are employing algorithms, the rest will have a strong motivation to do the same, lest they be left out of the market. The result is a certain sector of the market where all competitors use algorithms to constantly monitor the actions of other competitors, consumer choices and changes in the market environment in real time and respond to them, thus creating a transparent environment and one conducive to a kind of collusion.

However, it can be rather difficult to assess whether algorithms increase or reduce the prospect of collusion as they tend to modify the structure of market conditions and supply-side factors, which together could have a positive, negative or ambiguous impact on the sustainability of collusion.

The problem with algorithms that ultimately allow collusion is that this type of behaviour can happen without the developers of the algorithms even knowing it, let alone the companies that employ them; that is, the great challenge of a stricter legal measure would be the investigation of how the collusion occurred, whether it occurred as a result of the negligence of the owners of the algorithms to understand what their pricing tools were up to or whether it was just something that the algorithm developed by mimicking other algorithms.

In that sense, there are some traditional measures that antitrust authorities can put in place to address at least some of the competition concerns. These alternatives and less stringent approaches could include the use of market research, merger control enforcement, and the use of remedies.

Unlike the economic approach, which considers collusion to be a market outcome, the legal approach focuses on the means used by competitors to achieve such a collusive result. Competition law does not generally prohibit collusion as such but prohibits anti-competitive agreements. In that sense, the development of algorithms created solely for the purpose of creating collusion, and thereby achieving higher prices in the market, is illegitimate for being an anti-competitive practice.[35] CADE, however, has recently stated that the issue of collusion achieved by algorithms remains to be more thoroughly examined.[36]

If the circumstances are sufficiently clear that the algorithm was developed for this purpose, this would not exclude the possibility of excessive pricing based on established joint dominance. The challenge is to prove the existence of joint dominance since price collusion can be difficult to detect as algorithms do not take the

[34]D. Foster, Computer collusion, available at http://www.frontier-economics.com/uk/en/news-and-articles/articles/article-i2197-algorithms-and-price-collusion/. Accessed 3 September 2021.

[35]http://www.oecd.org/daf/competition/Algorithms-and-colllusion-competition-policy-in-the-digital-age.pdf - Access on 18 April 2020.

[36]Mercados de Plataformas Digitais, p. 133. Available at https://cdn.cade.gov.br/Portal/centrais-de-conteudo/publicacoes/estudos-economicos/cadernos-do-cade/plataformas-digitais.pdf. Accessed 31 August 2021.

form of a unified price that has been maintained for a long time, thus making it difficult to verify whether the price at any given time was really above its competitive level.

The emergence of algorithms and their application in the market has generated concern about the behavioural future of these algorithms since there is a possibility that, at some point, they may help to implement conventional cartel-like behaviour. This could be achieved because these algorithms are developed and become capable of processing all available and relevant pricing information, and therefore by monitoring and analyzing or anticipating their competitors' responses to current and future prices, competitors may be more easily able to find a supra-competitive price balance that they can agree on and then maintain it over time and through relevant changing circumstances.

A possible countermeasure to the risk created by such new types of cartels is the use of other algorithms developed to detect collusive proposals and, in general, possible cartel behaviour. Thus, the use of such algorithms may put the technology at service of antitrust and competition enforcement. Aside from that, any legal or regulatory initiative must take into account the possible negative impacts of a prohibition, especially the discouraging of development and the adoption of new and disruptive technologies.

14.3.5 Ethical Limits of Personalized Prices

Personalized prices and other tools provided by the use of artificial intelligence, as long as they are used in compliance with the law, represent *a priori* a combination of benefits to both the companies and the consumers as it is a way of facilitating purchase research mechanisms and helping the consumers to have access to products that match their profile. On the other hand, this may prevent them from easily finding all other products they may be interested in, regardless of the type.

It is important to highlight that, here, we are addressing the situation where the programming process of algorithms shows to the consumer the products whose prices are more compatible with his/her profile but not the distortion of this technology to change the price of the same product according to the profile of the person or even the denial to offer a product due to the consumer's profile, known as geo pricing and geo-blocking, which, whatever the area of life is, amounts to discrimination and other infringements of the Brazilian Consumer Protection Code, for instance of its Articles 4°, caput, items I and III; 6°, items II, III and IV; 37, §2°; and 39, items II, IX and X.

Due to such benefits of personalized prices, it would not be reasonable to prohibit their use in certain areas. However, this should be implemented with a view to making the use of this technology more compatible with the principles that protect consumers, data subjects and advertising guiding principles, such as transparency, promotional identification and the right to choose whether or not to use a given technology.

14.3.6 Incorporation of Legal Requirements and Ethical Values into the Programming Process of Algorithms to Prevent Discrimination

In the same line as the above arguments, there is a need for a specific regulation to guarantee that the programming process of algorithms complies with all objectives and principles of law, especially the most recent ones, such as Brazilian laws no. 12.965/2014 and no. 13.709/2018, and the recent alterations to consumer law, which arose exactly from the need to keep up with technological evolution, in order to prevent discrimination.

Both the GDPR and the LGPD state, quite clearly, that personal data processing activities must be developed—from the beginning—with an aim to comply with established legal requirements and principles, in order to protect the rights of data subjects, in addition to complying with the fundamental principles that guide privacy by design.

The GDPR Recital 71 establishes that treatment agents, specifically data controllers, must use appropriate procedures and apply technical organizational measures in order to prevent discriminatory effects against natural persons. In its Recital 78, the GDPR establishes that personal data processing activities must observe—from their conception—the principles that govern "privacy by design", which, in sum, establishes that any new development of services or products or a new personal data processing activity should always be guided by the search for the defence of individual rights and people's freedom, adopting or creating structured transactions with personal data in accordance with established principles and recognized good practices. In addition, such activities must also comply with the legal requirements listed by the privacy and personal data protection regulations.

Furthermore, privacy by design is based on seven principles,[37] among which is respect for the user's privacy, always keeping him or her at the centre of the development of activities, products and services.

Like the European Regulation, the LGPD, in Article 6, items IX and 46, paragraph 2, establishes, respectively, that the activities of personal data processing cannot be carried out with illicit or abusive discriminatory purposes—including in the case of automated decisions—and that security, technical and administrative measures capable of protecting personal data from any form of inappropriate or illicit treatment must be observed from the product or service design phase throughout its execution.

Although the LGPD does not employ "Recitals", legal requirements and socially and ethically recognized values must be incorporated into the algorithm programming process in Brazil to avoid discrimination.

Brazilian law, as can be seen from Article 20 of the LGPD, quoted many times in this report, also allows the data subject to request a review of decisions taken solely

---

[37] A. *Cavoukian, "7 Foundational Principles". Available at* https://www.ipc.on.ca/wp-content/uploads/Resources/7foundationalprinciples.pdf. Accessed 26 April *2020.*

on the basis of the automated processing of personal data that affect his or her interests, including decisions designed to define his/her profile, be it personal, professional, consumer or credit data or aspects of his/her personality, and the ANPD will be able to carry out an audit to ascertain possible discriminatory aspects in the automated processing of personal data.

Of course, this incorporation of legal requirements and ethical values would have to be updated and re-evaluated from time to time, given the changing and evolving landscape of consumer and data subject protection.

14.3.7 Human Tasks That Can Be Carried Out Automatically by AI: Future Developments

It is possible for advertising campaigns, also in the creative field, to be able to be carried out autonomously by artificial intelligence.

It is worth mentioning that in 2018, Lexus, the luxury vehicle division of the automaker Toyota, ran an advertising campaign whose script was created by artificial intelligence.

The artificial intelligence architecture of the Artificial Intelligence Creative Adversarial Network (AICAN) has already been tested in order to generate new artistic works, bearing in mind the minimum requirement of originality, that is, an element capable of differentiating intellectual work from others previously created by human beings. Through the use of AICAN, the so-called discriminatory neural network has access to a vast database labelled with various existing artistic styles, and then it uses such information to identify and differentiate the styles. The neural network called the generator does not have access to any existing intellectual work but generates new images based on random input data. In addition, the discriminator network emits two distinct signs, which are, by definition, contradictory: the first sign is the classification of "art or not art", and the second sign is "style ambiguity", that is, how confusing the discriminator network becomes or not in trying to identify the style of the artistic work generated by the generator network based on the labelled artistic styles. Then the generator network uses this style ambiguity to improve its ability to generate artistic works that do not follow existing artistic styles and therefore have an increasing level of style ambiguity. The input data can certainly include advertising campaigns.[38]

The project "Generating 'Art' by Learning About Styles and Deviating from Style Norms", carried out by the researcher Ahmed Elgammal of the Art and Artificial Intelligence Laboratory at Rutgers University in New Jersey, USA, proposes to maximize the deviation from existing artistic styles and the greatest possible framing of the pre-established art concept, from the point of view of artistic works created by human beings. In Brazil, the protection of intellectual work does

[38] V. da Silva Ferro *As obras artísticas geradas pela inteligência artificial*: considerações e controvérsias, p. 113. 1a. ed. Rio de Janeiro: Lumen Juris. 2020.

not depend on the value or criteria of classification as "art or not art". In order to assess especially the framing of the works generated by AI in the pre-established concept of art, Elgammal conducted experiments based on the following criteria: whether or not the art was intentional, visually structured, communicative and inspiring. The purpose of this experiment was to compare the response of humans with art generated by the AICAN model in relation to art generated by human artists. The results of the experiment showed that people were unable to satisfactorily make this distinction.[39]

As AICAN is a project developed in the United States, it is believed that the criterion adopted by researchers to measure creativity is different from the criterion used in civil law countries. In Brazil, in general, it is believed that to be endowed with creativity, the work needs to reflect the personality of its creator. Thus, even if a work generated by AI is produced by an AICAN architecture algorithm, it would not meet the originality requirement to be subject to protection by the Brazilian Copyright Law. This circumstance does not, however, exclude the possibility of creating an advertising campaign autonomously by the AI. The absence of copyright protection would therefore imply the immediate falling into the public domain of such work, except in the event of legislative changes, based on (1) the perception of the work's value and its consequent possibility of appropriation as a legal asset as well as (2) the justifying theories of copyright.[40]

Technology has improved several activities in society, increasing productivity among different sectors, besides the screenwriting of ad pieces. In this regard, it is possible to highlight the following tasks, which were previously performed only by humans and nowadays are provided or also performed entirely through the application of algorithms—associated or not with artificial intelligence:

1. Writing of books, articles, papers and other literary materials that might be created by algorithms programmed with specific purposes or themes
2. Searching for information (as in links to websites on search engines, content on social media, books in reading apps, songs and audiovisual materials in streaming platforms), which does not require a greater effort from humans to find since the algorithm already filters and ranks data;
3. Finding direction and routes, which is an activity usually calculated by Global Positioning System (GPS) and navigation apps and no longer by a human looking at a map;
4. Simple communication, given that some applications contain algorithms that, combined with artificial intelligence, can simulate entire conversations, without human intervention (for example, Siri and Alexa)
5. Photoshop and other design adjustments on photos and images online, which no longer requires human intervention;

[39] Idem, p. 114.

[40] Ibidem, p. 117–160.

6. Driving, considering the existence of automated vehicles that combine artificial intelligence and algorithms

14.3.8 Personalized Election Campaigns and Democracy

In the digital age, the use of data is essential to enhancing the dialogue between candidates and the electorate. However, there is a very narrow link between the correct use of data and its abusive and undue use, which consequently creates a negative impact on the final result of election campaign processes around the world.

The discussion on fake news and personalized campaigns has increased in the last few years, especially after the specific events during the US presidential election and the BREXIT poll in 2016. In Brazil, it was not different.

With big data, candidates can have access to the preferences and information of each Internet user, therefore getting to know which is the best way to attract their attention. Through the use of certain forms of content targeting, such as dark posts and micro-targeting, it is possible to shape an artificial scenario close to the one desired by the candidates, depending on how much they are willing to invest in the process.

However, what may seem common and harmless for users/voters in fact is not.

The use of such strategies, with excessive targeting of content and personalized information, can jeopardize the debate between the electorate. Besides promoting alienation and a superficial debate, such strategies can be considered to be manipulative and are, nonetheless, unfair.

In other words, there is a strong and dangerous political polarization, social demobilization and the so-called silencing effect of certain public issues, consequently affecting individuals' informational self-determination. Therefore, the undue use of digital strategies to obtain votes during an election shows, in the end, a manipulated result of the campaign, with false data arising from artificial moves.

The Brazilian authorities started to think about solutions to combat the negative impacts created by electoral strategies, mainly due to the legal omission in regard to those specific tactical manoeuvring.

According to Article 5, items X and XII, of the Brazilian Federal Constitution, the inviolability of intimacy, privacy and honour are fundamental rights, as well as the confidentiality of telegraph correspondence and communications, except in the case of a court order.

In view of such fundamental constitutional rights, the LGPD establishes that the use of personal data must follow the principles of Article 6 of the said law, such as purpose, adequacy, necessity and transparency. With respect to sensitive personal data, which consist of information regarding politics, affiliation and political organization, a special treatment is required by the law—such as its use linked to the consent provided by the data subject.

In addition to the LGPD, the Electoral Law[41] has been recently adjusted with respect to electoral digital content. Among its changes, the law sets forth that the candidate cannot use false profiles/users to propagate information, especially false information regarding other candidates (creating the so-called candidacy deconstruction), under penalty of payment of a fine.

Despite all the constant efforts of the authorities, the legislation regarding digital strategies for election campaigns is still superficial, while the referred strategies are increasingly accurate, representing a real impact on democracy and a real political polarization without, however, rendering a true portrait of the electorate's will.

Acknowledgements The rapporteur would like to thank his colleagues from the Study Committee in Competition from the Brazilian Intellectual Property Association—Anna Carolina Peixoto, Ana Caroline Silva, Cecília Cunha, Juliana Sydow, Letícia Carneiro, Maria Eduarda Junqueira, Marilia Baracat, Raphael Campos, Renata Soraia Luiz and Vanessa Ferro—for the inestimable aid with research and for responding to the questions.

[41] Federal Law N°. 9,504 of 30 September 1997.

15 Germany

Thomas Hoeren and Hanna Hoffmann

15.1 Introduction

The development of artificial intelligence (AI) leads to new challenges in all legal fields. The application possibilities of algorithms are becoming more and more extensive. While algorithms were once only used for recognising patterns in data sets,[1] today they are increasingly making forecasting decisions and thus taking on human tasks.[2] From a competition law perspective, it is worth taking a look at the use of AI for marketing purposes, which often includes native advertising and personal pricing, but the role of influencers is also in focus.

Firstly, this report examines the influence of algorithms on the customer's contractual freedom (Sect. 15.2). Secondly, the relationship between competition law and data protection law is presented (Sect. 15.3), referring in particular to current German and European case law. Thirdly, certain commercial practices that are currently of particular importance are examined, such as influencing, AI and marketing and personalised prices (Sect. 15.4). Fourthly, the impact of personalised political campaigns on elections and democracy as a whole is discussed (Sect. 15.5). Finally, future perspectives are outlined.

[1] Bundesverband Digitale Wirtschaft e.V., Anwendungsgebiete. Available at https://www.bvdw.org/themen/kuenstliche-intelligenz/anwendungsgebiete/. Accessed 23 August 2021.

[2] Fraunhofer Big Data AI, Potenzialanalyse "Künstliche Intelligenz". Available at https://www.bigdata.fraunhofer.de/de/big-data/kuenstliche-intelligenz-und-maschinelles-lernen/potenzialanalyse%2D%2Dkuenstliche-intelligenz-.html. Accessed 23 August 2021.

T. Hoeren (✉) · H. Hoffmann
Westfälische Wilhelms-Universität, Münster, Germany
e-mail: hoeren@uni-muenster.de; hanna.hoffmann@uni-muenster.de

B. Kilpatrick et al. (eds.), *Antitrust in Data Driven Markets & Legal Framework for Influencers, Native Advertising and Control over the Use of AI in Marketing*, LIDC Contributions on Antitrust Law, Intellectual Property and Unfair Competition, https://doi.org/10.1007/978-3-031-07422-6_15

15.2 Algorithms' Influence on the Customer's Contractual Freedom

The infinite flood of information on the Internet makes the use of certain tools, such as search engines, necessary. The mass of data can hardly be grasped by an individual and could not be processed without a preselection of search engines.[3] Therefore, the search algorithms of search engines determine which offers and thus potential contractual partners are suggested to the customer.[4] A search engine often acts as a "gatekeeper".[5] The software's suggestion finding process remains hidden from the user, which becomes problematic if external (third-party) factors, such as advertising, outweigh the user's interests when creating search results.[6] In order to get an objective selection of the best options, a user would have to do further research.[7] However, the average user only looks at the first two sides of results when using a search engine.[8] Therefore, it is no surprise that empirical studies show that ranking has a considerable influence on consumers' purchasing decisions.[9] However, it is questionable whether it can be said that the customer's contractual freedom is being eroded by the use of search engines.[10] There is no denying that the customer's rights are affected. However, it should be noted that similar situations have existed even before the emergence of search engines: if someone buys an item in a shop, one does not know whether it is possible to get it cheaper elsewhere. Search engines are the only way to sort content on the Internet, and the use of the Internet itself chiefly offers a huge advantage in the freedom of choice, which is greater than ever before.

However, in order to be fully informed, the customer needs to know which criteria search engines use to create ranking and how these criteria are rated. The primary concern is to prevent unfair influence using irrelevant ranking criteria,[11]

---

[3] C. Busch, Mehr Fairness und Transparenz in der Plattformökonomie?, GRUR 2019, pp. 788–796; J. Grapentin, Die Erosion der Vertragsgestaltungsmacht durch das Internet und den Einsatz Künstlicher Intelligenz, NJW 2019, pp. 181–185.

[4] J. Grapentin, Die Erosion der Vertragsgestaltungsmacht durch das Internet und den Einsatz Künstlicher Intelligenz, NJW 2019, pp. 181–185.

[5] S. Ott, Ich will hier rein! Suchmaschinen und das Kartellrecht, MMR 2006, pp. 195–202.

[6] J. Grapentin, Die Erosion der Vertragsgestaltungsmacht durch das Internet und den Einsatz Künstlicher Intelligenz, NJW 2019, pp. 181–185.

[7] J. Grapentin, Die Erosion der Vertragsgestaltungsmacht durch das Internet und den Einsatz Künstlicher Intelligenz, NJW 2019, pp. 181–185.

[8] S. Ott, Ich will hier rein! Suchmaschinen und das Kartellrecht, MMR 2006, pp. 195–202.

[9] R. Ursu, The power of rankings: quantifying the effect of rankings on the online consumer search and purchase decisions, Marketing Science 37 (4), 2018, pp. 530–552.

[10] With regard to the erosion of the free will in civil law by the internet see J. Grapentin, Die Erosion der Vertragsgestaltungsmacht durch das Internet und den Einsatz Künstlicher Intelligenz, NJW 2019, pp. 181–185.

[11] C. Busch, Mehr Fairness und Transparenz in der Plattformökonomie?, GRUR 2019, pp. 788–796.

which leads to worse results for consumers. Although ranking algorithms are "black box systems", which are not transparent to the customer,[12] there are already regulations to ensure some transparency. According to Art. 5 (2) of Regulation 2019/1150,[13] "[p]roviders of online search engines shall set out the main parameters, which individually or collectively are most significant in determining ranking and the relative importance of those main parameters, by providing an easily and publicly available description, drafted in plain and intelligible language, on the online search engines of those providers. They shall keep that description up to date." If, for example, commission payments are considered for the ranking, the provider must indicate this and has to explain how such payments affect the ranking.[14] The more transparency there is on the market, the more informed the consumer would be. However, disclosing too much information (e.g. the entire algorithm) would make it possible to manipulate the rankings[15] and might also affect trade secrets. Therefore, this is correctly not required by the regulation.

Internet services such as search engines affect the customer's contractual freedom not only by pre-arranging information but also by using personalised advertisements. Since there is an increasing appearance of "non-monetary markets" on the Internet,[16] some of those free service platforms have recently achieved very high market shares. Google, for example, has about 90% of the search engine market in most countries of the European Economic Area.[17] From a financial point of view, customers who are going to use services like search engines are going to receive information free of charge so that this business model is financed only through the advertising site, where advertisers pay a price to Google each time a search customer clicks on an ad.[18] The more information (data) about users is available, the more precisely this advertising can target the user.[19] If the advertisement fits the specific user well, the probability that he or she will click on it and ultimately purchase the advertised products increases.[20] In this respect, the user data has a value for platform companies

[12] C. Busch, Mehr Fairness und Transparenz in der Plattformökonomie?, GRUR 2019, pp. 788–796.

[13] Regulation 2019/1150 of 20 June 2019 on promoting fairness and transparency for business users of online intermediation services, OJ 2019 L 186, p. 57–79.

[14] C. Busch, Mehr Fairness und Transparenz in der Plattformökonomie?, GRUR 2019, pp. 788–796.

[15] C. Busch, Mehr Fairness und Transparenz in der Plattformökonomie?, GRUR 2019, pp. 788–796.

[16] T. Körber, Konzeptionelle Erfassung digitaler Plattformen und adäquate Regulierungsstrategien, ZUM 2017, pp. 93–101.

[17] European Commission, Antitrust: Commission sends Statement of Objections to Google on comparison shopping service, 15 April 2015. Available at https://ec.europa.eu/commission/presscorner/detail/en/MEMO_15_4781. Accessed 23 August 2021.

[18] T. Körber, Konzeptionelle Erfassung digitaler Plattformen und adäquate Regulierungsstrategien, ZUM 2017, pp. 93–101.

[19] P. Hacker, Daten als Gegenleistung: Rechtsgeschäfte im Spannungsfeld von DS-GVO und allgemeinem Vertragsrecht, ZfPW 2019, pp. 148–197.

[20] T. Körber, Konzeptionelle Erfassung digitaler Plattformen und adäquate Regulierungsstrategien, ZUM 2017, pp. 93–101.

such as Google or Meta.[21] Therefore, the customers "pay" for the service by paying attention to the advertising displayed on the search page and the disclosure of their data.[22]

It is obvious that the provision of data in certain constellations can have the character of a "fee" or "payment", and therefore the relevant regulations regarding paid services should be applied.[23] The new Directive 2019/770 on certain aspects concerning contracts for the supply of digital content and digital services recognises the disclosure of data as "remuneration" and treats it almost as other contracts for remuneration.[24]

However, while data can sometimes have the character of remuneration, there is still a difference (from a financial point of view) between free services and paid services. On the one hand, unlike money, data are neither scarce nor exclusive, and several companies can have access to the same data.[25] On the other hand, data can become outdated and obsolete.[26]

In December 2020, the EU Commission proposed its Digital Services Act (DSA).[27] The draft contains several rules and obligations concerning online advertising. According to Art. 24 DSA, online platforms that display advertising on their online interfaces shall ensure that the recipients of the service can identify, for each specific advertisement displayed to each individual recipient, in a clear and unambiguous manner and in real time that the information displayed is an advertisement, as well as the natural or legal person on whose behalf the advertisement is displayed and meaningful information about the main parameters used to determine the recipient to whom the advertisement is displayed. In addition, according to Art. 30 DSA, very large online platforms that display advertising on their online interfaces shall compile and make publicly available through application programming interfaces a repository containing certain information, until one year after the advertisement was displayed for the last time on their online interfaces. This repository shall include at least the content of the advertisement; the natural or legal person

[21] T. Körber, Konzeptionelle Erfassung digitaler Plattformen und adäquate Regulierungsstrategien, ZUM 2017, pp. 93–101.

[22] P. Hacker, Daten als Gegenleistung: Rechtsgeschäfte im Spannungsfeld von DS-GVO und allgemeinem Vertragsrecht, ZfPW 2019, pp. 148–197; T. Körber, Konzeptionelle Erfassung digitaler Plattformen und adäquate Regulierungsstrategien, ZUM 2017, pp. 93–101.

[23] T. Körber, Konzeptionelle Erfassung digitaler Plattformen und adäquate Regulierungsstrategien, ZUM 2017, pp. 93–101.

[24] G. Spindler and K. Sein, Die endgültige Richtlinie über Verträge über digitale Inhalte und Dienstleistungen, MMR 2019, pp. 415–420.

[25] T. Körber, Konzeptionelle Erfassung digitaler Plattformen und adäquate Regulierungsstrategien, ZUM 2017, pp. 93–101.

[26] T. Körber, Konzeptionelle Erfassung digitaler Plattformen und adäquate Regulierungsstrategien, ZUM 2017, pp. 93–101.

[27] EU Commission, Proposal for a Regulation of the European Parliament and of the Council on a Single Market For Digital Services (Digital Services Act) and amending Directive 2000/31, COM (2020) 825 final of 15.12.2020.

on whose behalf the advertisement is displayed; the period during which the advertisement was displayed; whether the advertisement was intended to be displayed specifically to one or more particular groups of recipients of the service and, if so, the main parameters used for that purpose and the total number of recipients of the service reached; and, where applicable, aggregate numbers for the group or groups of recipients to whom the advertisement was targeted specifically. Furthermore, according to Art. 26 DSA, those very large online platforms shall identify, analyse and assess at least once a year any significant systemic risks stemming from the functioning and use made of their services in the European Union. When conducting these risk assessments, the platforms shall consider, among other things, how their systems for selecting and displaying advertisements influence specific systemic risks, like the negative effects on the exercise of the fundamental rights to respect for private and family life, the freedom of expression and information, the prohibition of discrimination and the rights of the child. According to Art. 27 DSA, these very large online platforms shall additionally put in place reasonable, proportionate and effective mitigation measures aimed at limiting the display of advertisements in association with the service they provide. It remains to be seen whether these regulations will be adopted by the European Union.

Besides that, in April 2021, the EU Commission proposed a regulation for harmonised rules on AI, the so-called Artificial Intelligence Act (AIA).[28] The AIA draft also provides certain transparency requirements, which are discussed in more detail in the following sections.

15.2.1 Algorithmic Forecast and Individual Autonomy

The customer's contractual freedom could also be affected by algorithmic forecasts. Algorithms often undertake important forecast decisions and carry out contract negotiations, while the customer remains an observer. Depending on the program, artificial intelligence has sometimes a broad and sometimes a limited scope of action.[29] However, it is advised to often have a broad scope of action to take advantage of the benefits of artificial intelligence programs. Therefore, the customer cannot predict the outcome produced by the software and whether such outcome corresponds to his or her will, but he or she must allow himself or herself to be attributed any explanatory action of the software.[30] This is problematic because the result of the software may be objectively correct (based on mathematical calculations) but may not correspond to what the customer subjectively wanted.

[28] EU Commission, Proposal for a Regulation of the European Parliament and of the Council laying down harmonised rules on artificial intelligence (Artificial Intelligence Act) and amending certain union legislative acts, COM (2021) 206 final of 21 April 2021.

[29] J. Grapentin, Die Erosion der Vertragsgestaltungsmacht durch das Internet und den Einsatz Künstlicher Intelligenz, NJW 2019, pp. 181–185.

[30] J. Grapentin, Die Erosion der Vertragsgestaltungsmacht durch das Internet und den Einsatz Künstlicher Intelligenz, NJW 2019, pp. 181–185.

Besides opaque algorithmic forecast, the ability of data collectors to manipulate users based on information collected about them raises more and more public concerns.[31] The more information the data collector has, the better he or she can exercise a considerable influence on the affected person.[32] The data collector can understand how he or she can influence the affected person best by knowing his or her motives, weaknesses and vulnerabilities.[33] Some scholars say that digital surveillance and the use of information to influence our decision-making is not only about unfair commerce since it does not just diminish our interests but also threatens our autonomy.[34] Such interference in our autonomy can only be assumed if the algorithm acts in a manipulative way. Manipulation only occurs if it is a hidden influence. "Manipulating someone means intentionally and covertly influencing their decision-making, by targeting and exploiting their decision-making vulnerabilities."[35] Since "[t]he ruling idea behind the ideal of personal autonomy is that people should make their own lives",[36] we are only autonomous if we act freely and can face existing choices. And only if we can critically reflect on our goals and desires can we understand them authentically as our own.[37]

Nevertheless, this would not constitute a significant restriction of individual autonomy if the customer could still terminate a contract that does not correspond to his or her will. However, under section 119 (1) of the German Civil Code (Bürgerliches Gesetzbuch (BGB)), the right to contest is only available if the customer is in error. If a customer enters into a contract using a computer, a previous input error by the user can affect the declaration later made by the software. The error does not originate from the software: it can be traced back to the user.[38] The use of artificial intelligence shifts the decision-making process significantly to the software

[31] M. J. Abramowitz, Stop the Manipulation of Democracy Online, The New York Times of 11 December 2017; J. Doubek, How Disinformation and Distortions On Social Media Affected Elections Worldwide, 16 November 2017. Available at https://www.npr.org/sections/alltechconsidered/2017/11/16/564542100/how-disinformation-and-distortions-on-social-media-affected-elections-worldwide?t=1624346148419. Accessed 22 June 2021; M. I. Vayena, Cambridge Analytica and Online Manipulation, 2018. Available at https://blogs.scientificamerican.com/observations/cambridge-analytica-and-online-manipulation/. Accessed 22 June 2021.

[32] N. M. Richards, The Dangers of Surveillance, Harvard Law Review 2013, 126 (7), pp. 1934–1965.

[33] D. Susser, B. Roessler and H. Nissenbaum, Technology, autonomy and manipulation, Internet policy review 2019, 8 (2), pp. 1–22.

[34] B. Frischmann and E. Sellinger, Re-Engineering Humanity, 1st ed, 2018, p. 271; S. Zuboff, The Age of Surveillance Capitalism: The Fight for a Human Future at the New Frontier of Power, 1st ed., New York: Public Affairs 2019, p. 347.

[35] D. Susser, B. Roessler and H. Nissenbaum, Technology, autonomy and manipulation, Internet policy review 2019, 8 (2), pp. 1–22.

[36] J. Raz, The Morality of Freedom, Oxford: Clarendon Press 1986, p. 369.

[37] D. Susser, B. Roessler and H. Nissenbaum, Technology, autonomy and manipulation, Internet policy review 2019, 8 (2), pp. 1–22.

[38] BGH, Judgement of 16 October 2012 – X ZR 37/12, MMR 2013, pp. 296–298.

itself.[39] If the software determined the content of the declaration and the user only gave a blanket declaration,[40] the error must have occurred within the software. However, a software program generally does not make mistakes but makes decisions based on a mathematically correct calculation.[41] If an unwanted result occurs, it is not an error in the legal sense.[42] Therefore, a right to contest on the ground of error is generally precluded, and the user cannot terminate the contract. This further limits the autonomy of the customer.[43]

15.2.2 The Instrument of Informed Consent

Since there is an influence on the customer's contractual freedom, it can be compensated by the instrument of informed consent. The requirements for consent as a condition for lawfulness are present in various General Data Protection Regulation (GDPR) provisions. In addition to Art. 6 (1) lit. a GDPR, further requirements can be found in Art. 4 no. 11, Art. 7 and Art. 8 GDPR.[44]

The legitimacy of consent is a central requirement in data protection law.[45] Although the previous regulation, Art. 7 (a) of Directive 95/46, unlike Art. 6 (1) lit. a GDPR, used "unambiguously" when referring to the consent of the user, this requirement also applies to the new regulation since Art. 4 no. 11 and recitals 32, 42 and 43 have further specifications.[46] By asking for "one or more specific purposes", Art. 6 (1) lit. a GDPR prevents the possibility of extending the purpose of

---

[39] J. Grapentin, Die Erosion der Vertragsgestaltungsmacht durch das Internet und den Einsatz Künstlicher Intelligenz, NJW 2019, pp. 181–185.

[40] T. Nitschke and P. Sester, Software-Agent mit Lizenz zum…?, CR 2004, pp. 548–554; J. Grapentin, Die Erosion der Vertragsgestaltungsmacht durch das Internet und den Einsatz Künstlicher Intelligenz, NJW 2019, pp. 181–185.

[41] C. D. Müller-Hengstenberg and S. Kirn, Intelligente (Software-)Agenten: Eine neue Herausforderung unseres Rechtssystems – Rechtliche Konsequenzen der "Verselbstständigung" technischer Systeme, MMR 2014, pp. 307–313; J. Grapentin, Die Erosion der Vertragsgestaltungsmacht durch das Internet und den Einsatz Künstlicher Intelligenz, NJW 2019, pp. 181–185.

[42] C. D. Müller-Hengstenberg and S. Kirn, Intelligente (Software-)Agenten: Eine neue Herausforderung unseres Rechtssystems – Rechtliche Konsequenzen der "Verselbstständigung" technischer Systeme, MMR 2014, pp. 307–313.

[43] J. Grapentin, Die Erosion der Vertragsgestaltungsmacht durch das Internet und den Einsatz Künstlicher Intelligenz, NJW 2019, pp. 181–185.

[44] M. Albers and R.-D. Veit. In: Wolff and Brink (eds), BeckOK Datenschutzrecht, 31st ed, C.H. Beck 2020, Art. 6 DS-GVO, para. 23.

[45] J. Masing, Herausforderungen des Datenschutzes, NJW, 2012, pp. 2305–2311; M. Albers and R.-D. Veit. In: Wolff and Brink (eds), BeckOK Datenschutzrecht, 31st ed, C.H. Beck 2020, Art. 6 DS-GVO, para. 19.

[46] M. Albers and R.-D. Veit. In: Wolff and Brink (eds), BeckOK Datenschutzrecht, 31st ed, C.H. Beck 2020, Art. 6 DS-GVO, para. 7.

processing after consent has been given.[47] Recital 42 specifies the requirements for consent by saying that the declaration of consent pre-formulated by the controller should be written in a “clear and plain language and it should not contain unfair terms”. These requirements ensure that the user is informed in a clearly understandable way[48] and about the specific intention.

However, some voices question the adequacy of consent to ensure an appropriate balance between the data sovereignty of individuals and the economic interests of businesses. Even if the user is aware of all the circumstances of the data processing, he or she may be forced to accept the extensive processing of his or her personal data only because there are no comparable alternatives. The negligence of users represents another major risk. Some users may not take the time to read through data protection declarations but may carelessly consent to the use of their data by clicking on them.[49] This risk could be reduced by the use of warning symbols or the implementation of a European uniform certification system, which classifies a website or app according to its truthfulness in terms of data processing.[50] Furthermore, these proposals would not constitute a one-sided consumer protection regulation to the disadvantage of the digital economy: if data protection and data security were correctly understood as a competitive strength, the European digital economy could gain an advantage over international competition precisely through transparent and user-oriented business models.[51]

Another issue is the expiry date of consent. Especially in the case of non-written requests, it might be difficult for the company processing the data to know when the consent is invalid or has been revoked,[52] and it might be even more difficult for the user to know whether the company is complying.

Similar problems occur with the transparency requirements provided by the EU Commission’s proposal for an Artificial Intelligence Act.[53] For example, according to Art. 52 AIA, AI systems that interact with natural persons must inform the natural person that he or she is interacting with such a system unless it is obvious. This exception offers much room for interpretation, which will have to be further determined by the responsible supervisory authorities and courts. Furthermore, it remains unclear how the duty to inform should look in practice. On the one hand, a too simple

---

[47] M. Albers and R.-D. Veit. In: Wolff and Brink (eds), BeckOK Datenschutzrecht, 31st ed, C.H. Beck 2020, Art. 6 DS-GVO, para. 23.

[48] P.-L. Krüger, Datensouveränität und Digitalisierung, ZRP 2016, pp. 190–192.

[49] P.-L. Krüger, Datensouveränität und Digitalisierung, ZRP 2016, pp. 190–192.

[50] S. Kraska, Datenschutz-Zertifizierungen in der EU-Datenschutzgrundverordnung, ZD 2016, pp. 153–154; G. Hornung and K. Hartl, Datenschutz durch Marktanreize – auch in Europa? - Stand der Diskussion zu Datenschutzzertifizierung und Datenschutzaudit, ZD 2014, pp. 219–225.

[51] P.-L. Krüger, Datensouveränität und Digitalisierung, ZRP 2016, pp. 190–192.

[52] E. M. Frenzel. In Paal and Pauly (eds), Datenschutzgrundverordnung, Bundesdatenschutzkontakt, 2nd ed, C.H. Beck 2018, Art. 6 DS-GVO, para. 11.

[53] EU Commission, Proposal for a Regulation of the European Parliament and of the Council laying down harmonised rules on artificial intelligence (Artificial Intelligence Act) and amending certain union legislative acts, COM (2021) 206 final of 21 April 2021.

information to the affected person about the AI system does not help since many people will ignore the information. On the other hand, it is generally assumed that only an informed person is able to take responsibility for himself or herself and take necessary protective measures. In the context of AI systems, often the persons concerned are not in the position to recognise the indirect consequences, which may lie in the future. Instead of referring to the use of an AI system, it would be preferable to provide information about the criteria used by AI systems and their possible consequences.

15.2.3 Improvement of Transparency

Besides informed consent, the improvement of transparency reduces algorithms' influence on the customer's contractual freedom. The principle of transparency is included in several provisions of the GDPR (Arts. 5 (1) lit. a, 12, 13, 14, 15, 22 GDPR and recitals 58, 60, 71). According to Art. 15 (1) GDPR, the data subject shall have access to the personal data concerning him or her being processed, but at the same time, the specific amount of information to be communicated is highly controversial and equally relevant in practice.[54] The problem becomes obvious in the following example: according to Art. 15 (1) GDPR, a health insurance company must inform a policyholder requesting information of his or her "master data", such as name, date of birth, contact data, health data, medical statements and medical certificates. It is being discussed whether the internal notes of the insurance company about the policyholder, internal reports, internal statements, past correspondence or the premium account at the insurance tariff must also be provided.

In the literature, some voices say that Art. 15 (1) GDPR is not restricted within the limitations of section 826 BGB and therefore covers information on *everything* that contains personal data of the data subject, such as past correspondence from parties, internal memos on the data subject or internal opinions.[55] In individual cases, this can lead to a comprehensive provision of information.[56] The Cologne Higher Regional Court takes the same position: the term "personal data" under Art. 15 (1) GDPR is not limited to "master data" but also extends to electronically stored notes on telephone calls and other conversations with the customer.[57] In contrast, the Regional Court of Cologne takes the opposite view when it states that Art. 15 (1) GDPR does not refer to all internal transactions or correspondence but is intended

[54] F. Schmidt-Wudy, In: Wolff and Brink (eds), BeckOK Datenschutzrecht, 31st ed, C.H. Beck 2020, Art. 15 DS-GVO, para. 52.

[55] M. Riemer, Der Datenauskunftsanspruch als "discovery" im deutschen Zivilprozess, ZD 2019, pp. 413–415.

[56] V. Scheepers, Der Umgang mit einem Auskunftsersuchen nach der DSGVO, DStR 2019, pp. 1109–1111.

[57] OLG Köln, Judgement of 26 July 2019 – 20 U 75/18.

to ensure that the person concerned can assess the scope and content of the stored data.[58] This view could also be supported by Art. 15 (3) first sentence GDPR. According to this provision, the controller shall provide a copy of the personal data undergoing processing. It could be argued that the operator cannot be expected to make a copy of every piece of the data collected. In June 2021, the German Federal Court of Justice (Bundesgerichtshof (BGH)) followed a comprehensive understanding of Art. 15 (1) GDPR.[59] The court ruled that Art. 15 (1) GDPR covers all "personal data" within the meaning of the broad definition of Art. 4 no. 1 GDPR, and therefore the data subject shall even have access to internal notes.

In terms of time, it is not clear what the right covers. Art. 15 (1) GDPR does not address the problem raised in the "Rijkeboer" decision of the Court of Justice of the European Union (CJEU), according to which the right to information under data protection law covers not only the data currently held by the responsible party but also data from the past.[60] Some voices therefore say that the silence of the GDPR on this issue can only be understood to mean that there should be no claim to information on "past" data and only actually available data are to be considered.[61] Others say that the "Rijkeboer" principles would still apply.[62] It remains to be seen which view is supported by the case law.

Additionally, it becomes problematic for the consumers when the website operator does not know what data are collected from third parties on its website. In its judgement in the "Fashion ID" case, the CJEU disputed whether the operator of a website who adds to that website a social plug-in that causes the browser of the visitor to ask for the content of the provider of that plug-in and, therefore, to transmit the visitor's personal data to that provider can be treated as a controller within the meaning of Art. 2 (d) of Directive 95/46, even though that operator has no influence on the processing of the data that are transmitted to the provider. The data of the user are passed on to the provider of the plug-in regardless of whether the plug-in is clicked on or even whether the user is a member of the social network.[63]

The court found that the operator of a website placing the "Facebook-Like-button" (a social plug-in that causes the transmission of website users' personal data to Facebook) and Facebook can be qualified as jointly responsible under EU

[58] LG Köln, Judgement of 19 June.2019 – 26 S 13/18.

[59] BGH, Judgement of 15 June 2021 – VI ZR 576/19.

[60] CJEU, case C-553/07, College van burgemeester en wethouders van Rotterdam v M. E. E. Rijkeboer, Judgement of 7 May 2009, ECLI:EU:C:2009:293.

[61] F. Schmidt-Wudy. In: Wolff and Brink (eds), BeckOK Datenschutzrecht, 31st ed, C.H. Beck 2020, Art. 15 DS-GVO, para. 52.2.

[62] L. Franck. In: P. Gola (ed.), DS-GVO, 2nd ed, C.H. Beck 2018, Art. 15 DS-GVO, para. 8; A. Dix. In: Simitis, Hornung and Spiecker gen. Döhmann (eds), Datenschutzrecht, 1st ed, Nomos 2019, Art. 15 DS-GVO, para. 45.

[63] CJEU, case C-40/17, Fashion ID GmbH & Co. KG v Verbraucherzentrale NRW eV, Judgement of 29 July 2019, ECLI:EU:C:2019:629; Datenschutzrechtliche Verantwortlichkeit für Social Plugin auf Website, NJW 2019, pp. 2755–2760.

data protection law. The term "controller" in Art. 2 (d) of Directive 95/46 is widely defined in order to ensure the effective and comprehensive protection of data subjects.[64] The website operator is responsible within the meaning of Directive 95/46 even if, beyond the decision to use the social plug-in, he or she does not or cannot influence the processing of personal data. It is not necessary for each of the operators to have access to the personal data concerned. An operator may therefore be (jointly) responsible even if it does not process personal data itself as long as the controller decides on the purposes and means of processing personal data with another operator that processes the data.[65]

The CJEU's findings on shared responsibility in connection with social plug-ins can easily be transferred to the GDPR as the definition of (jointly) "responsible" in Art. 4 no. 7 GDPR has not changed compared to the Directive.[66] The GDPR contains specific legal consequences for joint controllership, such as the requirement of an agreement between the responsible parties, which defines who is responsible for compliance with which obligations under the GDPR (Art. 26 (1) first sentence GDPR), without this agreement having any external effect (Art. 26 (3) GDPR).[67] According to Art. 26 (2) second sentence GDPR, "the essentials" of this agreement have to be made accessible to the data subject, for example in the context of providing the information according to Arts. 13 and 14 GDPR.[68] The information that the operator of the website is required to provide to the data subject must relate only to operations involving the processing of personal data. In practice, those joint controllers must agree on who and how they will fulfil the relevant information obligations.[69]

Transparency is thus created due to the fact that the operator of a website that integrates social plug-ins from third-party providers such as Meta, Google, Twitter and Microsoft (LinkedIn) must inform the data subject and obtain a separate consent before collecting and transferring the data to the provider.[70] In addition, the shared

---

[64] CJEU, case C-131/12, Google Spain SL and Google Inc. v Agencia Española de Protección de Datos (AEPD) and Mario Costeja González, Judgement of 13 May 2014, EU:C:2014:317, para. 34; CJEU, case C-210/16, Unabhängiges Landeszentrum für Datenschutz Schleswig-Holstein gegen Wirtschaftsakademie Schleswig-Holstein GmbH, Judgement of 5 June 2018, ECLI:EU:C:2018:388, para. 28.

[65] CJEU, case C-40/17, Fashion ID GmbH & Co. KG v Verbraucherzentrale NRW eV, Judgement of 29 July 2019, ECLI:EU:C:2019:629, para. 85.

[66] J. Spittka and R. Mantz, Datenschutzrechtliche Anforderungen an den Einsatz von Social Plugins, NJW 2019, pp. 2742–2745.

[67] J. Spittka and R. Mantz, Datenschutzrechtliche Anforderungen an den Einsatz von Social Plugins, NJW 2019, pp. 2742–2745.

[68] J. Spittka and R. Mantz, Datenschutzrechtliche Anforderungen an den Einsatz von Social Plugins, NJW 2019, pp. 2742–2745.

[69] F. Moos and T. Rothkegel, „Gefällt mir"-Button von Facebook – Fashion ID, MMR 2019, pp. 579–587.

[70] A. Sattler, Gemeinsame Verantwortlichkeit – getrennte Pflichten, GRUR 2019, pp. 1023–1026.

responsibility operators are also both liable for the rights and compensation claims of the data subjects (Art. 26 (3), 82 (4) and (5) GDPR).[71]

Besides that, the AIA draft[72] also provides certain transparency requirements, but they only apply to AI systems within the meaning of Art. 3 no. 1 AIA. Following a risk-based approach, the AIA draft differentiates between four risk categories: the first category of AI involves unacceptable risks and therefore is prohibited (Art. 5 AIA).

The second category covers the so-called high-risk AI systems. These include AI systems with a safety component of a product or products, which are covered by the Union harmonisation legislation listed in Annex II or AI systems referred to in Annex III. In chapter 2 of the AIA draft, there are specific requirements that have to be fulfilled before or while using a high-risk AI system. According to Art. 13 AIA, AI systems should be designed and developed in such a way that their functioning is sufficiently transparent so that users can use the system appropriately and interpret its results correctly. The user is any natural or legal person, public authority, agency or other body using an AI system under its authority, except where the AI system is used in the course of a personal non-professional activity (Art. 3 no. 4 AIA). The call for transparency is a reminder of the discussion about the existence of a so-called "Right to Explanation" within the GDPR.[73] Also, here, the question of technical feasibility arises, as well as the resolution of possible conflicts with other rights. Technical barriers are encountered with deep learning, for example, because even if it is known what information the algorithm bases its decision on (input), there is no knowledge of how it reaches the result (output). According to the current state of research, in many fields of AI application, one has to decide between very well-functioning AI systems that hardly provide any insight into their functioning and more transparent but then also less well-functioning AI systems. Furthermore, transparency obligations can come into conflict with other rights, especially the protection of trade secrets. Both the algorithm itself and the functioning of the algorithm can be subsumed under the concept of a trade secret according to Art. 2 no. 1 of the Trade Secrets Directive.[74] The consequences are very extensive: the owner of the trade secret can object to almost any interference. The exceptions,

---

[71] J. Spittka and R. Mantz, Datenschutzrechtliche Anforderungen an den Einsatz von Social Plugins, NJW 2019, pp. 2742–2745.

[72] EU Commission, Proposal for a Regulation of the European Parliament and of the Council laying down harmonised rules on artificial intelligence (Artificial Intelligence Act) and amending certain union legislative acts, COM (2021) 206 final of 21.4.2021.

[73] B. Mittelstadt et al., Why a Right to Explanation of Automated Decision-Making Does Not Exist in the General Data Protection Regulation, Int. Data Priv. Law, 2017, 7 (2), pp. 76–99; B. Casey, A. Farhangi and R. Vogt, Rethinking explainable machines: The GDPR's "Right to Explanation" debate and the rise of algorithmic audits in enterprise, 34 Berkeley Tech. L. J. (2019), pp. 143–188; M. Kaminski, The Right to Explanation, Explained, 34 Berkeley Tech. L. J. (2019), pp. 189–218; S. Wachter and B. Mittelstadt, A Right to Reasonable Inferences: Re-thinking Data Protection Law in the Age of Big Data and AI, Colum. Bus. L. Rev. 2019 (2), pp. 494–620.

[74] Directive (EU) 2016/943 of 8 June 2016 on the protection of undisclosed know-how and business information (trade secrets) against their unlawful acquisition, use and disclosure.

regulated in Art. 5 of the Trade Secrets Directive, are not likely to apply, so that the enforcement of transparency requirements could fail due to not only technical but also legal barriers.[75]

In addition to high-risk AI systems, Art. 52 of the AIA draft also names three groups of AI systems to which certain weakened information and transparency obligations apply. As mentioned before, the first group focuses on AI systems that interact with natural persons unless it is obvious that they are facing an AI. The other two groups are emotion recognition and biometric categorisation systems as well as so-called deep fake applications. It is not quite clear why the Commission has decided to regulate these three groups in particular. Between the high-risk AI systems and the AI systems without a risk, there are likely to be other groups in addition to those listed in Art. 52 AIA. Therefore, a standard clause with sufficiently defined criteria would have been useful here as well.[76]

However, it should be noted that the transparency obligation only applies to the provider and user and not to the affected person.

15.2.4 The Right to Explanation

To improve transparency and therefore reduce algorithms' influence on the customer's contractual freedom, it is worth taking a deeper look into the so-called "Right to explanation".[77] There are three possible legal bases for a Right to Explanation in the GDPR: safeguards against automated decision-making, as required under Art. 22 (3) GDPR, commented upon by recital 71; notification duties under Art. 13 and Art. 14 GDPR, commented upon by recitals 60–62; or the right of access under Art. 15 GDPR, commented upon by recital 63.

According to Art. 13 (2) lit. f and Art. 14 (2) lit. g GDPR, the controller shall provide the data subject with the existence of an automated decision-making and, at least in those cases, meaningful information about the logic involved, to ensure fair and transparent processing. Art. 15 (1) lit. h GDPR grants the corresponding right to information. Contrary to the still ambiguous opinion of the data protection authorities,[78] the wording "at least in these cases" is to be understood in such a way that the information obligations or the right to information also exist in such cases in which there is merely a decision supported by algorithmic systems.[79]

[75] H. Hoffmann, Regulierung der Künstlichen Intelligenz, K & R 2021, pp. 369–374.

[76] H. Hoffmann, Regulierung der Künstlichen Intelligenz, K & R 2021, pp. 369–374.

[77] For a comprehensive overview, see H. Hoffmann and J. Kevekordes, Das Right to Explanation, DuD 2021, (forthcoming).

[78] See the contradictory statements of the Article 29 Data Protection Working Party in WP 251rev.01 v. 3.10.2017 on p. 25 and 31.

[79] M. Bäcker, In: J. Kühling and B. Buchner (eds), DSGVO, 3rd ed., C.H. Beck 2020, Art. 13 DSGVO paras 52 f.; A. Dix, In: S. Simitis, G. Hornung and I. Spiecker gen. Döhmann (eds), Datenschutzrecht, 1st ed., Nomos 2019, Art. 13 DSGVO, para. 16; B.A. Mester, In: J. Taeger and D. Gabel (eds), 3rd ed., Fachmedien R & W 2019, Art. 13 DSGVO paras 27 f.; a different view is

A limitation to only automated decision-making would tempt controllers to avoid the rule by having a person formally confirm the decision.[80] Since Art. 13 (1) GDPR requires that the controller shall provide the information at the time when personal data are obtained, the duty to inform cannot be focused on individual decision-making as this has not even taken place at this point but only on the general, global functioning of the system.[81] The same applies to Art. 14 (2) and Art. 15 (1) GDPR. On the one hand, they have the same wording; on the other hand, they both refer to "envisaged consequences".[82]

Another legal basis is Art. 22 (3) in conjunction with recital 71 fifth sentence GDPR. According to Art. 22 (1) GDPR, the data subject has the right not to be subject to a decision based solely on automated processing which produces legal effects concerning them or similarly significantly affects them. Paragraph 2 provides three exceptional cases in which such an automated decision is nevertheless permitted. In the case of Art. 22 (2) lit. a and c, the data controller shall implement suitable measures to safeguard the data subject's rights and freedoms and legitimate interests. These include at least the right to obtain human intervention on the part of the controller, to express his or her point of view and to contest the decision. If these requirements are not met, the data processing is prohibited, irrespective of the data subject's exercise of any right under Art. 22 of the GDPR.[83] Recital 71 fourth sentence GDPR requires an explanation of the decision reached after such an assessment. Although this obligation is anchored only in the recitals and not directly in the normative text, it must be considered when determining a Right to Explanation. Art. 22 (3) GDPR contains an exhaustive list of minimum measures, which is

---

taken by M. Martini, Blackbox Algorithmus, Springer 2019, p. 183; M. Martini, Algorithmen als Herausforderung für die Rechtsordnung, JZ 2017 (21), pp. 1017–1025; F. Schmidt-Wudy, In: S. Brink and H. A. Wolff (eds), BeckOK-Datenschutzrecht, 35th ed., C.H. Beck 2021, Art. 15 DSGVO para. 77.

[80] In result also S. Wachter, B. Mittelstadt and L. Floridi, Why a right to explanation of automated decision-making does not exist in the general data protection regulation, Int. Data Privacy Law 2017, Vol. 7(2), pp. 76–99; A. D. Selbst and J. Powles, Meaningful information and the right to explanation, Int. Data Privacy Law 2017. Vol. 7(4), pp. 233–424; see also Article 29 Data Protection Working Party, WP 251rev.01, p. 21.

[81] S. Wachter, B. Mittelstadt and L. Floridi, Why a right to explanation of automated decision-making does not exist in the general data protection regulation, Int. Data Privacy Law 2017, Vol. 7(2), pp. 76–99; M. Martini, Blackbox Algorithmus, Springer 2019, p. 191.

[82] See Article 29 Data Protection Working Party, WP 251rev.01, pp. 25–26; S. Wachter, B. Mittelstadt and L. Floridi, Why a right to explanation of automated decision-making does not exist in the general data protection regulation, Int. Data Privacy Law 2017, Vol. 7(2), pp. 76–99; M. Martini, Blackbox Algorithmus, Springer 2019, p. 192; M. Bäcker, In: J. Kühling and B. Buchner (eds), DSGVO, 3rd ed., C.H. Beck 2020, Art. 15 DSGVO para. 27.

[83] Article 29 Data Protection Working Party, WP 251rev.01, p. 19; B. Buchner, In: J. Kühling and B. Buchner (eds), DSGVO, 3rd ed., C.H. Beck 2020, Art. 22 DSGVO para. 12; M. Helfrich, In: G. Sydow (ed), Europäische Datenschutzgrundverordnung, 2nd editon, Nomos 2018, Art. 22 para. 38–41; M. Martini, In: B.P. Paal and D. A: Pauly (eds), DSGVO, 3rd ed., C.H. Beck 2021, Art. 22 para. 29b; the current state of the dispute: K. v. Lewinski, In: S. Brink and H. A. Wolff (eds), BeckOK-Datenschutzrecht, 35th ed. (1.2.2021), C.H. Beck 2021, Art. 22 DSGVO para. 2.1.

why in certain situations a Right to Explanation can still exist.[84] Since Art. 22 (3) GDPR uses the word "at least", it depends on the assessment of the individual case to decide whether a Right to Explanation exists.[85]

In addition, the AIA draft[86] also contains transparency obligations for high-risk AI systems (Art. 13) and mitigated transparency obligations for certain AI systems (Art. 52).

A comprehensive Right to Explanation, which the data subject can invoke, is currently not enshrined in the law. The existing provisions must be further specified by the legislator—or at least by the courts. However, Arts. 13–15 and 22 GDPR and Art. 13 of the draft AI Regulation indicate that there is a duty to explain algorithmic decisions. Nevertheless, transparency obligations and the Right to Explanation conflict with other rights, especially the protection of trade secrets, and face technical barriers.

15.3 Unfair Commercial Practices and the GDPR

Data privacy is discussed in the context of the German Act Against Restraints of Competition (Gesetz gegen Wettbewerbsbeschränkungen (GWB)) as well as in the context of the German Unfair Competition Act (Gesetz gegen den unlauteren Wettbewerb (UWG)). In the following section, we will address the question of when data protection can amount to a competition law issue. It is also being dealt with the question of whether the use of data requires consumer harm to become a competition law concern and if it requires the company to be in a dominant position.

The Higher Regional Court of Düsseldorf ruled in the "Facebook I" case[87] on the abuse of a dominant position on the German market for social networks. While the German Federal Cartel Office (Bundeskartellamt) justified the infringement against Facebook due to the abuse of a dominant market position by arguing that network users had to agree to contractual conditions which were inappropriate in terms of data protection law when registering and which enabled Facebook to collect, link

---

[84] M. E. Kaminski, The right to explanation, explained, 34 Berkeley Tech. L. J. 2019, Vol. 34(1), pp. 189–218.

[85] M. E. Kaminski, The right to explanation, explained, 34 Berkeley Tech. L. J. 2019, Vol. 34(1), pp. 189–218; S. Schulz, In: P. Gola (ed), DSGVO, 2nd ed., C.H. Beck 2018, Art. 22 para. 33; K. v. Lewinski, In: S. Brink and H. A. Wolff (ed), BeckOK-Datenschutzrecht, 35th ed. (1.2.2021), C.H. Beck 2021, Art. 22 DSGVO para. 47.; in result also S. Wachter, B. Mittelstadt and L. Floridi, Why a right to explanation of automated decision-making does not exist in the general data protection regulation, Int. Data Privacy Law 2017, Vol. 7(2).

[86] EU Commission, Proposal for a Regulation of the European Parliament and of the Council laying down harmonised rules on artificial intelligence (Artificial Intelligence Act) and amending certain union legislative acts, COM (2021) 206 final of 21.4.2021.

[87] OLG Düsseldorf, Judgement of 26 August 2019 – VI Kart 1/19 (V).

and use data generated outside the network,[88] the court found that the abuse of a dominant position by Facebook could not be clearly proven.

However, in its decision on 23 June 2020, the Federal Court of Justice (BGH) rejected this reasoning put forward by the Higher Regional Court and ruled that the exploitation of a dominant position does not always require a causal connection between market dominance and the disapproved conduct (conduct causality) in the case of an abuse of conditions under section 19 (1) GWB.[89] At least a causal connection between market dominance and the market result (causality of results) may be sufficient if, due to special market conditions, the conduct of the dominant company leads to market results which would not be expected in the case of functioning competition and the conduct not only constitutes exploitation but is at the same time also likely to hinder competition. By receiving an indispensable service (the use of the social network) only in conjunction with another undesirable service (the provision of a personalised experience based on data generated by the user's activity outside the network), users were forced into a service content and were no longer able to decide freely on the extent to which their data would be disclosed. The consent of the users was not actually voluntary within the meaning of Art. 6 (1) (a) GDPR. Thus, the disputed terms of use constituted a so-called forced extension of service. Contrary to the opinion of the Higher Regional Court, the assessment of such a forced extension of service remains unaffected by the fact that the provision of a personalised experience using data generated outside the network is free of charge in monetary ways. Rather, by imposing an undesirable service, the consideration for the desired service (namely the use of the social network) is increased in the form of disclosure of personal user data. Therefore, the Federal Court of Justice considered Facebook's exploitation of its dominant position to be abusive.

Furthermore, the Federal Court of Justice found no serious doubt that the disputed terms of use lead to an anti-competitive market result. This is underlined by the assumption that without Facebook's dominant position, there would be a usage offer linked to a lower level of data disclosure. Since there is no such an offer, the conditions used by Facebook lead to an impediment in competition both on the market for social networks dominated by the company and on the related market for online advertising. Besides the already existing direct network effects and the enormous economic value of personalised user data, the disputed terms of use further strengthened the market entry barriers for potential competitors and therefore hindered competition.

As a pioneering case worldwide, the Federal Court of Justice's decision states an important sign for competition on social platforms. However, the outcome of the principal proceedings remains to be seen. Currently, the proceedings have been

[88] German Federal Cartel Office, Case Report of 15 February 2019, B6-22/16 – NZKart 2019, pp. 178–184.

[89] BGH, Judgement of 23 June 2020, KVR 69/19 – NZKart 2020, pp. 473–483.

suspended due to questions of interpretation submitted to the European Court of Justice (ECJ).[90]

In the draft and explanatory memorandum of the tenth amendment of the GWB, the Federal Ministry of Economics and Energy and the German Parliament already referred to the comments of the Higher Regional Court of Düsseldorf in order to clarify that no qualified requirements in the sense of a "strict causality" could be derived from the previous wording of section 19 (1) GWB.[91] The Federal Ministry of Economics and Energy explicitly mentioned that if the unjustified disclosure of personal data required proof of strict causality, it would often encounter difficulties precisely in those cases where the exploitation of the market opponent is also favoured by other factors, such as existing information asymmetries or rational apathy on the part of the demand side, and where corresponding behaviour can therefore also be exhibited by non-dominant companies in individual cases.[92] Even after the decision of the Federal Court of Justice, the legislator wanted to create legal certainty as clarification was still pending for other constellations.[93] Therefore, the wording of section 19 (1) GWB no longer requires an abusive exploitation of a dominant position but rather considers sufficient a mere abuse of a dominant position, which can be exclusionary.

The tenth amendment of the GWB also recognises the increased importance of data and prepares for new challenges posed by advancing digitalisation. Therefore, section 18 (3) no. 3 GWB now states that, in particular, the access of a company to data relevant to competition shall be taken into account when assessing the market position of a company in relation to its competitors. Moreover, section 19 (2) no. 4 GWB now expressly mentions that abuse of market power may lie in refusing to grant access to data. The condition is that access must be objectively necessary for action on one of the previous or following markets and the refusal of access threatens to eliminate effective competition on that market, unless the refusal of access is objectively justified. Admittedly, numerous consequential problems exist behind

[90]Request for a preliminary ruling from the Oberlandesgericht Düsseldorf lodged on 22 April 2021 – Facebook Inc. and Others v Bundeskartellamt, C-252/21 (OJEU 2021/C 320/20).

[91]BMWi, Entwurf (RefE) eines Zehnten Gesetzes zur Änderung des Gesetzes gegen Wettbewerbsbeschränkungen für ein fokussiertes, proaktives und digitales Wettbewerbsrecht 4.0 (GWB-Digitalisierungsgesetz), pp. 70–71. Available at https://www.d-kart.de/wp-content/uploads/2019/10/GWB-Digitalisierungsgesetz-Fassung-Ressortabstimmung.pdf. Accessed 23 August 2021; BT-Drs. 19/23492 p. 71.

[92]BMWi, Entwurf (RefE) eines Zehnten Gesetzes zur Änderung des Gesetzes gegen Wettbewerbsbeschränkungen für ein fokussiertes, proaktives und digitales Wettbewerbsrecht 4.0 (GWB-Digitalisierungsgesetz), p. 71.

[93]BT-Drs. 19/23492 p. 71.

such a claim.[94] Nevertheless, it is a fallback situation if there are no sector-specific data access obligations.[95]

Regarding companies in the area of data and platform markets, the new centrepiece of abuse control is section 19a GWB. While the application of competition law usually requires a company to be in a dominant position, the prerequisite for the application of section 19a GWB is not a dominant position in traditional understanding (market power) but only a determination by the Federal Cartel Office that the company concerned has an "outstanding cross-market significance for competition". Among other things, the access of a company to competition-relevant data (section 19a (1) no. 4 GWB) and also the gatekeeper function of a company (section 19a (1) no. 5 GWB) must be considered while making this determination. It seems as if section 19a (2) no. 4 GWB is a direct reaction to the Facebook case as the provision specifically contains permission from the Federal Cartel Office to prohibit a company from using terms and conditions that make the use of services conditional on users consenting to the processing of data from other services of the company or a third-party provider without giving users sufficient choice as to the circumstance, purpose and manner of the processing.

Since the changes made to competition law with regard to digital companies entered into force on 19 January 2021, the Federal Cartel Office is extending the scope of its proceedings and is examining whether Facebook is subject to the new rules applying to undertakings of paramount significance for competition across markets (section 19a German Competition Act, GWB) and whether linking the Oculus and the Facebook network is to be assessed on this basis.[96]

In May 2021, the Federal Cartel Office initiated a proceeding against Amazon based on the new rules for large digital companies. In addition to inspecting tying or bundling strategies, the Federal Cartel Office examines whether Amazon creates or raises barriers to market entry by processing data relevant to competition.[97]

A few days later, a proceeding against Google was also initiated, where the Federal Cartel Office will undertake an in-depth analysis of Google's data processing terms.[98]

---

[94] R. Podszun and F. Brauckmann, GWB-Digitalisierungsgesetz: Der Referentenentwurf des BMWi zur 10. GWB-Novelle, GWR 2019, pp. 436–438.

[95] R. Podszun and F. Brauckmann, GWB-Digitalisierungsgesetz: Der Referentenentwurf des BMWi zur 10. GWB-Novelle, GWR 2019, pp. 436–438.

[96] Bundeskartellamt, Press release of 28 January 2021. Available at https://www.bundeskartellamt.de/SharedDocs/Meldung/DE/Pressemitteilungen/2021/28_01_2021_Facebook_Oculus.html. Accessed 23 August 2021.

[97] Bundeskartellamt, Press release of 18 May 2021. Available at https://www.bundeskartellamt.de/SharedDocs/Meldung/DE/Pressemitteilungen/2021/18_05_2021_Amazon_19a.html?nn=3591286. Accessed 23 August 2021.

[98] Bundeskartellamt, Press release of 25 May 2021. Available at https://www.bundeskartellamt.de/SharedDocs/Meldung/DE/Pressemitteilungen/2021/25_05_2021_Google_19a.html?nn=3591286. Accessed 23 August 2021.

In June 2021, the Federal Cartel Office initiated a proceeding against Apple, the last of the four GAFA companies (Google, Apple, Facebook and Amazon). Among other aspects, it will examine Apple's extensive integration across several market levels, the magnitude of its technological and financial resources and its access to data. A main focus of the investigations will be on the operation of the App Store as it enables Apple in many ways to influence the business activities of third parties.[99]

In May 2021, the French competition authority *Autorité de la concurrence* fined Google EUR 220 million for abusing its dominant position on the online advertising market. In the course of the settlement, Google also accepted a series of proposed commitments to settle the case, including promises to make it easier for competitors to make use of data and the online-ad tools.[100]

In June 2021, the European Commission opened a formal antitrust investigation to assess whether Facebook violated the rules on anti-competitive agreements between companies (Art. 101 TFEU) and on the abuse of a dominant position (Art. 102 TFEU) by using advertising data gathered, in particular, from advertisers in order to compete with them on markets where Facebook is active, such as classified ads, and by tying its online classified ad service "Facebook Marketplace" to its social network.[101]

15.4 Commercial Practices in the Debate

Since social media and AI-based systems become more and more important in online marketing, commercial practices have changed enormously. Firstly, the significance and legal consequences of influencing in modern marketing will be presented. Afterwards, the relationship between AI and marketing and particularly their influence on data protection is examined, as well as selected aspects of personalised prices. The current debate on the prohibition on selling products on platforms as well as advertising them on search engines will be discussed afterwards. At the end of the section, the upcoming challenges concerning product liability are outlined and a future perspective is given.

[99] Bundeskartellamt, Press release of 21 June 2021. Available at https://www.bundeskartellamt.de/SharedDocs/Meldung/DE/Pressemitteilungen/2021/21_06_2021_Apple.html. Accessed 23 August 2021.

[100] Autorité de la concurrence, Press Release of 7 June 2021. Available at https://www.autoritedelaconcurrence.fr/en/press-release/autorite-de-la-concurrence-hands-out-eu220-millions-fine-google-favouring-its-own. Accessed 23 August 2021.

[101] European Commission, Opening of Proceedings caseAT.40684Facebook leveraging. Available at https://ec.europa.eu/competition/antitrust/cases/dec_docs/40684/40684_1812_3.pdf. Accessed 23 August 2021.

15.4.1 Influencing

15.4.1.1 Introduction

Online marketing is a central aspect of corporate management. In the last few years, in addition to classic advertising, advertising on social media channels has changed tremendously.

Social media platforms offer companies the possibilities of more or less conventional Internet advertising. In addition to paid, static banner advertising or short video insertions, this can also be a sponsored posting.[102] Furthermore, a normal posting can also be an advertisement within the meaning of the relevant provisions.[103] According to section 5a UWG and section 6 of the German Telemedia Act (Telemediengesetz (TMG), advertising and normal postings must actually be strictly separated from each other. In order not to violate the separation rule, advertising must be labelled as such. However, a certain additional problem is posed by any mandatory or clarifying information, which often cannot be listed, or at least not completely, due to the limited size of such an electronic advertisement.[104]

So-called hashtags are used to tag articles on social media, such as Twitter or Instagram, and make them easier to find. These can contain characteristic terms of the posting as well as short sentences and are introduced by the diamond symbol "#'. In comparison to the mere use of buzzwords, this further emphasises their recognisability and thus reinforces their meaningfulness.[105] The use of hashtags in the context of commercial transactions may have consequences under trademark law or competition law.

In addition, on various social media platforms such as Facebook and Instagram, the photos uploaded by users can be tagged to products or companies.[106] This may also have consequences under trademark law or competition law.

Especially on social networks such as Instagram or YouTube, the so-called influencer marketing becomes more and more important.[107] People with many followers on social media advertise for third parties and receive a service in return. They use their reporting to present the products to their loyal followers in a positive way and thus induce them to make a purchase decision. Generally, the advertising is part of a normal posting, so consumers cannot distinguish whether it is the actual opinion of the influencer or whether the influencer is presenting the product only on the basis of a contract with a company. The problem is that the ads are often not marked as advertising. Often, the posts are only marked with "#ad", which is not

[102] F. Lichtnecker, Ausgewählte Werbeformen im Internet unter Berücksichtigung der neueren Rechtsprechung, MMR 2014, pp. 523–528.

[103] OLG Köln, Judgement of 19 May 2017 – 6 U 155/16.

[104] F. Lichtnecker, Neues aus dem Social Media-Marketing, MMR 2018, pp. 512–517.

[105] F. Lichtnecker, Neues aus dem Social Media-Marketing, MMR 2018, pp. 512–517.

[106] F. Lichtnecker, Neues aus dem Social Media-Marketing, MMR 2018, pp. 512–517.

[107] F. Suwelack, Schleichwerbung als Boombranche?, MMR 2017, pp. 661–665; U. Borsch, Der angemaßte Influencer – Markenpiraterie 2.0, MMR 2018, pp. 127–129.

really conclusive and does not solve the problem completely. National courts have already ruled on the first cases following this issue.[108]

If there is a contractual agreement between the company and the influencer, it is considered to be advertising, which in this respect is a commercial act pursuant to section 2 (1) no. 1 UWG.[109] If the contribution is (paid) advertising, it must also be recognisable as such and marked as such. Otherwise, there would be a violation of section 6 (1) nos. 1, 2 TMG; no. 11 of the annex to section 3 (3) UWG; section 5a (6) UWG; and section 22 (1) or section 74 in connection with section 8 (3) of the Interstate Media Treaty (Medienstaatsvertrag (MStV)).[110] If the advertising character is already obvious from the design of the post itself, there is usually no separate labelling necessary,[111] although one should be cautious here. As with other advertising campaigns, proper labelling includes the use of meaningful words such as "advertising" or "advertisement", possibly in combination with a hashtag.[112]

15.4.1.2 Recent Case Law

However, it has been disputed for some time whether a declaration is necessary if the influencer just presents a product without having an advertisement contract with a company and without receiving remuneration. The Higher Regional Court of Karlsruhe (OLG Karlsruhe) ruled that commercial interests are also pursued with gratuitous contributions.[113] After all, the contributions are intended to contribute to increasing the value of the influencer's image and thus to obtain advantages for any future paid advertising contracts.[114] Therefore, labelling is necessary, especially when "tab links" are used.[115] The Higher Regional Court of Hamburg (OLG Hamburg) ruled that there is no need for advertising labelling because with professionally designed posts and follower numbers in the millions, the commercial purpose is recognizable to consumers at first glance (section 5a (6) UWG).[116] In addition, there would also be a lack of consistency in comparison with the print media because there are also personal product recommendations by individual editors in print media that do not have to be labelled as advertising if no service in

[108]LG Hannover, Judgement of 8 June 2017 – 3 U 53/17, C. Sobottka, Schleichwerbung durch Influencer-Marketing in sozialen Medien, MMR 2017, pp. 769–772; KG Berlin, Ruling of 11 October 2017, 5 W 221/17, Kennzeichnungspflicht bei Influencer Marketing, MMR 2018, pp. 98–99.

[109]F. Lichtnecker, Neues aus dem Social Media-Marketing, MMR 2018, pp. 512–517.

[110]M. Gerecke, Kennzeichnung von werblichen Beiträgen im Online-Marketing, GRUR 2018, pp. 153–159.

[111]T. Fuchs and C. Hahn, Erkennbarkeit und Kennzeichnung von Werbung im Internet, MMR 2016, pp. 503–507.

[112]LG Hagen, Judgement of 13 September 2017 – 23 O 30/17, GRUR-RR 2017, pp. 510–512.

[113]OLG Karlsruhe, Judgement of 9 September 2020 – 6 U 38/19, MMR 2021, pp. 159–164.

[114]OLG Karlsruhe, Judgement of 9 September 2020 – 6 U 38/19, MMR 2021, pp. 159–164.

[115]OLG Karlsruhe, Judgement of 9 September 2020 – 6 U 38/19, MMR 2021, pp. 159–164.

[116]OLG Hamburg, Judgement of 2 July 2020 – 15 U 142/19, MMR 2020, pp. 767–772.

return has been provided.[117] The Higher Regional Court of Munich (OLG München) went even further and denied the existence of a commercial act (pursuant to section 2 (1) no. 1 UWG) because the fact that influencers indirectly pursue the goal of making themselves more interesting for future advertising contracts with gratuitous contributions is fundamentally immanent to the actions of the media industry, which is dependent on advertising revenues.[118]

It seems more convincing that if the influencer just presents a product without an agreement with a company and without using tab links, this cannot be defined as an advertisement, and a declaration is not necessary. For the consumer, it is crucial to know which products are recommended free of charge and which products are advertised in return for remuneration. However, if gratuitous contributions about a product also must be labelled as advertising, it becomes impossible for the consumer to distinguish between them. This would constitute over-labelling, which would be contrary to the protective purpose of section 5a (6) UWG. However, as the case law of the Higher Regional Courts is inconsistent, a decision by the Federal Court of Justice (BGH) remains to be seen. A decision of the Federal Court of Justice (BGH) is expected at the beginning of September 2021, where the court has the chance to provide clarity in three cases at once as it will have to rule on the question of whether influencers are obliged to label their gratuitous Instagram posts as advertising.

On 2 December 2020, another important act concerning online marketing and therefore also influencing came into force: the Act to Strengthen Fair Competition (Gesetz zur Stärkung des fairen Wettbewerbs), which modified the UWG. In particular, a provision was added to section 14 UWG stating that the court in whose district the infringement was committed does not have jurisdiction in disputes concerning infringements in electronic commerce or telemedia (section 14 (2) third sentence no. 1 UWG). In the case of infringements of unfair competition law committed on the Internet, the place of infringement is the whole of Germany due to the fact that the infringement can be retrieved from anywhere on the Internet, which means that under the old rules, any German regional court had jurisdiction and the competitor could choose a court (so-called flying jurisdiction). Thus, it could choose the court that, according to previous decision-making practice, was favourably disposed towards its application. The amendment to the law has severely restricted this option for competitors in Internet cases. Now, only the general place of jurisdiction at the defendant's registered address is to apply (section 14 (2) UWG). Since then, however, there has also been a decision by the Regional Court of Düsseldorf (LG Düsseldorf), which nevertheless assumed jurisdiction in the case of a YouTube advertisement, even though the defendant's registered address was not in its district.[119] According to the Regional Court of Düsseldorf, in order for jurisdiction to be excluded, the anti-competitive act must be an infringement that can only be committed in electronic commerce or telemedia and is therefore already linked to

[117] OLG Hamburg, Judgement of 2 July 2020 – 15 U 142/19, MMR 2020, pp. 767–772.

[118] OLG München, Judgement of 25 June 2020 – 29 U 2333/19, GRUR 2020, pp. 1096–1098.

[119] LG Düsseldorf, Judgement of 15 January 2021 – 38 O 3/21, GRUR-RS 2021, p. 402.

these media. It argues that the legislator's intention was only to abolish the "flying jurisdiction" in the case of abusive cease-and-desist letters, as in the case of violations of information and labelling obligations on the Internet.[120] This interpretation has already been the subject of much criticism, in particular because it contradicts the clear wording, and in the view of many, no contrary intention is apparent from the explanatory memorandum.[121] Moreover, the Higher Regional Court of Düsseldorf (OLG Düsseldorf) has already declared it unacceptable in an obiter dictum.[122] Since the Regional Court of Düsseldorf nevertheless continues to adhere to this interpretation and the Regional Court of Frankfurt (LG Frankfurt) has also followed suit, it remains to be seen how other courts will deal with this issue.[123]

15.4.1.3 German Bill to Strengthen Consumer Protection in Competition and Commercial Law

In addition to the court ruling, there were also legislative attempts to face the rising issues regarding online marketing, social media and influencing. On 4 November 2020, the German Federal Ministry of Justice and Consumer Protection published a ministerial draft bill to strengthen consumer protection in competition and commercial law.[124] The draft was followed by numerous comments from German associations as well as Instagram itself. Following on from this, a government draft for the modernisation of consumer protection law was published in order to comply with the requirements of Directive 2019/2161 and in order to provide legislative clarification with regard to influencer marketing.[125] On 10 June 2021, the German parliament (Bundestag) passed this bill that will enter into force on 28 May 2022.[126]

The explanatory memorandum to the bill specifically addresses the problems of influencer marketing. It requires regulations to clarify the scope of application of the UWG and other changes to distinguish non-commercial communication from commercial actions.

First, the bill clarifies in section 1 (2) UWG that provisions regulating specific aspects of unfair business acts take precedence over the UWG when assessing whether an unfair business act has been committed. However, pursuant to section 1

[120] LG Düsseldorf, Judgement of 15 January 2021 – 38 O 3/21, GRUR-RS 2021, p. 402.

[121] S. Wettig and G. Kiparski, Wiederaufleben des fliegenden Gerichtsstandes contra legem!?, CR 2021, pp. 177–182.

[122] OLG Düsseldorf, Judgement of 16 February 2021 – I-20 W 11/21, GRUR-Prax 2021, p. 158.

[123] LG Düsseldorf, Judgement of 26 February 2021 – 38 O 19/21, GRUR-RS 2021, p. 4044; LG Frankfurt, Judgement of 11 May 2021 – 3-06 O 14/21, GRUR-RS 2021, p. 11813.

[124] Referentenentwurf eines Gesetzes zur Stärkung des Verbraucherschutzes im Wettbewerbs- und Gewerberecht, Available at https://www.bmjv.de/SharedDocs/Gesetzgebungsverfahren/Dokumente/RefE_Staerkung_Verbraucherschutz_Wettbewerbs-_und_Gewerberecht.pdf?__blob=publicationFile&v=2. Accessed 23 August 2021.

[125] Regierungsentwurf eines Gesetzes zur Stärkung des Verbraucherschutzes im Wettbewerbs- und Gewerberecht, BT-Drs. 19/27873.

[126] BGBl. I p. 3504.

UWG, the precedence only extends as far as the aspect in question is conclusively regulated in the more specific provision.[127] If, for example, the more specific provision only contains supplementary information requirements, the general information requirements of section 5a and b UWG continue to apply.

At the same time, the UWG is applicable to regulations on specific aspects of unfair acts only with regard to those aspects that are not addressed in the specific regulations.

Additionally, it can sometimes be difficult to distinguish between commercial communication and private expression of opinion on social media or blogs. Therefore, the definition of the commercial act in section 2 (1) no. 2 UWG is supplemented to the effect that it must no longer only be objectively but also directly related to the promotion of sales.[128] In this regard, section 2 (2) no. 2 UWG explains that although the criterion "objectively" has to be interpreted in a way that it comprises "directly" in the case of actions towards consumers, a clarification is appropriate due to the importance of this central definition.[129] This is explained by the following example: in certain forms of promotion of one's own business, there is no direct connection to sales promotion, for example if an influencer recommends or mentions goods or services and has received no remuneration or similar consideration for doing so and the mention merely promotes his or her own notoriety.

Furthermore, section 5a (4) second sentence UWG clarifies that no commercial purpose is to be assumed in the case of an act in favour of another entrepreneur if the person acting does not receive any remuneration or similar consideration from the other entrepreneur or does not allow himself or herself to be promised such consideration. This new amendment is intended to provide a secure legal framework for the actions of influencers when they recommend goods and services without themselves profiting from them by remuneration or a similar consideration.[130]

According to the comments on section 5a (4) second sentence UWG, it appears unreasonable to require a labelling as "commercial" for such actions.[131] The term "similar consideration" also includes commissions, products sent by the third-party company that the trader may use or keep, as well as press trips, the provision of equipment or the assumption of costs. The consideration may also be of a temporary nature. The mere increase of the own awareness, for example of influencers, through such actions, however, cannot be considered as consideration. The consideration does not have to take place in a direct temporal connection. The consideration must have been initiated by the entrepreneur for whose benefit the action is performed. If the consideration is provided by a third party, such as an agency, this is attributed to the entrepreneur according to general principles. In contrast, consideration provided by independent third parties that is not initiated by the entrepreneur is not included.

[127] BT-Drs. 19/27873, p. 31.
[128] BT-Drs. 19/27873, p. 32.
[129] BT-Drs. 19/27873, p. 32.
[130] BT-Drs. 19/27873, p. 34.
[131] BT-Drs. 19/27873, p. 34.

This corresponds to the general assessment in the case of media companies from which employees also receive remuneration for the production of contributions.

However, according to the explanatory memorandum, it should be noted that sentence 2 is only applicable to the assessment of the question of whether there is an act for the benefit of a third-party entrepreneur.[132] The question of whether a recommendation of an influencer made free of charge also constitutes a business act insofar as it pursues the purpose of promoting one's own company is assessed solely in accordance with section 5a (4) first sentence UWG and the definition of a business act in section 2 (1) no. 2 UWG. The question of whether an act in favour of one's own company exists does not depend solely on the receipt of a fee since self-promotion is also generally subject to labelling requirements if it is not otherwise recognizable. In the case of recommendations made free of charge by influencers, therefore, it should be taken into account, in accordance with the addition in section 2 (1) no. 2 UWG, whether there is a direct connection to the sale of products and services. In this context, it will also be necessary to consider the fact that influencers in this area could be assessed like media companies, which are also regularly financed by advertising revenues and are also particularly attractive to their clients if they reach a large number of people.

In addition, the freedom of expression and media freedom under Art. 11 of the Charter of Fundamental Rights of the European Union must be included in the assessment. The effort to achieve advertising revenue according to a ruling of the Court of Appeal in Berlin (Kammergericht Berlin) does not justify providing each statement with a reference, with which a subordinate or inferior value of the contribution is connected.[133]

15.4.2 AI and Marketing

In ages where online marketing takes the leading role in commercial practices, the use of AI-based systems also increased and therefore became common practice. However, with the use of personal data as a foundation for forecast decisions, contract negotiations and personalised advertisements, the upcoming popularity of AI-based systems in the field of online marketing leads to higher requirements for data protection and the need for regulation. This is stressed even more by the fact that the frequent use of AI-based systems also causes a discriminatory aspect.

15.4.2.1 Data Quality

For the creation of personalised advertising, the high quality of the underlying data sets has to be ensured. High quality of data is crucial for sustainable and valuable

[132] BT-Drs. 19/27873, p. 35.

[133] Kammergericht (KG) Berlin, Judgement of 08.01.2019, Az. 5 U 83/18.

data use.[134] Therefore, guaranteeing high data quality is very important for the provision of algorithmic decisions. An incorrect data basis can lead to incorrect decisions, without the person concerned being aware of this. The evaluation of data quality is a complex task as the context of use, distortions in the data collection process or the structural disadvantages of individual groups may cause errors. Data quality can be understood as readable and processable data or can refer to the accuracy of the data.[135] So one can differentiate between technical data quality and the quality of the information itself.[136]

Art. 5 (1) lit. d GDPR requires the controller to ensure that the data processed are accurate and, where necessary for the purposes of the processing, are kept up to date. He or she should not just act when the data subject exercises his or her right of rectification;[137] rather, the controller must, on his or her own initiative, take appropriate measures to delete or correct inaccurate data without delay.[138] The appropriate steps to ensure accuracy and keep the data up to date will be determined on a case-by-case basis, taking into consideration the purpose of the processing.[139] This is also shown in the wording of the provision by the limitation of the obligation to update through the use of the phrase "if necessary": data that have become obsolete for their initial purpose may still be relevant for evidence purposes.[140]

As a general rule, higher standards must be set for the control of data in the context of their collection than for the control of the accuracy of stock data.[141] However, with regard to the data saved by the collector within the meaning of Art. 4 no. 7 GDPR, he or she must ensure that they take notice of information that calls into question the accuracy of the stored data and that subsequently, if necessary, the existing data are checked and corrected.[142] In addition, Art. 5 (1) lit. d GDPR says

[134] T. Hoeren, Data quality as a key issue for big data, MMR 2016, pp. 8–11; J. Stevens, Datenqualität bei algorithmischen Entscheidungen. In: David, Geihs, Lange and Stumme (eds), Informatik 2019, pp. 367–380.

[135] P. Bitter and S. Uphues. In: Kolany-Raiser, Heil, Orwat and Hoeren (eds), Big Data, 1st ed, C.H. Beck 2019, p. 81.

[136] P. Bitter and S. Uphues. In: Kolany-Raiser, Heil, Orwat and Hoeren (eds), Big Data, 1st ed, C.H. Beck 2019, p. 81.

[137] P. Schantz. In: Wolff and Brink (eds), BeckOK Datenschutzrecht, 31st ed, C.H. Beck 2020, Art. 5 DS-GVO, para. 28.

[138] H. Huber, Verbot der Diskriminierung aus Gründen der Staatsangehörigkeit – Ausländerzentralregister, NVwZ 2009, pp. 379–382.

[139] P. Schantz. In: Wolff and Brink (eds), BeckOK Datenschutzrecht, 31st ed, C.H. Beck 2020, Art. 5 DS-GVO, para. 29.

[140] H. Heberlein. In: Ehmann and Selmayr (eds), Datenschutzgrundverordnung, 2nd ed, C.H. Beck 2018, Art. 5, para. 24.

[141] P. Schantz. In: Wolff and Brink (eds), BeckOK Datenschutzrecht, 31st ed, C.H. Beck 2020, Art. 5 DS-GVO, para. 29.

[142] P. Schantz. In: Wolff and Brink (eds), BeckOK Datenschutzrecht, 31st ed, C.H. Beck 2020, Art. 5 DS-GVO, para. 29.

that the accuracy of the data must be guaranteed "in relation to the purposes for which they are processed".[143]

Data may also be inaccurate in the light of their purpose if they are incomplete and may, therefore, lead to false impressions or wrong decisions.[144]

The AIA draft[145] also provides requirements for data and data governance. According to Art. 10 (3) AIA, the training, validation and testing data sets of high-risk AI systems shall be relevant, representative, free of errors and complete. They shall have the appropriate statistical properties, including, where applicable, as regards the persons or groups of persons on which the high-risk AI system is intended to be used. Nevertheless, a precondition for the applicability of Art. 10 AIA is that it is a high-risk AI system. Therefore, the provision will only apply in certain cases.

15.4.2.2 The Use of Incorrect Data

Algorithms create the profile of the customer, so that advertisements correspond as closely as possible to the customer's wishes. Depending on the profile and method used, the customer will be sorted into categories that do not correspond to his or her interests. Particularly, this danger exists if algorithms use incorrect or outdated data or if they automatically combine several criteria. The customer may be confronted with an advertisement that subjectively does not necessarily match his or her wishes. He or she has no possibility to contradict the classification into certain criteria.[146]

Nevertheless, companies without algorithms display even more unsuitable ads. Whether this is a manipulation of contractual freedom depends on whether the measures are carried out secretly or not.[147] It must always be remembered that our subconscious is regularly influenced in such a way that we do not even notice it.

15.4.2.3 Data Management

Especially when large amounts of data are used, there must be strict rules on how they have to be encrypted, transmitted, stored and processed. The larger is the amount of data, the higher is the risk of abuse.

The accountability principle regulated in Art. 5 (2) GDPR is a real change from the previous legal situation: the data controller must prove compliance with the

[143] E. M. Frenzel. In Paal and Pauly (eds), Datenschutzgrundverordnung, Bundesdatenschutzkontakt, 2nd ed, C.H. Beck 2018, Art. 5 DS-GVO, para. 41.

[144] A. Roßnagel. In: Simitis, Hornung and Spiecker gen. Döhmann (eds), Datenschutzrecht, 1st ed, Nomos 2019, Art. 5 DS-GVO, para. 139.

[145] EU Commission, Proposal for a Regulation of the European Parliament and of the Council laying down harmonised rules on artificial intelligence (Artificial Intelligence Act) and amending certain union legislative acts, COM (2021) 206 final of 21 April 2021.

[146] P. Scholz. In: Simitis, Hornung and Spiecker gen. Döhmann (eds), Datenschutzrecht, 1st ed, Nomos 2019, Art. 4 Nr. 4 DS-GVO, paras 9 ff.

[147] D. Susser, B. Roessler and H. Nissenbaum, Technology, autonomy and manipulation, Internet policy review 2019, 8 (2), pp. 1–22.

regulation.[148] This means that the controller must be able to prove at any time that he or she is processing personal data lawfully. This has significantly aggravated the situation for controllers, which is linked to high documentation requirements that must be implemented in practice.[149]

Improvement can be achieved by using data management software that is designed for large amounts of data. In order to achieve the best application, one should also check if there is a specially developed software tool for the branch in question. Furthermore, the software must be up to date, and software updates must be carried out regularly.

Following on from this, Art. 15 of the AIA draft[150] requires that high-risk AI systems shall be designed and developed in such a way that they achieve an appropriate level of accuracy, robustness and cybersecurity and perform consistently in those respects throughout their life cycle. According to Art. 15 (4), the technical solutions to address AI-specific vulnerabilities shall include, where appropriate, measures to prevent and control attacks trying to manipulate the training data set (data poisoning), inputs designed to cause the model to make a mistake (adversarial examples) or model flaws.

15.4.2.4 National Restrictions on the Creation of User Profiles That Also Use Data from International Third-Party Sources

Before the implementation of the GDPR, section 15 (3) of the German Telemedia Act (Telemediengesetz (TMG)) was applicable. Section 15 (3) TMG allows the creation of pseudonymous user profiles for advertising purposes as long as the user has not objected. Section 15 (3) TMG does neither cover the use of data to create comprehensive personality profiles, nor does it cover the transmission of usage data to third parties.[151] However, it is precisely the creation of such a personality profile that AI often focuses on.

Since the GDPR became applicable, according to Art. 22 (1) GDPR, the data subject shall have the right not to be subject to a decision based solely on automated processing, including profiling, which produces legal effects concerning him or her or similarly significantly affects him or her. "Profiling" is defined in Art. 4 no. 4 GDPR as any form of automated processing of personal data consisting of the use of personal data to evaluate certain personal aspects relating to a natural person. This also applies to individual evaluations of usage profiles but not to the collective

[148] A. Jung, Datenschutz-(Compliance-)Management-Systeme – Nachweis- und Rechenschaftspflichten nach der DS-GVO, ZD 2018, pp. 208–213.

[149] A. Jung, Datenschutz-(Compliance-)Management-Systeme – Nachweis- und Rechenschaftspflichten nach der DS-GVO, ZD 2018, pp. 208–213.

[150] EU Commission, Proposal for a Regulation of the European Parliament and of the Council laying down harmonised rules on artificial intelligence (Artificial Intelligence Act) and amending certain union legislative acts, COM (2021) 206 final of 21.4.2021.

[151] A. Dix and P. Schaar. In: Roßnagel (ed), Beck'scher Komm. Zum Recht der Mediendienste, 1st ed, C.H. Beck 2013, § 15 TMG, paras 62 f.

evaluation of many usage profiles for statistical purposes.[152] However, the decision must have a legal effect or must affect the person in a similar way.

Art. 22 GDPR does not expressly regulate the special conditions under which a profile may be created and further processed unless it results in an exclusively automated decision. Instead, the general rules of justification, in particular Arts. 6 and 9 GDPR, regulate whether the data processing procedure is lawful. According to Art. 6 (1) lit. f GDPR, the creation of user profiles has to be "necessary for the purpose of the legitimate interests pursued by the controller or by a third party". It is therefore necessary to balance the interests in each individual case. This evaluation must consider the following factors: the responsible expectation of the data subjects and foreseeability/transparency, the possibilities for intervention by the data subjects, the chaining of data, the players involved, the duration of observation, the types of data and the scope of the data processing.[153] According to one view, as far as AI programs are used for the purpose of profiling measures that interact with cookies, there is a high risk for the rights of the data subject, so that his or her interests outweigh and Art. 6 (1) lit. f is not applicable.[154] Therefore, it would be necessary to obtain consent under Art. 6 (1) lit. a GDPR. Others argue using recital 47 of the GDPR, according to which direct marketing measures can be based on legitimate interests. If this is permitted, then tracking user behaviour beforehand should be even more so.[155] The interests of users are taken into account by precisely matching advertising and avoiding advertisement that is not in line with their interests. And even if this is AI based, Art. 21 (1) first sentence GDPR provides a right to object to the provisions of Art. 6 (1) lit. f GDPR.[156]

When user profiles use data from international third-party sources, the joint collectors have a shared responsibility according to Art. 26 GDPR. In this respect, the rights of the data subject are protected.

Whether more stringent restrictions are necessary depends on the understanding of the data protection authorities and courts. If they say that Art. 6 (1) lit. f GDPR is not applicable and it is rather necessary to obtain consent under Art. 6 (1) lit. a GDPR, there is no need for a stricter regulation.

[152] S. Schleipfer, Datenschutzkonformes Webtracking nach Wegfall des TMG, ZD 2017, pp. 460–466.

[153] DSK, Orientierungshilfe der Aufsichtsbehörden für Anbieter von Telemedien, Stand May 2019, p. 16. Available at https://www.datenschutzkonferenz-online.de/media/oh/20190405_oh_tmg.pdf. Accessed 23 August 2021.

[154] F. Schmidt, Dynamische und personalisierte Preise – datenschutz-, wettbewerbs- und kartellrechtliche Grenzen. In: Taeger (ed), Tagungsband DSRI-Herbstakademie 2016, Smart World – Smart Law? Weltweite Netze mit regionaler Regulierung, DSRITB 2016, pp. 1007–1022; M. Albers and R.-D. Veit. In: Wolff and Brink (eds), BeckOK Datenschutzrecht, 31st ed, C.H. Beck 2020, Art. 6 DS-GVO, para. 49.

[155] T. Gausling, Künstliche Intelligenz im digitalen Marketing, ZD 2019, pp. 335–341.

[156] T. Gausling, Künstliche Intelligenz im digitalen Marketing, ZD 2019, pp. 335–341.

The AIA draft[157] only prohibits AI-based social scoring for general purposes done by public authorities (Art. 5 (1) lit. c). According to the explanatory memorandum of the draft, other manipulative or exploitative practices affecting adults that might be facilitated by AI systems could be covered by the existing data protection, consumer protection and digital service legislation that guarantees that natural persons are properly informed and have the free choice not to be subject to profiling or other practices that might affect their behaviour.[158]

15.4.2.5 Unequal Competitive Conditions Due to the Use of AI

The use of AI as such does not constitute an infringement of competition. The right to free competition allows each individual company to decide for itself whether to use AI. Even before AI came on the market, companies could decide to implement different strategies. Nevertheless, companies with large databases can often use AI more effectively. Instead of challenging competition law, it should therefore be decided whether access to data needs to be (re-)regulated.

15.4.2.6 Discrimination Against Companies That Were Not Subject to Previous Purchase Decisions

The evaluation of previous purchase decisions does not constitute discrimination against other companies that were not yet subject to purchase decisions. The evaluation of previous purchase decisions is reflected in the profile created. Also, new companies can get access to this profile and thus benefit from it. The fact that a new company has not evaluated its own purchasing decision yet does not conflict with the principle of fair competition as it can also define the target group manually and still benefit from the profiling.

15.4.2.7 Legal, Social and Ethical Values to Prevent Discrimination

With the increasing digitalisation and widespread use of artificial intelligence, the value neutrality of the programs is of primary importance.[159] The algorithms make their decisions based on the underlying data sets and the programming of the code. As long as these data sets meet the wished social and ethical values, the algorithm will decide on the basis of these values. The question of what social and ethical standards the decision should be based on must nevertheless be answered as the data set must be selected on the basis of this since it has to reflect this. In practice, it is going to be problematic as it is hardly conceivable to find data sets that meet these standards. If the data set contains discriminatory elements, the decision of the

[157] EU Commission, Proposal for a Regulation of the European Parliament and of the Council laying down harmonised rules on artificial intelligence (Artificial Intelligence Act) and amending certain union legislative acts, COM (2021) 206 final of 21.4.2021.

[158] EU Commission, Proposal for a Regulation of the European Parliament and of the Council laying down harmonised rules on artificial intelligence (Artificial Intelligence Act) and amending certain union legislative acts, COM (2021) 206 final of 21 April 2021, p. 13.

[159] H. Steege, Algorithmenbasierte Diskriminierung durch Einsatz von Künstlicher Intelligenz, MMR 2019, pp. 715–721.

algorithm will also be discriminatory. In order to prevent this from happening, it is necessary to look carefully when selecting the training data.

In this context, Art. 10 of the AIA draft[160] provides certain requirements for training, validation and testing data. According to Art. 10 (3) AIA draft, the data shall be representative and free of errors. The AIA draft does not solve the problem outlined above.[161]

15.4.2.8 Prohibition of AI Marketing in Certain Areas of Life

Moreover, there should be certain areas where AI advertising campaigns should either be prohibited or highlighted and regulated. These areas are the most sensitive and vulnerable to abuse, such as political campaigns, religions, medical care or culture-related ads.

15.4.3 Personalised Prices

Personalised pricing as a kind of sub-category of marketing is the subject of the following section addressing selected aspects such as data protection issues, conflicts concerning competition law and consumer's individual disadvantages due to specific proposed higher prices. As a counterweight to pricing algorithms, the development of algorithms for consumers increases.

15.4.3.1 Interference with Art. 22 (1) GDPR

Personalised advertising and personalised pricing are exclusively created by automatically generated user profiles based on automated processing. However, Art. 22 (1) GDPR only prohibits an automated individual decision that has legal effects on the data subject or that has similarly significant effects on him or her.

In principle, personalised advertising and personalised prices have no direct legal effects or do not similarly significantly affect the data subject[162] since advertising is not an offer to buy but rather an invitation to make an offer. They have no impact on the potential risk to which the European legislator refers in the norm.[163] This is

---

[160] EU Commission, Proposal for a Regulation of the European Parliament and of the Council laying down harmonised rules on artificial intelligence (Artificial Intelligence Act) and amending certain union legislative acts, COM (2021) 206 final of 21 April 2021.

[161] H. Hoffmann, Regulierung der Künstlichen Intelligenz, K & R 2021, pp. 369–374.

[162] M. Martini. In: Paal and Pauly (eds), Datenschutzgrundverordnung, Bundesdatenschutzkontakt, 2nd ed, C. H. Beck 2018, Art. 22 DS-GVO, para. 23; S. Schulz. In: Gola (ed), Datenschutz-Grundverordnung, 2nd ed, C. H. Beck 2018, Art. 22 DS-GVO, para. 28; S. Gierschmann, Gestaltungsmöglichkeiten bei Verwendung von personenbezogenen Daten in der Werbung, MMR 2018, pp. 7–12; K.-U. Plath and J. M. Grages, "Let's Stay in Touch" - Direktwerbung unter der DSGVO, CR 2018, pp. 770–782; R. B. Abel, Automatisierte Entscheidungen im Einzelfall gem. Art. 22 DS-GVO, ZD 2018, pp. 304–307.

[163] M. Martini. In Paal and Pauly (eds), Datenschutzgrundverordnung, Bundesdatenschutzkontakt, 2nd ed, C.H. Beck 2018, Art. 22 DS-GVO, para. 23.

justified by a reference to Art. 21 (2) GDPR and recital 70.[164] By providing a right of objection to direct advertising to the data subject, they connect it to lawful processing operations and assume that such processing is in principle lawful and not subject to the prohibition of Art. 22. Therefore, the possibility to offer different services and prices to different customers is an expression of the free market economy and finds its limits only in the standards of anti-discrimination law.[165]

However, some voices in the literature doubt whether this argumentation can still be maintained in the current era of "behavioural advertising" and "price discrimination".[166] They argue that a similarly significant effect might occur in individual cases if there were enormous deviations from the market price, the limited availability of the product or the degree of necessity for the person concerned.[167]

If an AI system is involved, the EU Commission's proposal for an Artificial Intelligence Act (AIA)[168] might also be applied soon. Depending on the risk of the AI system, different measures may be necessary. Most likely, personalised prices created by an AI system only fall under Art. 52 AIA, which provides weakened transparency and information obligations.

15.4.3.2 Further Data Protection Requirements

As soon as personal data are processed, the processing has to comply with the principles of Art. 5 GDPR. Regarding the processing of personal data for the personalisation of content, advertising and prices, the principles of "purpose limitation" and "data minimization" are of particular relevance.

According to Art. 5 (1) lit. b GDPR, personal data shall be collected for specified, explicit and legitimate purposes and not further processed in a manner that is incompatible with those purposes (principle of "purpose limitation").

According to Art. 5 (1) lit. c GDPR, personal data shall be adequate, relevant and limited to what is necessary in relation to the purposes for which they are processed (principle of "data minimization"). This principle is a concretisation of the principle of necessity.[169]

[164] M. Martini. In: Paal and Pauly (eds), Datenschutzgrundverordnung, Bundesdatenschutzkontakt, 2nd ed, C.H. Beck 2018, Art. 22 DS-GVO, para. 23.

[165] S. Schulz. In: Gola (ed), Datenschutz-Grundverordnung, 2nd ed, C.H. Beck 2018, Art. 22 DS-GVO, para. 28.

[166] J. Hladjk. In: Ehmann and Selmayr (eds), Datenschutz-Grundverordnung, 2nd ed, C. H. Beck 2018, Art. 22, para. 9.

[167] P. Schantz and H. A. Wolff, Das neue Datenschutzrecht, C.H. Beck 2017, para. 737; T. J. Tillmann and V. Vogt, Personalisierte Preise im Big-Data-Zeitalter, VuR 2018, pp. 447–455.

[168] EU Commission, Proposal for a Regulation of the European Parliament and of the Council laying down harmonised rules on artificial intelligence (Artificial Intelligence Act) and amending certain union legislative acts, COM (2021) 206 final of 21 April 2021.

[169] A. Roßnagel, In: S. Simitis, G. Hornung and I. Spiecker gen. Döhmann (eds), Datenschutzrecht, 1st ed., Nomos 2019, Art. 5 DSGVO, para. 116.

At this point, the purpose of the data processing is particularly important. The purpose is the relevant point of reference for the principle of data minimisation.[170] In fact, it is even possible to process very large amounts of data as long as there is a sufficient purpose to be achieved.[171] The principle of data minimisation does not call for an absolute reduction or limitation of the amount of data, and therefore even "big data" analyses do not violate a priori the principle of data minimisation.[172] In order to comply with the above-mentioned principles, it is ultimately only a question of the permissible definition of a broad purpose and compliance with the principle of compatibility of purposes.[173] The principle of data minimisation therefore only restricts the collection and use of data for the personalisation of content/advertising/pricing to a limited extent. The Vienna Regional Court, for example, found that the processing of data by Facebook for the personalisation of content and advertising was lawful, and it did not address the plaintiff's argument that Facebook was violating the principle of data minimisation.[174]

While profiling for the purpose of the personalisation of advertising is not necessary to protect the legitimate interests of the controller, it is only lawful with the informed consent of the data subject.[175] However, consumers have long been aware of business models in which they "pay" for certain services with their data for the purpose of personalised advertising.[176] Accordingly, obtaining consent for processing for the purposes of personalised advertising is not such an obstacle in practice.

Profiling for the purpose of personalising prices is also not necessary to protect the legitimate interests of the controller.[177] Such legitimate interests may also be of an economic nature, but at least when weighing the interests, one will have to conclude that the fundamental rights of consumers outweigh the profit maximisation interests of the controllers.[178] Therefore, with regard to personalised prices, data

---

[170] T. Herbst, In: J. Kühling and B. Buchner (eds), DSGVO, 3rd ed., C.H. Beck 2020, Art. 5 DSGVO para. 22.

[171] T. Herbst, In: J. Kühling and B. Buchner (eds), DSGVO, 3rd ed., C.H. Beck 2020, Art. 5 DSGVO para. 56.

[172] T. Herbst, In: J. Kühling and B. Buchner (eds), DSGVO, 3rd ed., C.H. Beck 2020, Art. 5 DSGVO para. 56.

[173] T. Herbst, In: J. Kühling and B. Buchner (eds), DSGVO, 3rd ed., C.H. Beck 2020, Art. 5 DSGVO para. 56.

[174] LGZ Wien, Judgement of 30 June 2020 – 3 Cg 52/14k-91, ZD 2021, pp. 25–32.

[175] A. Golland, Das Kopplungsverbot in der Datenschutz-Grundverordnung, MMR 2018, pp. 130–135.

[176] L. Mischau, Daten als „Gegenleistung" im neuen Verbrauchervertragsrecht, ZEuP 2020, pp. 335–365.

[177] T. J. Tillmann and V. Vogt, Personalisierte Preise im Big-Data-Zeitalter, VuR 2018, pp. 447–455.

[178] T. J. Tillmann and V. Vogt, Personalisierte Preise im Big-Data-Zeitalter, VuR 2018, pp. 447–455.

processing is also only lawful with the informed consent of the data subject.[179] This would require an explicit query as to whether the data subject consents to his or her data being used for the purpose of individual pricing.[180] In contrast to consent for personalised advertising, many companies are likely to shy away from this, especially since it is to be expected that such clauses would attract media attention.[181]

In December 2020, the EU Commission addressed this issue by proposing its Digital Markets Act (DMA).[182] According to Art. 5 lit. a DMA, a gatekeeper shall refrain from combining personal data sourced from their core platform services with personal data from any other services offered by the gatekeeper or with personal data from third-party services and from signing in end users to other services of the gatekeeper in order to combine personal data, unless the end user has been presented with a specific choice and provided consent in the sense of the GDPR.

15.4.3.3 Price Collusion Due to the Use of the Same or Similar Algorithms

Intentional or unintentional price agreements may arise when companies use the pricing algorithms and data sets of one and the same third-party supplier.[183] This can lead to the situation where all suppliers using the same system offer the same price to the individual, so that the customer has hardly any choice.[184] In this case, prices increase, and the entire price acceptance of the customer can be absorbed.[185] The same scenario may also occur if many companies use personal prices and their algorithms are becoming very similar.[186] This is even likely since there can be just one "correct" price that matches the customers' willingness to pay.[187]

[179] F. Hofmann and F. Freiling, Personalisierte Preise und das Datenschutzrecht - Anforderungen an die datenschutzrechtliche Einwilligung ZD 2020, pp. 331–335; T. J. Tillmann and V. Vogt, Personalisierte Preise im Big-Data-Zeitalter, VuR 2018, pp. 447–455.

[180] F. Hofmann and F. Freiling, Personalisierte Preise und das Datenschutzrecht - Anforderungen an die datenschutzrechtliche Einwilligung ZD 2020, pp. 331–335.

[181] F. Hofmann and F. Freiling, Personalisierte Preise und das Datenschutzrecht - Anforderungen an die datenschutzrechtliche Einwilligung ZD 2020, pp. 331–335.

[182] EU Commission, Proposal for a Regulation of the European Parliament and of the Council on contestable and fair markets in the digital sector (Digital Markets Act), COM(2020) 842 final of 15 December 2020.

[183] M. Ebers, Dynamic Algorithmic Pricing: Abgestimmte Verhaltensweise oder rechtmäßiges Parallelverhalten?, NZKart 2016, pp. 554–555; CJEU, case C-74/14, "Eturas" UAB and Others v Lietuvos Respublikos konkurencijos taryba, Judgement of 21 January 2016; Monopolkommission, XXII. Hauptgutachten der Monopolkommission gemäß § 44 Abs. 1 Satz 1 GWB, Wettbewerb 2018, para. 167.

[184] T. J. Tillmann and V. Vogt, Personalisierte Preise im Big-Data-Zeitalter, VuR 2018, pp. 447–455.

[185] M. Schleusener, Personalisierte Preise im Handel – Chancen und Herausforderungen. In: Stüber and Hudetz (ed), Praxis der Personalisierung im Handel, 1st ed, Springer 2017, pp. 71–89.

[186] BGH, Judgement of 3 July 2003 – I ZR 211/01; BGHZ 155, 301, 305.

[187] BGH, Judgement of 3 July 2003 – I ZR 211/01; BGHZ 155, 301, 305.

With regard to possible behavioural coordination in the context of pricing algorithms, intentional collusion with the aid of algorithms would be the case if algorithms are deliberately used by companies to achieve behavioural coordination.[188] This must be distinguished from unintentional algorithmic collusion, in which coordination is brought about independently by the algorithms in the sense of "artificial intelligence" without any direct human intervention.[189]

Nowadays, an urgent need for legislative action regarding intentional price collusion cannot be identified. The current cases of intentional collusion using price algorithms can be handled adequately with the applicable competition law, as any form of explicit collusion based on communication using algorithms is legally treated in the same way as human collusion.[190] One example is the case pursued by the British antitrust authority (Competition and Markets Authority (CMA)) in which two Amazon Marketplace retailers implemented a price agreement for posters with the help of automatic pricing software (dynamic pricing algorithms). The companies themselves had clearly entered into a cartel agreement, and the use of algorithms therefore only served to implement it.[191]

Moreover, under the circumstances of the current state of the art and the existing market conditions, the occurrence of unintentional "algorithmic collusion" still appears to be comparatively remote.[192] Nevertheless, unintentional price agreements, like the ones described above, are not covered by section 1 GWB. If the circumstances in the future allow the occurrence of unintentional "algorithmic collusion", there should be a stricter regulation of the conditions of use of certain forms of algorithms to prevent companies from circumventing normal competition and thereby harming consumers.

In any case, it must be considered that algorithms significantly simplify price agreements and make it much more difficult to detect them.[193] Algorithms can thus promote collusion.[194]

Furthermore, it might become difficult to decide whether price fixing was intentional or unintentional if price algorithms are able to observe each other and are able

---

[188] BeckOK InfoMedienR/Paal, 31st ed, C.H. Beck 1.2.2021, AEUV Art. 101 para. 133a.

[189] BeckOK InfoMedienR/Paal, 31st ed, C.H. Beck 1.2.2021, AEUV Art. 101 para. 133a.

[190] OECD, Directorate for Financial and Enterprise Affairs, Algorithms and Collusion – Note from the European Union, 14 June 2017, DAF/COMP/WD(2017)12; BeckOK InfoMedienR/Paal, 31st ed, C.H. Beck 1.2.2021, AEUV Art. 101 para. 137.

[191] Competition and Markets Authority (CMA) 30 September 2016, case no. 50223, 19 (25 ff.) – Online sales of posters and frames. Available at https://assets.publishing.service.gov.uk/media/57ee7c2740f0b606dc000018/case-50223-final-non-confidential-infringement-decision.pdf. Accessed 23 August 2021.

[192] BeckOK InfoMedienR/Paal, 31st ed, C.H. Beck 1.2.2021, AEUV Art. 101 para. 133a.

[193] W. Kerler, Illegale Preisabsprachen: Drohen uns bald Algorithmus-Kartelle?. Available at https://www.wired.de/article/preisabsprachen-durch-kuenstliche-intelligenz-drohen-uns-algorithmus-kartelle. Accessed 23 August 2021; Monopolkommission, XXII. Hauptgutachten der Monopolkommission gemäß § 44 Abs. 1 Satz 1 GWB, Wettbewerb 2018, para. 171.

[194] BeckOK InfoMedienR/Paal, 31st ed, C.H. Beck 1.2.2021, AEUV Art. 101 para. 133a.

to match each other without leaving any traces.[195] As part of a stricter regulation, some people suggest a reversal of the burden of proof: the operator of the algorithm would have to demonstrate to the verifying authority that the use of the algorithm did not contribute to infringements of competition law.[196] However, this is usually rejected with the argument that such proof would be very difficult or even impossible for many companies with dynamic price algorithms.[197] This may be due to, for example, the structure of the algorithm or because the company would have to disclose its trade secrets.

15.4.3.4 Collusion of Algorithms to Achieve Higher Prices

Any form of explicit collusion based on communication using algorithms is legally treated in the same way as human collusion.[198] Entrepreneurs coordinate their prices or quantities with each other using algorithms. This enables companies to achieve higher prices and possibly higher profits than it would be possible in free competition.[199] As already explained above, intentional collusion with the aid of algorithms is covered by section 1 GWB, while unintentional price agreements, in which coordination is brought about independently by algorithms, are not covered by section 1 GWB. The described scenario (deliberate use of algorithms to coordinate the behaviour of the entrepreneur involved) is therefore not lawful.

In order to prevent such actions, several countermeasures are requested: it is proposed that the competition authorities are better equipped with specialist staff and technical resources, that algorithms for consumers and interference algorithms are developed, that algorithms are used to detect collusion at an early stage, that transparency obligations are introduced for users and manufacturers of the algorithms and that the speed of price changes get limited.[200] Other voices suggest not overestimating the possibilities of algorithmic collusion. Instead, antitrust and competition agencies should address the issue of blockchain-based collusion.[201]

15.4.3.5 Promotion of the Stability of Cartels Through the Use of Price Algorithms

Although supervision and assembly are not new problems, they are becoming more relevant as a result of the increased use of algorithms and the greater availability of

[195] Monopolkommission, XXII. Hauptgutachten der Monopolkommission gemäß § 44 Abs. 1 Satz 1 GWB, Wettbewerb 2018, para. 185.

[196] L. Bernhardt, Algorithmen, Künstliche Intelligenz und Wettbewerb, NZkart 2019, pp. 314–317.

[197] B. Paal, Missbrauchstatbestand und Algorithmic Pricing, GRUR 2019, pp. 43–53.

[198] OECD, Directorate for Financial and Enterprise Affairs, Algorithms and Collusion – Note from the European Union, 14.6.2017, DAF/COMP/WD(2017)12.

[199] M. Gal, Algorithms as Illegal Agreements, Berkeley Technology Law Journal 2018.

[200] L. Bernhardt, Algorithmen, Künstliche Intelligenz und Wettbewerb, NZkart 2019, pp. 314–317.

[201] T. Schrepel, Collusion by Blockchain and Smart Contracts, Harvard Journal of Law & Technology Vol. 33 (2019), pp. 117–166.

market data.[202] If prices are set with the help of algorithms, possible agreements can be made more quickly. In addition, collusive behaviour can be stabilised through greater price transparency. This applies in particular to the possibility of quick price adaptations if deviations from the collusive agreement are found. This would make short-term deviations unprofitable as companies would be able to analyse data on the price changes of their competitors in real time. As a result, mutual monitoring and sanctions are easier to enforce among cartel members, and the companies involved have less incentives to break the agreement.[203]

As a result, the increasing utilisation of price algorithms can help cartels achieve greater stability.[204]

15.4.3.6 Combating Consumer Discrimination Through Consumer, Competition and Unfair Competition Regulations

The prohibition of misleading advertising in section 5 UWG does not constitute a real obstacle to personalised pricing because of the principle of price freedom.[205] The amount of the price, price fairness and the underlying calculation of the price are not subject to any legal control and do not have to be disclosed.[206] Therefore, prices set by the entrepreneur can be increased or decreased at any time if he or she considers it appropriate.[207]

Nevertheless, personalised prices can result in interaction with the German General Equal Treatment Act (Allgemeines Gleichbehandlungsgesetz (AGG)).[208] The provision in section 19 of the AGG constitutes a market conduct rule within the meaning of section 3a UWG, so that an infringement would be sanctionable under unfair competition law.[209] However, there is no general principle of equal treatment, so that prices do not have to be the same for everyone,[210] nor does an offer say that the goods will be offered to every customer for the same price.[211] Certain discriminations, such as those based on age, gender and religion, are prohibited (section 19 AGG). Although section 20 AGG contains the possibility of justifying

---

[202] M. Gal, Algorithms as Illegal Agreements, Berkeley Technology Law Journal 2018.

[203] L. Bernhardt, Algorithmen, Künstliche Intelligenz und Wettbewerb, NZkart 2019, pp. 314–317.

[204] Monopolkommission, XXII. Hauptgutachten der Monopolkommission gemäß § 44 Abs. 1 Satz 1 GWB, Wettbewerb 2018.

[205] T. J. Tillmann and V. Vogt, Personalisierte Preise im Big-Data-Zeitalter, VuR 2018, pp. 447–455.

[206] F. Hofmann, Der maßgeschneiderte Preis, WRP 2016, pp. 1074 - 1081; D. Tietjen and F. Flöter, Dynamische und personalisierte Preise: Welche lauterkeitsrechtlichen Schranken gelten für Unternehmen?, GRUR-Prax 2017, pp. 546–548.

[207] BGH, Judgement of 13 March 2003 – I ZR 212/00, BGHZ 154, 306.

[208] T. J. Tillmann and V. Vogt, Personalisierte Preise im Big-Data-Zeitalter, VuR 2018, pp. 447–455.

[209] BT-Drs. 16/1780, 48/49.

[210] E. I. Obergfell, Personalisierte Preise im Lebensmittelhandel – Vertragsfreiheit oder Kundenbetrug?, ZLR 2017, pp. 290–301.

[211] F. Hofman, Der maßgeschneiderte Preis, WPR 2016, pp. 1074–1081.

different treatment, the higher willingness to pay does not justify different pricing.[212] In the case of personalised prices, the categories listed in section 19 AGG matter, but basically the aim is to charge the highest possible price: women, for example, are not generally confronted with higher prices than men.[213] Thus, a man and a woman can be shown the same high price, although women would generally be more willing to pay for the underlying product.[214]

In general, the General Equal Treatment Act does not cover personalised prices in principle, although discrimination may well occur in individual cases, which are comprised. For this reason, at least the criteria on which pricing is based should be disclosed.[215] This is essential in order to identify and complain about any discrimination.[216] Such obligation does not currently exist and would have to be regulated by law.[217] However, the problems mentioned above with regard to the disclosure of an algorithm would then occur.[218]

15.4.3.7 Prohibition of the Use of Price Algorithms in Certain Areas of Life

Since personalised prices can be used by companies to maximise profits by making the consumer pay higher prices, the use of personalised pricing is particularly controversial in certain areas of life. Particularly, that applies in those areas that are a central component of basic services.

In the health sector, for example, there is a risk that personalised prices may undermine the principle of solidarity of statutory health and long-term care insurance.[219] Another area where the use of personalised pricing is highly problematic is food supply. The purchase of food is a key part of the basic supply. It is essential for everyone. Therefore, it is extremely important to provide consumers with maximum transparency in pricing policy.[220]

---

[212] F. Hofman, Der maßgeschneiderte Preis, WPR 2016, pp. 1074–1081.

[213] T. J. Tillmann and V. Vogt, Personalisierte Preise im Big-Data-Zeitalter, VuR 2018, pp. 447–455.

[214] T. J. Tillmann and V. Vogt, Personalisierte Preise im Big-Data-Zeitalter, VuR 2018, pp. 447–455.

[215] H. Zander-Hayat, L.A. Reisch and C. Steffen, Personalisierte Preise – Eine verbraucherpolitische Einordnung, VuR 2016, pp. 403–410.

[216] H. Zander-Hayat, L.A. Reisch and C. Steffen, Personalisierte Preise – Eine verbraucherpolitische Einordnung, VuR 2016, pp. 403–410.

[217] T. J. Tillmann and V. Vogt, Personalisierte Preise im Big-Data-Zeitalter, VuR 2018, pp. 447–455.

[218] T. J. Tillmann and V. Vogt, Personalisierte Preise im Big-Data-Zeitalter, VuR 2018, pp. 447–455.

[219] Verbraucherzentrale Bundesverband e.V., Personalisierte Preise Diskussionspapier des Verbraucherzentrale Bundesverbands, 23. September 2016, p. 7. Available at https://www.vzbv.de/sites/default/files/vzbv_position_preisdifferenzierung_16-09-21_pdf.pdf. Accessed 23 August 2021.

[220] E. I. Obergfell, Personalisierte Preise im Lebensmittelhandel – Vertragsfreiheit oder Kundenbetrug?, ZLR 2017, pp. 290–301.

Sectors of basic need should therefore remain free of personalised prices. In order to achieve this, appropriate regulations should be laid down in special laws.[221]

15.4.3.8 Consumer-Friendly Algorithms

The development of algorithms for consumers, so-called algorithmic consumers, is suggested in order to provide a counterweight to the pricing algorithms used by companies. Examples of such applications are digital assistants such as Alexa (Amazon), Siri (Apple) or Cortana (Microsoft). Algorithmic consumers could compensate for competitive disadvantages for consumers and prevent collusion. Algorithmic consumers could also reduce search and transaction costs, help consumers avoid prejudice and make more rational and intelligent decisions, and create or strengthen buyer power. This of course presupposes that the algorithms act in the interest of the users and do not consider the interests of the companies or other third parties.

However, such a possibility requires a high level of available data. Since companies would always have an information advantage, such an algorithmic consumer solution would be difficult to implement. Another problem is the cost that consumers would have to pay for the use of such algorithms. If there is also implicit or explicit coordination by algorithms, it is unclear how a consumer algorithm is supposed to protect consumers from the negative effects of this collusion, especially in terms of higher prices. Furthermore, it is questionable which institution should have the competence to support and verify consumer algorithms.

However, the common use of platforms for price comparison and price forecasting is a first approach. The German price regulation (Preisabgabenverordnung) is intended to strengthen the position of the consumer by creating an optimal price comparison and to promote competition. According to the legislator, correct and complete consumer information guarantees price clarity, price transparency and price truthfulness. The consumer must be able to obtain full information on the price level by comparing prices. In the end, this allows the consumer to decide on the most favourable purchase or service offer, thereby promoting competition and helping to reduce the increase in prices. Finally, there is also the question of how to ensure that they are independent and therefore neutral.

15.4.4 Prohibition of Online Distribution

In the past few years, rising online marketing faces increasing concerns about online distribution, which leads to a debate on the prohibition of the sale of products on certain platforms as well as the prohibition of advertising them on search engines.

[221] T. J. Tillmann and V. Vogt, Personalisierte Preise im Big-Data-Zeitalter, VuR 2018, pp. 447–455.

15.4.4.1 Prohibition of the Sale of Products on Certain Platforms

The admissibility of marketplace bans has been the subject of a large number of court and authority decisions in recent years.[222] The German Federal Cartel Office (Bundeskartellamt) has considered such prohibitions to be generally inadmissible.[223] However, the EJC recently clarified in its *Coty* ruling that platform bans do not in principle constitute a restriction of the customer group and also do not constitute a restriction on passive sales to end users within the meaning of Art. 4 lit. b and lit. c of Commission Regulation 330/2010.[224]

The court has ruled that a selective distribution network does not fall within the prohibition of Art. 101 (1) TFEU if the selection of resellers is made based on objective criteria, which are determined uniformly for all eligible resellers and applied without discrimination. However, the characteristics of the product in question have to require such a distribution network in order to preserve its quality, and the criteria cannot go beyond what is necessary. Regarding the question of whether selective distribution may be necessary in respect of luxury goods, the court held that the quality of such goods is based not only on their physical characteristics but also on their prestigious character, which gives them an aura of luxury. That aura is an essential element in enabling consumers to distinguish the luxury good from other similar products; therefore, any damage to that aura also affects the quality of the goods themselves. A selective distribution system helps preserve the aura of luxury. Therefore, a selective distribution system for luxury goods with the primary purpose of ensuring the luxury image of those goods is compatible with Art. 101 (1) TFEU provided that the conditions set out before are fulfilled.

15.4.4.2 Prohibition of Product Advertising on Search Engines

Fundamental legal questions remain unresolved for both platform bans and AdWords restrictions.[225] In the 2017 annual report of the German Federal Cartel Office, they assumed that "the discussion on the antitrust assessment of distribution restrictions on the Internet is far from complete".[226]

---

[222] T. Reimers, S. Brack and C. Modest, Kartellrechtliche Compliance in Zeiten der Digitalisierung, NZKart 2018, pp. 453–459.

[223] Bundeskartellamt, Case Summary B2-98/11, Unlawful restrictions of online sales of ASICS running shoes, 25 January 2016. Available at https://www.bundeskartellamt.de/SharedDocs/Entscheidung/EN/Fallberichte/Kartellverbot/2016/B2-98-11.pdf?__blob=publicationFile&v=2. Accessed 23 August 2021; Bundeskartellamt, Case Summary B3-137/12, Adidas abandons ban on sales via online marketplaces,19 August 2014. Available at https://www.bundeskartellamt.de/SharedDocs/Entscheidung/EN/Fallberichte/Kartellverbot/2014/B3-137-12.pdf?__blob=publicationFile&v=2. Accessed 23 August 2021.

[224] CJEU, case C-230/16, Coty Germany GmbH v Parfümerie Akzente GmbH, Judgement of 6 December 2017, ECLI:EU:C:2017:941.

[225] T. Reimers, S. Brack and C. Modest, Kartellrechtliche Compliance in Zeiten der Digitalisierung, NZKart 2018, pp. 453–459.

[226] Bundeskartellamt, Tätigkeitsbericht 2017, p. 11.

Only those constellations in which the respective distribution partner is a commercial agent in the sense of the EU Guidelines on Vertical Restraints should be largely unproblematic: in this case, the ban on cartels does not apply to the relationship between suppliers and distributors.[227]

The ASICS distribution system prohibited dealers from using various forms of search engine advertising, including the use of ASICS trademarks as keywords. Adidas' distribution conditions also contained a ban on search engine advertising.[228] The German Federal Cartel Office considered the respective restrictions to be hardcore restrictions not exempted by Regulation 330/2010. However, following the *Coty* judgement, this view may no longer be sustainable. It follows from the ruling that restrictions on online distribution can no longer easily be classified as restrictions on passive sales—at least not if advertising on the Internet is not completely prohibited and customers continue to have access to the Internet offer of the distributors by means of "normal" searches.[229]

15.4.5 Product Liability

With regard to digital marketing, issues related to product liability have to be examined too. In a recent decision, the European Court of Justice has again held that Product Liability Directive 85/347 does not apply to the defectiveness of intellectual content.[230] However, in the future, at least software marketed and sold digitally might fall under the Directive. At their conference on May 7, 2021, the German consumer protection ministers unanimously decided that the current Product Liability Act (Produkthaftungsgesetz), which is a result of the Directive, no longer offers sufficient protection in an increasingly digitalised and technically interconnected world.[231] In their opinion, consumers need to be better protected against damage caused by software errors, and therefore the provisions of product liability law should cover software that is not permanently integrated into a physical product.[232] This opinion was also already expressed by the German federal government.[233] In addition, product liability law should also cover digital damages, including data loss, in the future.[234] Furthermore, since digital products can change as a

[227] T. Reimers, S. Brack and C. Modest, Kartellrechtliche Compliance in Zeiten der Digitalisierung, NZKart 2018, pp. 453–459.

[228] T. Reimers, S. Brack and C. Modest, Kartellrechtliche Compliance in Zeiten der Digitalisierung, NZKart 2018, pp. 453–459.

[229] T. Reimers, S. Brack and C. Modest, Kartellrechtliche Compliance in Zeiten der Digitalisierung, NZKart 2018, pp. 453–459.

[230] CJEU, case C-65/20, Judgement of 10 June 2021, ECLI:EU:C:2021:471.

[231] Ergebnisprotokoll der 17. Verbraucherschutzministerkonferenz am 7. Mai 2021, p. 84.

[232] Ergebnisprotokoll der 17. Verbraucherschutzministerkonferenz am 7. Mai 2021, p. 84.

[233] Stellungnahme der Bundesregierung der Bundesrepublik Deutschland zum Weißbuch zur Künstlichen Intelligenz, pp. 24 f.

[234] Ergebnisprotokoll der 17. Verbraucherschutzministerkonferenz am 7. Mai 2021, p. 85.

result of updates or machine learning, the ministers consider it necessary to review the definition of the term "placing on the market" in the context of product liability.[235] In view of the increasing complexity of "digital" products, they also believe that it is necessary to readjust the rules on the burden of proof as, in many cases, it remains unclear whether the cause of damage lies in the hardware, the software, faulty data or an application error.[236] In this respect, they reiterated the Federal Council's (Deutscher Bundesrat) statement that such uncertainties should not be to the detriment of consumers, who should be granted a simplification of the burden of proof.[237] The clarification of such questions and whether the Directive should be transformed into a regulation had also already been demanded by the European Parliament.[238]

15.5 The Impact of Personalised Political Campaigns on Democracy

Through personalised political campaigns, also called microtargeting, a party can match its message to the specific interests and vulnerabilities of the voters. In terms of democracy, there are both advantages and disadvantages.

For citizens, such microtargeting may lead to more relevant advertising. It can reach citizens who ignore traditional media, and it can interest people in politics through tailored messages. Microtargeting might thus increase information, interest in politics and electoral turnout.[239] For politicians, microtargeting can be efficient, effective and—in some cases—cheap.[240] Microtargeting can benefit public opinion as it can lead to more diverse political campaigns and to more knowledge among voters about certain issues.[241]

At the same time, microtargeting also brings with it major disadvantages and risks for democracy. Unconscious manipulation and an undermining of democratic decision-making are the biggest dangers of big data applications.[242] It can invade people's privacy and can be used to exclude or manipulate people. Microtargeting

235 Ergebnisprotokoll der 17. Verbraucherschutzministerkonferenz am 7. Mai 2021, p. 84.

236 Ergebnisprotokoll der 17. Verbraucherschutzministerkonferenz am 7. Mai 2021, p. 85.

237 BR-Drs. 95/20 (B), p. 6.

238 European Parliament resolution of 20 October 2020 with recommendations to the Commission on a civil liability regime for artificial intelligence (2020/2014(INL)), P9_TA(2020)0276.

239 F. J. Zuiderveen Borgesius et al., Online Political Microtargeting: Promises and Threats for Democracy, Utrecht Law Review 2018, pp. 82–96.

240 F. J. Zuiderveen Borgesius et al., Online Political Microtargeting: Promises and Threats for Democracy, Utrecht Law Review 2018, pp. 82–96.

241 F. J. Zuiderveen Borgesius et al., Online Political Microtargeting: Promises and Threats for Democracy, Utrecht Law Review 2018, pp. 82–96.

242 C. Pentzold and L. Fölsche, Die öffentliche Verhandlung von Big Data in politischen Kampagnen, p. 52. http://www.abida.de/sites/default/files/ABIDA%20Gutachten%20Digitaler%20Demos.pdf. Accessed 23 August 2021.

can enable a political party to misleadingly present itself to various people as a different party with a different political program.[243] For politicians, microtargeting also brings with it threats. Some types of microtargeting are so expensive that large parties with more financial resources gain an unfair advantage. And new intermediaries, such as online marketing companies, become more powerful.[244] One risk for public opinion is that the priorities of political parties could become opaque. In addition, the political debate can be split if different groups of voters focus on different issues.[245]

These risks must be taken seriously, and if they occur, they threaten democracy. Nevertheless, microtargeting could have less influence in Europe than in the United States because of differences in the legal and electoral systems. The GDPR makes microtargeting more difficult in the European Union than in the United States.[246] In order to minimise the risks, the following approaches should be adopted: expanding the political education of citizens, expanding and strengthening journalistic expertise, information obligation for data-based activities of parties and reviewing legal conditions for data ownership, data transfer and data consolidation.[247]

Depending on the specific design of the AI systems, it is possible that the systems will be prohibited under Art. 5 (1) lit. a, b, c AIA draft or will be classified as high-risk AI systems and therefore be subject to special safeguards. Due to the partially very vague formulations in the AIA draft,[248] no clear statements can be made today. In the recital 15 AIA draft, however, it is pointed out that the aim is to counteract the dangers to democracy.

15.6 Future Perspectives

After highlighting the problems that society, the authorities and the legal system must deal with in the face of rapidly changing commercial practices and the implementation of algorithms in the marketing sectors, future perspectives are presented.

[243] F. J. Zuiderveen Borgesius et al., Online Political Microtargeting: Promises and Threats for Democracy, Utrecht Law Review 2018, pp. 82–96.

[244] F. J. Zuiderveen Borgesius et al., Online Political Microtargeting: Promises and Threats for Democracy, Utrecht Law Review 2018, pp. 82–96.

[245] F. J. Zuiderveen Borgesius et al., Online Political Microtargeting: Promises and Threats for Democracy, Utrecht Law Review 2018, pp. 82–96.

[246] B. Kolany-Raiser and T. Radtke, Microtargeting, Gezielte Wähleransprache im Wahlkampf, 2018, p. 6 ff. http://www.abida.de/sites/default/files/16_Microtargeting.pdf. Accessed 23 August 2021.

[247] C. Pentzold and L. Fölsche, Die öffentliche Verhandlung von Big Data in politischen Kampagnen, p. 65 ff. Available at http://www.abida.de/sites/default/files/ABIDA%20Gutachten%20Digitaler%20Demos.pdf. Accessed 23 August 2021.

[248] H. Hoffmann, Regulierung der Künstlichen Intelligenz, K & R 2021, pp. 369–374.

15.6.1 Specific Regulatory Instruments

As already shown, it is difficult to detect the use of algorithms if they are not labelled as such. A labelling requirement is therefore the first step towards creating transparency. In addition, transparency is required with regard to the understandability and verifiability of the algorithms used.[249] In order to control the application and use under the current legislation, public authorities must therefore be better provided with personnel and technical equipment.

Besides that, there are various points where regulatory instruments could be deployed:[250] an independent technical supervisory association could be established to verify the appropriate and correct use of decision-making systems and to address shortcomings at different stages. Input monitoring could check whether training data are adequate and of good quality. An incorrectly designed decision system can be detected by black box experiments, which can be used to test the functionality of such systems. Professional ethics for data scientists could ensure that they are aware of certain principles in their work. A guidance document that helps users interpret the results can contribute to more competent use of decision systems. Last but not least, a better external researchability of algorithmic decision systems would generally ensure the independent evaluation of the overall process.

However, before considering the use of other regulatory instruments, one should ask where exactly these should be used. To do this, one must first clarify which algorithms should be made transparent, for whom and why.[251]

Some of these aspects are already mentioned in the EU Commission's proposal for an Artificial Intelligence Act (AIA).[252] Even though this is a good approach, the legal requirements must also take into account technical developments. Therefore, the progress of the so-called *Explainable AI* (XAI) in particular must be considered.

[249] C. Husted, Algorithmen-Transparenz. Was steckt hinter dem Buzzword?, 6 May 2019. Available at https://algorithmenethik.de/2019/05/06/algorithmen-transparenz-was-steckt-hinter-dem-buzzword/. Accessed 23 August 2021.

[250] K. Zweig, Wo Maschinen irren können, Bertelsmann Stiftung 2018. Available at https://www.bertelsmann-stiftung.de/fileadmin/files/BSt/Publikationen/GrauePublikationen/WoMaschinenIrrenKoennen.pdf. Accessed 23 August 2021.

[251] C. Husted, Algorithmen-Transparenz. Was steckt hinter dem Buzzword?, 6 May 2019. Available at https://algorithmenethik.de/2019/05/06/algorithmen-transparenz-was-steckt-hinter-dem-buzzword/. Accessed 23 August 2021.

[252] EU Commission, Proposal for a Regulation of the European Parliament and of the Council laying down harmonised rules on artificial intelligence (Artificial Intelligence Act) and amending certain union legislative acts, COM (2021) 206 final of 21 April 2021.

15.6.2 No Need to Establish Special Units

In our view, it is not necessary to establish special units within the authorities. Special knowledge can be obtained by consulting experts and internally training employees.

However, if this evaluation changes, consideration may be given to creating special departments within the supervisory authorities. The establishment of special chambers at the courts does not appear to make sense as there will be interfaces with algorithms in almost all areas of life.

15.6.3 Codes of Conduct

Up until now, codes of conduct have played a minor role in data protection. The GDPR wants to help the instruments for self-regulation gather steam: for this purpose, the legal basis of the codes of conduct, Art. 40 GDPR, was substantially expanded and is more detailed than section 38a of the former German Federal Data Protection Act (Bundesdatenschutzgesetz (BDSG)).[253]

Codes of conduct are guidelines for good data protection practice, which are developed by experts in the relevant field and are similar to technical standards: they standardise data processing in connection with typical services and products offered in an industry.[254] According to Art. 40 (2) GDPR, codes of conduct should "specify" the application of the regulation. The need for such a clarification arises from the fact that the regulation is in numerous places undefined and contains general clauses.[255] Codes of conduct can be used as interpretative aids and thus serve to create and maintain legal certainty.[256] It also follows that codes of conduct are not a legal basis for the processing of personal data.[257] Codes of conduct are intended to make it easier to apply the regulation effectively. Recital 98 states that account should be taken of the specific nature of processing operations carried out in certain sectors and the special needs of micro, small and medium-sized enterprises.

[253] C. Herfurth and F. Engel, Codes of Conduct im Konzern?, ZD 2017, pp. 367–372.

[254] N. Raschauer. In: Sydow (ed), Europäische Datenschutzgrundverordnung, 2nd ed, Nomos 2018, Art. 40 DS-GVO, para. 3.

[255] V. Jungkind. In: Wolff and Brink (eds), BeckOK Datenschutzrecht, 31st ed, C.H. Beck 2020, Art. 6 DS-GVO, para. 23.

[256] B. Paal. In: Paal and Pauly (eds), Datenschutzgrundverordnung, Bundesdatenschutzkontakt, 2nd ed, C.H. Beck 2018, Art. 6 DS-GVO, para. 11.

[257] P. Laue. In: Laue and Kremer, Das neue Datenschutzrecht in der betriebsrechtlichen Praxis, 2nd ed, Nomos 2019, § 8, para. 7.

Codes of conduct can help avoid uncertainty and legal disputes to a certain extent,[258] but the legal basis for data processing remains the regulation.[259] Self-regulation and codes of conduct can therefore not replace legal regulations.[260]

The EU Commission's DSA draft contains provisions for a framework for the development of specific codes of conduct for online advertising. According to Art. 36 DSA, the Commission shall encourage and facilitate the drawing up of codes of conduct at the Union level between online platforms and other relevant service providers, such as providers of online advertising intermediary services or organisations representing recipients of the service and civil society organisations or relevant authorities, to contribute to further transparency in online advertising. The codes of conduct shall pursue an effective transmission of information, in full respect for the rights and interests of all the parties involved, and a competitive, transparent and fair environment in online advertising, in accordance with Union and national law, in particular on competition and the protection of personal data. The codes of conduct shall address at least the transmission of information held by providers of online advertising intermediaries to recipients of the service and the transmission of information held by providers of online advertising intermediaries to repositories containing certain information regarding the displayed advertisements.

15.6.4 A Vigilant and Learning Approach

New social media platforms with fresh or at least newly interpreted functions are constantly emerging. They offer special possibilities for advertising use to justify their special features and stand out from other offers. Although new problems are always associated with this, the legal principles known from other, comparable constellations are usually transferable to these new types of questions.[261] However, the existing regulations must be regularly reviewed in light of upcoming changes. Due to changing user habits and greater familiarity with new social media offerings, a certain adjustment of the requirements would be desirable in some constellations.[262]

---

[258] N. Raschauer. In: Sydow (ed), Europäische Datenschutzgrundverordnung, 2nd ed, Nomos 2018, Art. 40 DS-GVO, para. 7.

[259] V. Jungkind. In: Wolff and Brink (eds), BeckOK Datenschutzrecht, 31st ed, C.H. Beck 2020, Art. 6 DS-GVO, para. 23; Landesbeauftragte für Datenschutz und Informationsfreiheit NRW, Verhaltensregeln und Akkreditierung von Überwachungsstellen nach der DS-GVO. https://www.ldi.nrw.de/mainmenu_Datenschutz/submenu_Datenschutzrecht/Inhalt/Verhaltensregeln-_-Code-of-Coduct/Inhalt/Verhaltensregeln-und-Akkreditierung-von-Ueberwachungsstellen-nach-der-DS-GVO/Verhaltensregeln-und-Akkreditierung-von-Ueberwachungsstellen-nach-der-DS-GVO.html. Accessed 23 August 2021.

[260] A. Roßnagel. In: Simitis,Hornung and Spiecker, Datenschutzrecht, 1st ed, Nomos 2019, Art. 40 DS-GVO, para. 20.

[261] F. Lichtnecker, Neues aus dem Social Media-Marketing, MMR 2018, pp. 512–517.

[262] F. Lichtnecker, Neues aus dem Social Media-Marketing, MMR 2018, pp. 512–517.

The DSA and DMA drafts seem to try to learn from prior complications. A frequently raised criticism of antitrust supervision is that it is not fast enough to effectively stop abusive practices. Therefore, the DMA provides an ex ante regulatory tool to control the behaviour of gatekeepers under Arts. 5 and 6 DMA.[263] Furthermore, the findings of the evaluation report on the GDPR,[264] especially regarding enforcement, have a direct impact on the DSA and DMA drafts: to address the different levels of enforcement across EU member states under the GDPR, the DSA and DMA provide a modified enforcement regime. By involving the Commission, the lack of enforcement is supposed to be prevented.

15.7 Conclusion

In an overall assessment, Germany and the European Union are taking different measures that positively contribute to the addressing of new challenges. Nevertheless, AI regulation can only go hand in hand with technical feasibility and the consideration of ethical and social circumstances and cannot be dealt with on a legal level alone. Therefore, the circumstances must regularly be assessed from not only a legal but also an ethical perspective. In addition to economic aspects, the focus must be on the autonomy of the persons concerned, especially their informational self-determination, which is protected by Art. 8 CFR. Both economic and self-determination interests have to be fairly balanced.

The new legislative procedures at the EU and national levels provide an insight into the new policy directions. With the help of harmonising legislation, the European Union wants to make the EU internal market more competitive. In order to be able to effectively implement this goal, it is useful to seek solidarity with US partners.[265] Furthermore, the European Commission's proposal for an Artificial Intelligence Act as the world's first response to the issue of regulating AI is a step in the right direction but needs to be adopted in the further regulation process.

---

[263] See N. Gielen and S. Uphues, Digital Markets Act und Digital Services Act, EuZW 2021, pp. 627–637.

[264] EU Commission, COM(2020) 264 final.

[265] See U. von der Leyen, Speech at the World Economic Forum in Davos 2021, "I want to invite our friends in the United States to join our initiatives. Together, we could create a digital economy rulebook that is valid worldwide." Available at https://ec.europa.eu/commission/presscorner/detail/en/speech_21_221. Accessed 23 August 2021.

Hungary

16

Adam Liber, Lili Albert, and Judit Firniksz

16.1 Introduction

The emergence of data-driven business models, data analytics and algorithms provide unprecedented opportunities to gain and maintain a competitive advantage on the part of big tech service providers whose main income is from the analysis, processing and selling of data. However, this also triggers high exposure by individuals through technology because there seems to be a significant imbalance between providers and users. The use of digital services and the utilisation of algorithms and artificial intelligence (AI) is now essential in the enforcement activities of various regulatory agencies prosecuting consumer protection, competition and data protection infringements, and it seems that it is necessary to provide a coordinated answer to the issues raised by new business models, considering the interplay between privacy law, competition law and consumer protection law issues.

This contribution will explore the main aspects of this interplay under EU and Hungarian laws.

Adam Liber acted as working group lead. Lili Albert and Judit Firniksz acted as working group members.

A. Liber (✉)
PROVARIS Varga & Partners, Budapest, Hungary
e-mail: Liber.Adam@provaris.hu

L. Albert
Deloitte Legal, Budapest, Hungary
e-mail: lialbert@deloittece.com

J. Firniksz
Firniksz Law Firm, Budapest, Hungary
e-mail: judit.firniksz@ecommercelawyer.eu

B. Kilpatrick et al. (eds.), *Antitrust in Data Driven Markets & Legal Framework for Influencers, Native Advertising and Control over the Use of AI in Marketing*, LIDC Contributions on Antitrust Law, Intellectual Property and Unfair Competition,
https://doi.org/10.1007/978-3-031-07422-6_16

16.2 Privacy Implications

Regarding the use of algorithms or other AI-based applications, chatbots, customer tracking and similar means, the implementation of such technologies inevitably involves the processing of personal data and therefore triggers various obligations under the General Data Protection Regulation (GDPR). These obligations include securing an appropriate legal basis and transparency; accuracy (data quality); data security; compliance with purpose limitation, data minimisation, and storage limitation; documentation requirements; and the requirement to conduct data protection impact assessments as part of these efforts with a view to the high risks that the use of such technologies may cause to the data subject.

In this context, data protection by design and by default (DPbDD) is of particular importance when designing applications or algorithms because the data controller must assess DPbDD requirements, which cover the effective implementation of data protection principles under Article 5 (1) GDPR and data subjects' rights and freedoms by design and by default, which controllers must take into account when designing the processing. The Hungarian Data Protection Authority (DPA) has not yet released any guidance on application design issues; however, the European Data Protection Board (EDPB) has released Guidelines 4/2019 on Article 25 Data Protection by Design and by Default ("DPbDD Guidelines"). The addressees of DPbDD obligations are data controllers. However, controllers (especially small data controllers) seldom have the opportunity to build and design their own systems, and the GDPR does not seem to address software developers and manufacturers that play a key role in DPbDD implementation.

This part of the country report explores the privacy implications with reference to the applicable data protection principles under Article 5 (1) GDPR and related accountability obligations under Article 5(2) GDPR.

16.2.1 Lawfulness and Purpose Limitation

Under the GDPR, the lawfulness principle requires that the controller identify a valid legal basis for the processing of personal data. Moreover, to comply with the purpose limitation principle, the controller must collect data for specified, explicit, and legitimate purposes and not further process the data in a manner that is incompatible with the purposes for which they were collected. To comply with these requirements, the EDPB in its DPbDD Guidelines expects that the whole processing life cycle must be in line with the relevant legal grounds of processing and that the design of the processing should therefore be shaped by what is necessary to achieve these legitimate purposes.

To the extent *consent* is used to legitimise data processing (Article 6 (1) (a) GDPR), the goal would be to enable the data subjects (consumers) to make informed decisions so they could understand what they are agreeing to when providing their personal data, which protects the consumer. According to the

Hungarian Advertising Act,[1] *advertisements may be conveyed to natural persons by way of direct contact (...), only upon the express prior consent of the person to whom the advertisement is addressed. The statement of consent may be made out in any way or form, on condition that it contains the name of the person providing it, (...), furthermore, any other personal data authorized for processing by the person providing the statement, including an indication that it was given freely and in possession of the necessary legal information.*[2]

If the data controller relies on *contract* as a legal basis, then the controller must predetermine what personal data are objectively necessary for the performance of the contract or necessary to take steps at the request of the data subject prior to entering into the contract. In this context, Article 6(1)(b) GDPR covers the preliminary processing of necessary personal data before entering into a contract in order to facilitate the actual entering into that contract by the data subject, such as responding to an enquiry of the data subject; however, it may not cover processing activities that are carried out solely on the initiative of the data controller, such as service improvement, fraud prevention or behavioural advertising, because such processing operations require reliance on other legal bases. Article 6(1)(b) GDPR can also cover processing operations objectively necessary for the performance of the contract and delivery of the contractual services, i.e. processing reasonably foreseen and necessary within a normal contractual relationship, such as payment processing, sending formal reminders about outstanding payments or correcting errors or delays in the performance of the contract, contractual warranty, or contract termination and returning of goods. The processing options, by default, may not allow the retrieval of data directly from other sources than the data subject itself because this will also require a different legal basis. However, special categories of data cannot be processed by solely relying on Article 6(1)(b) GDPR.

The use of advertising tracking cookies can only be legitimised based on consent (pursuant to Section 13/A (4) of the Act CVIII of 2001 on certain issues of electronic commerce services and information society services), whereas consent may not be combined with other legal bases, such as legitimate interests (see item 123 of Guidelines 05/2020 on consent under Regulation 2016/679 and item 40 of Opinion 5/2019 on the interplay between the ePrivacy Directive[3] and the GDPR). This is because Article 5 (3) of the ePrivacy Directive takes precedence over Article 6 of the GDPR. The GDPR mentions cookies only once, in recital 30. If the cookies are used to identify users, then they qualify as personal data so they are subject to the GDPR. Under the GDPR, controllers have the right to process their users' data (the cookies) as long as they receive consent or if they have a legitimate interest.

---

[1] Act XLVIII of 2008 on the Basic Requirements and Certain Restrictions of Commercial Advertising Activities.

[2] Section 6 of the Hungarian Advertising Act.

[3] Directive 2002/58 of the European Parliament and of the Council of 12 July 2002 concerning the processing of personal data and the protection of privacy in the electronic communications sector (Directive on privacy and electronic communications), OJ 2002 L 201, p. 37.

The legality of processing may be challenged if the data controller relies on the use of cookie walls, i.e. when consumers may not use a website without consenting to the use of personal data or when cookies also likely violate the tie-in ban under Article 7 (4) GDPR. Article 7 (4) GDPR says: *when assessing whether consent is freely given, utmost account shall be taken of whether, inter alia, the performance of a contract, including the provision of a service, is conditional on consent to the processing of personal data that is not necessary for the performance of that contract.* This provision makes a distinction between processing activities necessary for the performance of a contract and services conditioned on certain processing activities that are not in fact necessary for the performance of the contract. If consent is given in the latter situation, it is presumed not freely given. The legal basis of consent and contract cannot be merged and blurred.[4] In practice, data controllers in Hungary seldomly respect the tie ban under Article 7 (4) GDPR; therefore, data subjects erroneously get the impression that they are giving their consent in line with Article 6 (1) (a) when signing a contract or accepting terms of service. This may occur when data controllers request data subjects to "accept" a privacy notice. Indeed, if the data subject must explicitly "accept" the privacy notice, it could be considered bundling of consent with contractual requirements. When consumers are precluded from using a website without consenting to the use of their personal data, the recently updated consent guidelines of the EDPB also provided clarification saying that "In order for consent to be freely given, access to services and functionalities must not be made conditional on the consent of a user to the storing of information, or gaining of access to information already stored, in the terminal equipment of a user (so called cookie walls)". Practically, if there is no possibility to access the website content without accepting the cookies and the data subject is not presented with a genuine choice, this will not constitute valid consent, and this practice will be illegal.

Under the GDPR, processing must be limited to what the legal basis strictly gives grounds for. Under the GDPR's opening clauses, algorithm operators must consider other restrictions under member state law when creating user profiles. Under Hungarian law, the relevant limitations are laid down by the Act XLVII of 1997 on Processing and Protection of Medical and Other Related Personal Data, which limits the processing purposes of data concerning health and requires the explicit consent of the data subject if the processing purpose does not comply with the Act. Moreover, Act No XXI of 2008 on the protection of human genetic data and the regulation of human genetic studies, research and biobanks limits the export of genetic data form third countries that do not secure the adequate level of protection for personal data.

[4] Guidelines 05/2020 on consent under Regulation 2016/679, Version 1.1, adopted on 4 May 2020, item 26.

16.2.2 Transparency

Anyone who spends a lot of time on the Internet can hardly afford to read the numerous data protection regulations that apply to advertising, prices, etc., and in practice, data protection declarations are often extensive and difficult to understand.

According to the GDPR's transparency principle, from the start, the controller must be clear and open with the data subject about how they will collect, use and share personal data. In this context, the controller must act transparently and present information about the processing in such a manner that makes it easy for the data subject to understand what processing is necessary to enter into or perform the contract, whether the data subject is obliged to provide personal data and what the possible consequences are if the data subject does not provide such data.

If the website operator does not know what data are collected from third parties on his or her website, it is likely that the operator will be unable to comply with the transparency requirements under the GDPR.

On 17 February 2017, the Hungarian DPA released cookie guidance relating to website and online shop operations. This guidance says that before implementing cookies or similar technologies placing information on the users' end device, the website operator must (1) map the cookies that it wishes to use on its site and (2) determine whether notice or consent is required for the use of each. The guidance also confirms that the website operator is liable for the use of third-party cookies that transmit information—such as user behaviour data—to third parties. Accordingly, the website operator must have the user's consent to allow the use of cookies that collect and transmit user information to third parties. The guidance says that special attention must be given to the use of social plug-in modules monitoring user behaviour or tracking other user activities. The website operator must be aware of the scope of the data collected by third-party cookies, including the data categories collected and the relevant processing purposes, such as analytics, advertising or market research. The operator also must be transparent about data collection practices relative to the use of its website.

Due to the inherent complexity of the black box operation of modern computing systems, it is likely that data subjects (or even experts) will be unable to understand how artificial intelligence works. However, even if a passenger may be unable to explain the technical operation of a tram or airplane, there is no reason to exclude someone from using these means of travel because only the travel destination and his safety will be of utmost importance for a passenger.

Algorithmic transparency is a principle that expects that the factors influencing the decisions made by algorithms should be visible, or transparent, to the people who use, regulate and are affected by systems that employ those algorithms. The EU regulatory model of transparency (Article 12 GDPR) focuses on "meaningful transparency", which means that in case of automated decision-making, including profiling, the data controller must provide meaningful information that the average data subject is able to understand at least *about the logic involved, as well as the significance and the envisaged consequences of such processing for the data subject*" (Article 13 (2)(f) GDPR). This also means that information of a technical nature that data subjects are unable to understand must be omitted from the disclosure. Within the context of AI operations, transparency should cover information on how the algorithm was taught, how its vulnerabilities were addressed, what the data source is, how its results have been validated and what impact it may have on the individual. When being transparent about AI, the data controller must disclose the risks and possible outcomes of the decision-making process rather than its technical functionality because the latter will be of little value for individuals. Understanding what data are used by the algorithm to determine an outcome can be useful in securing algorithmic transparency. However, because non-personal data (such as statistical data) might also play an important role in the decision, the specific amount of personal data used may not always be informative with regard to the opacity of the algorithm's operation.[5]

16.2.3 Accuracy and Data Quality

According to the accuracy principle, personal data shall be accurate and kept up to date. The controller shall take every reasonable step to ensure that personal data that are inaccurate are erased or rectified without delay, having regard to the purposes for which they are processed. Accuracy is of utmost importance because decisions about personalised advertising, content or prices are made by algorithms based on the customer's existing data sets. In order to comply with the GDPR, the advertiser must secure that the quality of the underlying data sets is ensured for the creation of personalised advertising, content or prices.

Data quality means that personal data shall be accurate, relevant, complete, up to date and consistent. The data accuracy principle is one of the key data protection principles under the GDPR. Article 5 (1) (c) GDPR also requires that personal data are adequate and relevant in relation to the purposes for which they will be processed. Under the accountability principle (Article 5 (2) GDPR), the data controller must be able to demonstrate compliance with these requirements. The main data quality issues may be incompleteness, inaccuracy (wrong field entered), inconsistency, invalidity, redundancy and non-standard data.

Artificial intelligence can also enhance the use of large amounts of data sensibly and enable extended applications in real time by employing real-time data analytics, which improves response times and can make the data management process more robust. This technology requires processing power and extensive infrastructure but provides an increased degree of accuracy and efficiency for data processing. There are readily available, state-of-the-art technologies on the market.

[5] EPRS, A governance framework for algorithmic accountability and transparency, April 2019. Available at https://www.europarl.europa.eu/RegData/etudes/STUD/2019/624262/EPRS_STU(2019)624262_EN.pdf item 3.6.2. Accessed 25 August 2021.

The Hungarian DPA has not yet released any guidance on the handling of data quality issues. At the same time, the EDPB has released its DPbDD Guidelines, which cover compliance requirements regarding data accuracy.

In practice, to secure data quality, the controller may use tools that enable entry field validations; detect and handle misspellings; have email (such as disposable email) and IP blacklist filtering; allow data cleansing and testing by using static analysis tools, which check the data against quality issues, such as random white spaces or unwanted character removal; use predefined choices in input fields or drop-down lists rather than free-text entries; provide measurement and metrics regarding data quality; and permit data subjects to correct the data.

16.2.4 Fairness

The "fairness" principle requires that personal data shall not be processed in a way that is detrimental, discriminatory, unexpected or misleading to the data subject. In that regard, the ethical aspects of data processing are of utmost importance. There are strong arguments[6] that developers and implementers of AI have a social responsibility to incorporate socially and ethically recognised values into the programming design process to avoid unfairness and discrimination that the functioning of the algorithm may cause. As part of this process, developers and implementers must secure the diagnosis and correction of any biases that the algorithm may cause.

It is likely that advertising campaigns, also in the creative field, are able to be carried out autonomously by artificial intelligence in the future. Algorithms evaluate previous purchase decisions and then place new advertisements. This may constitute discrimination against other companies that were not subject to previous purchase decisions, because if the data used for the training of the algorithm is selected in a discriminatory way, then the machine learning algorithm will also likely continue the discriminatory practice. The Hungarian Competition Authority (HCA) and the Hungarian DPA have no practice in that regard. However, depending on the specific circumstances of the case, this may constitute an abuse of a dominant position *via-á-vis* companies that were not subject to previous purchase decisions; moreover, data subjects who are exposed to discriminatory data processing may claim that the data controller did not comply with the fairness principle when processing personal data in a discriminatory way.

The creation of personalised advertising or personalised prices may fall under Article 22 (1) GDPR provided that data processing activities have serious consequences for individuals. According to the EDPB Guidelines on automated individual decision-making and profiling for the purposes of Regulation 2016/679, only serious impactful effects will be covered by Article 22. According to the GDPR, "*profiling means any form of automated processing of personal data consisting of*

[6] The Institute of Electrical and Electronics Engineers (IEEE) - 'Ethically Aligned Design' First Edition, March 2019.

the use of personal data to evaluate certain personal aspects relating to a natural person, in particular to analyse or predict aspects concerning that natural person's performance at work, economic situation, health, personal preferences, interests, reliability, behaviour, location or movements". There are potentially three ways in which profiling may be used, namely (1) general profiling; (2) decision-making based on profiling; and (3) solely automated decision-making, including profiling, which produces legal effects or similarly significantly affects the data subject.[7] When the decision is based only on an algorithm, we can speak about the process described in point (3). This is what happens exactly in the case of personalised advertising and personalised prices; that is, an algorithm assesses the personal data and decides on the basis of the assessed personal data (for example location data or an online identifier).

The creation of personalised advertising or personalised prices may qualify as automated individual decision-making. According to the GDPR, the data subject shall have the right not to be subject to a decision based solely on automated processing, including profiling, which produces legal effects concerning him or her or similarly significantly affects him or her. Hungarian law does not explicitly require that the consumer shall be given a choice between personalised advertisements or personalised prices on the one hand and non-personalised on the other; however, this can be derived from the fairness principle of the GDPR.

16.2.5 Integrity and Confidentiality

Data security principle means that controllers and processors must implement appropriate technical and organisational measures to ensure a level of security appropriate to the risk of data processing. To comply with data security requirements, the controller is expected to perform a risk analysis, which assesses the risks against the security of personal data and counter-identified risks, such as accidental or unlawful destruction, loss, alteration, unauthorised disclosure of, or access to, personal data transmitted, stored or otherwise processed, which may in particular lead to physical, material or non-material damage.

Considering the privacy risks that the use of artificial intelligence may pose for the individual, it will likely trigger the requirement of a data protection impact assessment (DPIA). Under Article 35 (1) GDPR, where data processing operations are likely to result in a high risk to the rights and freedoms of natural persons, the data controller must carry out a DPIA before the commencement of data processing. Regarding the assessment of such risks, the controller must consider the type of processing when using new technologies and the nature, scope, context and purposes

---

[7] Article 29 Data Protection Working Party Guidelines on Automated individual decision-making and Profiling for the purposes of Regulation 2016/679, available at https://ec.europa.eu/newsroom/article29/item-detail.cfm?item_id=612053. As last Revised and Adopted on 6 February 2018. Last accessed 19 September 2020.

of the processing that may occur. The use of artificial intelligence for processing personal data will likely meet the legal requirement for completing a DPIA under Article 35 (1) GDPR because AI is an innovative technological solution that may pose risks to individuals' privacy and data protection rights, such as discrimination, bias or inaccuracy in the algorithms or unexpected outcome by data subjects.

Under Article 35 (4) GDPR, the Hungarian DPA established a list of processing operations that are subject to the requirement for a data protection impact assessment (NAIH backlist).[8] Under the NAIH blacklist, the use of new technologies for data processing or automated decision-making producing legal effects or similarly significant effects requires a DPIA. The relevant processing operations cover the processing of large amounts of data obtained via sensor-equipped devices (e.g. smart televisions, smart household appliances, smart toys, etc.) and transferred through the Internet or other channels, and such devices providing data on the characteristics of the financial status, health condition, personal interests, trustworthiness or conduct, residence or movement of the natural person, and such data form the basis of profiling.

16.2.6 Data Minimisation

The GDPR also expects the controller to comply with data minimisation. Within the context of the use of algorithms, data minimisation means that only personal data that are adequate, relevant and limited to what is necessary for the purpose may be processed. According to the DPbDD Guidelines, this means that the controller must predetermine which features, parameters or supporting functions are permissible in their systems regarding data processing. In designing and using algorithms, the controller is expected to avoid data processing when this is possible for the relevant purpose and must also limit the amount of personal data collected to what is necessary for the purpose.

16.2.7 Storage Limitation

The storage limitation principle says that the controller must ensure that personal data are kept in a form that permits the identification of data subjects for no longer than is necessary for the purposes for which the personal data are processed. Practically, the algorithm's operator must determine what data and length of storage are necessary for the purpose of data processing, and the operator must be able to justify why the period of storage is necessary for a given purpose.

---

[8] Available at https://naih.hu/list-of-processing-operations-subject-to-dpia-35-4%2D%2Dgdpr.html. Last accessed 19 September 2020.

16.3 Consumer Protection and Competition

When shopping on the Internet, consumers have access to a bigger range of products and potential contract partners. However, when using search engines, it is the search engine and not the user who decides which products are proposed to the consumer. The proposals often come about through artificial intelligence and do not necessarily correspond to the interest of the user.

The use of search engines, digital comparison tools, personalised advertisements and personalised offers (personalised targeting) raise the question of whether the consumer needs to understand the logic used by search engines and direct-to-consumer (DTC) business models and, if the answer is positive, to what extent the consumer must have an insight into the operation of search engines and DTCs. Under the Hungarian regulatory framework, search engines, DCTs and personalised targeting based on consumer profiles would be assessed as a factor influencing the consumers' decision-making process by limiting the range of available choices; therefore, this issue would be treated in the framework of the Prohibition of Unfair Commercial Practices Against Consumers Act ("UCP Act).

The current regulatory framework includes the fundamental legal requirements to enable enforcement as regards personalised targeting. Competent authorities, however, meet the challenge of how to investigate the technicalities of personalisation and provide proper evidence for the legal assessment. Beyond proceedings focused on particular market behaviours, authorities may play a significant role in facilitating awareness of consumers and self-regulation of market players.

16.3.1 Market Power and the Use of AI

The interplay between data, artificial intelligence and market power has not yet been assessed in the Hungarian legal practice. General market tendencies show that digital sectors tend to be leading sectors also in AI, and the use of AI has the potential to accelerate shifts in market share, revenue and profit. As for the effect on competitive conditions, the role of algorithms cannot be separated from access to the data serving as an input to algorithmic analysis (e.g. health care data). Further, AI algorithms might change the nature of the concerned goods and services (e.g. new fields of personalised health care services are becoming available as a result also of AI-based research and analytics methods), and the transformation may reach and change the general mindset of the given industry as well. Therefore, due to the complexity of the issue, a case-by-case analysis is required.

Autonomous algorithmic agents or autonomous software agents being present in the economy raise several questions, which can be linked to the contracting parties' private autonomy: (1) how to interpret in legal context the autonomy of software agents—whether they can take individual, stand-alone decisions (which do not fall into the category of automatised decision-making); (2) what the potential risks and benefits of cooperation are between people and software agents in economic decision-making; (3) how network risks rooting in the integrated operation of computer systems can be treated.

Consequently, it has to be assessed whether the contracting party is in a position to decide whether or not any software agent should be involved in decision-making; if the decisions are left to algorithms, is there any requirement or benchmark about their accuracy? Further, human supervision and algorithm cooperation are key issues: whether and to what extent should humans be able to intervene in the decision-making process? These issues may even let us draw the conclusion that the focus of investigation should be shifted from the restriction of private autonomy to the redefinition of the concept of private autonomy itself to adapt to the changing economic environment.

16.3.2 Algorithmic Collusion

Algorithms often set prices of products on the Internet in a flexible way. If different companies are using the same or similar algorithms to set their prices, it might lead to intentional or unintentional price collusion. It is questionable whether companies should be required to disclose the functionality of their algorithms to make the process of price setting more transparent because transparency in price setting may have anti-competitive effects, such as price fixing.

Digital businesses increasingly use algorithms in their price setting, and such pricing algorithms can monitor the pricing of competitors. Increasing the transparency of such tools (its structure, workings etc.) and sharing publicly such competitively sensitive information may be considered illegal if this may lead to tacit collusion regarding price setting. However, if the user of the algorithm must provide an explanation that no collusion occurred, then the company may be required by the Competition Authority to disclose the functionality of the algorithm and that its price setting complies with the law.

Up to now, Hungarian legal practice did not lay down clear principles on how to differentiate between algorithm's collusion on their own (i.e. algorithmic collusion) and the use of code to enforce collusion (i.e. collusion by code). Market tendencies indicate that intermediary players, such as platforms, may also be involved in algorithmic collaboration. Existing cartel rules might apply to such relationships.

If the use of algorithms deliberately allows undertakings to collude with the aim of raising the prices, such behaviour is considered unlawful from the antitrust point of view. For instance, sharing pricing algorithms with competitors is a rather clear-cut infringement, but there may be more complex situations where parallel behaviour is taking place without the market players' performing any explicit coordination or collusion.

According to the general antitrust logic, a low level of market transparency may offer an opportunity to the cartel members to "cheat" the cartel system and entice consumers away from their competitors. Thus, cartels usually monitor competitors' actions to shorten the period in which the cheating cartel member can use its strategy, and if revealed, deviations from the cartel rules are followed by sanctions of the cartel system. The use of price algorithms as tools can contribute to the growth and/or sustenance of market transparency since deviation from agreed prices can be identified within seconds, and the cheating member can face real-time sanctioning interventions.

16.3.3 Consumer Discrimination in Pricing

Online platforms often price discriminate among consumers; platforms also apply dynamic pricing structures based on customer segmentation. In practice, this means that the consumers will see different prices depending on the browser type and browsing history, type of end-user device, number and timing of website visits, geographic location and similar user behaviour data. There are common strategies to avoid such price discrimination and dynamic pricing, such as the following:

- By blocking the IP address and geographic location data or by using a virtual private network (VPN), which permits the change of IP/geographic location
- By blocking the placement of third-party tracking cookies on the end-user device
- By using price comparison websites and conducting a search for more information
- By changing the end-user device or browser type to do a search for a product and to perform a purchase, considering that first-time customers likely become more benefits than returning buyers

Further, the use of personalised prices may also depend on other product characteristics. Among others, a potentially relevant categorisation can be based on the level of difficulty of consumers' information search prior to taking their transactional decisions. Concerning goods, at which consumers can typically rely on with a view to their prior experience or consumers can collect direct information on all essential features before taking their transactional decisions (i.e. search goods), personalised pricing can be subject to further assessment, provided that unjustified discrimination between the consumers is not taking place. As regards experience goods and credence goods, where an accurate prior evaluation of the product as compared to personal needs is more difficult or even impossible for the consumers, consumers are obviously not in the position to compare and assess personalised prices.

In some countries, some markets (e.g. health products, health care services) belong to the regulated sectors, and for some products, regulated prices are defined. In such markets, for instance, it can be worth considering whether the personalising of prices can be introduced if the aspects of personalisation can be properly regulated and the consumer can have relevant and comprehensible information on personalisation methods.

16.3.4 Online Sale Restrictions

According to the Court of Justice of the European Union's (CJEU's) recent practice,[9] manufacturers can prohibit the sale of their products on certain platforms in order to protect the product's image. *Article 101(1) TFEU must be interpreted as meaning that a selective distribution system for luxury goods designed, primarily, to preserve the luxury image of those goods complies with that provision to the extent that resellers are chosen on the basis of objective criteria of a qualitative nature that are laid down uniformly for all potential resellers and applied in a non-discriminatory fashion and that the criteria laid down do not go beyond what is necessary. Article 101 (1) TFEU must be interpreted as not precluding a contractual clause, such as that at issue in the main proceedings, which prohibits authorised distributors in a selective distribution system for luxury goods designed, primarily, to preserve the luxury image of those goods from using, in a discernible manner, third-party platforms for the internet sale of the contract goods, on condition that that clause has the objective of preserving the luxury image of those goods, that it is laid down uniformly and not applied in a discriminatory fashion, and that it is proportionate in the light of the objective pursued, these being matters to be determined by the referring court.* One of the conclusions of the judgment is that in the case of luxury goods, the protection of a brand image can sufficiently require a selective distribution system. The CJEU also stressed that its judgment in the *Pierre Fabre* case is in line with the *Coty* judgment: the prohibition on all online sales of (even luxury) products remains prohibited as it would be detrimental for the final consumers. However, there is no relevant Hungarian legislation or practice in this matter.

Manufacturers would also like to prohibit distributors from advertising the manufacturer's products on search engines, but this practice is very likely to restrict competition. The European Commission fined Guess EUR 39,821,000 for restricting retailers from online advertising and selling cross-border to consumers in the other Member States (geo-blocking), in breach of EU competition rules.[10] The Commission has found in its investigation—among others—that Guess' distribution agreements restricted authorised retailers from using the Guess brand names and trademarks for the purposes of online search advertising. Guess revealed this infringement during its cooperation with the Commission. The Commission decided that this was a restriction by object; however, the retailers of Guess were able to sell online. Based on the Commission's decision, manufacturers cannot prohibit distributors from advertising the manufacturer's products on search engines because

[9] CJEU, case C-230/16, *Request for a preliminary ruling under Article 267 TFEU from the Oberlandesgericht Frankfurt am Main (Higher Regional Court, Frankfurt am Main, Germany), made by decision of 19 April 2016, received at the Court on 25 April 2016, in the proceedings*, ECLI:EU:C:2017:941, pt 6.

[10] EU Commission Comp. DG, Case AT.40428, *Commission Decision of 17.12.2018 relating to proceedings under Article 101 of the Treaty on the Functioning of the European Union and Article 53 of the Agreement on the European Economic Area.*

the distributors will not be able to effectively generate traffic to their own websites by means of online search advertising. There is no relevant Hungarian legislation or practice in this matter.

16.3.5 Practice in Hungary as to Influencer Marketing

In Hungary, the practice of influencer marketing is developing more slowly than in other EU member countries, for example in Germany. No decision was published by the courts on this matter so far. The possible reason for this is that the HCA has only dealt with influencers in six cases up to this point. In the first three out of six cases, the authority has accepted the commitments from the influencers and companies; i.e., the authority has not imposed a fine. In two cases,[11] the authority found that the influencers and companies had not sufficiently implemented their commitments; therefore, the HCA imposed a fine. In the most recent case[12] from June 2020, the authority punished the influencer and also the company for unmarked advertisements. In all three cases, the authority fined both the influencers and the companies. The highest fine was imposed in case VJ/49/2018: 5 million HUF (approx. 14,000 EUR).

At the same time as publishing its first decision in connection with influencers, the HCA also published a guideline[13] that provides useful advice on how influencers should publish sponsored content in order to make it legally compliant.

16.4 Enforcement

In Hungary, the HCA is in charge of the enforcement of the UCP Act (i.e. the Hungarian implementation of Directive 2005/29 of the European Parliament and of the Council of 11 May 2005 concerning unfair business-to-consumer commercial practices in the internal market) if the unfair commercial practice materially distorts competition. Pursuant to the Hungarian Constitution, the enforcement of data protection rules is the exclusive competence of the Hungarian DPA, being the supervisory authority of Hungary under Article 51 (1) GDPR. The HCA and the Hungarian DPA have a cooperation agreement,[14] and these authorities discuss cooperation between these two authorities.

The HCA generally recognises that data protection enhances "consumer welfare" because consumers consider the privacy aspects of online products as a significant

[11] VJ/49/2018; VJ/5/2019.

[12] VJ/44/2019.

[13] https://www.gvh.hu/pfile/file?path=/en/actualities/aktualis_hirek_gvh_megfeleles_velemenyvezer_2017_11_30_a&inline=true.

[14] Available at https://www.gvh.hu/pfile/file?path=/gvh/egyuttmukodesi_megallapodasok/gvh_egyuttmuk_NAIH_GVH_2015_03_13&inline=true. Last accessed 19 September 2020.

product characteristic. In the *Google Allo* case (Vj/88/2016, closed with Commitment Decision), the HCA investigated whether consumers received the information necessary for an informed decision on Google's data processing activity. In its decision, the HCA considered that data processing is an essential aspect of Google products and that the consumer informational self-determination right has not only privacy but also market and competition aspects. Accordingly, the HCA acknowledged that the non-transparent nature of data processing may be a relevant aspect of the consumer's transactional decision.

Even if data protection law enforcement is the competence of the Hungarian DPA, the HCA adopted a policy to intervene and enforce unfair competition rules to the extent that data protection is part of consumer or customer welfare, and the breach of data protection provisions (such as privacy-related transparency/information provisions) constitutes an unfair commercial practice against consumers. The HCA has not yet applied the antitrust rules to social networks' and similar online providers' data processing activities.